Clifford Algebras and their Applications in Mathematical Physics

Fundamental Theories of Physics

An International Book Series on The Fundamental Theories of Physics: Their Clarification, Development and Application

Volume 55

Clifford Algebras and their Applications in Mathematical Physics

Proceedings of the Third Conference held at Deinze, Belgium, 1993

edited by

F. Brackx

R. Delanghe

and

H. Serras

Department of Mathematical Analysis,
University of Ghent,
Ghent, Belgium

SPRINGER-SCIENCE+BUSINESS MEDIA, B.V.

Library of Congress Cataloging-in-Publication Data

Clifford algebras and their applications in mathematical physics : proceedings of the third conference held at Deinze, Belgium, 1993 / edited by F. Brackx, R. Delanghe, and H. Serras.
p. cm. -- (Fundamental theories of physics)
ISBN 0-7923-2347-5 (alk. paper)
1. Clifford algebras--Congresses. 2. Mathematical physics--Congresses. I. Brackx, F. II. Delanghe, Richard. III. Serras, H. IV. Series.
QC20.7.C55C55 1993
530.1'5157--dc20 93-31536

ISBN 978-94-010-4886-6 ISBN 978-94-011-2006-7 (eBook)
DOI 10.1007/978-94-011-2006-7

Printed on acid-free paper

TABLE OF CONTENTS

CLASSICAL MECHANICS

MATHEMATICAL PHYSICS

PHYSICAL MODELS

1. Mrs. Krüger
2. Artibano Micali
3. Waldyr Rodrigues
4. Mrs. Nancy Hestenes
5. Mrs. Alan Common
6. Mrs. Paulina Mc Ewan
7. Sören Sprössig
8. Eiman Taha Abou el Dahab
9. W.E. Baylis
10. Claude Daviau
11. Roy Chisholm
12. Richard Delanghe
13. Fred Brackx
14. Nono Kiyoharu
15. Julien Van Hamme
16. Ron Shaw
17. Andrzej Trautman
18. David Larner
19. Franco Piazzese
20. Jaime Vaz
21. Alan Common
22. Josep Manel Parra Serra
23. Klaus Gürlebeck
24. Jose Ricardo De Resende Zeni
25. Roger Boudet
26. Jacub Rembielinski
27. Steve Gull
28. Wolfgang Sprössig
29. W.S. Leng
30. Garret Sobczyk
31. Heinz Krüger
32. Chris Doran
33. Jarolim Bureš
34. David Hestenes
35. Michael V. Shapiro
36. Frank Sommen
37. Mrs. Sprössig
38. Nadine Van Acker
39. Marius Mitrea
40. Peter Van Lancker
41. Richard C. Pappas
42. John Mc Ewan
43. Anthony Lasenby
44. Swanhild Bernstein
45. Jan Cnops
46. Herman Serras
47. Jan de Graaf
48. Guy Laville
49. William M. Pezzaglia
50. Ruth Farwell
51. Bart Klein Obbink
52. Matthew Watkins
53. Helmuth Malonek
54. Rafal Ablamowicz
55. Klaus Habetha
56. Bernhelm Booss-Bavnbek
57. James P. Crawford
58. Vladimir Souček
59. Manfred Stein
60. Vladimir V. Kisil
61. Nikolai L. Vasilevski
62. Juri I. Karlovich
63. Enrique Ramirez de Arellano
64. Laurence L. Boyle

List of Participants

Ablamowicz R. (USA)
Abou el Dahab E.T. (Egypt)
Baylis W.E. (Canada)
Bernstein S. (Germany)
Booss-Bavnbek B. (Denmark)
Boudet R. (France)
Boyle L.L. (UK)
Brackx F. (Belgium)
Bureš J. (Czechia)
Chisholm J.S.R. (UK)
Cnops J. (Belgium)
Common A.K. (UK)
Crawford J.P. (USA)
Daviau C. (France)
de Graaf J. (the Netherlands)
Delanghe R. (Belgium)
Doran C. (UK)
Farwell R. (UK)
Gürlebeck K. (Germany)
Gull S. (UK)
Habetha K. (Germany)
Hestenes D. (USA)
Karlovich Y.I. (Ukraina)
Kisil V.V. (Ukraina)
Klein Obbink B. (the Netherlands)
Krüger H. (Germany)
Larner D. (the Netherlands)
Lasenby A. (UK)
Laville G. (France)
Lawrynowicz J. (Poland)
Leng W.S. (UK)
Malonek H. (Portugal)
McEwan J. (UK)
Micali A. (France)
Mitrea M. (USA)
Nôno K. (Japan)
Pappas R.C. (USA)
Parra Serra J.M. (Spain)
Pezzaglia W.M. (USA)
Piazzese F. (Italy)
Porteous I.P. (UK)
Porter R.M. (Mexico)
Ramírez de Arellano E. (Mexico)
Rembieliński J. (Poland)
Rodrigues W. (Brazil)
Roos P. (the Netherlands)
Serras H. (Belgium)
Shapiro M.V. (Mexico)
Shaw R. (UK)
Sobczyk G. (Mexico)
Sommen F. (Belgium)
Souček V. (Czechia)
Sprössig W. (Germany)
Stein M. (Germany)
Trautman A. (Poland)
Van Acker N. (Belgium)
Van hamme J. (Belgium)
Van Lancker P. (Belgium)
Vasilevski N.I. (Mexico)
Vaz J. (Brazil)
Watkins M. (UK)
Zeni J.R. (Brazil)

Preface

This *International Conference on Clifford Algebras and Their Applications in Mathematical Physics* is the third in a series of conferences on this theme, which started at the *University of Kent* in Canterbury in 1985 and was continued at the *Université des Sciences et Techniques du Languedoc* in Montpellier in 1989.

Since the start of this series of Conferences the research fields under consideration have evolved quite a lot. The number of scientific papers on Clifford Algebra, Clifford Analysis and their impact on the modelling of physics phenomena have increased tremendously and several new books on these topics were published. We were very pleased to see old friends back and to wellcome new guests who by their inspiring talks contributed fundamentally to tracing new paths for the future development of this research area.

The Conference was organized in Deinze, a small rural town in the vicinity of the University town Gent. It was hosted by *De Ceder*, a vacation and seminar center in a green area, a typical landscape of Flanders's "plat pays".

The Conference was attended by 61 participants coming from 18 countries; there were 10 main talks on invitation, 37 contributions accepted by the Organizing Committee and a poster session. There was also a book display of Kluwer Academic Publishers.

As in the Proceedings of the Canterbury and Montpellier conferences we have grouped the papers accordingly to the themes they are related to: Clifford Algebra, Clifford Analysis, Classical Mechanics, Mathematical Physics and Physics Models. Needless to say that this classification is far from absolute. We are grateful to the authors who respected the deadline for submission of their paper; they helped us and Kluwer Academic Publishers substantially in providing the scientific community with the results presented at the Conference as quickly as possible.

This Conference was made possible by the generous support of the *Belgian National Fund for Scientific Research NFWO*. Thanks are also due to *Kluwer Academic Publishers*, the *Bank Brussel Lambert BBL* and the Public Relations Office of the *Universiteit Gent RUG* for their sponsorship. We also like to thank the management of *De Ceder* for their hospitality, their cuisine and their assistance with the organization. Last but not least we are grateful to the staff of our Department of Mathematical Analysis for their continuous concern and co-operation.

In their foreword to the Proceedings of the first Conference in Canterbury, the organizers J. S. R. Chisholm and A. K. Common wrote that they were pleased that "...the participants at the Workshop see it as the first in a regular series of meetings". We in our turn are very glad that Prof. K. Habetha, Rector of the *RWTH Aachen*, has consent to organize the next meeting in Germany in 1996.

F. Brackx R. Delanghe H. Serras

CLIFFORD ALGEBRAS
and
APPLICATIONS

QUANTUM CLIFFORD ALGEBRAS

T. BRZEZIŃSKI *
Department of Applied Mathematics and Theoretical Physics
University of Cambridge
Silver St., CB3 9EW Cambridge, U.K.

L.C. PAPALOUCAS
Institute of Mathematics
University of Athens
106 79 Athens, Greece

and

J. REMBIELIŃSKI
Department of Mathematical Physics
University of Łódź
ul. Pomorska 149/153, 90–236 Łódź, Poland

Abstract. Quantum multiparameter deformation of real Clifford algebras is proposed. The corresponding irreducible representations are found.

Key words: Clifford algebras, quantum deformations

1. Introduction

As is well known there exists a direct relation between the exterior and the Grassmann algebras. Let us consider a unital algebra $C(x)$ over C freely generated by the elements $x^1, \ldots, x^N$ and a two-sided ideal $J_0 \subset C(x)$ generated by $x^i x^j - x^j x^i$, with $i, j = 1, \ldots, N$. Now, the quotient algebra $M_0 = C(x)/J_0$ is freely generated by $x^1, \ldots, x^N$ subject to the commutativity relation

$$x^i x^j = x^j x^i. \tag{1}$$

Introducing a new set of generators dx^k, $k = 1, \ldots, N$, where d is the exterior differential operator, we can extend this algebra first to an A_0-module and then to the exterior algebra Ω, with the standard product given by

$$x^i\, dx^j = dx^j\, x^i, \tag{2}$$

$$dx^i\, dx^j = -\, dx^j\, dx^i. \tag{3}$$

The generators dx^i define a finite dimensional subalgebra of Ω with the multiplication given by the exterior product (3). This algebra is a differential realisation of the abstract Grassmann algebra generated by the set $\hat{\gamma}^i$, $i = 1, \ldots, N$ subject to the

* On leave from Institute of Mathematics, University if Łódź, ul. Banacha 22, 90–238 Łódź, Poland

F. Brackx et al. (eds.), *Clifford Algebras and their Applications in Mathematical Physics*, 3–8.

relations

$$\hat{\gamma}^i\hat{\gamma}^j + \hat{\gamma}^j\hat{\gamma}^i = 0. \tag{4}$$

Clifford algebra is defined as the central extension of the algebra (4)

$$\gamma^i\gamma^j + \gamma^j\gamma^i = 2g^{ij}I, \tag{5}$$

with $g^{ij} = g^{ji}$, $\overline{g}^{ij} = g^{ij}$.

Now, it is easy enough to apply this procedure to the Manin's hyperplane. In this case we choose the ideal J [1] leading to the following reordering rules

$$x^ix^k = q_{ik}x^kx^i, \tag{6}$$

with $q_{ik} \in C - \{0\}$, $q_{ik} = q_{ki}^{-1}$, $q_{kk} = 1$. Consequently, we obtain [2]

$$[Bdx^i\, dx^k = -q_{ik}\, dx^k\, dx^i \tag{7}$$

as basic rules for the two-form sector of the corresponding twisted exterior algebra. Similarly as in the standard case, identifying dx^i with $\hat{\Gamma}^i$, we obtain

$$\hat{\Gamma}^i\hat{\Gamma}^k + q_{ik}\hat{\Gamma}^k\hat{\Gamma}^i = 0, \tag{8}$$

as a generalisation of the Grassmann algebra multiplication rules.

In the next sections of the paper we construct central extensions of the algebra (8).

2. Clifford Algebra $C^{p,q}$ in the Witt Basis

The *canonical basis* $\{\gamma^\mu\}$ for the Clifford algebra $C^{p,q}$ is defined as follows

$$\gamma^\mu\gamma^\nu + \gamma^\nu\gamma^\mu = 2g^{\mu\nu}I, \tag{9}$$

where

$$g = \begin{pmatrix} I_p & 0 \\ 0 & -I_q \end{pmatrix}, \tag{10}$$

$p \geq q$, $\mu, \nu = 1, \ldots, (p+q)$.

For $p+q = 2n$ there exists another standard basis, the so called *Witt basis* [3, 4]. As we will see later this basis is suitable for a non-commutative generalisation of the Clifford algebra. In the Witt basis metric tensor takes the form

$$G = \begin{pmatrix} 0 & I_n \\ I_n & 0 \end{pmatrix}, \tag{11}$$

while the new generators are defined by

$$\gamma_N^a = \frac{1}{\sqrt{2}}(\gamma^a + i\gamma^{a+n}), \qquad \gamma_N^b = \frac{1}{\sqrt{2}}(\gamma^b - \gamma^{b+n}), \tag{12}$$

$$\gamma_P^a = \frac{1}{\sqrt{2}}(\gamma^a - i\gamma^{a+n}), \qquad \gamma_P^b = \frac{1}{\sqrt{2}}(\gamma^b + \gamma^{b+n}), \tag{13}$$

where $a = 1, \ldots, (n-q)$, $b = (n-q+1), \ldots, n$. Using Eq. (9) we obtain the following commutation relations

$$\gamma_N^\alpha \gamma_N^\beta = -\gamma_N^\beta \gamma_N^\alpha, \tag{14}$$
$$\gamma_P^\alpha \gamma_P^\beta = -\gamma_P^\beta \gamma_P^\alpha, \tag{15}$$
$$\gamma_N^\alpha \gamma_P^\beta = -\gamma_P^\beta \gamma_N^\alpha + \delta^{\alpha\beta} I, \tag{16}$$

$\alpha, \beta = 1, \ldots, n$. In particular, we notice that ${\gamma_N^\alpha}^2 = {\gamma_P^\alpha}^2 = 0$.

We see that γ_N^α and γ_P^α span two totally isotropic n-dimensional subspaces in the generating sector of the Clifford algebra $C^{p,q}$, $p + q = 2n$.

The hermitian conjugation can always be chosen as [5]

$$\gamma^{\mu\dagger} = g^{\mu\mu}\gamma^\mu, \tag{17}$$

so that

$$\gamma_N^{\alpha\dagger} = \gamma_P^\alpha. \tag{18}$$

Therefore γ_N^α (γ_P^α) behave like *fermionic annihilation (creation) operators.*

For $p+q = 2n+1$, it is necessary to add an extra generator γ^0, $\gamma^{0\dagger} = \gamma^0$ satisfying

$$\gamma^{0^2} = I, \quad \{\gamma^0, \gamma_N^\alpha\} = \{\gamma^0, \gamma_P^\alpha\} = 0. \tag{19}$$

For $p + q = 2n$, real forms $C^{p,q}$ can be reconstructed via

$$A = \sum_{\alpha=1}^{n-q} (z_\alpha \gamma_N^\alpha + \bar{z}_\alpha \gamma_P^\alpha) + \sum_{\beta=n-q-1}^{n} (a_\beta \gamma_N^\beta + b_\beta \gamma_P^\beta), \tag{20}$$

where $z_\alpha \in \boldsymbol{C}$, $a_\beta, b_\beta \in \boldsymbol{R}$, while for $p + q = 2n + 1$ we have

$$A' = A + a_0\gamma^0, \quad a_0 \in \boldsymbol{R}. \tag{21}$$

3. Quantum Deformation of Clifford Algebras $C^{p,q}$

According to the standard procedure we try to deform the Clifford algebras in the Witt basis via the following Ansatz

$$\Gamma_N^\alpha \Gamma_N^\beta = -q_{\alpha\beta}^N \Gamma_N^\beta \Gamma_N^\alpha, \qquad \Gamma_P^\alpha \Gamma_P^\beta = -q_{\alpha\beta}^P \Gamma_P^\beta \Gamma_P^\alpha, \tag{22}$$
$$\Gamma_N^\alpha \Gamma_P^\beta = -q_{\alpha\beta} \Gamma_P^\beta \Gamma_N^\alpha + 2\delta^{\alpha\beta}\Delta, \tag{23}$$
$$\Delta \Gamma_N^\alpha = \xi_N^\alpha \Gamma_N^\alpha \Delta, \qquad \Delta \Gamma_P^\beta = \xi_P^\beta \Gamma_P^\beta \Delta, \tag{24}$$

where $\Gamma_N^{\alpha\dagger} = \Gamma_P^\alpha$, $\Delta^\dagger = \Delta$.

The parameters $q_{\alpha\beta}^N$, $q_{\alpha\beta}^P$, $q^{\alpha\beta}$, ξ_N^α, ξ_P^α should satisfy a number of conditions following from the consistency with associativity and the hermitian conjugation rules. Solving these constraints, we obtain (for $q_{\alpha\alpha}^{N(P)} \neq -1$)[1]

$$\xi_N^\alpha = q_{\alpha\alpha}^{-1}, \quad \xi_P^\alpha = q_{\alpha\alpha}, \quad \bar{q}_{\alpha\beta} = q_{\beta\alpha}, \tag{25}$$
$$q_{\alpha\beta}^N = q_{\alpha\alpha}/q_{\alpha\beta}, \quad q_{\alpha\beta}^P = q_{\beta\beta}/q_{\alpha\beta}, \quad q_{\alpha\beta}q_{\beta\alpha} = q_{\alpha\alpha}q_{\beta\beta}. \tag{26}$$

---

[1] The case $q_{\alpha\alpha}^{N(P)} = -1$ leads to deformations of the symplectic or Crummeyrolle Clifford algebra.

Now, demanding the extension be *central*, we are led to

$$q_{\alpha\alpha} = 1, \tag{27}$$

so that

$$|q_{\alpha\beta}| = 1, \quad q^N_{\alpha\beta} = q_{\beta\alpha}, \quad q^P_{\alpha\beta} = q_{\beta\alpha}. \tag{28}$$

and Δ can be chosen as $+I$ (notice that $\gamma_N\gamma_P$ is normal, so positive definite).

Resulting is the following deformation of the Clifford algebra, obtained as a central extension of the q-deformed Grassmann algebra:

$$\Gamma^\alpha_N \Gamma^\beta_N = -q_{\beta\alpha}\Gamma^\beta_N\Gamma^\alpha_N, \tag{29}$$

$$\Gamma^\alpha_P \Gamma^\beta_P = -q_{\beta\alpha}\Gamma^\beta_P\Gamma^\alpha_P, \tag{30}$$

$$\Gamma^\alpha_N \Gamma^\beta_P = -q_{\alpha\beta}\Gamma^\beta_P\Gamma^\alpha_N + 2\delta^{\alpha\beta}I, \tag{31}$$

with $\Gamma^{\alpha\dagger}_N = \Gamma^\alpha_P$ and $\alpha, \beta = 1, \ldots, n$, where

$$|q_{\alpha\beta}| = 1, \qquad \overline{q}_{\alpha\beta} = q_{\beta\alpha}. \tag{32}$$

The odd case ($p+q = 2n+1$) can be treated by the extending the above algebra by

$$\Gamma^{0\dagger} = \Gamma^0, \qquad \Gamma^{0^2} = I \tag{33}$$

together with the Ansatz

$$\Gamma^0\Gamma^\alpha_N = -q^N_\alpha \Gamma^\alpha_N \Gamma^0, \qquad \Gamma^0\Gamma^\alpha_P = -q^P_\alpha \Gamma^\alpha_P \Gamma^0. \tag{34}$$

If we additionally demand the existence of the classical limit, we will obtain

$$q^{N(P)}_\alpha = 1, \tag{35}$$

i.e.

$$\Gamma^0\Gamma^\alpha_{N(P)} = -\Gamma^\alpha_{N(P)}\Gamma^0. \tag{36}$$

4. Representations

Here we construct a Fock space of representations of the deformed Clifford algebra. We consider even and odd cases separately and we conclude this section with the example of the deformed Dirac matrices in four dimensions.

4.1. The Even Case ($p+q=2n$)

We define the vacuum state

$$\Gamma^\alpha_N|0\rangle = 0, \qquad \langle 0|0\rangle = 1. \tag{37}$$

The basis of the Fock space can be defined via

$$|\sigma_1, \ldots, \sigma_n\rangle = \left(\frac{\Gamma^n_P}{\sqrt{2}}\right)^{\sigma_n} \cdots \left(\frac{\Gamma^1_P}{\sqrt{2}}\right)^{\sigma_1} |0\rangle, \tag{38}$$

where $\sigma_\alpha = 0, 1$.

Consequently

$$\Gamma_N^\alpha|\ldots,\sigma_\alpha,\ldots\rangle = \sqrt{2}\delta_{\sigma_\alpha 1}\prod_{\beta=1+\alpha}^{n}(-q_{\alpha\beta})^{\sigma_\beta}|\ldots,0,\ldots\rangle, \tag{39}$$

$$\Gamma_P^\alpha|\ldots,\sigma_\alpha,\ldots\rangle = \sqrt{2}\delta_{\sigma_\alpha 0}\prod_{\beta=1+\alpha}^{n}(-q_{\beta\alpha})^{\sigma_\beta}|\ldots,1,\ldots\rangle. \tag{40}$$

4.2. The Odd Case $(p+q=2n+1)$

When $p+q=2n+1$, the representations of $C^{p,q}$ can be easily derived from the representations of even $C^{p,q}$ by the following procedure. First, we replace the even case vacuum $|0\rangle$ by $|0\pm\rangle$, defined by Eqs. (37) and

$$\Gamma^0|0\pm\rangle = \pm|0\pm\rangle. \tag{41}$$

The Fock space is generated from the vacuum by the actions of the raising operators Γ_P^α, precisely as in the even case. The action of Γ_N^α, Γ_P^α on a standard state $|\sigma_1,\ldots,\sigma_n\pm\rangle$ is given by Eqs (39- 40) and

$$\Gamma^0|\sigma_1,\ldots,\sigma_n\pm\rangle = \pm(-)^{\Sigma\sigma_\alpha}|\sigma_1,\ldots,\sigma_n\pm\rangle. \tag{42}$$

4.3. Example

Let us consider the simplest non-trivial example: $n=2$, $p+q=4$, $q_{12}=\kappa$, $|\kappa|=1$. Explicitly, we have the following algebra

$$\begin{aligned}\Gamma_N^1\Gamma_N^2 &= -\kappa^{-1}\Gamma_N^2\Gamma_N^1, \qquad \Gamma_P^1\Gamma_P^2 = -\kappa^{-1}\Gamma_P^2\Gamma_P^1,\\ \Gamma_N^1\Gamma_P^2 &= -\kappa\Gamma_P^2\Gamma_N^1, \qquad \Gamma_N^2\Gamma_P^1 = -\kappa\Gamma_P^1\Gamma_N^2,\\ &\{\Gamma_N^1,\Gamma_P^1\} = \{\Gamma_N^2,\Gamma_P^2\} = 2I.\end{aligned}$$

and

$$(\Gamma_N^1)^2 = (\Gamma_N^2)^2 = (\Gamma_P^1)^2 = (\Gamma_P^2)^2 = 0.$$

According to the Eqs. (39–40) we obtain the following deformation of the Dirac matrices

$$\Gamma_P^1 = \sqrt{2}\begin{pmatrix}0&0&0&0\\1&0&0&0\\0&0&0&0\\0&0&-\kappa^{-1}&0\end{pmatrix}, \qquad \Gamma_N^1 = \sqrt{2}\begin{pmatrix}0&1&0&0\\0&0&0&0\\0&0&0&-\kappa\\0&0&0&0\end{pmatrix},$$

$$\Gamma_P^2 = \sqrt{2}\begin{pmatrix}0&0&0&0\\0&0&0&0\\1&0&0&0\\0&1&0&0\end{pmatrix}, \qquad \Gamma_N^2 = \sqrt{2}\begin{pmatrix}0&0&1&0\\0&0&0&1\\0&0&0&0\\0&0&0&0\end{pmatrix}.$$

5. Related Topics

Let us conclude with pointing out a number of interesting problems. One may investigate relations of the above twisted Clifford algebras to:

– multiparameter q-deformations of Spin and pseudo-orthogonal groups (corresponding to quantum groups);
– q-spinors;
– exotic statistics (anyons, etc.).

Acknowledgements

This work is supported by the KBN grant No. 2 0218 91 01. We would thank to University of Athens, Department of Mathematics for the hospitality.

References

1. Manin, Yu.I.: 1988, *Quantum Groups and Non-Commutative Geometry*, RIMS, Montreal
2. Brzeziński, T., Dąbrowski, H., Rembieliński, J.: 1992, 'On the Quantum Differential Calculus and the Quantum Holomorphicity', *Journal of Matheamtical Physics* Vol. no. **33**, pp. 19–24
3. Chevalley, C.C.: 1955, *The Algebraic Theory of Spinors*, Columbia Press, New York
4. Giller, S., Kosiński, P., Maślanka, P., Rembieliński, J.: 1986, 'On $SO(\nu,\nu)$ Pure Spinors', *Acta Physica Polonica* Vol. no. **B17**, pp. 985–993
Giller, S., Kosiński, P., Maślanka, P., Rembieliński, J.: 1986, 'On $SO(p,q)$ Pure Spinors', *Acta Physica Polonica* Vol. no. **B18**, pp. 713–727
5. Rembieliński, J.: 1992, 'Classification, Properties and Applications of the Majorana Representations of the Real Clifford Algebras $C^{p,q}$', in ed. A. Micali et al. *Workshop on Clifford Algebras and Their Applications in Mathematical Physics, Montpelier, 1989*

RELATIONS BETWEEN WITT RINGS AND BRAUER GROUPS

JACQUES HELMSTETTER
Institut Fourier
Université Grenoble I
B.P. 74
38402 Saint-Martin-d'Hères, France

and

ARTIBANO MICALI
Département des Sciences Mathématiques
Université Montpellier II
Place Eugène Bataillon
34095 Montpellier, France

1. The main results

Let R be a commutative ring with unit element. To R we can associate the *Witt ring* $W(R)$ which classifies the nondegenerate quadratic forms Q on finitely generated projective R-modules M and the *Brauer group* $Br(R)$ which classifies the *Azumaya algebras* A over R, that is, A is a finitely generated projective R-module and, if A^{rev} denotes the *reversed algebra* with multiplication $(x, y) \mapsto yx$, the algebra $A \otimes_R A^{rev}$ is canonically isomorphic to the algebra $End_R(A)$. Let us recall that each nondegenerate quadratic form Q on a R-module M has an image in $W(R)$, here denoted by $w(M, Q)$ or $w(Q)$, and that this image fullfils the following properties: $w(Q) + w(Q') = w(Q \oplus Q')$ (*orthogonal sum*), $-w(Q) = w(-Q)$ and $w(Q)w(Q') = w(Q \otimes Q')$ (*tensor product*), for all nondegenerate quadratic forms Q and Q'. Besides, each Azumaya algebra A has an image $b(A)$ in the Brauer group $Br(R)$ and $b(A)b(A') = b(A \otimes_R A')$, $b(A)^{-1} = b(A^{rev})$, for all R-Azumaya algebras A and A'. Moreover, the unit element in the Brauer group $Br(R)$ is the image of the ring R.

By means of Clifford algebras, we can define a mapping β from the Witt ring $W(R)$ into the Brauer group $Br(R)$, which is the purpose of this paper. Now we shall first assume that 2 is inversible in the ring R; further works are necessary to get rid of this assumption. To define and study β, we need some other objects.

The idempotents of the ring R build a Boole ring denoted $Ip(R)$; the *sum* of two idempotents i and j is $i + j - 2ij$ and their *product* is the ordinary product ij. This Boole ring operates on the Brauer group $Br_2(R)$ consisting of all elements b of

F. Brackx et al. (eds.), Clifford Algebras and their Applications in Mathematical Physics, 9–12.

order 2 in $Br(R)$, that is, such that $b^2 = 1$. This operation is denoted by exponents: $b(A)^i = b(iA \oplus (1-i)R)$. We must understand that $b^{i+j-2ij} = b^i b^j$, $b^{ij} = (b^i)^j$ and $(bc)^i = b^i c^i$, whenever b and c are in $Br_2(R)$ and i and j are in $Ip(R)$.

The R-module M (as above) has a local rank (or dimension) when localized at any prime ideal of R. Let i be the greatest idempotent in R such that the R-module iM has even rank when localized at any prime ideal of R. The idempotent i only depends on $w(M,Q)$ and by setting $\pi(w(M,Q)) = 1-i$, we get a ring homomorphism π, the *parity homomorphism*, from $W(R)$ into $Ip(R)$.

We need also the group $Q(R)$ of quadratic R-modules D of constant rank 1, that is, the modules D provided with an isomorphism $D \otimes_R D \to R$, which defines a nondegenerate quadratic form on D. Elements of the group $Q(R)$ are the isomorphy classes $q(D)$ of such quadratic modules D and they are multiplied according to the rule $q(D)q(D') = q(D \otimes_R D')$ which implies that $q(D)^{-1} = q(D)$. There is a natural mapping ρ from $W(R)$ into $Q(R)$. In fact, let $x \mapsto \hat{x}$ be the automorphism of the Clifford algebra $C(M,Q)$ associated with the *parity grading*; this means that $\hat{x}$ is x or $-x$ depending on x being even or odd; the submodule D of $C(M,Q)$ consisting of all x such that $xm = -m\hat{x}$ for all m in M is a quadratic module of constant rank 1 (indeed x^2 is in R for all x in D). Therefore, we set $\rho(w(M,Q)) = q(D)$.

Every invertible element λ in R has an image in $Q(R)$ which is denoted by $q(\lambda)$. This element is the isomorphy class of R provided with the mapping $s \otimes t \mapsto \lambda st$ for all s and t in R; notice that $q(\lambda^2) = q(1)$ for all invertible element λ in R. We shall also write 1 instead of $q(1)$, the unit element of $Q(R)$, and we set $\varepsilon = q(-1)$.

The Boole ring $Ip(R)$ operates on $Q(R)$ too and we write $q(D)^i = q(iD \oplus (1-i)R)$, with the canonical mapping $s \otimes t \mapsto st$ for all s and t in $(1-i)R$. Moreover there is a natural mapping σ from $Q(R) \times Q(R)$ into the Brauer group $Br(R)$. In fact, if D and D' are quadratic modules of rang 1, the Clifford algebra of the orthogonal sum $D \oplus D'$ is a quaternion algebra, therefore an Azumaya algebra, and its Brauer class is $\sigma(q(D), q(D'))$, by definition of σ.

Now we define the mapping β from $W(R)$ into $Br(R)$. If M has everywhere even rank (or odd rank), then the Clifford algebra $C(M,Q)$ (or the even subalgebra of $C(M,Q)$) is an Azumaya algebra. This allows us to define $\beta(w(M,Q))$. If M has not constant rank parity, we shall write $M = iM' \oplus (1-i)M''$ with M' (or M'') of even (or odd) rank and the definition of $\beta(w(M,Q))$ follows. The image of β lies in the subgroup $Br_2(R)$ of the Brauer group $Br(R)$ of elements of order 2, since every Clifford algebra is isomorphic to its reversed algebra. These main facts can be summarized in the following two theorems:

Theorem 1.1 *Let $V(R)$ be the set $Ip(R) \times Q(R) \times Br_2(R)$ provided with the following addition and multiplication:* $(i,q,b) + (j,r,c) = (i+j-2ij, qr\varepsilon^{ij}, bc\sigma(q,r)$ $\sigma(q,\varepsilon)^{(1-i)j}\sigma(\varepsilon,r)^{i(1-j)})$ *and* $(i,q,b)(j,r,c) = (ij, q^j r^i, b^j c^i \sigma(q,r)^{1-ij})$. *With these operations, $V(R)$ becomes an associative ring with zero equal to* $(0,1,1)$ *and with unit equal to* $(1,1,1)$. *Moreover,* $8v = 0$ *for all v in $V(R)$.*

The proof only needs the following facts: the Boole ring $Ip(R)$ operates in the groups $Q(R)$ and $Br_2(R)$; $\sigma(q,r)$ is a symmetric function of q and r such that $\sigma(q,rr') = \sigma(q,r)\sigma(q,r')$, $\sigma(q,r^i) = \sigma(q,r)^i$ and $\sigma(q,q) = \sigma(q,\varepsilon)$, for all q, r, r' in $Q(R)$ and i in the ring $Ip(R)$. All this would be clearer if $Q(R)$ and $Br_2(R)$ were additive instead of multiplicative groups (see next section). With additive notations, $Q(R)$ and $Br_2(R)$ would become $Ip(R)$-modules and σ would be an $Ip(R)$-bilinear symmetric mapping; the zero element of $V(R)$ would be $(0,0,0)$.

Theorem 1.2 *The mapping (π,ρ,β) which associates $(\pi(w),\rho(w),\beta(w))$ to all w in $W(R)$, is a ring homomorphism from $W(R)$ into $V(R)$ and its image is $Ip(R) \times Q(R) \times \beta(W(R))$.*

Example 1.3. Assume that R is the field of real numbers; then R contains two idempotents, 0 and 1, $Q(R)$ has two elements, $1 = q(1)$ and $\varepsilon = q(-1)$, and $Br(R) = Br_2(R)$ has also two elements, the class 1 of the R-algebra R and the class $\sigma(\varepsilon,\varepsilon)$ of the field of quaternions as Azumaya R-algebra; then $V(R)$ is isomorphic to $Z/8Z$ and $W(R)$ to Z. The homomorphism $W(R) \to V(R)$ is surjective and accounts for several 8-periodicity theorems. Such periodicity, with period 8, 4 or 2, occurs for every ring R.

2. The general setting

Let A be a Boole ring, M and N two A-modules and $\sigma : M \times M \to N$ a A-bilinear symmetric mapping such that there exists an element ε in M that verifies $\sigma(q,q) = \sigma(q,\varepsilon)$ for all q in M. We considerer on the set $A \times M \times N$ the structure of associative ring defined by the following addition and multiplication: $(i,q,b) + (j,r,c) = (i+j, q+r+ij\varepsilon, b+c+\sigma(q,r) + (1-i)j\sigma(q,\varepsilon) + i(1-j)\sigma(\varepsilon,r))$ and $(i,q,b)(j,r,c) = (ij, ir+jq, ic+jb+(1-ij)\sigma(q,r))$, for all i,j in A, q and r in M and b and c in N. This ring will be denoted by $(A,M,N,\sigma,\varepsilon)$ and he has the following properties: (i) his zero element is $(0,0,0)$ and his unit element is $(1,0,0)$; (ii) the characteristic of the ring $(A,M,N,\sigma,\varepsilon)$ is a divisor of 8 for the reason that $8(i,q,b) = 0$ for all (i,q,b) in $(A,M,N,\sigma,\varepsilon)$. Hence the ring $(A,M,N,\sigma,\varepsilon)$ has a structure of algebra over the ring $Z/8Z$; (iii) the invertible elements of the ring $(A,M,N,\sigma,\varepsilon)$ are those that we can write in the form $(1,q,b)$ for all q in M and b in N; (iv) the nilpotent elements of the ring $(A,M,N,\sigma,\varepsilon)$ are those that we can write in the form $(0,q,b)$, therefore the A-module $M \times N$ is isomorphic to an ideal of the ring $(A,M,N,\sigma,\varepsilon)$, the *nilradical* of this ring, denoted by $Nil(A,M,N,\sigma,\varepsilon)$; the ring isomorphism $(A,M,N,\sigma,\varepsilon)/Nil(A,M,N,\sigma,\varepsilon) \simeq A$ shows that *if $A = Z/2Z$, then $(A,M,N,\sigma,\varepsilon)$ is a local ring whose maximal ideal is its nilradical.*

A *morphism*, in the category of the rings $(A,M,N,\sigma,\varepsilon)$, is a triplet (α,f,g) : $(A,M,N,\sigma,\varepsilon) \to (A',M',N',\sigma',\varepsilon')$ were $\alpha : A \to A'$ is a homomorphism of Boole rings, $f : M \to M'$ and $g : N \to N'$ are additive mappings verifying $f(iq) = \alpha(i)f(q)$ and $g(ib) = \alpha(i)g(b)$ for all i in A, q in M and b in N; moreover, $f(\varepsilon) = \varepsilon'$ and

$\sigma' \circ (f \times f) = g \circ \sigma$. The last condition implies that $h' \circ f = g \circ h$ where $h : M \to N$ (or $h' : M' \to N'$) is the A-linear (or A'-linear) mapping defined by $q \mapsto \sigma(q, \varepsilon)$ (or $q' \mapsto \sigma'(q', \varepsilon')$). This means that $g(\sigma(q, \varepsilon)) = \sigma'(f(q), \varepsilon')$ for all q in M.

Let $GL_A(M, N, \sigma, \varepsilon)$ be the subgroup of the product group $GL_A(M) \times GL_A(N)$ (linear groups) consisting of the pairs (f, g) with f in $GL_A(M)$ and g in $GL_A(N)$ such that $f(\varepsilon) = \varepsilon$ and $\sigma \circ (f \times f) = g \circ \sigma$. If $Aut(A)$ (or $Aut(A, M, N, \sigma, \varepsilon)$) denotes the automorphism group of the Boole ring A (or of the ring $(A, M, N, \sigma, \varepsilon)$), the above considerations lead to the following result:

Proposition 2.1 *For every ring $(A, M, N, \sigma, \varepsilon)$ we have a group isomorphism $Aut(A, M, N, \sigma, \varepsilon) \simeq Aut(A) \times GL_A(M, N, \sigma, \varepsilon)$.*

We remark that the case where A is the Boole ring of idempotents of a commutative ring R was quoted in section 1.

REFERENCES

[1] J. Helmstetter. "Anneau de Witt et groupe de Brauer", in Séminaire d'Algèbre 1975/1976, Université de Provence, Marseille.

[2] A. Micali and Ph. Revoy. "Modules Quadratiques", Bull. Soc. Math. France, Mémoire n. 63, 1979, 144 p.

CLIFFORD ALGEBRA TABLES

IAN R. PORTEOUS
Department of Pure Mathematics
The University of Liverpool, PO Box 147, Liverpool L69 3BX, U.K.

Abstract. The paper is a commentary on Tables 13.66 of the author's book [3]. These classify the conjugation anti-involutions of the *five* Clifford algebras naturally associated to the orthogonal space $\mathbf{R}^{p,q}$, for each signature p, q.

Key words: Clifford Algebra, Spin group

1. Introduction

Tables 13.66 of [3] classify the conjugation anti-involutions of the *five* Clifford algebras naturally associated to the orthogonal space $\mathbf{R}^{p,q}$,for each signature p, q. It is because there are five and not just one of these algebras that the notation $\mathbf{R}_{p,q}$ for the standard real Clifford algebra for $\mathbf{R}^{p,q}$ is employed there rather than $Cl(p,q)$, or some such. Additional algebras $\overline{\mathbf{C}}_{p,q}$ and ${}^2\mathbf{R}^s_{p,q}$ are obtained in the first instance by tensoring $\mathbf{R}_{p,q}$ with the *superfield* or involuted field $\overline{\mathbf{C}}$, that is $\mathbf{C}$ assigned the conjugation involution, which fixes the field $\mathbf{R}$, and in the second instance with the superfield ${}^2\mathbf{R}^s$ (denoted in [3] not very happily by $hb\mathbf{R}$), that is the double field ${}^2\mathbf{R}$ of diagonal 2×2 real matrices, with *swapping* the slots as involution, which also fixes the field $\mathbf{R}$. The other two, namely $\mathbf{C}_n$ and ${}^2\mathbf{C}^s{}_n$, where $n = p+q$, are obtained by tensoring $\mathbf{R}_{p,q}$ with $\mathbf{C}$, or with the superfield ${}^2\mathbf{C}^s (= hb\mathbf{C})$, which fixes the field $\mathbf{C}$.

In each of these cases minus the identity on $\mathbf{R}^{p,q}$ extends both to an involution, the *main* or *grade* involution on the algebra and also to an anti-involution of the algebra, namely *conjugation*. In this way each of the five should be regarded as a *superalgebra*, that is as a $\mathbf{Z}_2$-graded algebra, equipped also with the conjugation anti-involution as an integral part of its structure. Roughly speaking $\mathbf{C}_n$ encodes the rank information in $\mathbf{R}_{p,q}$ while ${}^2\mathbf{R}^s_{p,q}$ encodes the signature information. It is of course the case that as superalgebras $\overline{\mathbf{C}}_{p,q}$ and $\mathbf{C}_n$ are isomorphic, but with their assigned conjugations they are not at all the same.

Since various conventions are current in the literature, all for the best of reasons, it is necessary here to say that for us $\mathbf{R}^{p,q}$ denotes the real linear space $\mathbf{R}^n$, where $n = p+q$, assigned a non-degenerate quadratic form with p negative eigenvalues and q positive ones, while for any $x \in \mathbf{R}^{p,q}$, $x^2 = -x \cdot x$ in $\mathbf{R}_{p,q}$. One could alternatively and equivalently take p, q to describe a non-degenerate quadratic form with p positive and q negative eigenvalues, with $x^2 = x \cdot x$ in $\mathbf{R}_{p,q}$, for any $x \in \mathbf{R}^{p,q}$.

Each Clifford algebra of any the five types described above is isomorphic to a full matrix algebra over $\mathbf{R}, {}^2\mathbf{R}, \mathbf{C}, {}^2\mathbf{C}, \mathbf{H}$, or ${}^2\mathbf{H}$, the module on which the matrix acts,

F. Brackx et al. (eds.), Clifford Algebras and their Applications in Mathematical Physics, 13–22.

identifiable with a minimal left ideal of the algebra, being the *spinor* space for the algebra. Now any anti-involution of such an algebra (in particular conjugation) may be regarded as the anti-involution adjoint to some symmetric or skew sesquilinear form on the spinor space. There are *ten* (*not* seven, *not* eight, *not* nine!) such classes and to appreciate Tables 13.66 one must be familiar with all of them. These, and their associated classical groups, were coded in [3] by the digits 0, 1, 2, ..., 9 as follows.

Code	Spinor space	Classical group
0	$\mathbf{R}$-module with $\mathbf{R}$ symmetric form	$O_m = O_m(\mathbf{R})$
1	${}^2\mathbf{R}$-module with ${}^2\mathbf{R}^{s}$ symmetric or skew form	$GL_m(\mathbf{R})$
2	$\mathbf{R}$-module with $\mathbf{R}$ skew form	$Sp_m(\mathbf{R})$
3	$\mathbf{C}$-module with $\mathbf{C}$ skew form	$Sp_m(\mathbf{C})$
4	$\mathbf{H}$-module with $\tilde{\mathbf{H}}$ skew or $\overline{\mathbf{H}}$ symmetric form	$Sp_m = Sp_m(\mathbf{H})$
5	${}^2\mathbf{H}$-module with ${}^2\tilde{\mathbf{H}}^{s}$ or ${}^2\overline{\mathbf{H}}^{s}$ symmetric or skew form	$GL_m(\mathbf{H})$
6	$\mathbf{H}$-module with $\tilde{\mathbf{H}}$ symmetric or $\overline{\mathbf{H}}$ skew form	$O_m(\mathbf{H})$
7	$\mathbf{C}$-module with $\mathbf{C}$ symmetric form	$O_m(\mathbf{C})$
8	$\mathbf{C}$-module with $\overline{\mathbf{C}}$ symmetric or skew form	U_m
9	${}^2\mathbf{C}$-module with ${}^2\mathbf{C}^{s}$ or ${}^2\overline{\mathbf{C}}^{s}$ symmetric or skew form	$GL_m(\mathbf{C})$

where in each case m denotes the dimension of the module, m necessarily being even in the real and complex symplectic cases 2 and 3.

The logic behind the numbering of these ten types derives from the order in which most of them appear in Table 3 below. Short notations for the classical groups, such as $GL_m(\mathbf{C})$ rather than the more usual $GL(m;\mathbf{C})$, are preferred here simply to save some space in some of the tables and hopefully make them easier to read. The notation $\tilde{\mathbf{H}}$ denotes the quaternion algebra $\mathbf{H}$ assigned the anti-involution $q = a+bi+cj+dk \mapsto \tilde{q} = a+bi-cj+dk$, while $\overline{\mathbf{H}}$ denotes $\mathbf{H}$ assigned the standard conjugation anti-involution $q = a+bi+cj+dk \mapsto \overline{q} = a-bi-cj-dk$. Geometrically the first of these involves reflection of the space of pure quaternions in a plane, which for definiteness we take to be that with equation $y = 0$ (any two such reflections are conjugate), while the second involves reflection of the space of pure quaternions in the origin.

Note that the three families of general linear groups in this context all turn up as *'unitary'* groups. In the simplest case, that of a symmetric ${}^2\mathbf{R}^{s}$ form, the elements of the group preserving the form actually occur as matrices of the form

$$\begin{pmatrix} a & 0 \\ 0 & (a^{\tau})^{-1} \end{pmatrix}$$

where a is any invertible $m \times m$ matrix and a^{τ} denotes its transpose, the group in this way being isomorphic to the general linear group $GL_m(\mathbf{R})$. In the skew real case or in any of the complex or quaternionic cases the automorphism in the bottom right-hand slot may be different, but that is irrelevant to the nature of the group.

2. The Tables 13.66 in more detail

The Clifford algebras $\mathbf{R}_{p,q}$ for $0 \le p,q < 7$ and $\mathbf{C}_n$ for $0 \le n < 7$ are shown in Tables 1 and 2, where $\mathbf{K}(m)$ denotes the real or complex algebra of $m \times m$ matrices over $\mathbf{K} = \mathbf{R}, {}^2\mathbf{R}, \mathbf{C}, {}^2\mathbf{C}, \mathbf{H}$, or ${}^2\mathbf{H}$.

$p \downarrow$ \ $q \rightarrow$								
±1	R	C	H	^{2}H	H(2)	C(4)	R(8)	^{2}R(8)
R	^{2}R	R(2)	C(2)	H(2)	^{2}H(2)	H(4)	C(8)	R(16)
C	R(2)	^{2}R(2)	R(4)	C(4)	H(4)	^{2}H(4)	H(8)	C(16)
H	C(2)	R(4)	^{2}R(4)	R(8)	C(8)	H(8)	^{2}H(8)	H(16)
^{2}H	H(2)	C(4)	R(8)	^{2}R(8)	R(16)	C(16)	H(16)	^{2}H(16)
H(2)	^{2}H(2)	H(4)	C(8)	R(16)	^{2}R(16)	R(32)	C(32)	H(32)
C(4)	H(4)	^{2}H(4)	H(8)	C(16)	R(32)	^{2}R(32)	R(64)	C(64)
R(8)	C(8)	H(8)	^{2}H(8)	H(16)	C(32)	R(64)	^{2}R(64)	R(128)

Table 1

$q \rightarrow$								
±1	C	^{2}C	C(2)	^{2}C(2)	C(4)	^{2}C(4)	C(8)	^{2}C(8)

Table 2

Table 1 extends indefinitely either way with period 8, while Table 2 has period 2. In either case one obtains the even Clifford algebra in any location by moving one square to the left, if necessary into the additional column on the left.

Tables 13.66 of [3] for the algebras $\mathbf{R}_{p,q}$, and $\mathbf{C}_n$, to be overlaid on the relevant part of Tables 1 and 2, are shown as Tables 3 and 4.

p mod 8 $\downarrow$ \ q mod 8 $\rightarrow$								
^{2}0	0	8	4	^{2}4	4	8	0	^{2}0
0	1	2	3	4	5	6	7	0
8	2	^{2}2	2	8	6	^{2}6	6	8
4	3	2	1	0	7	6	5	4
^{2}4	4	8	0	^{2}0	0	8	4	^{2}4
4	5	6	7	0	1	2	3	4
8	6	^{2}6	6	8	2	^{2}2	2	8
0	7	6	5	4	3	2	1	0

Table 3

n mod 8 $\rightarrow$								
^{2}7	7	9	3	^{2}3	3	9	7	^{2}7

Table 4

By contrast the table for $\overline{\mathbf{C}}_{p,q}$ is given in Table 5.

		$q \bmod 2 \rightarrow$	
$p \bmod 2$	$^{2}8$	8	$^{2}8$
$\downarrow$	8	9	8

Table 5

To complete the set we give as Tables 6 and 7 the tables for $^{2}\mathbf{R}^{s}_{p,q}$ and $^{2}\mathbf{C}^{s}_{n}$, the algebras for these, as algebras, being just the doubles of the algebras $\mathbf{R}_{p,q}$ and $\mathbf{C}_{n}$.

$-p+q \bmod 8 \rightarrow$								
$^{2}1$	1	9	5	$^{2}5$	5	9	1	$^{2}1$

Table 6

$n \bmod 2 \rightarrow$		
$^{2}9$	9	$^{2}9$

Table 7

For the codes 0, 4 and 8 in Tables 3 and 5 there is a further classification by signature. The choice along the top row or down the extra column on the left is the positive definite one. Elsewhere the choice is the neutral one.

Tables 3 to 7 may be appreciated the more if the various code numbers are replaced by the classical groups that preserve the sesquilinear forms on the spinor spaces. We give them here for $0 \le p, q < 8$ as Tables 8 to 12, contenting ourselves with the tables up to $n = p + q = 7$.

$p \downarrow$ \ $q \rightarrow$							
O_1	U_1	Sp_1	$^{2}Sp_1$	Sp_2	U_4	O_8	$^{2}O_8$
$GL_1(\mathbf{R})$	$Sp_2(\mathbf{R})$	$Sp_2(\mathbf{C})$	$Sp_{1,1}$	$GL_2(\mathbf{H})$	$O_4(\mathbf{H})$	$O_8(\mathbf{C})$	
$Sp_2(\mathbf{R})$	$^{2}Sp_2(\mathbf{R})$	$Sp_4(\mathbf{R})$	$U_{2,2}$	$O_4(\mathbf{H})$	$^{2}O_4(\mathbf{H})$		
$Sp_2(\mathbf{C})$	$Sp_4(\mathbf{R})$	$GL_4(\mathbf{R})$	$O_{4,4}$	$O_8(\mathbf{C})$			
$Sp_{1,1}$	$U_{2,2}$	$O_{4,4}$	$^{2}O_{4,4}$				
$GL_2(\mathbf{H})$	$O_4(\mathbf{H})$	$O_8(\mathbf{C})$					
$O_4(\mathbf{H})$	$^{2}O_4(\mathbf{H})$						
$O_8(\mathbf{C})$							

Table 8

$n \rightarrow$							
$O_1(\mathbf{C})$	$GL_1(\mathbf{C})$	$Sp_2(\mathbf{C})$	$^{2}Sp_2(\mathbf{C})$	$Sp_4(\mathbf{C})$	$GL_4(\mathbf{C})$	$O_8(\mathbf{C})$	$^{2}O_8(\mathbf{C})$

Table 9

$q \rightarrow$

$p \downarrow$							
U_1	${}^2U_1$	U_2	${}^2U_2$	U_4	${}^2U_4$	U_8	${}^2U_8$
$GL_1(\mathbf{C})$	U_2	$GL_2(\mathbf{C})$	U_4	$GL_4(\mathbf{C})$	U_8	$GL_8(\mathbf{C})$	
U_2	${}^2U_2$	U_4	${}^2U_4$	U_8	${}^2U_8$		
$GL_2(\mathbf{C})$	U_4	$GL_4(\mathbf{C})$	U_8	$GL_8(\mathbf{C})$			
U_4	${}^2U_4$	$U8$	${}^2U_8$				
$GL_4(\mathbf{C})$	U_8	$GL_8(\mathbf{C})$					
U_8	${}^2U_8$						
$GL_8(\mathbf{C})$							

Table 10

$q \rightarrow$

$p \downarrow$							
$GL_1(\mathbf{R})$	$GL_1(\mathbf{C})$	$GL_1(\mathbf{H})$	${}^2GL_1(\mathbf{H})$	$GL_2(\mathbf{H})$	$GL_4(\mathbf{C})$	$GL_8(\mathbf{R})$	${}^2GL_8(\mathbf{R})$
${}^2GL_1(\mathbf{R})$	$GL_2(\mathbf{R})$	$GL_2(\mathbf{C})$	$GL_2(\mathbf{H})$	${}^2GL_2(\mathbf{H})$	$GL_4(\mathbf{H})$	$GL_8(\mathbf{C})$	$GL_{16}(\mathbf{R})$
etc.							

Table 11

$n \rightarrow$

$GL_1(\mathbf{C})$	${}^2GL_1(\mathbf{C})$	$GL_2(\mathbf{C})$	${}^2GL_2(\mathbf{C})$	$GL_4(\mathbf{C})$	${}^2GL_4(\mathbf{C})$	$GL_8(\mathbf{C})$	${}^2GL_8(\mathbf{C})$

Table 12

The dimensions of the groups in Table 8 are shown in Table 13.

0	1	3	6	10	16	28	56
1	3	6	10	16	28	56	
3	6	10	16	28	56		
6	10	16	28	56			
10	16	28	56				
16	28	56					
28	56						
56							

Table 13

These depend only on the rank $n = p + q$ and not on the index p, q.

The table of differences of this sequence is

0	1	3	6	10	16	28	56	120	256
1	2	3	4	6	12	28	64	136	
1	1	1	2	6	16	36	72		
0	0	1	4	10	20	36			
0	1	3	6	10	16 ,				

where the fifth line repeats the first one. The dimensions of the groups in Table 9 are twice those of the groups in Table 8.

3. Tables of Spin groups

For each n the *norm* $g^{-}g$ of any element g of the group $Spin(n) = Spin(0,n) = Spin(n,0)$ is equal to $+1$, the group being a subgroup of the *even* classical group associated in Table 8 to the index $0,n$ or $n,0$. That group, being the part of the classical group that lies in the even Clifford algebra for the given index, lies in the Table either in the position $0, n-1$ or in the position $n, -1$ (in an extra column on the left that matches the first row). For any p, q with neither p nor q equal to 0 the norm of an element of $Spin(p,q)$ may be equal either to $+1$ or to -1. It is then the subgroup $Spin^{+}(p,q)$, consisting of those elements of $Spin(p,q)$ with norm $+1$, that is a subgroup of the even classical group for the index, namely the classical group in the position $p, q-1$.

The group $Spin(n)$ or $Spin^{+}(p,q)$, with $n = p+q$, has dimension $\frac{1}{2}n(n-1)$. It is the whole group for $n = p+q \leq 5$, but is of dimension one less than this, namely of dimension 15 , rather than 16, for $n = p+q = 6$. In that case it happens that each algebra has a real-valued determinant and lowering the dimension by 1 corresponds to taking the determinant equal to 1. In the case of $GL_2(\mathbf{H})$ the determinant is defined by representing each quaternionic 2×2 matrix as a 4×4 complex matrix and then taking the determinant of that matrix, this necessarily being a positive real number.

The groups $Spin(n)$ for $n \leq 6$, and the groups $Spin^{+}(p,q)$ for $p+q \leq 6$, in the case that both p and q are non-zero, are shown in Table 14.

	$q \rightarrow$						
p	± 1	O_1	U_1	Sp_1	${}^2Sp_1$	Sp_2	SU_4
$\downarrow$	O_1	$GL_1(\mathbf{R})$	$Sp_2(\mathbf{R})$	$Sp_2(\mathbf{C})$	$Sp_{1,1}$	$SL_2(\mathbf{H})$	
	U_1	$Sp_2(\mathbf{R})$	${}^2Sp_2(\mathbf{R})$	$Sp_4(\mathbf{R})$	$SU_{2,2}$		
	Sp_1	$Sp_2(\mathbf{C})$	$Sp_4(\mathbf{R})$	$SL_4(\mathbf{R})$			
	${}^2Sp_1$	$Sp_{1,1}$	$SU_{2,2}$				
	Sp_2	$SL_2(\mathbf{H})$					
	SU_4						

Table 14

The groups $Spin(n;\mathbf{C})$, for $n \leq 6$, are shown in Table 15.

$n \rightarrow$						
± 1	$O_1(\mathbf{C})$	$GL_1(\mathbf{C})$	$Sp_2(\mathbf{C})$	${}^2Sp_2(\mathbf{C})$	$Sp_4(\mathbf{C})$	$SL_4(\mathbf{C})$

Table 15

Most recent writers do not give the whole of these Tables and, as Pertti Lounesto has remarked [2], at least two authors are in error with regard to $Spin^{+}(3,3)$, claiming that it is isomorphic to $SL_4(\mathbf{R})/\mathbf{Z}_2$. In [3] the isomorphism $Spin(6) \cong SU_4$ is proved as Proposition 13.61, but based on Exercise 11.65, while in the second (1981) edition of [3] there is a second proof, based on Diagram 21.6, as a lead up to the discussion of triality. The first of these proofs is the more explicit. The discovery of similar isomorphisms for the groups $Spin^{+}(p,q)$, where $p+q=6$ and neither p nor q is equal to 0, is then set as part of Exercise 13.83. In fact the proof of Proposition 13.61 of [3] is easily adapted to provide proofs of the isomorphisms $Spin^+(3,3) \cong SL_4(\mathbf{R})$ and $Spin(6,\mathbf{C}) \cong SL_4(\mathbf{C})$, as we now show. These are of interest as exemplifying the point stressed at the outset that in the context of Clifford algebras the general linear groups are best thought of as generalised unitary groups, where the relevant conjugation involves the swapping of components of a module over a double field.

The proof involves the following diagram of "left coset exact" sequences of maps

$$\begin{array}{ccccc}
Sp_2(\mathbf{R}) = SL_2(\mathbf{R}) & \longrightarrow & SL_3(\mathbf{R}) & \longrightarrow & S(\mathbf{R}^3 \times \mathbf{R}^3) \\
\downarrow & & \downarrow & & \downarrow ? \\
Sp_4(\mathbf{R}) & \longrightarrow & SL_4(\mathbf{R}) & \xrightarrow{\pi} & Q \subset SL_4(\mathbf{R}) \\
\downarrow & & \downarrow & & \\
S(\mathbf{R}^4 \times \mathbf{R}^4) & \xrightarrow{1} & S(\mathbf{R}^4 \times \mathbf{R}^4), & &
\end{array}$$

where, for any n, $S(\mathbf{R}^n \times \mathbf{R}^n) = \{(x.y) \in \mathbf{R}^n \times \mathbf{R}^n : x \cdot y = 1\}$.

What we prove is that the second row of this diagram may be identified with the left-coset exact sequence

$$Spin^+(3,2) \xrightarrow{\iota} Spin^+(3,3) \xrightarrow{\pi} S(\mathbf{R}^3 \times \mathbf{R}^3),$$

and hence that $Spin^+(3,3)$ may be identified with $SL_4(\mathbf{R})$, the identification of $Spin^+(3,2)$ with $Sp_4(\mathbf{R})$ having been previously established. Explicitly the second map π of this sequence is the obvious action of the group $Spin^+(3,3)$, through the group $SO^+(3,3)$, on the point

$$\left(\begin{pmatrix} 0 \\ 0 \\ 1 \end{pmatrix}, \begin{pmatrix} 0 \\ 0 \\ 1 \end{pmatrix} \right)$$

of the quadric $S(\mathbf{R}^3 \times \mathbf{R}^3)$ in $\mathbf{R}^3 \times \mathbf{R}^3$ with equation $x \cdot y = 1$, that is

$$-(x_1 - y_1)^2 - (x_2 - y_2)^2 - (x_3 - y_3)^2 + (x_1 + y_1)^2 + (x_2 + y_2)^2 + (x_3 + y_3)^2 = 4.$$

The subgroup of the group $Spin^+(3,3)$ that leaves that point fixed is identifiable with $Spin^+(3,2)$, that identification being the map ι, and the fibres of the map π are all the left cosets of this subgroup in $Spin^+(3,3)$.

To return to the diagram, the map $SL_3(\mathbf{R}) \to S(\mathbf{R}^3 \times \mathbf{R}^3)$ is defined by

$$\begin{pmatrix} a_{00} & a_{01} & a_{02} \\ a_{10} & a_{11} & a_{12} \\ a_{20} & a_{21} & a_{22} \end{pmatrix} \mapsto \left(\begin{pmatrix} a_{02} \\ a_{12} \\ a_{22} \end{pmatrix}, \begin{pmatrix} A_{02} \\ A_{12} \\ A_{22} \end{pmatrix} \right),$$

the first component of the image being the last column of the matrix a and the second component the last column of the the matrix $(a^{\tau})^{-1}$, with

$$a_{02}A_{02} + a_{12}A_{12} + a_{22}A_{22} = \det a = 1.$$

The fibres of this map are the left cosets in $SL_3(\mathbf{R})$ of the subgroup $SL_2(\mathbf{R})$, identified with the inverse image of

$$\left(\begin{pmatrix}0\\0\\1\end{pmatrix},\begin{pmatrix}0\\0\\1\end{pmatrix}\right).$$

The map $SL_4(\mathbf{R}) \to S(\mathbf{R}^4 \times \mathbf{R}^4)$ is analogously defined.

The map $\pi : SL_4(\mathbf{R}) \to Q \subset SL_4(\mathbf{R})$ is defined, as is the map π in Exercise 11.65 of [3], by

$$a \mapsto \begin{pmatrix} a_{00} & a_{01} & a_{02} & a_{03}\\ a_{10} & a_{11} & a_{12} & a_{13}\\ a_{20} & a_{21} & a_{22} & a_{23}\\ a_{30} & a_{31} & a_{32} & a_{33}\end{pmatrix}\begin{pmatrix} a_{11} & -a_{01} & a_{31} & -a_{21}\\ -a_{10} & a_{00} & -a_{30} & a_{20}\\ a_{13} & -a_{03} & a_{33} & -a_{23}\\ -a_{12} & a_{02} & -a_{32} & a_{22}\end{pmatrix},$$

with fibres the left cosets in $SL_4(\mathbf{R})$ of the group $Sp_4(\mathbf{R})$ identified with the inverse image by π of the unit matrix. Note that the inverse of any element of $Sp_4(\mathbf{R})$ is then explicitly given.

The map $Sp_4(\mathbf{R}) \to S(\mathbf{R}^4 \times \mathbf{R}^4)$ is defined by

$$a \mapsto \begin{pmatrix} a_{00} & a_{01} & a_{02} & a_{03}\\ a_{10} & a_{11} & a_{12} & a_{13}\\ a_{20} & a_{21} & a_{22} & a_{23}\\ a_{30} & a_{31} & a_{32} & a_{33}\end{pmatrix} \mapsto \left(\begin{pmatrix}-a_{12}\\ a_{02}\\ -a_{32}\\ a_{22}\end{pmatrix},\begin{pmatrix}a_{03}\\ a_{13}\\ a_{23}\\ a_{33}\end{pmatrix}\right),$$

with

$$a_{02}a_{13} - a_{12}a_{03} + a_{22}a_{33} - a_{32}a_{03} = (-a_{12}, a_{02}, -a_{32}, a_{22}) \cdot (a_{03}, a_{13}, a_{23}, a_{33}) = 1.$$

That the map between the two copies of $S(\mathbf{R}^4 \times \mathbf{R}^4)$ is the identity is then easily verified. By elementary diagram chasing (as in Proposition 5.24 of [3]) it follows that there is a linear map $\mathbf{R}^3 \times \mathbf{R}^3 \to \mathbf{R}(4)$, restricting to a bijection between $S(\mathbf{R}^3 \times \mathbf{R}^3)$ and the image Q of π, that makes the diagram commute. Explicitly this is the map

$$\left(\begin{pmatrix}x_0\\x_1\\x_2\end{pmatrix},\begin{pmatrix}y_0\\y_1\\y_2\end{pmatrix}\right) \mapsto \begin{pmatrix} y_2 & 0 & x_0 & y_1\\ 0 & y_2 & x_2 & -y_0\\ -y_0 & -y_1 & x_2 & 0\\ -x_1 & x_0 & 0 & x_2\end{pmatrix}.$$

The image of this linear map may be taken to be the space of *paravectors*, as they have recently come to be known, in the even Clifford algebra $\mathbf{R}^0_{3,3}$. Identification of $SL_4(\mathbf{R})$ with $Spin^+(3,3)$ follows directly.

The whole of the above argument goes through unchanged if the field $\mathbf{R}$ is replaced by the field $\mathbf{C}$. Accordingly $SL_4(\mathbf{C})$ may be identified with $Spin(6;\mathbf{C})$.

As a matter of fact it is enough simply to prove this last result. Then all the real cases follow by restriction.

4. Tables for Anti-involutions other than Conjugation

To return to Table 3, this table classifies conjugation for each of the algebras $\mathbf{R}_{p,q}$. The *reversion* anti-involution of $\mathbf{R}_{p,q}$, induced by the identity endomorphism of the orthogonal space $\mathbf{R}^{p,q}$, is classified by a similar table, namely Table 16.

$q \bmod 8$ $\rightarrow$								
$p \bmod 8$	0	7	6	5	4	3	2	1
$\downarrow$	^{20}	0	8	4	^{24}	4	8	0
	0	1	2	3	4	5	6	7
	8	2	^{22}	2	8	6	^{26}	6
	4	3	2	1	0	7	6	5
	^{24}	4	8	0	^{20}	0	8	4
	4	5	6	7	0	1	2	3
	8	6	^{26}	6	8	2	^{22}	2

Table 16

In this case positive-definiteness of the spinor product and compactness of the associated classical group reigns down the left-hand column, with neutrality everywhere else where there is a choice. Otherwise Table 16 is just Table 3 translated one square to the South-East.

More generally any orthogonal involution of $\mathbf{R}^{p,q}$ induces an anti-involution of $\mathbf{R}_{p,q}$, and one would like to classify all such. The relevant table for the algebras $\mathbf{R}_{0,n}$ just looks like Table 3 turned upside down, as displayed in Table 17.

0	8	4	^{24}	4	8	0	^{20}
7	6	5	4	3	2	1	0
6	^{26}	6	8	2	^{22}	2	8
5	6	7	0	1	2	3	4
4	8	0	^{20}	0	8	4	^{24}
3	2	1	0	7	6	5	4
2	^{22}	2	8	6	^{26}	6	8
1	2	3	4	5	6	7	0

Table 17

Here the first row classifies conjugation and the first column classifies reversion. The transition from conjugation to reversion as fewer and fewer in turn of the standard basis vectors of $\mathbf{R}^{p,q}$ have their signs changed by the involution is made by traversing this table steadily from the first row to the first column in a South-Westerly direction. At all intermediate stages neutrality is the rule.

For the algebras $\mathbf{R}_{p,n-p}$ comprising the pth row of Table 3 there is a similar table. For example in the case that $p = 2$ we have, for the algebras $\mathbf{R}_{2,n-2}$, Table 18, which is just Table 17, bordered by *two* additional rows on the top and *two* additional rows on the left that respect the periodicity 8.

		2	$^{2}2$	2	8	6	$^{2}6$	6	8
	0	1	2	3	4	5	6	7	0
0	$^{2}0$	0	8	4	$^{2}4$	4	8	0	$^{2}0$
1	0	7	6	5	4	3	2	1	0
2	8	6	$^{2}6$	6	8	2	$^{2}2$	2	8
3	4	5	6	7	0	1	2	3	4
4	$^{2}4$	4	8	0	$^{2}0$	0	8	4	$^{2}4$
5	4	3	2	1	0	7	6	5	4
6	8	2	$^{2}2$	2	8	6	$^{2}6$	6	8
7	0	1	2	3	4	5	6	7	0

Table 18

Here also the transition from conjugation to reversion as fewer and fewer in turn of the standard basis vectors of $\mathbf{R}^{p,q}$ have their signs changed by the involution is made by traversing the Table steadily from the first row to the first column in a South-Westerly direction. The spinor product is positive-definite if the involution changes the sign of all vectors of square -1 while all those of square $+1$ remain fixed. It is neutral otherwise.

The reversion classification of the algebras $\mathbf{C}_n$ coincides with the conjugation classification.

5. Historical note

The explicit classifications of the conjugation and reversion anti-involutions in the tables of Clifford algebras are in a 1969 Liverpool MSc thesis by Tony Hampson [1]. On obtaining the answers Hampson and I wrote to my colleague Prof. Terry Wall who was at that time on a visit to Mexico. He replied with two letters, written a couple of weeks apart, drawing our attention to his paper [4] which we had not read, and which presented the entire theory very succinctly! It was only later that we learned that between these two letters he had been driven 100 miles by his wife Sandra to the hospital in Acapulco to have his appendix out! The original question which led to our undertaking this work was posed by my colleague Dr R. (Bob) H. Boyer, who was tragically the victim of a madman's bullet in Austin, Texas, on August 1st, 1966. To his memory I dedicate this paper.

References

1. Hampson, A. (1969) *On the Table of Clifford Algebras*, M.Sc. Thesis, University of Liverpool.
2. Lounesto, P. (1992) Private communication.
3. Porteous, I.R. (1969) *Topological Geometry*, 1st Edition, Van Nostrand, (1981) 2nd Edition, with additional material on Triality, Cambridge University Press.
4. Wall, C.T.C. (1968) 'Graded algebras, antiinvolutions, simple groups and symmetric spaces,' *Bull. Am. Math Soc.*74, 198-202.

FINITE GEOMETRY AND THE TABLE OF REAL CLIFFORD ALGEBRAS

R. SHAW
School of Mathematics
University of Hull
Hull HU6 7RX
England

Abstract. Associated with the real Clifford algebra $Cl(p, q)$, $p+q=n$, is the finite Dirac group $G(p, q)$ of order 2^{n+1}. The group $V_n = G(p, q)/\{\pm 1\}$, viewed additively, is an n-dimensional vector space over GF(2) = {0, 1} which comes equipped with a quadratic form Q and associated alternating bilinear form B. The finite geometry of V_n, B, Q, in part familiar, in part less so, is described, and is then used in conjunction with the representation theory of $G(p, q)$ to give a pleasantly clean derivation of the well-known table, [9], of the algebras $Cl(p, q)$. In particular the finite geometry highlights the "antisymmetry" of the table about the column $q-p=1$.

Key words: Finite geometry, Clifford algebra, Caps on quadrics

1. Introduction: the GF(2) Connection

Let $Cl(p, q)$ denote the universal Clifford algebra for a real n-dimensional orthogonal space $R(p, q)$ of signature (p, q). If $\{e_1, e_2, ..., e_n\}$ is a fixed choice of orthonormal basis for $R(p, q) \subset Cl(p, q)$, then the e_i (together with -1, if $n=p=1$) generate, under Clifford multiplication, a finite group $G(p, q)$, called the *Dirac group* of $Cl(p, q)$, whose order is 2^{n+1} : $G(p, q) = \{\pm 1, \pm e_i, \pm e_i e_j, ..., \pm w\}$, $i<j< ...$, where $w = e_1 e_2 \ldots e_n$. Observe that every commutator $ghg^{-1}h^{-1}$ and every square g^2 in $G(p, q)$ is equal to $+1$ or -1. Consequently the quotient group $G(p, q)/\{\pm 1\}$ by the central subgroup $\{1, -1\}$ is an elementary abelian 2 - group V_n of order 2^n. We will therefore view V_n additively, as a vector space of dimension n over the finite field $\mathrm{GF}(2) = \mathbf{F}_2 = \{0, 1\}$. If, for the central extension $\{\pm 1\} \to G(p, q) \to V_n$, we choose a section $e : V_n \to G(p, q) : x \mapsto e_x$, we can display the Dirac group as $G(p, q) = \{\pm e_x : x \in V_n\}$. Then $e_x e_y = (-1)^{f(x,y)} e_{x+y}$, for some function $f : V_n \times V_n \to GF(2)$ which depends on our choice of section. However in

$$(e_x)^2 = (-1)^{Q(x)}, x \in V_n, \tag{1.1}$$

the function $Q : V_n \to GF(2)$ is independent of our choice of section. Since $(e_x e_y)^2 = (e_{x+y})^2$, it follows from (1.1) that we have the commutator result

$$e_x e_y (e_x)^{-1} (e_y)^{-1} = (-1)^{B(x,y)} \tag{1.2}$$

where $B : V_n \times V_n \to GF(2)$ is given by $B(x, y) = Q(x+y) + Q(x) + Q(y)$ and is thus also independent of our choice of section. Clearly $B(x, x) = 0$; also an easy check shows that B is a bilinear form on V_n. Consequently, as after eq. (2.1) below, *Q is a quadratic form on V_n*, with B its associated alternating bilinear form.

F. Brackx et al. (eds.), Clifford Algebras and their Applications in Mathematical Physics, 23–31.

Is B non-degenerate? The answer depends on whether n is even or odd. If n is even then $G(p, q)$ has centre $\{\pm 1\}$, which fact entails that the only vector $k \in V_n$ which satisfies $B(k, x) = 0$ for all $x \in V_n$ is the zero vector – i.e. B is non-degenerate. If $n = 2m + 1$ is odd then B, being alternating, is necessarily degenerate. *However it is still of maximal rank*, namely 2m. For $G(p, q)$ now has centre $\{\pm 1, \pm w\}$, which fact entails, on writing $\pm w = \pm e_k$, that B has 1-dimensional kernel spanned by k.

REMARK Clifford algebraists and finite geometers frequently carry out the same computations without realizing it — the former proceeding multiplicatively using commutativity / anticommutativity and the latter proceeding additively using orthogonality / non-orthogonality. As an elementary example, in Clifford algebra one argues that if e_x anticommutes with both e_y and e_z, then e_x commutes with their product $e_y e_z$; in finite geometry (over GF(2)) one argues that if x is non-orthogonal to both y and z, then x is orthogonal to their sum $y + z$. (For from $B(x, y) = 1$ and $B(x, z) = 1$ it follows by bilinearity that $B(x, y + z) = 1 + 1 = 0$.) Similarly all Clifford algebraists will (time and time again!) have computed the square of the volume element $w = e_1 e_2 \ldots e_n$ and arrived at the result $w^2 = (-1)^{q+m}$, where $m = [n/2]$. For the finite geometry computation, see the proof of lemma 4.6. As a third example, the results of Dye [3] could have been deduced from the table (theorem 5.2) of real Clifford algebras. Also much of [5] is essentially Sp(4, 2) geometry.

Sections 2 – 4 below deal with relevant finite geometry, but in outline only, with many proofs omitted, due to lack of space. (A fuller account, [12], will be available on request.) It should be remarked that the geometry in question is of a rather peculiar kind, due to the ground field being GF(2). Moreover the results in Sections 3 and 4, and in particular theorem 4.5, which are crucial as far as Clifford algebra applications are concerned, appear to be little known even by finite geometry specialists. There are, see [12], many interesting interactions of this area of finite geometry with real Clifford algebras, but in this paper we chiefly confine ourselves, in Section 5, to explaining how the finite geometry result in theorem 4.5 translates into the well-known table of real Clifford algebras, as given in [9], table 13.26.

Of course the fact that the Dirac group $G(p, q)$, although finite, nevertheless captures much of importance concerning the Clifford algebra $Cl(p, q)$ has been appreciated previously by many authors, see e.g. [4], [2]. That the study of $G(p,q)$ profitably involves finite geometry over GF(2) is also well-known: accounts of relevant mathematics can be found dotted about the literature, cf. [10], [7], [6]. Nevertheless the finite geometry derivation in Section 5 of the table of real Clifford algebras has not, it seems appeared previously in the literature.

2. Finite Geometry over GF(2)

2.1. SYMPLECTIC GEOMETRY

Let $V = V_n = V(n, 2)$ denote any n-dimensional vector space over GF(2). Due to the peculiar nature of GF(2) a one-dimensional subspace $\prec x \succ \subset V$ is spanned by a *unique* nonzero vector $x \in V$. Consequently the points of the projective geometry $PG(n-1, 2) = \mathbf{P}(V)$ associated with V can be identified with the nonzero vectors of V. Of course, relative to a choice of basis for V, we have $V_n \cong (\mathbf{F}_2)^n$, and so $|V_n| = 2^n$ and $|PG(n-1, 2)| = 2^n - 1$.

Suppose now that V_n is equipped with a preferred alternating bilinear form $B : V \times V \to GF(2)$. Note that the alternating condition $B(x, x) = 0$ for all $x \in V$, in conjunction with bilinearity, implies (since $-1 = +1$) that B is symmetric. At times it proves convenient to abbreviate $B(x, y)$ to $x.y$. Two vectors x, y are said to be orthogonal, or perpendicular, whenever $x.y = 0$. However it seems clear enough that *over GF(2) non-orthogonality of vectors should be considered an equally important property.* For the "negative" information $x.y \neq 0$ implies, over GF(2), the "positive" fact that $x.y = 1$. This insight is followed up in sections 3, 4 below.

Since B is alternating its rank is even. If V has even dimension $n = 2m$ and if B is non-degenerate, i.e. of rank 2m, then the invariance group Sp(V) of B is (isomorphic to) the symplectic group Sp(2m, 2). It is well-known, [1], that two alternating forms on a vector space V (over any field) lie on the same GL(V)-orbit if and only if they have the same rank.

2.2. ORTHOGONAL GEOMETRY

Suppose now that $V = V(n, 2)$ is equipped with a preferred quadratic form $Q : V \to GF(2)$. Then Q determines an associated alternating bilinear form B via the relation

$$Q(x+y) + Q(x) + Q(y) = B(x,y). \tag{2.1}$$

In fact, over GF(2), a quadratic form Q on V is nothing more than *any* solution Q of (2.1) for some choice of alternating bilinear form B.

LEMMA 2.1 A given non-degenerate alternating form B on V_{2m} supports precisely two distinct Sp(2m, 2)-orbits Quad$^{\epsilon}(B)$, $\epsilon = \pm 1$, of quadratic forms, with Quad$^{\epsilon}(B)$ consisting of $N^{\epsilon} = 2^{2m-1} + \epsilon 2^{m-1}$ quadratic forms Q^{ϵ}, each having N^{ϵ} zeros. ∎

If $Q \in$ Quad$^{\epsilon}$, we write $\epsilon(Q) = \epsilon$ and refer to $\epsilon(Q)$ as the type of Q. Quadratic forms of types +1, −1 are also termed *hyperbolic*, *elliptic*, respectively. In projective language the nonzero vectors x satisfying $Q(x) = 0$ are referred to as the points on a projective quadric in PG(2m — 1, 2). This quadric (still assuming we are dealing with the case when the associated alternating form B is nondegenerate) is said to be of the type $\mathcal{H}_{2m-1}$ or $\mathcal{E}_{2m-1}$ according as $\epsilon(Q) = +1$ or −1, see [8]. Of course for Q of type ϵ the *projective* quadric has $N^{\epsilon} - 1$ points on it.

Consider also the odd-dimensional case in which $V_n = V_{2m+1}$ is equipped with an alternating form B having rank 2m and one-dimensional kernel $\prec k \succ$. Solutions Q of (2.1) such that $Q(k) = 1$ are said to be of *parabolic* type. In projective language, see [8], such a Q defines a $\mathcal{P}_{2m}$ quadric with *nucleus* k. Solutions Q of (2.1) such that $Q(k) = 0$ fall into two kinds: in projective language they define *hyperbolic* and *elliptic cones*, $\Pi_0\mathcal{H}_{2m-1}$ and $\Pi_0\mathcal{E}_{2m-1}$, whose *vertex* Π_o is k. Unlike the parabolic quadrics, the latter are *singular* in that their equations can be expressed using fewer than the $2m+1$ coordinates. In the odd-dimensional case under consideration we define the type $\epsilon(Q)$ of Q to be +1, 0 or −1 according as Q is of the foregoing hyperbolic, parabolic or elliptic types, respectively. By making use of the knowledge gained in lemma 2.1 one easily derives the next lemma.

LEMMA 2.2 The three kinds $\epsilon(Q) = +1, 0, -1$ of quadratic form Q on $V(2m+1, 2)$ which are supported by an alternating form B of maximal rank 2m are distinguished by the property that $Q(x) = 0$ has $2^{2m} + \epsilon(Q)2^m$ solutions. A given B supports 2^{2m} Q's of the parabolic type $\epsilon = 0$ and $2^{2m-1} + \epsilon 2^{m-1}$ Q's of type $\epsilon = \pm 1$. ∎

In the case of a quadratic form Q on V_n whose alternating form B is of maximal rank, let $\delta(Q) =$ (no. of zeros of Q) – (no. of non-zeros of Q). From lemmas 2.1, 2.2 we see that

$$\delta(Q) = \begin{cases} \epsilon(Q)2^m & \text{if } n = 2m \\ \epsilon(Q)2^{m+1} & \text{if } n = 2m+1. \end{cases} \tag{2.2}$$

The following lemma, whose proof is immediate, will also prove useful.
LEMMA 2.3 Let B be of maximal rank on V_n. For $Q_0 \in \text{Quad}(B)$ and $a \in V_n$, let $Q_a \in \text{Quad}(B)$ be defined by $Q_a(x) = Q_0(x) + B(a, x) = Q_0(a + x) + Q_0(a)$. Then $\epsilon(Q_a) = +\epsilon(Q_0)$, if $Q_0(a) = 0$, and $= -\epsilon(Q_0)$, if $Q_0(a) = 1$. ∎

3. Non-orthogonality and Symplectic Geometry

Let $V = V(n, 2)$ be equipped with a preferred alternating bilinear form B. Initially we make no restrictions on the rank of B. Consider a set $S = \{a_1, a_2, ..., a_s\}$ of vectors which are mutually non-perpendicular: $a_i.a_j \equiv B(a_i, a_j) = 1, i \neq j$. We will refer to such a set as an OS (=Off-diagonal Set). Perhaps the "O" can remind one that S is off-diagonal in an Outstanding manner, and the "S" can remind one that the geometry is Symplectic. Our terminology leads to pronounceable acronyms — unlike alternatives such as PNPS (=Pair-wise Non-Perpendicular Set) or MNCS (= Mutually Non-Conjugate Set). An OS S in V_n will be termed

a BOS for V_n if S is a Basis for V_n,

a COS if S is Complete — in the sense that there does not exist $x \in V$ such that $x.a_i = 1$, each $i = 1, 2, ..., s$,

a DOS if S is a linearly Dependent set,

a MOS if $s = |S|$ is Maximal for the given $n = \dim V$.

Given the OS $S = \{a_1, ..., a_s\} \subset V_n$, let S denote the subspace $\prec S \succ$ and put $u = \sum a_i \in S$. Then lemmas 3.1 – 3.5 below hold. Their proofs are all very easy, depending on very little more than that a sum of t $1's$ is, over GF(2), equal to 0 or 1 according as t is even or odd. Theorem 3.6 is an immediate consequence of these lemmas.

LEMMA 3.1 Let $0 \neq x \in S = \prec a_1, ..., a_s \succ$. Then

(i) $x.a_i = 0, i = 1, ..., s \Leftrightarrow s = 2r + 1$ and $x = u$,

(ii) $x.a_i = 1, i = 1, ..., s \Leftrightarrow s = 2r$ and $x = u$. ∎

LEMMA 3.2 Suppose that $s = |S| = 2r$ is even. Then

(i) S is nonsingular (i.e. $S \cap S^{\perp} = \{0\}$),

(ii) S is a BOS for S (and so even-sized DOS's do not exist),

(iii)$S^* = S \cup \{u\}$ is the unique extension of S to a larger OS for S (and then $S^* - \{a\}$ is a BOS for S for any $a \in S$ *),

(iv) S^* is a COS for V_n. ∎

LEMMA 3.3 Suppose that $s = 2r + 1 (r \geq 1)$ is odd and that S is a DOS. Then

(i) $\sum_1^s a_i = 0$, this being the only linear dependence amongst the $a_i \in S$ (and so, for each i, $S - \{a_i\}$ is a BOS for S and $\dim\ S = 2r$),

(ii) S is non-singular,

(iii) S is a COS for V_n. ∎

LEMMA 3.4 Suppose that $s = 2r + 1$ is odd and that $\mathcal{S}$ is a BOS for S (and so $\dim S = 2r + 1$). Then

(i) rad $S(= S \cap S^{\perp}) = \prec u \succ$,

(ii) $\mathcal{S}$ is a COS for S. ∎

LEMMA 3.5 (i) Suppose that B is non-degenerate on V_{2m}. Given an orthogonal decomposition $V_{2m-2} \perp V_2$ let $\{b_1, ..., b_{2m-2}\}$ be any BOS for V_{2m-2} and $\{u, v\}$ any basis for V_2. Then $\{a_1, ..., a_{2m-2}, u, v\}$ is a BOS for V_{2m} if $a_i = b_i + u + v$.

(ii) Suppose that B has rank 2m on V_{2m+1}. Given a decomposition $V_{2m+1} = V_{2m} \perp \prec k \succ$, let $\{a_1, ..., a_{2m}\}$ be any BOS for V_{2m}. Then $\{a_i, ..., a_{2m}, k + u\}$ is a BOS for V_{2m+1} if $u = \sum a_i$. ∎

THEOREM 3.6 If V_n is equipped with an alternating bilinear form B of maximal rank, then

(i) BOS's exist for (V_n, B),

(ii) a MOS for V_{2m} has size $2m + 1$ and a MOS for V_{2m+1} has size $2m + 1$,

(iii) if $n = 2m$ every MOS is the extension of a BOS for V_n,

(iv) if $n = 2m + 1$ there are two kinds of MOS:

a) the BOS's for V_n,

b) the MOS's for the nonsingular hyperplanes $V_{2m} \subset V_n$;

moreover, for any BOS $\{a_1, ..., a_{2m+1}\}$, $\sum a_i$ spans rad(V_{2m+1}),

(v) COS's exist of any odd size $2r + 1$, $1 \le r \le m = [n/2]$. ∎

4. BOS's, MOS's and Quadratic Forms

Let $\mathcal{S} = \{a_1, \dots , a_n\}$ be a BOS for (V_n, B). Given non-negative integers p, q, with $p + q = n$, there exists a unique quadratic form Q on V_n supported by B such that $Q(a_1) = ... = Q(a_p) = 0$, and $Q(a_{p+1}) = ... = Q(a_n) = 1$. For equation (2.1) fixes Q up to a linear form, and the preceding conditions fix the linear form. Indeed by varying the order of the basis vectors a_i, and varying (p, q), we obtain in this way all 2^n quadratic forms Q satisfying (2.1). We will say that Q is of type (p, q) with respect to the BOS $\mathcal{S}$ for (V_n, B). We will also say that $\mathcal{S}$ is a BOS of type (p, q) with respect to Q. *In fact we shall only be interested in the case of quadratic forms Q whose B is of maximal rank.* We wish to define a function $\epsilon(p, q)$ of the non-negative integers p, q by $\epsilon(p, q) = \epsilon(Q)$, for any Q of type (p, q). This definition is allowed on account of the following lemma.

LEMMA 4.1 If also Q' is of type (p, q) with respect to a BOS $\mathcal{S}' = \{a'_n, ..., a'_n\}$ for (V_n, B') (with B' of maximal rank), then $\epsilon(Q') = \epsilon(Q)$.

PROOF If $T \in \mathrm{GL}(V_n)$ is defined via $Ta_i = a'_i$, then $B' = {}^T B$ and $Q' = {}^T Q$. ∎

LEMMA 4.2 $\epsilon(p + 1, q + 1) = \epsilon(p, q)$ ("mod (1,1) periodicity")

PROOF Use $Q_n \oplus Q_2$ on $V_n \perp V_2$, with BOS's as in lemma 3.5(i); see [12]. ∎

LEMMA 4.3 (i) $\epsilon(p + 1, q) = \epsilon(q + 1, p)$

(ii) $\epsilon(p, q + 1) = -\epsilon(q, p + 1)$.

PROOF In lemma 2.3 take (i) $a = a_1$ (ii) $a = a_{p+q+1}$, where $\{a_1, \dots, a_{p+q+1}\}$ denotes a BOS with respect to which Q_o has type (i) $(p + 1, q)$ (ii) $(p, q + 1)$. ∎

COROLLARY 4.4 (i) $\epsilon(p, q + 4) = \epsilon(p + 4, q)$ ("mod (4, −4) periodicity")

(ii) $\epsilon(p + 8, q) = \epsilon(p, q) = \epsilon(p, q + 8)$ ("mod 8 periodicity") ∎

THEOREM 4.5

$$\epsilon(p,q) = \begin{cases} +1 & \text{if } p-q \equiv 0,1,2 \pmod 8 \\ 0 & \text{if } p-q \equiv 3,7 \pmod 8 \\ -1 & \text{if } p-q \equiv 4,5,6 \pmod 8. \end{cases}$$

PROOF Let the values $\epsilon(p, q)$ be set out as a table in which the columns are labelled by $q-p$ and the rows by $p+q$. By lemma 4.2 the values are constant down each column. By lemma 4.3 *the table is symmetric about the column* $q-p=-1$, *and skew symmetric about the column* $q-p=+1$ (which last is therefore a column of 0's). The whole table now follows from the trivial values $\epsilon(1,0)=1$, $\epsilon(1,1)=1$ upon repeatedly reflecting about the two columns $q-p=\pm 1$ – leading to the stated values for $\epsilon(p, q)$. ∎

LEMMA 4.6 For Q of type (p,q) w.r.t. the BOS $\{a_1, ..., a_n\}$, and $u=\Sigma a_i$, we have

$$Q(u) = \begin{cases} 0 & \text{if } p-q \equiv 0,1 \pmod 4 \\ 1 & \text{if } p-q \equiv 2,3 \pmod 4. \end{cases}$$

PROOF Let $a_{ij}=a_i+a_j$, and note that, for example $a_{12}.a_{34}=1+1+1+1=0$ and $Q(a_{12})=Q(a_1)+Q(a_2)+1$. Consequently $Q(u)=\Sigma_i Q(a_i)+[n/2]=q+[n/2]$. ∎

Let us now adopt projective language. Given a quadric Q^ϵ in $PG(n-1,\ 2)$, of type ϵ, then it supports BOS's of type (p,q) – meaning p points of the BOS on Q^ϵ, and q points off Q^ϵ – precisely for those $(p,\ q)$ satisfying $\epsilon(p,\ q)=\epsilon$, where $\epsilon(p,\ q)$ is as given by theorem 4.5. It is also of interest to know the type $(r,\ s)$, say, of MOS's supported by Q^ϵ. Of course, by theorem 3.6(ii), $r+s=2m+1$ where $m=[n/2]$.

THEOREM 4.7 A quadric Q^ϵ, $\epsilon=\pm 1$, in $PG(n-1,\ 2)$ supports a MOS of type $(r,\ s)$ if and only if $r-s+1\equiv 2\epsilon \pmod 8$. A parabolic quadric ($\epsilon=0$) supports MOS's of any type $(r,\ s)$, such a MOS being a BOS if and only if $r-s+1\equiv 0,\ 4 \pmod 8$.

PROOF Use theorem 4.5 in conjunction with lemma 4.6 and theorem 3.6(iii),(iv). ∎

For example a $\mathcal{H}_5$ quadric supports MOS's only of type (4,3) or (0,7), and a $\mathcal{E}_5$ only those of type (6,1) or (2,5). Similarly a $\mathcal{H}_7$ supports only the types (9,0), (5,4), (1,8), and a $\mathcal{E}_7$ only the types (7,2), (3,6). Of special interest to geometers are OS's of type $(k,0)$, called, see [3], k - *caps* for the quadric Q, for maximal k, and OS's of type $(0,l)$, called l - *non-secant sets*, for maximal l. The maximal sizes for k and l on the whole follow immediately from theorem 4.7. For example $\mathcal{H}_7$ supports a 9-cap, while the maximal size of a cap for $\mathcal{E}_7$ is 7. However in some instances an extra consideration enters. For example consider $\mathcal{E}_7$ again. We know that a non-polar set of type (3,6) exists, and hence an OS of type (0,6); but by lemmas 3.2,4.6 the latter extends to a COS of type (0,7); we thus see that the maximal size of a non-secant set for $\mathcal{E}_7$ is 7, not 6. In this manner *our results are seen to subsume all those of Dye* [3], but are more general in two respects: (i) we treat OS's of type $(r,\ s)$, not just those of type $(k,0)$ or $(0,l)$; (ii) our results apply also to the cones $\Pi_o\mathcal{H}_{2m-1}$ and $\Pi_0\mathcal{E}_{2m-1}$. Our approach via theorem 4.5 was motivated by knowing the importance of the corresponding Clifford algebra theorem 5.2, and appears to us to be simpler that that of [3].

5. Finite Geometry Derivation of the Table of Real Clifford Algebras

From now on the finite geometry of (V_n, B, Q) will be that arising from the Dirac group $G(p,q) \subset Cl(p,q)$ via the central extension $\{\pm 1\} \rightarrow G(p,q) \rightarrow^{\pi} V_n$. If $x \in V_n$ then $\pi^{-1}(x) = \{e_x, -e_x\}$ for a section $e : V_n \rightarrow G(p,q)$. Observe that $x.y = 0$ or 1 according as $e_x e_y = +e_y e_x$ or $- e_y e_x$. Consequently the notion in section 3 of an OS $\{a_1, ..., a_s\}$ translates into that of an AS (=Anticommuting Set) $\{u_1, ..., u_s\} \subset G(p,q)$, where $u_i u_j = -u_j u_i$, $i \neq j$, and $\pi^{-1}(a_i) = \{u_i, -u_i\}$. Similarly the notion of a BOS, COS, DOS, MOS translate into corresponding notions, say BAS, CAS, DAS, MAS, for $G(p,q)$. For example the orthonormal basis $\{e_1, ..., e_n\}$ for $R(p,q)$ is a BAS $\subset G(p,q)$ which corresponds to a BOS for V_n of type (p,q) with respect to Q; also, if n is even, $\{e_1, ..., e_n, w\}$ is a MAS for $G(p,q)$, cf. theorem 3.6(iii). At this point it would be a pleasant exercise, see [12], to compile a dictionary of translation between concepts and results for (V_n, B, Q) and their counterparts in $G(p,q)$.

Instead we proceed directly to a finite geometry proof of theorem 5.2. To this end we need those complex irreducible representations (irreps) of $G(p,q)$, $p+q=n$, which faithfully represent $\{\pm 1\}$. The class structure of G entails that in the even case $n = 2m$ there is *one* such irrep D, of dimension 2^m (the remaining 2^n irreps of G being the 1-dimensional ones arising from V_n) and that in the odd case $n = 2m+1$ there are *two* such irreps D_+, D_-, each of dimension 2^m. By Schur's lemma $D_+(w) = \alpha I$ (with $\alpha = \pm 1$ *or* $\pm\, i$ according as $w^2 = +1$ or -1). D_- can be taken to agree with D_+ on the even elements of $G(p,q)$, but differ by a sign on the odd elements; in particular $D_-(w) = -\alpha I$. We now use (a) Burnside's theorem, applied to the complex irreps of $G(p,q)$ of dimension 2^m, $m = [n/2]$, and (b) the fact that for $n = 2m+1$ the complexified algebra $Cl(p,q)^c$ possesses the central idempotents e_+, $e_- = 1 - e_+$, where $e_+ = (1+w)/2$, if $w^2 = +1$, and $e_+ = (1+iw)/2$, if $w^2 = -1$. We thereby derive the (very well-known) result.

$$Cl(p,q)^c \cong \begin{cases} \mathbf{C}(2^m) & \text{if } n = 2m \\ \mathbf{C}(2^m) \oplus \mathbf{C}(2^m) & \text{if } n = 2m+1. \end{cases} \tag{5.1}$$

However we are interested in the *real* algebras $Cl(p,q)$, and so in order to progress further we need to know the reality type of the irrep D of $G(p,q)$, as given by its ***Frobenius-Schur indicator*** $\eta(p,q)$, which by definition equals $+1$, 0 or -1 according as D is of real, strictly complex or quaternionic type, respectively. Now $\eta(p,q)$ is determined either (see [11]) by

$$Av_G D(g^2) = \eta(p,q)/dim\ D \tag{5.2}$$

or, on taking the trace, by the well-known classical criterion

$$Av_G \chi(g^2) = \eta(p,q) \quad (\chi = \text{ character of } D). \tag{5.3}$$

THEOREM 5.1 $\eta(p,q) = \epsilon(p,q)$.

PROOF Use equation (5.2) in conjunction with equations (1.1) and (2.2). ∎

Now it is easy to see that the Frobenius-Schur type $\epsilon(p,q)$ of the complex irrep D of $G(p,q)$ determines the real Clifford algebra $Cl(p,q)$, as follows. If $n = 2m$, then

$$Cl(p,q) \cong \begin{cases} \mathbf{R}(2^m) & \text{if } \epsilon(p,q) = +1 \\ \mathbf{H}(2^{m-1}) & \text{if } \epsilon(p,q) = -1. \end{cases} \tag{5.4}$$

while if $n = 2m + 1$, then

$$Cl(p,q) \cong \begin{cases} \mathbf{R}(2^m) \oplus \mathbf{R}(2^m) & \text{if } \epsilon(p,q) = +1 \\ \mathbf{C}(2^m) & \text{if } \epsilon(p,q) = 0 \\ \mathbf{H}(2^{m-1}) \oplus \mathbf{H}(2^{m-1}) & \text{if } \epsilon(p,q) = -1. \end{cases} \tag{5.5}$$

(If $n = 2m$ we know from theorem 4.5 that $\epsilon(p, q) = \pm 1$. If $n = 2m+1$, and if $\epsilon = 0$, then $e_+ = (1 + iw)/2$ does not lie in the real algebra, but w itself acts as imaginary unit, and so $Cl(p, q)$ is isomorphic to the complexification of the even subalgebra of $Cl(p, q)$, i.e. to $\mathbf{C}(2^m)$.) In conclusion note that the finite geometry table of $\epsilon(p, q)$, as provided by theorem 4.5, translates via (5.4) and (5.5) into the well-known table of real Clifford algebras (see Table 13.26 in [9]):

THEOREM 5.2 If $p + q = 2m$, then

$$Cl(p,q) \cong \begin{cases} \mathbf{R}(2^m) & \text{if } p - q \equiv 0, 2 \pmod 8 \\ \mathbf{H}(2^{m-1}) & \text{if } p - q \equiv 4, 6 \pmod 8. \end{cases}$$

If $p + q = 2m + 1$ then

$$Cl(p,q) \cong \begin{cases} \mathbf{R}(2^m) \oplus \mathbf{R}(2^m) & \text{if } p - q \equiv 1 \pmod 8 \\ \mathbf{C}(2^m) & \text{if } p - q \equiv 3, 7 \pmod 8 \\ \mathbf{H}(2^{m-1}) \oplus \mathbf{H}(2^{m-1}) & \text{if } p - q \equiv 5 \pmod 8. \end{cases}$$

REMARK The five kinds of algebra in theorem 5.2 classify the *ungraded* real Clifford algebras. As we have shown, they arise from the five kinds of projective quadrics over GF(2), $\mathcal{H}_{2m-1}$, $\mathcal{E}_{2m-1}$, $\Pi_0\mathcal{H}_{2m-1}$, $\mathcal{P}_{2m}$, $\Pi_0\mathcal{E}_{2m-1}$, whose bilinear form is of maximal rank. However, as pointed out at the conference by Professor Trautmann, one ought to classify the *graded* Clifford algebras, which arise out of the $\mathbf{Z}_2$-grading $Cl(p,q)_0 \oplus Cl(p,q)_1$ of $Cl(p,q)$, where $Cl(p,q)_0$ denotes the even subalgebra (consisting of elements fixed under the main involution). For example the ungraded algebras $Cl(4,2)$ and $Cl(3,3)$ are isomorphic, both being $\cong \mathbf{R}(8)$, but as graded algebras they are not isomorphic, since $Cl(4,2)_0 \cong \mathbf{C}(4)$ and $Cl(3,3)_0 \cong \mathbf{R}(4) \oplus \mathbf{R}(4)$. Now it is easy to see (by Clifford algebra or finite geometry) that we have the well-known isomorphisms $Cl(p, q+1)_0 \cong Cl(p, q)$ and $Cl(p+1, q)_0 \cong Cl(q, p)$. Upon using theorem 5.2 one then easily sees that there are *eight* kinds of graded real Clifford algebras. It is straightforward to give a finite geometry interpretation of these eight kinds. For now V_n comes along with a preferred "even hyperplane" H, whose equation is $\sum x_i = 0$ in coordinates with respect to a (p,q)-BOS $\{a_1, \ldots, a_n\}$. Corresponding to the Dirac group $G(p,q)$ possessing a privileged subgroup of index 2, namely the even Dirac group $G(p,q)_0 = G(p,q) \cap Cl(p,q)_0$, the quadric Q has a privileged "subquadric" Q_0, the restriction of Q to the hyperplane H. For a quadric Q of the three types $(i)\mathcal{H}_{2m-1}$, $(ii)\mathcal{E}_{2m-1}$, $(iii)\mathcal{P}_{2m}$, there are two types of section Q_0 by H, namely $(i)\Pi_0\mathcal{H}_{2m-3}$ or $\mathcal{P}_{2m-2}$, $(ii)\Pi_0\mathcal{E}_{2m-3}$ or $\mathcal{P}_{2m-2}$, $(iii)\mathcal{H}_{2m-1}$ or $\mathcal{E}_{2m-1}$, respectively. (In the last case the section can not be of type $\Pi_0\mathcal{P}_{2m-2}$, since the nucleus $k = \sum a_i$ of $\mathcal{P}_{2m}$ does not lie in H.) On the other hand for Q of the remaining two types $(iv)\Pi_0\mathcal{H}_{2m-1}$, $(v)\Pi_0\mathcal{E}_{2m-1}$, there is only one type of section Q_0, namely $(iv)\mathcal{H}_{2m-1}$, $(v)\mathcal{E}_{2m-1}$, since H does not pass through the vertex k of these cones. So we see that there are indeed precisely eight possibilities for the pair Q_0, Q.

References

1. Artin, E. (1957), Geometric Algebra. Interscience, New York.
2. Braden, H.W. (1985) "N-dimensional spinors: their properties in terms of finite groups". J. Math. Phys. 26, 613-620.
3. Dye, R.H. (1992) "Maximal sets of non-polar points of quadrics and symplectic polarities over GF(2)". Geom. Ded. 44, 281-293.
4. Eckmann, B. (1942) "Gruppentheoretischer Beweis des Satzes von Hurwitz-Radon über die Komposition quadratishcher Formen". Comment. Math. Helv. 15, 358-366.
5. Eddington, A.S. (1936) Relativity Theory of Protons and Electrons. Cambridge Univ. Press, Cambridge.
6. Frenkel, I., Lepowsky, J. & Meurman A. (1988), Vertex Operator Algebras and the Monster. Academic Press, San Diego.
7. Griess, R.L. (1973) "Automorphisms of extra special groups and non vanishing degree 2 co-homology". Pacific. J. Math. 48, 403-422.
8. Hirschfeld, J.W.P. & Thas, J.A. (1991), General Galois Geometries, Clarendon, Oxford.
9. Porteous, I.R. (1981), Topological Geometry. Cambridge Univ. Press, Cambridge.
10. Quillen, D. (1971) "The mod 2 cohomology ring of extra-special 2-groups and the spinor groups". Math. Ann. 194, 197-212.
11. Shaw, R. (1986) "The ten classical types of group representations". J. Phys. A: Math. Gen. 19, 35-44.
12. Shaw, R. (1993) "Finite geometry, Dirac groups, and the table of real Clifford algebras". Hull Math. Res. Repts. VI, No 8.

JORDAN FORM IN CLIFFORD ALGEBRA

GARRET SOBCZYK
Universidad Nacional Autónoma de México, FES-C
Apartado Postal #25, Cuautitlan Izcalli 54700, Estado de México
garret@unamvm1.bitnet

Abstract. The Jordan form of an element in a algebra over the complex field is uniquely determined by special generators of the factor algebra of its minimal polynomial. Examples are considered in Clifford algebra.

Key words: Jordan form - analytic function - Lagrange-Sylvester polynomial - Clifford algebra

1. Introduction

The Jordan form of a matrix is uniquely determined by its minimal polynomial. We generalize this result to an element in an associative $\mathbb{C}$-algebra.

Let D be a subset of $\mathbb{C}$ and $f : D \to \mathbb{C}$ be an analytic function on D. The Lagrange-Sylvester interpolation polynomial makes it possible to extend the domain D of f to a subset D_M of square matrices, $f : D_M \to M$, provided that the spectrum of the matrices in D_M is contained in D (Gantmacher 1967, p.97).

Let $\mathcal{A}$ be any associative $\mathbb{C}$-algebra with unity. We show that the domain of the function f can be extended to a domain $D_{\mathcal{A}} \subset \mathcal{A}$, $f : D_{\mathcal{A}} \to \mathcal{A}$, if the roots of the minimal polynomials of each $x \in D_{\mathcal{A}}$ are in D.

Examples are considered in Clifford algebra.

Let $\mathbb{R}[\lambda]$ and $\mathbb{C}[\lambda]$ denote the rings of polynomials over the real and complex numbers. Augustin Cauchy observed in 1847 that the factor ring $\mathbb{R}[\lambda]/<\lambda^2+1>$ of the principal ideal $<\lambda^2+1>$ is isomorphic to the algebra of complex numbers. We define for each $\psi \in \mathbb{C}[\lambda]$ a $\mathbb{C}$-algebra $\mathbb{C}\{m_1, \ldots, m_r\}$ which is isomorphic to $\mathbb{C}[\lambda]/<\psi>$. In (Sobczyk 1993), we define a $\mathbb{R}$-algebra $\mathbb{R}\{m_1, \ldots, m_r; q_1, \ldots, q_s\}$ which is isomorphic to $\mathbb{R}[\lambda]/<\psi>$.

2. Definitions

Let $\{m_1, \ldots, m_r\}$ be a set of r nondecreasing postive integers with only the first h of them $\equiv 1$.

Definition 1 *The set of elements*

$$\{u_j, n_k^t |, 1 \le j \le r,\ h+1 \le k \le r,\ and\ 1 \le t < m_k\},$$

define an associative and commutative $(m_1 + \ldots + m_r)$-dimensional $\mathbb{C}$-algebra

$$\mathbb{C}\{m_1, \ldots, m_r\} \equiv\ span\{u_j, n_k^t\},$$

F. Brackx et al. (eds.), *Clifford Algebras and their Applications in Mathematical Physics*, 33–41.

where the operations of addition and multiplication of the basis elements are determined by

$$\{u_1 + \ldots + u_r = 1,\ u_i u_j = \delta_{ij} u_i,\ n_k^{m_k-1} \neq 0 \text{ but } n_k^{m_k} = 0,\ n_k u_k = n_k\}.$$

The elements u_j make up a *partition of unity* and are *mutually annihilating idempotents.* The elements n_j are *nilpotents* with the respective *indexes* m_j. The nilpotents n_j are *projectively* related to the corresponding idempotents u_j. We adopt the convention that $n_j \equiv 0$, for $j = 1, \ldots, h$.

The complex r-dimensional algebra $\mathbb{C}\{1, \ldots, 1\} \simeq \mathbb{C}^r$.

For the set $\{m_1, \ldots, m_r\}$ of non-decreasing positive integers and a set of distinct complex numbers $\{\lambda_j \in \mathbb{C};\ j = 1, 2, \ldots, r\}$, let

$$\psi \equiv (\lambda - \lambda_1)^{m_1} \ldots (\lambda - \lambda_r)^{m_r} \equiv \psi_j \tilde{\psi}_j,\ \in\ \mathbb{C}[\lambda] \tag{1}$$

where

$$\psi_j \equiv (\lambda - \lambda_j)^{m_j} \text{ and } \tilde{\psi}_j \equiv (\lambda - \lambda_1)^{m_1} \ldots \check{j} \ldots (\lambda - \lambda_r)^{m_r}.$$

Given ψ, ψ_j, *and* $\tilde{\psi}_j$, we now define a set consisting of $2r - h$ polynomials. These polynomials, as shown in Theorem 1, generate the factor ring $\mathbb{C}[\lambda]/<\psi>$.

Definition 2 *The set consisting of $2r - h$ polynomials*

$$\{u_j(\lambda), n_k(\lambda);\quad j = 1, 2, \ldots r, \text{ and } k = h+1, \ldots, r\},$$

where each polynomial has degree $= -1 + \deg \psi$, is specified by:

i) For each $m_j > 1$,

$$n_j(\lambda) \equiv \Big\{\frac{1}{\tilde{\psi}_j(\lambda_j)} + \Big(\frac{1}{\tilde{\psi}_j(\lambda_j)}\Big)'(\lambda - \lambda_j) + \ldots$$

$$+ \frac{1}{(m_j - 2)!}\Big(\frac{1}{\tilde{\psi}_j(\lambda_j)}\Big)^{(m_j-2)}(\lambda - \lambda_j)^{m_j-2}\Big\}(\lambda - \lambda_j)\tilde{\psi}_j(\lambda),$$

where the indicated derivatives of $\big(\frac{1}{\tilde{\psi}_j(\lambda)}\big)$ are taken with respect to λ and evaluated at $\lambda = \lambda_j$.

ii) The polynomials $u_j(\lambda)$ are uniquely defined in terms of the polynomial $w(\lambda) \equiv \lambda - n_{h+1}(\lambda) - n_{h+2}(\lambda) - \ldots - n_r(\lambda)$ by

$$u_j(\lambda) \equiv \frac{\tilde{\psi}_j(w(\lambda))}{\tilde{\psi}_j(\lambda_j)} = \frac{(w(\lambda) - \lambda_1) \ldots \check{j} \ldots (w(\lambda) - \lambda_r)}{(\lambda_j - \lambda_1) \ldots \check{j} \ldots (\lambda_j - \lambda_r)} \mod \psi$$

and the requirement that $\deg u_j(\lambda) = -1 + \deg \psi$.

We adopt the convention that $n_j(\lambda) \equiv 0$, for $j = 1, \ldots, h$.

3. Jordan Form

The following Lemma gives a *Jordan partition* of $\lambda \in \mathbb{C}[\lambda]$, with respect to $\psi \in \mathbb{C}$.

Lemma 1 *The polynomials* $\{u_j(\lambda), n_j(\lambda)\}$ *satisfy the identities*
i) $u_1(\lambda) + \ldots + u_r(\lambda) \equiv 1$,
and
ii) $\lambda \equiv (\lambda_1 u_1(\lambda) + n_1(\lambda)) + \ldots + (\lambda_r u_r(\lambda) + n_r(\lambda))$.

PROOF: The identities are a consequence of the fact that $n_k(\lambda_j) = 0$, $w(\lambda_j) = \lambda_j$ and $u_j(\lambda_k) = \delta_{jk}$.

Q.E.D.

The theorem below gives the relationship between the $\mathbb{C}[\lambda]/<\psi>$ and $\mathbb{C}\{m_1, \ldots, m_r\}$. The case when all $m_j = 1$ can be found in (Turnbull and Aitken, 1955, p.163), and in (I.N. Herstein, 1969, p.48).

Theorem 1 *i) The cosets of polynomials* $\{u_j(\check{\lambda}), n_j(\check{\lambda})\}$ *generate* $\mathbb{C}[\lambda]/<\psi>$.
ii) The algebra $\mathbb{C}[\lambda]/<\psi>$ *is isomorphic to* $\mathbb{C}\{m_1, \ldots, m_r\}$.

PROOF: *i)* Let $\{u_j(\lambda), n_j(\lambda)\}$ be given as in Definition 2. For $j \neq k$, $u_j(\lambda)u_k(\lambda) = 0 \ mod\ \psi$. Furthermore, for $w \equiv w(\lambda)$

$$u_j^2(\lambda) = \frac{(w-\lambda_1)^2 \ldots \check{j} \ldots (w-\lambda_r)^2}{(\lambda_j-\lambda_1)^2 \ldots \check{j} \ldots (\lambda_j-\lambda_r)^2} \ mod\ \psi$$

$$= \frac{(w-\lambda_1)^2 \ldots \check{j} \ldots (w-\lambda_r)(w-\lambda_j+\lambda_j-\lambda_r)}{(\lambda_j-\lambda_1)^2 \ldots \check{j} \ldots (\lambda_j-\lambda_r)^2} \ mod\ \psi$$

$$= \frac{(w-\lambda_1)^2 \ldots \check{j} \ldots (w-\lambda_{r-1})^2(w-\lambda_r)}{(\lambda_j-\lambda_1)^2 \ldots \check{j} \ldots (\lambda_j-\lambda_{r-1})^2(\lambda_j-\lambda_r)} \ mod\ \psi,$$

$$\ldots$$

$$= \frac{(w-\lambda_1) \ldots \check{j} \ldots (w-\lambda_r)}{(\lambda_j-\lambda_1) \ldots \check{j} \ldots (\lambda_j-\lambda_r)} \ mod\ \psi = u_j(\lambda),$$

so they are idempotents.

The polynomials $[n_k(\lambda)]^s \ mod\ \psi$, together with the $u_j(\lambda)$ make up a basis of $\mathbb{C}[\lambda]/<\psi>$.

ii) Let $u_j, n_k \subset \mathbb{C}\{m_1, \ldots, m_r\}$ be given as in Definition 1. We *solve* the equation

$$\lambda \equiv \sum_{j=1}^{r}(\lambda_j u_j + n_j) \in \mathbb{C}\{m_1, \ldots, m_r\} \qquad (*)$$

for each n_k and u_j and find that $n_k \equiv n_j(\lambda)$ and $u_j \equiv u_j(\lambda)$ as given in Definition 2.

For any term where $m_j > 1$, note that

$$\lambda - \lambda_j = (\lambda_1 - \lambda_j + n_1)u_1 + \ldots + n_j + \ldots + (\lambda_r - \lambda_j + n_r)u_r,$$

since $\lambda_j \equiv \lambda_j u_1 + \ldots + \lambda_j u_r$, and use this to get

$$(\lambda - \lambda_j)^{m_j} = (\lambda_1 - \lambda_j + n_1)^{m_j} u_1 + \ldots \check{j} \ldots + (\lambda_r - \lambda_j + n_r)^{m_j} u_r$$

by applying the algebraic relations in Definition 1. We evaluate $\widetilde{\psi}_j(\lambda)$, getting

$$\widetilde{\psi}_j(\lambda) = (\lambda_j - \lambda_1 + n_j)^{m_1} \ldots \check{j} \ldots (\lambda_j - \lambda_r + n_j)^{m_r} u_j \ \in \ \mathbb{C}\{m_1, \ldots, m_r\}.$$

We now evaluate the product $(\lambda - \lambda_j)^{m_j - 1}\widetilde{\psi}_j(\lambda)$, getting

$$(\lambda - \lambda_j)^{m_j - 1}\widetilde{\psi}_j(\lambda) = \widetilde{\psi}_j(\lambda_j) n_j^{m_j - 1} \ \in \ \mathbb{C}\{m_1, \ldots, m_r\}.$$

Repeating this process, we find the recursive relationships

$$(\lambda - \lambda_j)^{m_j - 2}\widetilde{\psi}_j(\lambda) = \widetilde{\psi}_j(\lambda_j) n_j^{m_j - 2} + \widetilde{\psi}_j'(\lambda_j) n_j^{m_j - 1} \ \in \ \mathbb{C}\{m_1, \ldots, m_r\},$$

and

$$(\lambda - \lambda_j)^{m_j - 3}\widetilde{\psi}_j(\lambda) = \widetilde{\psi}_j(\lambda_j) n_j^{m_j - 3} + \widetilde{\psi}_j'(\lambda_j) n_j^{m_j - 2} + \widetilde{\psi}_j''(\lambda_j) n_j^{m_j - 1}$$

$$= \ldots .$$

The process terminates when we get down to n_j. If we solve these recursive relationships explicitly for the nilpotents n_j in terms of polynomials in λ, we find that each $n_j \equiv n_j(\lambda)$ as given in Definition 2.

In terms of the polynomial $w(\lambda) \equiv \lambda - n_1(\lambda) - \ldots - n_r(\lambda)$, we can write the equation $(*)$ in the form

$$w(\lambda) = \lambda_1 u_1 + \ldots + \lambda_r u_r,$$

which we now solve for $u_j \equiv u_j(\lambda)$.

From $u_1 + \ldots + u_r = 1$, it follows that

$$w(\lambda) - \lambda_j = (\lambda_1 - \lambda_j)u_1 + \ldots \check{j} \ldots + (\lambda_r - \lambda_j)u_r.$$

We then find that

$$(w(\lambda) - \lambda_1) \ldots \check{j} \ldots (w(\lambda) - \lambda_r) = (\lambda_j - \lambda_1) \ldots \check{j} \ldots (\lambda_j - \lambda_r) u_j,$$

from which it follows that the $u_j \equiv u_j(\lambda)$ are given as in Definition 2.

Q.E.D.

Lemma 1 and Theorem 1 can be applied to any element $\lambda = x$ in an associative $\mathbb{C}$-algebra $\mathcal{A}$, provided that x has the minimal polynomial (1). The set of elements $\{u_j \equiv u_j(x), n_j \equiv n_j(x)\}$ generates a $\mathbb{C}$-subalgebra $\mathcal{A}_{\{m_1,\ldots,m_r\}} \subset \mathcal{A}$, and x can be written in the Jordan form

$$x = \sum_{j=1}^{r} (\lambda_j u_j + n_j).$$

If $M \in \mathcal{M}_n$ the algebra of complex $(n \times n)$-matrices, and M has the minimal polynomial (1), then M can be written in the Jordan form

$$M = \sum_{j=1}^{r} (\lambda_j u_j + n_j),$$

which can be used to derive the Jordan normal form.[1] See (Sobczyk 1993).

4. Analytic Functions

Let $\mathcal{A}$ be an associative $\mathbb{C}$-algebra, and let $\mathcal{A}_{\{m_1,\ldots,m_r\}}$ be the subalgebra of $\mathcal{A}$ generated by an element $x \in \mathcal{A}$ having the minimal polynomial ψ. Any $p \in \mathcal{A}_{\{m_1,\ldots,m_r\}}$ can be expressed in terms of the partition $\{u_j, n_j\}$ by

$$p = \sum_{j=1}^{r} p_j u_j, \tag{2}$$

where

$$p_j \equiv \lambda_j + N_j, \quad N_j \equiv \sum_{k=1}^{m_j-1} a_{jk} n_j^k, \quad a_{jk} \in \mathbb{C}.$$

Thus, p is uniquely specified by $m_1 + \ldots + m_r$ complex scalars. The $\lambda_j \in \mathbb{C}$ are called the *characteristic values* and make up the *spectrum* of p.

The *j-parts* $p_j u_j$ of p are mutually annihilating because $u_j u_k = 0$ for $j \neq k$. This greatly simplifies the multiplication in $\mathcal{A}_{\{m_1,\ldots,m_r\}}$:

$$p \equiv \sum p_j u_j, \ \textit{and}\ q \equiv \sum q_j u_j \ \Rightarrow\ pq = \sum p_j q_j u_j \ \in \mathbb{C}\{m_1, \ldots, m_r\}.$$

We can now extend the domain of the analytic function $f(z)$ to any element $p \in \mathcal{A}$ provided that f is analytic on the spectrum of p.

Definition 3 *For $p \in \mathcal{A}_{\{m_1,\ldots,m_r\}}$ as given in equation (2),*

$$f(p) \equiv \sum_{j=1}^{r} f(p_j) u_j \quad \in \quad \mathcal{A}_{\{m_1,\ldots,m_r\}},$$

[1] This has been shown in an unpublished manuscript by the author, *The Eigenprojector Problem: Oversight of the Century?*

where

$$f(p_j) = f(\lambda_j + N_j) = f(\lambda_j) + f'(\lambda_j)N_j + \dots$$

$$+\frac{1}{(m_j-1)!}f^{(m_j-1)}(\lambda_j)N_j^{m_j-1}.$$

The justification of Definition 3 is the Taylor series expansion of f around points in the spectrum of $p \in \mathcal{A}$.

Definition 3 liberates the Classical Lagrange-Sylvester Interpolation polynomial, from its bondage in a matrix (Turnbull and Aitken, p. 76), (Gantmacher, p. 101), and it extends the rich theory of complex analysis (Ahlfors, 1966) to an even richer $\mathbb{C}\{m_1, \dots, m_r\}$ analysis (Sobczyk, DGM 1992). On the more mundane side, p has a unique *inverse* $p^{-1} \equiv 1/p$ provided that its characteristic values are all non-zero. Its inverse is found by applying Definition 3 to the analytic function $f(z) = 1/z$.

5. The Clifford Algebra $Cl_{2,3}$

Let $Cl_{2,3}$ $(++---)$ denote the Clifford algebra of De Sitter space. Any Clifford algebra $Cl_{q,q+1}$ can be considered to be the algebraic extension of a $\mathbb{C}$-algebra $\mathbb{C}\{m_1, \dots, m_r\}$ (Sobczyk 1992).

We denote the k-vector part of $g \in Cl_{2,3}$ by $< g >_k$. We also write $< g >_{j+k} \equiv < g >_j + < g >_k$. An arbitrary element $g \in Cl_{2,3}$ can be written in the form

$$g = \alpha + v + b, \tag{3}$$

where $\alpha =< g >_{0+5}$ is a complex scalar, $v =< g >_{1+4}$ is a complex vector, and $b =< g >_{2+3}$ is a complex bivector. The *reversion* of g, defined by

$$g^\dagger = \alpha + v - b,$$

changes the sign of the complex bivector part of g. We also need the operation g^v, defined by

$$g^v \equiv \alpha - v + b,$$

which changes the sign of the complex vector part of g.

The *determinant* of the geometric number g is given by

$$det(g) \equiv (gg^\dagger)(gg^\dagger)^v. \tag{4}$$

Some care must be exercised in extending the *inner* and *outer* products (Hestenes and Sobczyk, 1984, p.8) to complex elements (3); we define

$$v \cdot b \equiv \frac{1}{2}(vb - bv) =< vb >_{1+4},$$

and

$$v \wedge b \equiv \frac{1}{2}(vb + bv) = < vb >_{2+3},$$

so that $vb = v \cdot b + v \wedge b$. For complex bivectors b_1 and b_2, we define the products

$$b_1 b_2 = b_1 \cdot b_2 + b_1 \wedge b_2 + [b_1, b_2],$$

where $b_1 \cdot b_2 \equiv < b_1 b_2 >_{0+5}$, $b_1 \wedge b_2 \equiv < b_1 b_2 >_{1+4}$, and $[b_1, b_2] \equiv < b_1 b_2 >_{2+3}$.

Theorem 2 *Each $g \in Cl_{2,3}$ satisfies its characteristic equation*

$$g^4 - 4\alpha g^3 + (6\alpha^2 - 2\beta)g^2 - 4(\alpha^3 - \alpha\beta + \delta)g +$$

$$(\alpha^4 - 2\beta\alpha^2 + 4\delta\alpha + \beta^2 - \gamma) \equiv 0, \tag{5}$$

where the constants β, γ, δ are defined by

$$\beta = v^2 + b \cdot b, \quad \gamma = (b \wedge b)^2 + 4(v \wedge b) \cdot (v \wedge b), \quad \delta = (v \wedge b) \cdot b.$$

PROOF: We calculate

$$(g - \alpha)^2 = (v + b)^2 = (v^2 + b \cdot b) + b \wedge b + 2v \wedge b.$$

and

$$[(g - \alpha)^2 - \beta]^2 = (b \wedge b + 2v \wedge b)^2$$

$$= (b \wedge b)^2 + 4(b \wedge b) \wedge (v \wedge b) + 4(v \wedge b)^2.$$

Finally, we calculate

$$[(g - \alpha)^2 - \beta]^2 - \gamma$$

$$= 4(b \wedge b) \wedge (v \wedge b) + 4(v \wedge b) \wedge (v \wedge b)$$

$$= 4 < (b \wedge v)\ (v \wedge b) >_{1+4} + 4 < (b \wedge b)\ (v \wedge b) >_{2+3}$$

$$= 4 < bvb >_{0+5} (v + b) = 4[(v \wedge b) \cdot b](g - \alpha).$$

The last step requires use of the definitions of the complex inner and outer products given earlier.

The proof is completed by multiplying out the above expressions to find the form of the characteristic equation of g given in the statement of the theorem.

Q.E.D.

The coefficients of the characteristic equation of g determine the *invariants* of g in terms of the complex scalars $\alpha, \beta, \gamma, \delta$. For example, the *trace* of g is the negative of the coefficient of g^3, or

$$tr\ g \equiv 4\alpha.$$

We get another expression for the $det(g)$, defined in (4), by taking the constant term of the characteristic equation (5),

$$det\ g = \alpha^4 - 2\alpha^2\beta + \beta^2 - \gamma + 4\alpha\delta.$$

6. Evaluation of Functions in $Cl_{2,3}$.

Let $g = \alpha + v + b \in Cl_{2,3}$, and let $f : \mathbb{C} \to \mathbb{C}$ be analytic on the spectrum of g. We wish to evaluate $f(g) \in Cl_{2,3}$. The minimal polynomial of g can be found from the characteristic equation (5). By appealing to Theorem 1, we can construct a partition of idempotents and nilpotents. The problem is completed by using the definition of $f(g)$ given in Definition 3.

We can often evaluate $f(g)$ without explicitly having to find its mimimal polynomial. We shall indicate the procedure by way of the following examples. CLICAL (Lounesto, Mikkola, and Vierros, 1987) proved to be very useful for checking these examples.

Case i) $g = \alpha + v$.

If $v^2 \neq 0$, define $\beta = \sqrt{v^2}$, and $\hat{v} \equiv v/\beta$. We also need the idempotents

$$v_+ \equiv \frac{1}{2}(1+\hat{v}) \text{ and } v_- \equiv \frac{1}{2}(1-\hat{v}).$$

Noting that $g = (\alpha+\beta)v_+ + (\alpha-\beta)v_-$, and appealing to Definition 3, we find that

$$f(g) = f(\alpha+\beta)v_+ + f(\alpha-\beta)v_-.$$

When $v^2 = 0$, we appeal directly to Definition 3 and get

$$f(g) = f(\alpha) + f'(\alpha)v.$$

Case ii) $g = \alpha + b$.

If $b^2 = 0$, we appeal directly to Definition 3 and get

$$f(\alpha + b) = f(\alpha) + f'(\alpha)b.$$

If $b^2 \neq 0$, then $b^2 = b \cdot b + b \wedge b$. We can now apply *Case i)* to b^2 with $f(z) = \sqrt{z}$ and $f(z) = 1/\sqrt{z}$ in order to find $\beta \equiv \sqrt{b^2}$ and $1/\beta$, which we use to define

$$\hat{b} \equiv \frac{1}{\beta}b.$$

We then find the reduction

$$f(g) = f(\alpha + \beta)b_+ + f(\alpha - \beta)b_-$$

in terms of the idempotents b_+ and b_-, which are defined in terms of $\hat{b}$. But we are still not done since β is a geometric number of the type discussed in *Case i)*. Appealing once again to *Case i)*, $f(\alpha+\beta)$ and $f(\alpha-\beta)$ can be evaluated in terms of the idempotents of β.

Case iii) $g = \alpha + v + b$, where $v \wedge b = 0$.

In this case $(v+b)^2 = v^2 + b^2 = \omega + v'$, where ω is a complex scalar and v' is a complex vector. By again appealing to *Case i)*, we can calculate $\beta \equiv \sqrt{\omega + v'}$, and use this to define

$$\hat{S} \equiv \frac{1}{\beta}(v+b), \quad S_+ \equiv 1/2(1+\hat{S}) \text{ and } S_- \equiv 1/2(1-\hat{S}).$$

We can then evaluate $f(g)$ in terms of the idempotents S_+ and S_- and reduce the problem to one of the previous cases.

Case iv) $g = \alpha + v + b$, where $v \cdot b = 0$.

In this case $vb = bv$, so that

$$(v+b)^2 = v^2 + 2vb + b.$$

Just as for *Case iii)*, this case can be reduced to a *Case ii)* using up to four idempotents, depending upon whether or not $v^2 \neq 0$.

Case v) $g = \alpha + v + b$, where $b^2 = 0$.

Acknowledgements

I greatly benefited from discussions with Professor Jaime Keller and want to thank him for inviting me to Universidad Nacional Autónoma de México where this work was written. Professor Zbigniew Oziewicz read an earlier version of this manuscript suggesting many improvements. The author wishes to thank the Organizers for an excellent Conference.

References

Ahlfors, Lars: 1966, *Complex Analysis*, 2nd ed., McGraw-Hill Book Company, New York.

Gantmacher, F.R.: 1960, *Matrix Theory*, Vol. 1, Chelsea Publishing Company, New York.

Herstein, I.N.: 1976 *Rings with Involution*, Chicago Lecture Series in Mathematics, University of Chicago Press, Chicago.

Hestenes David and Garret Sobczyk: 1984, *Clifford Algebra to Geometric Calculus: A Unified Language for Mathematics and Physics*, D. Reidel Publishing Company, Dordrecht.

Lounesto Pertti, Risto Mikkola and Vesa Vierros: 1987 *Clical User's Manual*, Research Reports A248, Helsinki University of Technology.

Sobczyk Garret: 1993, in Oziewicz, Z., A. Borowicz, and B. Jancewicz, ed(s)., *Twistors, Spinors, Clifford Algebra, and Quantum Deformations*, , Kluwer Publishing Company, Dordrecht.

Sobczyk Garret: 1992, in Catto, S. and A. Rocha, ed(s)., *Differential Geometry Methods in Theoretical Physics*, , World Scientific Publ. Co., Singapore, pp.397-407.

Sobczyk Garret: 1992, 'Unipotents, Idempotents, and a Spinor Basis for Matrices', *Advances in Applied Clifford Algebras* [2], No. 1.

Turnbull H.W. and A.C. Aitken: 1969, *An Introduction to the Theory of Canonical Matrices*, Dover Publications, Inc..

CLIFFORD ANALYSIS:
Local and global theory for Dirac-type operators

ELLIPTIC BOUNDARY VALUE PROBLEMS IN UNBOUNDED DOMAINS

S. BERNSTEIN
Freiberg University of Mining and Technology
Department of Mathematics
Institute of Applied Mathematics I
Bernhard-von-Cotta-Str. 2
D-09599 Freiberg

Abstract. By $T_G\mathbf{Q}MT_G$ is constructed an isomorphism between $\overset{o}{w}{}^1_{2,0}(G)$ and its dual space. We introduce the notion of uniform regularity and use weighted spaces for solving linear boundary value problems of Dirichlet-type in unbounded domains with unbounded boundary.

Key words: unbounded domains, unbounded boundary, quaternionic analysis, representation of solution, Dirichlet problems

1. Introduction

A central point in the theory of elliptic boundary value problems are continuous isomorphisms between the Sobolev space $\overset{o}{W}{}^1_2(G)$ and its dual space. Considering special systems created by quaternionic valued functionspaces, we will prove that $T_G\mathbf{Q}MT_G$ is such an isomorphism.

If we consider unbounded domains there arises some further problems. It is well known that the Dirichlet problem for the Laplacian for functions of $L_2(G)$ need not has any solution in $L_2(G)$ if G is an unbounded domain. In the same case the weakly singular Teodorescu transform T_G is unbounded in $L_2(G)$. These problems are solved by using weighted spaces with weight of type $(1+|x|^2)^{1/2}$.

Usually, there are considered outer domains, i.e. unbounded domains thats complement is bounded. But we don't want to exclude unbounded domains with unbounded boundary. That's why we introduce the notion uniform regularity that is the usual regularity if the domain is bounded.

Using all these results, we prove that $T_G\mathbf{Q}MT_G$ is a continuous isomorphism between $\overset{o}{w}{}^1_{2,0}(G)$ (if G is bounded $\overset{o}{w}{}^1_{2,0}(G) \equiv \overset{o}{W}{}^1_2(G)$) and its dual space. Furthermore, we solve linear boundary value problems of Dirichlet-type.

2. Preliminaries

We use the following weighted Sobolev-Slobodezkij-spaces (cf. [8]): Let $k = 0, 1, \ldots$, $\mu \in \mathbf{R}$, then we have

F. Brackx et al. (eds.), Clifford Algebras and their Applications in Mathematical Physics, 45–53.

$$w^k_{2,\mu}(\mathbf{R}^n) = \left\{ f \in \mathcal{D}'(\mathbf{R}^n) : \|f\|_{w^k_{2,\mu}(\mathbf{R}^n)} = \left[\int_{\mathbf{R}^n} \sum_{|\alpha| \le k} (1+|x|^2)^{\frac{\mu - 2(k-|\alpha|)}{2}} |\partial^\alpha f(x)|^2 \, dx \right]^{\frac{1}{2}} < \infty \right\}$$

and

$$w^{k-1/2}_{2,\mu}(\mathbf{R}^n) = \left\{ f \in \mathcal{D}'(\mathbf{R}^n) : \|f\|_{w^{k-1/2}_{2,\mu}(\mathbf{R}^n)} = \|f\|_{w^{k-1}_{2,\mu-1}(\mathbf{R}^n)} + \left[\int_{\mathbf{R}^n \times \mathbf{R}^n} \sum_{|\alpha| = k-1} \frac{|(1+|x|^2)^{\mu/4} \partial^\alpha f(x) - (1+|y|^2)^{\mu/4} \partial^\alpha f(y)|^2}{|x-y|^{n+1}} dx dy \right]^{\frac{1}{2}} < \infty \right\},$$

and

$$L_{2,\mu}(\mathbf{R}^n) = w^0_{2,\mu}(\mathbf{R}^n).$$

Let $G \subset \mathbf{R}^3$ be a domain then

$$w^s_{2,\mu}(G) = \{ f|_G : f \in w^s_{2,\mu}(\mathbf{R}^n) \}$$

with the norm

$$\|g\|_{w^s_{2,\mu}(G)} = \inf_{f \in w^s_{2,\mu}(\mathbf{R}^n),\ f|_G = g} \|f\|_{w^s_{2,\mu}(\mathbf{R}^n)}.$$

$\overset{o}{w}{}^k_{2,\mu}(G)$ denotes the closure of $C^\infty_c(G)$(set of function that have derivates of all order and compact support in G) in the norm of $w^k_{2,\mu}(G)$.

If G is a bounded domain the spaces $w^k_{2,\mu}(G)$ and $W^k_2(G)$ coincide.

Let $\mathbf{H}$ be the quaternionic algebra with the basic elements e_0, e_1, e_2, e_3 that fulfills the relations

$$e_0^2 = 1,\ e_0 e_i = e_i e_0 = e_i,\ e_i^2 = -1,\ e_i e_j + e_j e_i = 0,\ i,j = 1,2,3,$$
$$e_1 = e_2 e_3,\ e_2 = e_3 e_1,\ e_3 = e_1 e_2.$$

A function u will be called $\mathbf{H}$-valued or quaternionic valued if u is a linear combination $\sum_{i=0}^{3} u_i e_i$ with scalar valued functions u_i. The conjugated function to u is $\bar{u} = u_0 e_0 - \sum_{i=1}^{3} u_i e_i = u_0 e_0 - \underline{u} = \Re u - \Im u$. In the same way we consider $\mathbf{H}$- valued functionspaces:

$$u \in F_{\mathbf{H}} \iff u_i \in F,\ i = 0,1,2,3,$$

where F is a scalar valued functionspace. Thus $F_{\mathbf{H}}$ is a bi-H-module ([2]). We identify $x \in \mathbf{R}^3$ with $x = \sum_{i=0}^{3} x_i e_i \in \mathbf{H}$.

Furthermore, we use a quaternionic calculus that was introduced in [4]: Let $u \in C^1_{\mathbf{H}}(\bar{G})$(by $\bar{G}$ is denoted the closure of G) then the Dirac operator D is defined by

$$Du = \sum_{i=1}^{3} \frac{\partial}{\partial x_i} e_i u,$$

the Teodorescu transform by

$$T_G u = -\int_G e(x-y)u(y)dG_y, \quad e(x) = -\frac{1}{4\pi}\frac{x}{|x|^3},$$

and

$$F_\Gamma u = \int_\Gamma e(x-y)n(y)u(y)d\Gamma_y, x \notin \Gamma,$$

where $\Gamma = \partial G$ is the boundary of the domain G and $n(y) = \sum_{i=1}^{3} n_i(y)e_i$ with $(n_1(y), n_2(y), n_3(y))$ is the outer normal at the point $y \in \Gamma$.

3. Regularity of unbounded domains

As mentioned above we also want to consider domains with unbounded boundary. For this case we further develop the notion of uniform regularity that was first used by F.E. Browder for usual Sobolev spaces ([3]).

Definition 3.1 *Let G be a domain in $\mathbf{R}^n$ with boundary Γ. Then G is said to be uniformly C^m-regular, $m \geq 1$, if there exists a family of open sets $\{N_k\}_{k=1}^{\infty} \subset \mathbf{R}^n$ and homeomorphisms $\{\Phi_k : N_k \to B_1(y^k), y^k \in \mathbf{R}^{n-1}\}$, where $B_1(y^k)$ denotes the unit ball with centre in y^k, and an integer $E > 0$ such that the following conditions are satisfied:*

(1) Any $(E+1)$ distinct open sets of $\{N_k\}$ have an empty intersection.

(2) Let $\tilde{N}_k = \Phi_k^{-1}(\{y \in B_1(y^k) : |y - y_k| < \frac{1}{2}\})$. Then $\cup_k \tilde{N}_k$ contains the E^{-1}-neighborhood of Γ in G.

(3) Let $\Phi_k^{-1} : B_1(y^k) \to N_k$ be the inverse mapping to Φ_k , Φ_{jk} and Φ_{jk}^{-1} are the j-th components of Φ_k and Φ_k^{-1} respectivly. Then Φ_k and Φ_k^{-1} are mappings of class C^m, i.e. $\Phi_k \in C^m(N_k)$ and $\Phi_k^{-1} \in C^m(B_1(y^k))$, and further exists a constant M independent of x, y or k such that $|\partial^\beta \Phi_{jk}(x)| \leq M$, $x \in N_k$, $|\partial^\beta \Phi_{jk}^{-1}(y)| \leq M$, $y \in B_1(y^k)$, $|\beta| = 1$.

(4) For each k : $\Phi_k(N_k \cap G) = \{y \in B_1(y^k) : y_1 > 0\}$ and $\Phi_k(N_k \cap \Gamma) = \{y \in B_1(y^k) : y_1 = 0\} = B_1^0(y^k) \subset \mathbf{R}^{n-1}$.

(5) The weight-condition is fulfilled.

The weight-condition is caused by the special structure of our spaces, if there used other spaces this condiotion will be changing.

Definition 3.2 *The homeomorphisms $\Phi_k \in C^m(N_k)$ fulfills the weight-condition if there exists a positive constant K (independent of x, y or k) such that*
$|\partial^\beta \Phi_{jk}(x)| \leq K(1+|x|^2)^{\frac{1-|\beta|}{2}}$, $x \in N_k$, *and*
$|\partial^\beta \Phi_{jk}^{-1}(y)| \leq K(1+|y|^2)^{\frac{1-|\beta|}{2}}$, $y \in B_1(y^k)$, $0 \leq |\beta| \leq m$.

An important conclusion of the uniform C^m-regularity is:

Lemma 3.1 *Let G be a uniformly C^m-regular domain. Then there exists a countable covering $\{N_k\}_{k=1}^\infty$ of the closure of $\{x \in G : dist(x, \Gamma) < E^{-1}\}$ by open subsets of $\mathbf{R}^n$ and a family of functions $\{\eta_k\}$ with $\eta_k \in C_c^m(N_k)$ such that*

(a) *For every $x \in G$, we have $0 \leq \eta_k(x) \leq 1$, and* $\sum_k [\eta_k(x)]^2 = 1$.

(b) *There exist constants K_s, K_s', $m \geq [s]+1$, such that*
$$||u||^2_{w^s_{2,\mu}(G)} \leq K_s \sum_k ||\eta_k^2 u||^2_{w^s_{2,\mu}(G)} \leq K_s' ||u||^2_{w^s_{2,\mu}(G)}.$$

For the proof see [1].
We use this result to define the spaces $w^s_{2,\mu}(\Gamma)$:

Definition 3.3 *Let G be a uniformly C^m-regular domain and $m \geq [s]+1$ then*
$u \in w^s_{2,\mu}(\Gamma)$ *if* $(\eta_k^2 u) \circ \Phi_k^{-1} \in \overset{\circ}{w}{}^s_{2,\mu}(B_1^0(y^k))$ $\forall k$ *and*
$$||u||^2_{w^s_{2,\mu}(\Gamma)} = \sum_{k=0}^{\infty} ||(\eta_k^2 u) \circ \Phi_k^{-1}||^2_{w^s_{2,\mu}(B_1^0(y^k))} < \infty.$$

Then we obtain usual trace and extension theorems:

Theorem 3.1 *(trace theorem) Let G be a uniformly C^m-regular domain, $m \geq 1$. Then there exists a linear, continuous trace operator*
$tr_\Gamma : w^s_{2,\mu}(G) \to w^{s-1/2}_{2,\mu}(\Gamma)$, $m \geq [s]+1 > 3/2$,
such that $tr_\Gamma u = u\,|_\Gamma$ *if* $u \in C_c^s(\bar{G})$.

Theorem 3.2 *(extension theorem) Let G be a uniformly C^m-regular domain, $m \geq 1$. Then there exists a linear, continuous extension operator*
$Z_G : w^{s-1/2}_{2,\mu}(\Gamma) \to w^s_{2,\mu}(G)$, $m \geq [s]+1 > 3/2$,
such that $tr_\Gamma Z_G u = u$ $\forall u \in w^{s-1/2}_{2,\mu}(\Gamma)$.

For the proofs see [1]

4. The space $w^{-1}_{2,0,\mathbf{H}}(G)$

We want to describe the dual space to $\overset{\circ}{w}{}^1_{2,0}(G)$. It is easy to see that the norm

$$||u||^2_{w^1_{2,0,\mathbf{H}}(G)} = \int_G (1+|x|^2)^{-1}|u(x)|^2 dx + \sum_{|\alpha|=1} \int_G |\partial^\alpha u|^2 dx$$

of the normed bi-H-module $w^{1}_{2,0,\mathbf{H}}(G)$ may be created by the inner product

$$[u,v] = \sum_{|\alpha|\leq 1} \int_G (1+|x|^2)^{|\alpha|-1} \partial^\alpha \bar{u} \partial^\alpha v dx$$

and also by the inner product

$$(u,v) = \sum_{|\alpha|\leq 1} \int_G (1+|x|^2)^{|\alpha|-1} \partial^\alpha u \partial^\alpha \bar{v} dx.$$

With the inner product $[.,.]$ $w^{1}_{2,0,\mathbf{H}}(G)$ becomes a right H-Hilbert-module and with the inner product $(.,.)$ a left H-Hilbert-module. Using quaternionicaly functional analysis ([2], [4]) and results of scalar valued function spaces ([9]) we may define:

Definition 4.1 *Let G be an uniformly C^2-regular domain. Then the normed bi-H-module $w^{-1}_{2,0,\mathbf{H}}(G)$ is the space of all distributions S that can be represented as*

$$S = \sum_{|\alpha|\leq 1} (-1)^{|\alpha|} \partial^\alpha \bar{s_\alpha}, \quad \textit{where } s_\alpha \in L_{2,2(1-|\alpha|),\mathbf{H}}(G),$$

and the norm is $$||S||_{w^{-1}_{2,0,\mathbf{H}}(G)} = \inf_{\textit{all repres.}} \left\{ \sum_{|\alpha|\leq 1} ||s_\alpha||^2_{L_{2,2(1-|\alpha|),\mathbf{H}}(G)} \right\}^{1/2}.$$

It is easy to see that using one of the above mentioned inner products we obtain:

Theorem 4.1 *The dual space to the normed bi-H-module $\overset{\circ}{w}{}^{1}_{2,0}(G)$ is given by the normed bi-H-module $w^{-1}_{2,0,\mathbf{H}}(G)$.*

This is caused by the fact that all considered spaces are bi-H-moduls and we need not to distinguish "right" and "left" due to the inner products.

5. The Teodorescu transform T_G

In this section we prove some properties of the Teodorescu transform which is a weakly singular integral operator.

Theorem 5.1 *The operator $T_{\mathbf{R}^3} : L_{2,2\alpha,\mathbf{H}}(\mathbf{R}^3) \to L_{2,2(\alpha-1),\mathbf{H}}(\mathbf{R}^3)$, $-\frac{1}{2} < \alpha < \frac{3}{2}$, is continuous.*

Proof: Let $B_i(0)$, $i = 1,2,3$, be balls with radius i and the centre in the origin. Then we have

$$||4\pi Tu||^2_{L_{2,2(\alpha-1),\mathbf{H}}} = \int_{\mathbf{R}^3} (1+|x|^2)^{\alpha-1} \left| \int_{\mathbf{R}^3} \frac{x-y}{|x-y|^3} u(y) dy \right|^2 dx$$

$$\leq 2\left\{\int\limits_{\mathbf{R}^3\setminus B_2(0)} (1+|x|^2)^{\alpha-1}\left|\int\limits_{\mathbf{R}^3\setminus B_1(0)} \frac{x-y}{|x-y|^3}u(y)dy\right|^2 dx\right.$$

$$+\int\limits_{\mathbf{R}^3\setminus B_2(0)} (1+|x|^2)^{\alpha-1}\left|\int\limits_{B_1(0)} \frac{x-y}{|x-y|^3}u(y)dy\right|^2 dx$$

$$+\int\limits_{B_2(0)} (1+|x|^2)^{\alpha-1}\left|\int\limits_{\mathbf{R}^3\setminus B_3(0)} \frac{x-y}{|x-y|^3}u(y)dy\right|^2 dx$$

$$\left.+\int\limits_{B_2(0)} (1+|x|^2)^{\alpha-1}\left|\int\limits_{B_3(0)} \frac{x-y}{|x-y|^3}u(y)dy\right|^2 dx\right\}$$

$= 2\{I_1 + I_2 + I_3 + I_4\}$

On $\mathbf{R}^3\setminus B_1(0)$ and $\mathbf{R}^3\setminus B_2(0)$ the weights $(1+|x|^2)^{\alpha/2}$ and $|x|^\alpha$ are equivalent. Thus by using [7] we get

$$2I_1 \leq C_1\|u\|^2_{L_{2,2\alpha},\mathbf{H}(\mathbf{R}^3)}$$

Because weakly singular integral operators in bounded domains in L_2 are compact ([5]) we obtain

$$2I_4 \leq C_4\|u\|^2_{L_{2,2\alpha},\mathbf{H}(\mathbf{R}^3)}$$

Furthermore,

$$2I_2 \leq 2\int\limits_{\mathbf{R}^3\setminus B_2(0)} (1+|x|^2)^{\alpha-1}\int\limits_{B_1(0)} \frac{(1+|y|^2)^{-\alpha}}{|x-y|^4}dy\int\limits_{B_1(0)} (1+|y|^2)^{\alpha}|u(y)|^2 dydx$$

$$\leq 2\|u\|^2_{L_{2,2\alpha},\mathbf{H}(\mathbf{R}^3)}\int\limits_{B_1(0)} (1+|y|^2)^{-\alpha}\int\limits_{\mathbf{R}^3\setminus B_2(0)} \frac{(1+|x|^2)^{\alpha-1}}{|x-y|^4}dxdy$$

$$\leq C_2\|u\|^2_{L_{2,2\alpha},\mathbf{H}(\mathbf{R}^3)},\ \alpha < \frac{3}{2},$$

$$2I_3 \leq 2\int\limits_{B_2(0)} (1+|x|^2)^{\alpha-1}\int\limits_{\mathbf{R}^3\setminus B_3(0)} \frac{(1+|y|^2)^{-\alpha}}{|x-y|^4}dy\int\limits_{\mathbf{R}^3\setminus B_3(0)} (1+|y|^2)^{\alpha}|u(y)|^2 dydx$$

$$\leq 2\|u\|^2_{L_{2,2\alpha},\mathbf{H}(\mathbf{R}^3)}\int\limits_{\mathbf{R}^3\setminus B_3(0)} (1+|y|^2)^{-\alpha}\int\limits_{B_2(0)} \frac{(1+|x|^2)^{\alpha-1}}{|x-y|^4}dxdy$$

$$\leq C_3\|u\|^2_{L_{2,2\alpha},\mathbf{H}(\mathbf{R}^3)},\ -\frac{1}{2} < \alpha.$$

This completes the proof.

Theorem 5.2 *The operator* $\int\limits_G \frac{\partial}{\partial x_i} e(x-y)u(y)dG_y$, $i = 1, 2, 3$, *is a Calderon-Zygmund operator.*

The proof is analogous to that in [6] for the differentation of weakly singular integral operators.

Theorem 5.3 *The operator* $T_G : w^{-1}_{2,0,\mathbf{H}}(G) \to L_{2,0,\mathbf{H}}(G)$ *is continuous.*

Proof:Let $u \in w^{-1}_{2,0,\mathbf{H}}(G)$ then $u = f_{0,\mathbf{H}} + \frac{\partial}{\partial x_1} f_{1,\mathbf{H}} + \frac{\partial}{\partial x_2} f_{2,\mathbf{H}} + \frac{\partial}{\partial x_3} f_{3,\mathbf{H}}$, where $f_{0,\mathbf{H}} \in L_{2,2,\mathbf{H}}(G)$ and $f_{i,\mathbf{H}} \in L_{2,0,\mathbf{H}}(G)$, $i = 1, 2, 3$. We consider $T_G v$, where $v = v_{0,\mathbf{H}} + \sum\limits_{i=1}^{3} \frac{\partial}{\partial x_i} v_{i,\mathbf{H}}$ and $v_{j,\mathbf{H}} \in C^{\infty}_{c,\mathbf{H}}$, $j = 0, 1, 2, 3$. Then we get

$$T_G v = \int\limits_G e(x-y) \left\{ v_{0,\mathbf{H}}(y) + \sum_{i=1}^{3} \frac{\partial}{\partial x_i} v_{i,\mathbf{H}}(y) \right\} dy$$

$$= \int\limits_G e(x-y) v_{0,\mathbf{H}}(y) dy - \frac{1}{4\pi} \sum_{i=1}^{3} \int\limits_G \frac{\partial}{\partial x_i} e(x-y) v_{i,\mathbf{H}} dy.$$

The first integral is weakly singular and the second is a Calderon-Zygmund operator thus we obtain:

$$||T_G v||_{L_{2,0,\mathbf{H}}(G)} \le c_0 ||v_{0,\mathbf{H}}||_{L_{2,2,\mathbf{H}}(G)} + \sum_{i=1}^{3} c_i ||v_{i,\mathbf{H}}||_{L_{2,0,\mathbf{H}}(G)}$$

$$\text{and finally } ||T_G v||_{L_{2,0,\mathbf{H}}(G)} \le C ||u||_{w^{-1}_{2,0,\mathbf{H}}(G)}.$$

6. Orthogonal decomposition of $L_{2,0,\mathbf{H}}(G)$

The results of this section are simple generalisations of [4] to the case of unbounded domains. Thus, all proofs are analogous to [4]. Let $Mu = \sum\limits_{i=0}^{3} m_i(x) u_i e_i$, where $m_i \in C^{\infty}(\bar{G})$ and $m_i(x) > 0$, then $M^{-1}u = \sum\limits_{i=0}^{3} m_i^{-1} u_i e_i$ and $[u, v]_M = \int\limits_G M^{-1} u \overline{M^{-1} v} dG$ defines an inner product on the bi-**H**-module $L_{2,0,\mathbf{H}}(G)$.

Theorem 6.1 *Let G be an uniformly C^2-regular domain, then*

$$L_{2,0,\mathbf{H}}(G) = M\ ker D(G) \cap L_{2,0,\mathbf{H}}(G) \oplus_M D(\overset{o}{w}{}^{1}_{2,0}(G)),$$

where $\oplus_M$ denotes the orthogonal sum in sense of $[.,.]_M$.

We denote by $\mathbf{P}$ the orthoprojectior on $M\ ker D(G) \cap L_{2,0,\mathbf{H}}(G)$ and by $\mathbf{Q} = I - \mathbf{P}$ the orthoprojector on $D(\overset{o}{w}{}^{1}_{2,0}(G))$.

Lemma 6.1 *Let G be an uniformly C^2-regular domain and $g \in w^{1/2}_{2,0,\mathbf{H}}(\Gamma)$. Then there exists a uniquely determined solution $u \in w^{1}_{2,0,\mathbf{H}}(G)$ of the Dirichlet problem*

$$\begin{aligned} DM^{-1}Du &= 0 \quad \text{in } G, \\ u &= g \quad \text{on } \Gamma, \end{aligned}$$

where $u = F_\Gamma g + T_G \mathbf{P} Dh$ and $h \in w^{1}_{2,0,\mathbf{H}}(G)$ is an extension of g into G.

7. The $T_G \mathbf{Q} M T_G$ isomorphism

Theorem 7.1 *Let G be an uniformly C^2-regular domain. Then the operator*

$$T_G \mathbf{Q} M T_G : \; w^{-1}_{2,0,\mathbf{H}}(G) \to \overset{\circ}{w}{}^{1}_{2,0}(G)$$

is a continuous isomorphism.

Proof: We have the continuous mappings:

$$\begin{aligned} T_G : \; w^{-1}_{2,0,\mathbf{H}}(G) &\to L_{2,0,\mathbf{H}}(G), \\ M : \; L_{2,0,\mathbf{H}}(G) &\to L_{2,0,\mathbf{H}}(G), \\ \mathbf{Q} : \; L_{2,0,\mathbf{H}}(G) &\to L_{2,0,\mathbf{H}}(G) \cap \; im\, \mathbf{Q}, \\ T_G : \; L_{2,0,\mathbf{H}}(G) \cap \; im\, \mathbf{Q} &\to \overset{\circ}{w}{}^{1}_{2,0}(G). \end{aligned}$$

Further, let $u, v \in C^\infty_{c,\mathbf{H}}(G)$ and $T_G \mathbf{Q} M T_G u = T_G \mathbf{Q} M T_G v$ then $DM^{-1} D T_G \mathbf{Q} M T_G (u - v) = DM^{-1} \mathbf{Q} M T_G (u - v) = 0$ and $\mathbf{Q} M T_G (u - v) = Dw$, $w \in \overset{\circ}{w}{}^{1}_{2,0}(G)$, thus $DM^{-1}Dw = 0$ and it follows $M^{-1}Dw \in \; ker\, D$, $Dw \in \; im\, \mathbf{P}$ but we have also $Dw \in \; im\, \mathbf{Q}$, so we obtain $Dw = \mathbf{Q} M T_G (u - v) = 0$. Thus $M T_G (u - v) \in \; im\, \mathbf{P}$ and we get $D T_G (u - v) = u - v = 0$.

If $v \in C^\infty_{c,\mathbf{H}}(G)$ then there exists an $u \in C^\infty_{c,\mathbf{H}}(G)$ such that $T_G \mathbf{Q} M T_G u = v$. We take $u = DM^{-1}Dv$ and obtain by using Borel-Pompeiu formula ([4])
$T_G \mathbf{Q} M T_G D M^{-1} D v = T_G \mathbf{Q} M F_\Gamma (M^{-1} D v) + M^{-1} D v = T_G \mathbf{Q} M M^{-1} D v =$
$T_G \mathbf{Q} D v = T_G D v = v$.

8. Elliptic boundary value problems

Theorem 8.1 *Let G be an uniformly C^2-regular domain. Then the Dirichlet problem*

$$\begin{aligned} DM^{-1}Du &= f \;\; \text{in } G, \;\; f \in w^{-1}_{2,0,\mathbf{H}}(G), \\ u &= g \;\; \text{on } \Gamma, \;\; g \in w^{1/2}_{2,0,\mathbf{H}}(\Gamma), \end{aligned}$$

has the uniquely determined solution

$$u = T_G \mathbf{Q} M T_G f + T_G \mathbf{P} D h + F_\Gamma g \in w^{1}_{2,0,\mathbf{H}}(G),$$

where h is an extension of g into $w^{1}_{2,0,\mathbf{H}}(G)$.

Examples:

1) Equations of linear elasticity, $m \in \mathbf{R}\backslash[1,2]$,

$$\Delta\underline{u} + \frac{m}{m-2}\,\text{grad div}\,\underline{u} = -\underline{f} \text{ in } G,\ \underline{f} \in w^{-1}_{2,0,\mathbf{H}}(G),$$
$$\underline{u} = \underline{g},\ \text{on } \Gamma,\ \underline{g} \in w^{1/2}_{2,0,\mathbf{H}}(\Gamma)$$

is equivalent to

$$DM^{-1}Du = f \text{ in } G,\ f \in w^{-1}_{2,0,\mathbf{H}}(G),\ f_0 \equiv 0,$$
$$u = g \text{ on } \Gamma,\ g \in w^{1/2}_{2,0,\mathbf{H}}(\Gamma),\ g_0 \equiv 0,$$

where $m_0 = \frac{m-2}{2m-2}$, $m_1 = m_2 = m_3 \equiv 1$.

2) Time-independent Maxwell equations with constant permeability μ, the dielectric constant ϵ, $\epsilon > 0$, the electrical conductivity κ, $\kappa = \kappa(x) > 0$, the density ρ, the electric field $\underline{E}$ and the magnetic field $\underline{H}$.

We have the system

$$\begin{array}{ll} \text{div}\,\epsilon\underline{E} = \rho, & \text{div}\,\mu\underline{H} = 0 \\ \text{curl}\,\epsilon\underline{E} = \text{grad}\,\epsilon \times \underline{E}, & \text{curl}\,\mu\underline{H} = \mu\kappa\underline{E} + \text{grad}\,\mu \times \underline{H}. \end{array}$$

Set $a = \epsilon^{-1}\,\text{grad}\,\epsilon$, $b = \mu^{-1}\,\text{grad}\,\mu$, $f = -\rho e_0$, $\gamma = \epsilon^{-1}\mu\kappa$, $E = \epsilon\underline{E}$, $H = \mu\underline{H}$, $g = \text{tr}_\Gamma H$ thus

$$DE = f + \Im(aE) = f,$$
$$DH = \gamma E + \Im(bH) = \gamma E.$$

and we obtain

$$D\gamma^{-1}DH = f \text{ in } G,$$
$$H = g \text{ on } \Gamma.$$

References

1. Bernstein, S.: *Analytische Untersuchungen in unbeschränkten Gebieten mit Anwendungen auf quaternionische Operatortheorie und elliptische Randwertprobleme*, Dissertation, Freiberg University of Mining and Technology, 1993
2. Brachx, F., Delanghe, R., Sommen, F.: *Clifford analysis*, Pitman Advanced Publishing Program, Boston-London-Melbourne 1982
3. Browder, F.E.: *On the spectral theory of elliptic differential operators*, Math. Ann. 142 (1961), 22-130
4. Gürlebeck, K., Sprößig, W.: *Quaterionic Analysis and Elliptic Boundary Value Problems*, Akademie-Verlag, Berlin, 1989
5. Michlin, S.G.: *Lehrgang der mathematischen Physik*, Akademie-Verlag Berlin 1972
6. Michlin, S.G., Prößdorf, S.: *Singuläre Integraloperatoren*, Akademie-Verlag Berlin 1980
7. Stein, E.M., Weiss, G.: *Fractional integrals on n-dimensional Euclidean space*, J. Math. Mech. 7(1958), 503-514
8. Triebel, H.: *Interpolation Theory, Function Spaces, Differential Operators*, VEB Deutscher Verlag der Wissenschaften, Berlin 1978
9. Wloka, J.: *Partielle Differentialgleichungen*, B.G. Teubner Stuttgart 1982

DIRAC OPERATORS AND MANIFOLDS WITH BOUNDARY

B. BOOSS-BAVNBEK
Institut for matematik og fysik
Roskilde University
DK-4000 Roskilde, Denmark
e-mail: booss@jane.ruc.dk

and

K. P. WOJCIECHOWSKI*
Dept. Math., IUPUI
Indianapolis IN 46202, U.S.A.
e-mail: kwojciec@indyvax.iupui.edu

Abstract. We deal with various elliptic analogies to the classical Dirac equation; we explain our main analytical tools: invertible extension, Calderón projector, and twisted orthogonality of Cauchy data spaces; we investigate natural spaces of global elliptic boundary value problems for Dirac operators; and we develop an index theory for transmission problems and give additivity and non-additivity theorems for the index and the η-invariant under cutting and pasting of Dirac operators over partitioned manifolds. The explicit formulas rely on Clifford multiplication with vectors normal to the cutting submanifold.

Key words: Calderón projector, Clifford modules, Dirac operator, elliptic boundary problems, eta-invariant, index theory, partitioned manifolds, pseudo-differential Grassmannian, spectral flow, surgery

1. Begin With Clifford Modules

There are many different concepts of a Dirac operator in global analysis: classical and twisted Dirac operators on spin manifolds; operators of Dirac type with a square with scalar principal symbol; generalized (or compatible) Dirac operators defined by arbitrary (or compatible) connections on bundles of Clifford modules over Riemannian manifolds; full and split (odd-parity) Dirac operators; boundary Dirac operators; etc. The concepts depend on various geometrical features like dimension parity, orientation and chirality, almost complex structure, and suitable boundary. Each definition has its own merits and range of application.

Let X be a compact smooth oriented manifold (with or without boundary) with Riemannian metric g. Let $\dim X = n$. Let S be a complex vector bundle over X of Clifford modules; i.e. we have a representation

$$\mathbf{c} : C\ell(X) \longrightarrow \mathrm{Hom}(S, S)$$

with

$$\mathbf{c}(v)^2 = -\|v\|^2 Id_{S_x} \qquad \text{for } v \in TX_x \text{ and } x \in X. \tag{1}$$

* Work supported in part by NSF grant no. DMS-9105057.

F. Brackx et al. (eds.), Clifford Algebras and their Applications in Mathematical Physics, 55–66.

Recall that the Clifford bundle $C\ell(X)$ consists of the Clifford algebras $C\ell(TX_x, g_x)$, $x \in X$, which are associative algebras with unit generated by TX_x and subject to the relation $v \cdot w + w \cdot v = -2g_x(v, w)$. We shall call $\mathbf{c}$ *left Clifford multiplication* and occasionally write

$$\mathbf{c} : C^\infty(X; TX \otimes S) \longrightarrow C^\infty(X; S).$$

We may assume that S is equipped with a Hermitian metric which makes Clifford multiplication skew-adjoint, i.e. $\mathbf{c}(v)^* = -\mathbf{c}(v)$ for all $v \in TX_x$.

Definition 1 *A connection* $D : C^\infty(X; S) \longrightarrow C^\infty(X; T^*X \otimes S)$ *for* S *will be called* compatible *with the Clifford module structure of* S*, if it is* Leibnizian, *i.e. it satisfies the product rule*

$$\partial_v \langle s, s' \rangle = \langle D_v s, s' \rangle + \langle s, D_v s' \rangle, \tag{2}$$

and if $D\mathbf{c} = 0$*, i.e* D *is a* module derivation *with*

$$(D\mathbf{c})(v)(s) = D(\mathbf{c}(v)s) - \mathbf{c}(D^g v)s - \mathbf{c}(v)(Ds) = 0, \tag{3}$$

where D^g *denotes the Levi-Civita connection on* X.

Patching locally constructed spin connections together proves

Theorem 2 *(Branson, Gilkey [12]) There exist compatible connections on* S *which extend the Riemannian connection on* X *to* S.

Definition 3 *Let* $A : C^\infty(X; S) \longrightarrow C^\infty(X; S)$ *be a linear differential operator of first order operating on smooth sections of a* $C\ell(X)$*-module* S.
(a) We call A *an* operator of Dirac type, *if the principal symbol of its square is defining the Riemannian metric:*

$$\sigma_{A^2}(x, \xi) = \sum_{\mu,\nu=1}^{n} g^{\mu\nu}(x)\xi_\mu \xi_\nu. \tag{4}$$

(b) We call A *a* generalized Dirac operator, *if it can be written as* $A = \mathbf{c} \circ J \circ D$*, where* D *is a (not necessarily compatible) connection and*

$$J : C^\infty(X; T^*X \otimes S) \cong C^\infty(X; TX \otimes S)$$

denotes the canonical identification. In terms of a local orthonormal frame $v_1, \ldots, v_n$ *of* TX *we then have*

$$As|_x = \sum_{\nu=1}^{n} \mathbf{c}(v_\nu)(D_{v_\nu} s)|_x.$$

(c) We call A *a* (compatible) Dirac operator, *if it can be written as* $A = \mathbf{c} \circ J \circ D$*, where* D *is a compatible connection.*

Note. In this article we deal with compatible Dirac operators. However, most of the arguments remain valid for generalized non-compatible Dirac operators like the Dolbeault complex or, even more general, operators of Dirac type.

Clearly all (total) Dirac operators are elliptic and formally self-adjoint with a *Green's formula*

$$(As, s') - (s, As') = -\int_Y G(y)\langle s|_Y, s'|_Y\rangle, \tag{5}$$

where $G(y) := \mathbf{c}(n)$ denotes Clifford multiplication by the inward unit tangent vector.

For even n the splitting $\mathrm{C}\ell(X) = \mathrm{C}\ell^+(X) \oplus \mathrm{C}\ell^-(X)$ of the Clifford bundles induces a corresponding splitting of $S = S^+ \oplus S^-$ and a *chiral decomposition*

$$A = \begin{pmatrix} 0 & A^- \\ A^+ & 0 \end{pmatrix}.$$

The *partial Dirac operators* $A^\pm$ are especially interesting in index theory since they are also elliptic, but in general not self-adjoint and provide interesting integer-valued invariants as their indices. Like the Cauchy-Riemann operator $\bar\partial = \frac{1}{2}(\partial_x + i\partial_y) = \frac{1}{2}e^{i\varphi}(\partial_r + \frac{i}{r}\partial_\varphi)$ on the punctured two-disc all partial Dirac operators A^+ can be written in product form

$$A^+ = G(u,y)(\partial_u + B_u) \tag{6}$$

close to the boundary, where u denotes the inward oriented normal coordinate. Notice that the Clifford multiplication $G(u,y)$ defines a unitary morphism $S^+|_Y \to S^-|_Y$ and that $\{B_u\}$ is a family of self-adjoint (total) Dirac operators over Y. In the cylindrical case of a product metric close to Y the operators B_u and the morphisms $G(u,y)$ are independent of u.

2. Three Analysis Tools

We shall build our analysis on three basic properties which are not widely known and seem partly overlooked (2.1 and 2.3), partly insufficiently exploited (2.2) in the literature on partial differential equations.

2.1. Invertible Extension

Clifford multiplication by the inward normal vector gives a natural clutching of S^+ over one copy of X with S^- over a second copy of X to a smooth bundle $\widetilde{S^+}$ over the closed double $\widetilde{X}$. As observed in [39], the product forms of A^+ and A^- fit together over the boundary and provide a Dirac operator

$$\widetilde{A^+} := A^+ \cup A^- : C^\infty(\widetilde{X}; \widetilde{S^+}) \longrightarrow C^\infty(\widetilde{X}; \widetilde{S^-}).$$

Clearly $(A^+ \cup A^-)^* = A^- \cup A^+$; hence $\operatorname{index} \widetilde{A^+} = 0$. It turns out that $\widetilde{A^+}$ is invertible with a pseudo-differential elliptic inverse $(\widetilde{A^+})^{-1}$. Of course A^+ is not invertible and $r^+(\widetilde{A^+})^{-1}e^+A^+ \neq Id$, where $e^+ : L^2(X; S^+) \to L^2(\widetilde{X}; \widetilde{S^+})$ denotes the extension by zero operator and $r^+ : \mathcal{H}^t(\widetilde{X}; \widetilde{S^+}) \to \mathcal{H}^t(X; S^+)$ the natural restriction operator for Sobolev spaces, t real.

2.2. Calderón Projector

The next piece is the Calderón projector. It is a pseudo-differential projection onto the Cauchy data spaces (announced in Calderón [14] and proved in Seeley [34] in great generality). We define the *Cauchy data spaces*

$$H_+(A^+) := \{s|_Y \mid s \in C^\infty(X;S^+) \text{ and } A^+s = 0 \text{ in } X \setminus Y\}$$

and, for real t,

$$H_+(A^+,t) := \text{ closure of } H_+(A^+) \text{ in } \mathcal{H}^{t-\frac{1}{2}}(S^+|_Y);$$

and the *null spaces*

$$\ker_+(A^+,t) := \{s \in \mathcal{H}^t(X;S^+) \mid A^+s = 0 \text{ in } X \setminus Y\}.$$

The null spaces consist of sections which are distributional for negative t; which by elliptic regularity are smooth in the interior; and which by a Riesz operator argument can be shown to possess a trace $\gamma_0(s)$ over the boundary in $\mathcal{H}^{t-\frac{1}{2}}(Y;S^+|_Y)$.

First we construct a *Poisson type operator*

$$K_+ := r^+(\widetilde{A^+})^{-1}(\gamma_0^-)^*G : C^\infty(S^+|_Y) \longrightarrow C^\infty(S^+|_{X\setminus Y}).$$

It extends to a continuous mapping $\mathcal{H}^{t-\frac{1}{2}}(S^+|_Y) \to \ker_+(A^+,t)$ which is a bijection, if restricted to $H_+(A^+,t)$. Then we obtain the *Calderón projector* by taking the traces

$$\mathcal{P}_+ := \lim_{u \to 0_+} \gamma_u K_+$$

and similarly $\mathcal{P}_- := -\lim_{u\to 0_-} \gamma_u K_+$. Then $\mathcal{P}_+$ is a pseudo-differential projection with range$(\mathcal{P}_+) = H_+(A^+)$ and $\mathcal{P}_+ + \mathcal{P}_- = Id$.

The principal symbol p_+ of the Calderón projector is the projection onto the eigenspaces of the principal symbol $b(y,\zeta)$ of B_0 corresponding to non-negative eigenvalues. Hence it coincides with the principal symbol of the spectral projection $P_{\geq}(B_0)$. We call the space of pseudo-differential projections with the same principal symbol p_+ the *Grassmannian* Gr_{p_+}. It has enumerable many connected components; two projections P_1, P_2 belong to the same component, if and only if the *virtual codimension*

$$\mathrm{i}(P_2,P_1) := \mathrm{index}\{P_2P_1 : \mathrm{range}P_1 \to \mathrm{range}P_2\} \tag{7}$$

of P_2 in P_1 vanishes; the higher homotopy groups of each connected component are given by Bott periodicity.

2.3. Twisted Orthogonality of Cauchy Data Spaces

It is a nice feature of the Clifford multiplication G, first observed in [9], that it describes the orthogonal complement of the Cauchy data space of A^+ by

$$G^{-1}(H_+(A^-)) = (H_+(A^+))^\perp \tag{8}$$

and provides a short exact sequence

$$0 \to G^{-1}(H_+(A^-,t)) \hookrightarrow \mathcal{H}^t(Y;S^+|_Y) \xrightarrow{K_+} \ker_+(A^+,t) \to 0.$$

3. Some Basic Results

To get a closed operator, full regularity of the solutions, and a finite integer-valued index for Dirac operators over manifolds with boundary, we must impose global elliptic boundary conditions.

Definition 4 *A pseudo-differential operator* $R : C^\infty(Y; S^+|_Y) \to C^\infty(Y; V)$ *of order 0 defines an* elliptic boundary condition *for* A^+, *if* $\text{range} R^{(t)}$ *is closed in* $\mathcal{H}^t(Y; V)$ *and* $\text{range}(r) = \text{range}(rp_+) = \text{range}(p_+)$. *Here* $R^{(t)}$ *denotes the continuous extension of* R *to the* t*-th Sobolov space and* r *the principal symbol of* R.

Examples 5 *Typical examples are the* Atiyah-Patodi-Singer boundary condition *defined by the spectral projection* $P_{\geq}$ *of the boundary Dirac operator* B_0*;* generalized Atiyah-Patodi-Singer boundary conditions *defined by projections belonging to the Grassmannian* Gr_{p_+}*; and* local elliptic boundary conditions *characterized by the additional condition that the range of* r *can be written as the lifting of the vector bundle* V *under the natural projection* $T^*Y \setminus 0 \to Y$. *For even-dimensional* X *the Clifford multiplication becomes non-trivial and excludes the existence of local elliptic boundary conditions for* A *and* A^+ *(though not for systems); for odd-dimensional* X *we have natural local elliptic boundary conditions* $\Pi_\pm$ *defined by the orthogonal projection of* $S|_Y$ *onto* $(S|_Y)^\pm$.

Theorem 6 *The operator* $A_R^+ : \text{dom} A_R^+ \to L^2(X; S^-)$, *which acts like* A^+ *and is determined by*

$$\text{dom}\, A_R^+ := \{s \in \mathcal{H}^1(X; S^+) \mid R^{(0)}(\gamma_0 s) = 0\},$$

is a Fredholm operator from $L^2(X; S^+)$ *to* $L^2(X; S^-)$ *with index* $A_R^+ = i(R, \mathcal{P}_+)$.

Proof A first proof was sketched in Seeley [35]. We show how easy it is in our context: The subspace $\ker A_R^+$ of $\mathcal{H}^1(X; S^+)$ consists of smooth sections, since $A^+ s = 0$ and $R(\gamma_0 s) = 0$ imply that $h := \gamma_0 s$ belongs to the kernel of the boundary integral $R\mathcal{P}_+ : H_+(A^+, 1) \to \text{range} R^{(1)}$ which is contained in the kernel of the 'fan' $(Id - \mathcal{P}_+) + \mathcal{P}_+ R^* R \mathcal{P}_+$; the fan is elliptic by the symbol compatibility condition of Definition 4; hence h is smooth; hence also $s = K_+ h$ is smooth. (The concept of an *elliptic fan* is due to M. Birman and A. Solomyak [6]). The argument establishes the isomorphism

$$\ker A_R^+ \cong \ker\{R\mathcal{P}_+ : H_+ \longrightarrow \text{range}(R)\}$$

and the finite dimension of the kernel. That A_R^+ is a closed L^2 realization can be deduced from the explicit description of a left parametrix for A^+ by

$$(r^+ (\widetilde{A^+})^{-1} e^+) A^+ = Id - K_+ \gamma_0,$$

which is a direct consequence of the Calderón construction. Now all is simple: Since we have an explicit description of the adjoint operator

$$(A_R^+)^* = A^-_{G(Id-R)G^*}$$

(if R is a projection; otherwise replace R by the orthogonal projection onto the range of R^*) and since the range of A_R^+ is closed in L^2, we also have an explicit description of the cokernel of A_R^+ which, by the Clifford rotation of Cauchy data spaces of (8), can be identified with the cokernel of the boundary integral. □

It follows from the topology of Gr_{p_+} that for generalized Atiyah-Patodi-Singer boundary conditions index A_R^+ vanishes, if and only if the projection R belongs to the same connected component of Gr_{p_+} as the Calderón projector $\mathcal{P}_+$. More generally, we obtain two explicit versions of the classical *Agronovič-Dynin formula* (see [1]).

Theorem 7 *For two projections R_1 and $R_2 \in Gr_{p_+}$ we have*

$$\mathrm{index}(A_{R_1}^+) - \mathrm{index}(A_{R_2}^+) = \mathbf{i}(R_1, R_2); \tag{9}$$

and for two local elliptic boundary conditions $R_j : C^\infty(Y; S^+|_Y) \to C^\infty(Y; V_j)$, $j = 1, 2$, we have

$$\mathrm{index}(A_{R_1}^+) - \mathrm{index}(A_{R_2}^+) = \mathrm{index}\{R_1\mathcal{P}_+R_2 : C^\infty(Y; V_2) \to C^\infty(Y; V_1)\}. \tag{10}$$

From (9) we get a generalization of the *Atiyah-Patodi-Singer index formula* of [3]:

Theorem 8 *For $R \in Gr_{p_+}$ we have*

$$\mathrm{index}\ A_R^+ = \int_X \alpha(x) - \frac{1}{2}(\eta_B(0) + \dim\ker B) + \mathbf{i}(R, P_\geq).$$

Here $\alpha(x)$ denotes the locally defined *index density* of A^+ and

$$\eta_B(z) := \sum_{\lambda \in \mathrm{spec}B\setminus\{0\}} \mathrm{sign}\lambda|\lambda|^{-z} = \frac{1}{\Gamma(\frac{z+1}{2})}\int_0^\infty t^{\frac{z-1}{2}}\,\mathrm{tr}(Be^{-tB^2})\,dt \tag{11}$$

denotes the *η-function* of B. It is (i) well defined through absolute convergence for $\Re(z)$ large; (ii) it extends to a meromorphic function in the complex plane with isolated simple poles; (iii) its residues are given by a local formula; and (iv) it has a finite value at $z = 0$ (see e.g. Gilkey [16]).

Theorem 8 separates the contributions to the index from the whole manifold, from the structure on the boundary, and from the boundary condition in relation to the structure on the boundary. A special feature is that the correction term $\mathbf{i}(R, P_\geq)$ is not a homotopy invariant. It can change, e.g. under smooth deformations of the Riemannian metric, and is therefore a good candidate for more refined geometrical invariants.

To look at the geometrical aspects more closely, it is, as usual when working with Clifford algebras, appropriate to distinguish between the even- and the odd-dimensional case; and to alternate the focus between the index, the spectral flow, and the η-invariant.

4. Even Dimension. Index Theory

Let $M = X_1 \cup X_2$ be an even-dimensional closed partitioned manifold with $\partial X_1 = \partial X_2 = X_1 \cap X_2 = Y$. To illustrate the twisting of Cauchy data spaces by Clifford multiplication we prove

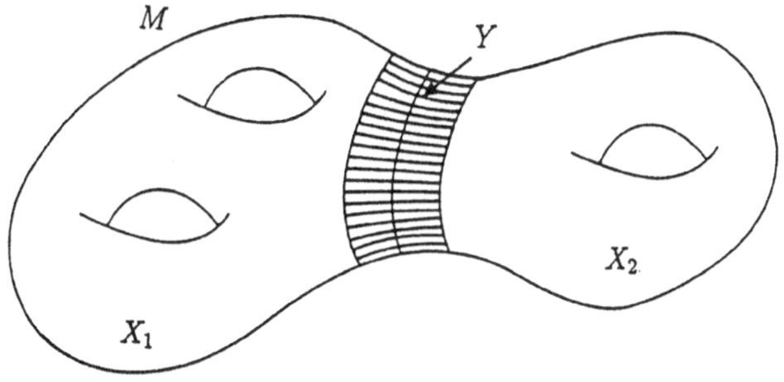

Fig. 1. Partitioned manifold

Theorem 9 *(Bojarski's conjecture [7]) Let A be a Dirac operator over M, let A^j denote its restriction to X_j, and let $\mathcal{P}(A^j)$ and $H(A^j)$ denote the corresponding Calderón projectors and L^2 closures of the Cauchy data spaces, $j = 1, 2$. Then*

$$\text{index}\, A = i(Id - \mathcal{P}(A^2), \mathcal{P}(A^1)) = \text{index}(H(A^1), H(A^2)).$$

Note. Notice that $Id - \mathcal{P}(A^2) = G\mathcal{P}(A^1)G^*$ by (8) and recall that

$$\mathrm{i}(Id - \mathcal{P}(A^2), \mathcal{P}(A^1)) := \text{index}\{(Id - \mathcal{P}(A^2))\mathcal{P}(A^1) : H(A^1) \longrightarrow H(A^2)^\perp\}$$

and

$$\text{index}(H(A^1), H(A^2)) := \dim(H(A^1) \cap H(A^2)) - \dim(L^2(Y; S|_Y)/(H(A^1) + H(A^2)).$$

Note. Closed subspaces for which the two dimensions in the preceding definition are finite are called ***Fredholm pairs*** of subspaces.

Proof It follows from the unique continuation property for Dirac operators and Green's formula (5) that $\ker A \cong H(A^1) \cap H(A^2)$. Now we apply the Clifford rotation formula for Cauchy data spaces (8) to get $\text{coker} A \cong \ker A^* \cong G(H(A^1)^\perp) \cap G(H(A^2)^\perp)$. The last space is isomorphic to $H(A^1)^\perp \cap H(A^2)^\perp$ which is the orthogonal complement of $H(A^1) + H(A^2)$ in $L^2(Y; S|_Y)$. □

Replacing the Clifford multiplication with an arbitrary unitary automorphism Φ of $S|_Y$, which is *consistent* (i.e. Φ commutes with p_+), leads us to the *general linear conjugation problem*, a generalization of the classical Riemann-Hilbert problem: We are looking for couples (s_1, s_2) with

$$A^j s_j = 0 \text{ in } X_j \setminus Y \quad \text{for } j = 1, 2 \qquad \text{and} \qquad s_2|_Y = \Phi(s_1|_Y). \tag{12}$$

Let $\text{index}_{LCP}(A, \Phi)$ denote the difference between the dimensions of the solution spaces of the original problem (12) and of the corresponding adjoint problem. We obtain

Theorem 10 *([8], [9])*

$$\text{index}_{LCP}(A, \Phi) = \text{index}\, (Id - \mathcal{P}(A^2))\Phi\mathcal{P}(A^1) = \text{index}\, A + \text{index}\, \mathcal{P}(A^1)\Phi\mathcal{P}(A^1).$$

Instead of the index of the generalized Toeplitz operator $\mathcal{P}(A^1)\Phi\mathcal{P}(A^1)$ we can calculate the index of the elliptic pseudo-differential operator $(Id-\mathcal{P}(A^1))-\Phi\mathcal{P}(A^1)$ of order 0 over Y applying the Atiyah-Singer index theorem. Or we determine the spectral flow $\mathrm{sf}\{B_u\}$ of any smooth family of elliptic self-adjoint operators connecting the Dirac operator B_0 with its gauge transform $\Phi^{-1}B_0\Phi$. Recall that the *spectral flow* is the difference between the number of eigenvalues, which change the sign from $-$ to $+$ as u goes from 0 to 1, and the number of eigenvalues, which change the sign from $+$ to $-$. It can be described as the index of the *suspension* $\{-\frac{\partial}{\partial u}+B_u\}$ which is an elliptic operator over $Y\times S^1$.

It is worth mentioning that for Dirac operators there is a one-one correspondence between the linear conjugation problems and the *cutting and pasting* of Dirac operators; and we obtain the same list of 'correcting' operators and explicit 'error terms' as above. By the decomposition of manifolds and operators into 'elementary' pieces, the cutting and pasting procedure also provides a direct inductive proof of the Atiyah-Singer index theorem for elliptic pseudo-differential operators on closed manifolds.

We close our discussion of the even-dimensional case with a *non-additivity theorem*. Consider two Dirac operators A^j over X_j, $j=1,2$, and assume that A^1 and A^2 are *consistent* with regard to Clifford multiplication, i.e. if A^1 takes the form $G(\partial_u+B)$ close to Y, then A^2 takes the form $G^{-1}(\partial_v-GBG^{-1})$ close to Y, where u denotes the inward normal on X_1 and v the inward normal on X_2. Then a Dirac operator $A^1\cup A^2$ is well defined over M and we obtain as a corollary to Theorem 8:

Theorem 11 *Let E_λ denote the eigenspace of the boundary Dirac operator B_0 to the eigenvalue λ, and $P_{\geq\lambda}$ the spectral projection onto the direct sum of all E_α with $\alpha\geq\lambda$. Then*

$$\mathit{index}\, A^1\cup A^2=\mathit{index}(A^1)_{P^1_{\geq\lambda}}+\mathit{index}(A^2)_{P^2_{\geq-\lambda}}+\dim E_\lambda .$$

Note. For $\lambda=0$ we obtain

$$\mathrm{index}(A^1\cup A^2)=\mathrm{index}(A^1)_{P_\geq}+\mathrm{index}(A^2)_{G^*(Id-P_\geq)G}+\dim\ker B,$$

which corresponds exactly to the Novikov additivity of the signature since $\operatorname{sign} X=\operatorname{index} A^+_{P_\geq}+\frac{1}{2}\dim\ker B$.

5. Odd Dimension

If n is odd, the total Dirac operator takes the form $A=G(\partial_u+B)$ near Y. Since G is a unitary bundle automorphism with $G^2=-Id_{S|_Y}$, it defines a decomposition of $S|_Y$ into the direct sum $S^+\oplus S^-$ of the subbundles of the $\pm i$-eigenvalues of $\{G_y\}_{y\in Y}$. With respect to this decomposition the operator A takes the following form near Y:

$$A=\begin{pmatrix} i & 0\\ 0 & -i\end{pmatrix}\left(\partial_u+\begin{pmatrix} 0 & B^-=(B^+)^*\\ B^+ & 0\end{pmatrix}\right). \tag{13}$$

Notice that the formal self-adjointness of $A=G(\partial_u+B)$ implies $GB=-BG$.

The product form (13) has various far reaching consequences. We begin with the classical *cobordism theorem*:

Theorem 12 *(Atiyah, Singer [4]) The index of a Dirac operator*

$$B^+ : C^\infty(Y;S^+) \longrightarrow C^\infty(Y;S^-)$$

over a closed even-dimensional manifold Y vanishes, if the couple (Y, S^+) is a 'boundary', i.e. if there exists a manifold X with boundary Y and a bundle of Clifford modules over X which, restricted to Y, is equal to $S^+ \oplus S^-$.

Proof From (10) we get for any A of the form (13) close to the boundary Y of an odd-dimensional manifold X $\operatorname{index} A_{\Pi_-} - \operatorname{index} A_{\Pi_+} = \operatorname{index} B^+$, where $\Pi_\pm$ denotes the local elliptic boundary conditions introduced in Example 5. But $\operatorname{index} A_{\Pi_\pm}$ vanishes by Green's formula (5). □

It follows that $\dim\ker B = 2\dim\ker B^+$. Moreover, the spectrum of B is symmetric with respect to 0; hence $\eta_B(z) \equiv 0$; and the index density α vanishes for odd-dimensional X. This reduces Theorem 8 to the simple formulas

$$\begin{aligned} \operatorname{index} A_{P_\geq(B)} &= -\dim\ker B_+ = \mathrm{i}(P_\geq(B), \mathcal{P}_+(A)) \qquad \text{and} \\ \operatorname{index} A_R &= -\dim\ker B_+ + \mathrm{i}(R, P_\geq(B)). \end{aligned}$$

That a kernel dimension appears indicates the non-homotopy invariance, see the example in Hitchin [21].

Now we discuss the true *odd* situation, i.e. we are given an odd-dimensional closed Riemannian manifold M which is partitioned in two manifolds X_1, X_2 with boundary by a hypersurface Y, and a bundle S of Clifford modules over M. We fix a bicollar neighbourhood N of Y in M. We assume that the metrics of M and S are product near Y.

The invariants we study on odd-dimensional manifolds are *spectral invariants*: the spectral flow, the η-invariant, and the analytic torsion. They are defined by the spectra of self-adjoint Dirac operators over M or Y; by the spectra of self-adjoint boundary problems; by the spectra of associated Laplacians; or by the spectra of naturally associated families. We shall begin with the spectral flow introduced above after Theorem 12. It plays a prominent role in the recent work of 'three-dimensional' topologists where it is used to determine the value of *Casson's invariant* and other invariants introduced recently (see Taubes [37], Witten [38]).

Let $\mathcal{D}$ denote the space of *cylindrical* Dirac operators acting on S with fixed principal symbol $G(x,\xi)$. Cylindrical means that the operator has the form (13) in the fixed cylinder N; i.e. elements of $\mathcal{D}$ differ over N only by endomorphisms of the bundle S. Let $\mathcal{D}^\times$ denote the subspace of invertible Dirac operators. We consider families

$$\mathcal{A} = \{A_r\} : (I;\{0,1\}) \longrightarrow (\mathcal{D};\mathcal{D}^\times).$$

It is not difficult to see that $\mathrm{sf}\{A_r\}$ is an integer-valued homotopy invariant of such families over M. In order to obtain an *odd* variant of Bojarski's theorem we need an invariant of families defined over Y. This was worked out recently by L. Nicolaescu [31]. Once again, the key is the 'twisted' orthogonality of Cauchy data of (8) which in this case says

$$\mathcal{P}(A) = -G(Id - \mathcal{P}(A^*))G = -G(Id - \mathcal{P}(A))G.$$

This equation has the following nice interpretation. Recall that $G|_Y$ defines a symplectic structure on $L^2(Y; S|_Y)$

$$\{s_1, s_2\} := (Gs_1, s_2) = \int_Y \langle Gs_1(y); s_2(y)\rangle dy.$$

Theorem 13 *([9]) For $A \in \mathcal{D}$, the Cauchy data space $H(A^1)$ is a Lagrangian subspace of $L^2(Y; S|_Y)$ with respect to G.*

Here $H(A^1)$ denotes the Cauchy data space of the operator $A^1 := A|_{X_1}$. *Lagrangian* means that $G(H(A^1))$ and $H(A^1)$ are orthogonal to each other and span the whole space. It is an easy consequence of Kuiper's theorem that the space of all Lagrangian subspaces is contractible (in the operator topology of the corresponding projections). To get a situation which is interesting from a topological point of view we consider the space $\mathcal{F}_2^G$ of all Fredholm pairs of Lagrangian subspaces of $L^2(Y; S|_Y)$. We have the following result proved by Nicolaescu [31] (for the definition of the K-group KR^{-7} see Karoubi [22]):

Theorem 14 *(Nicolaescu [31]) The space $\mathcal{F}_2^G$ has the homotopy type of a classifying space for KR^{-7}; in particular*

$$\pi_1(\mathcal{F}_2^G) \simeq Z.$$

It follows that to any loop $\{H_r^1, H_r^2\}$ of elements of $\mathcal{F}_2^G$ corresponds an integer $\mu\{H_r^1, H_r^2\}$, which is an obvious generalization of the standard *Maslov index* defined in finite dimensions (see Arnold [2]). Following standard procedures we can define the Maslov index for any path in $\mathcal{F}_2^G$ and formulate the 'odd' variant of Bojarski's theorem.

Theorem 15 *(Nicolaescu [31]) Let $\{A_r\} : (I; \{0,1\}) \to (\mathcal{D}; \mathcal{D}^\times)$ be a smooth family of cylindrical Dirac operators on M acting on S. Then*

$$sf\{A_r\} = \mu\{H(A_r^1), H(A_r^2)\}.$$

This theorem leads to a decomposition formula for the spectral flow. Let $\mathcal{L}_1$, $\mathcal{L}_2$ denote subspaces of $\ker B$ such that range$(P_>) + \mathcal{L}_j$ are Lagrangian subspaces of $L^2(Y; S|_Y)$ (or, equivalently, $\mathcal{L}_j$ is a Lagrangian subspace of $\ker B$ which is a finite-dimensional symplectic space). We denote by $A^j_{\mathcal{L}_j}$, $j = 1, 2$ the operator

$$\begin{cases} A^j_{\mathcal{L}_j} = A^j := A|_{X_j} \\ \operatorname{dom} A^j_{\mathcal{L}_j} := \{s \in \mathcal{H}^1(X_j; S|_{X_j}) \mid (P_> + \pi_j)(s|_Y) = 0\}, \end{cases}$$

where π_j denotes the projection onto $\mathcal{L}_j$. Then $A^j_{\mathcal{L}_j}$ is a closed self-adjoint operator in $L^2(X_j; S|_{X_j})$. Varying A (and so also its 'tangential' Dirac operator B) and choosing families $\{\mathcal{L}_{j,r}\}$, the following result follows from the work of Nicolaescu.

Theorem 16 $sf\{A_r\} = sf\{(A_r^1)_{\mathcal{L}_{1,r}}\} + sf\{(A_r^2)_{\mathcal{L}_{2,r}}\} + \mu\{\mathcal{L}_{1,r}, \mathcal{L}_{2,r}\}$.

This theorem has already been used in the work of Kirk and Klassen [23] in their computation of topological invariants of 3-manifolds.

We have a similar result for the η-invariant. The operator $A^1_{\mathcal{L}}$ has a discrete spectrum and the η-invariant of such operators is well defined (see [15]). The next result follows from the work of Lesch and Wojciechowski [27] and an observation made by W. Müller [30] that, modulo the integers, the η-invariant on manifolds of the form $M = X_1 \cup [-R, R] \times Y \cup X_2$ with cylindrical A does not depend on the length R of the cylinder.

Theorem 17 *Let $\eta(\mathcal{L}_1, \mathcal{L}_2)$ denote the η-invariant of A on the cylinder* $[-1,1] \times Y$ *with boundary condition $\mathcal{L}_1$ at $u = -1$ and $\mathcal{L}_2$ at $u = 1$. Then we have*

$$\eta(A) \equiv \eta((A^1)_{\mathcal{L}_1}) + \eta((A^2)_{\mathcal{L}_2}) + \eta(\mathcal{L}_1, \mathcal{L}_2) \quad mod\ Z,$$

Note. The integer contribution has been computed recently by U. Bunke [13].

6. History and Perspectives

For learning the basic relations between Dirac operators and global analysis on manifolds without boundary we refer to Palais [32], Karoubi [22], Gilkey [16], [17], Lawson and Michelsohn [26], and Berline, Getzler, and Vergne [5]; for details of the calculus on manifolds with boundary as sketched in the sections 1-4 of this article see Booß and Wojciechowski [11]; for other approaches than presented here in Section 5 to the 'odd' problem and to the cutting and pasting of η-invariants and analytic torsion over partitioned manifolds see Gilkey and Smith [19], [20], Roe [33], Singer [36], Booß and Wojciechowski [10], Gilkey [18], Mazzeo and Melrose [29], Klimek and Wojciechowski [24], [25], and Lück [28].

References

1. Agranovič, M.S., and Dynin, A.S.: 1962, 'General boundary value problems for elliptic systems in an n-dimensional domain', *Dokl. Akad. Nauk SSSR 146*, 511-514. (Russian; English translation *Soviet Math. Dokl. 3* (1962/63), 1323-1327).
2. Arnold, V.I.: 1967, 'Characteristic class entering in quantization conditions', *Funkcional. Anal. i Priložen. 1*, 1-14. (Russian; English translation *Functional Anal. Appl. 1*, 1-13; French translation Complément 1 to V.P. Maslov, *Théorie des Perturbations et Méthodes Asymptotiques*, Dunod, Gauthier-Villars, Paris 1972, 341-361).
3. Atiyah, M.F., Patodi, V.K., and Singer, I.M.: 1975, 'Spectral asymmetry and Riemannian geometry. I', *Math. Proc. Cambridge Phil. Soc. 77*, 43-69.
4. Atiyah, M.F., and Singer, I.M.: 1963, 'The index of elliptic operators on compact manifolds', *Bull. Amer. Math. Soc. 69*, 422-433.
5. Berline, N., Getzler, E., and Vergne, M.: 1992, *Heat Kernels and Dirac Operators*, Springer, Berlin.
6. Birman, M., and Solomyak, A.: 1982, 'On subspaces which admit pseudodifferential projections', *Vestnik Leningrad Univ. Nat. Mekh. Astronom. 82, no. 1*, 18-25 (Russian).
7. Bojarski, B.: 1979, 'The abstract linear conjugation problem and Fredholm pairs of subspaces', in: *In Memoriam I.N. Vekua* Tbilisi Univ., Tbilisi, pp. 45-60 (Russian).
8. Booß, B., and Wojciechowski, K.P.: 1985, 'Desuspension of splitting elliptic symbols I', *Ann. Global Anal. Geom. 3*, 337-383.
9. —, —: 1986, 'Desuspension of splitting elliptic symbols II', *Ann. Global Anal. Geom. 4*, 349-400.

10. —, —: 1989, 'Pseudo-differential projections and the topology of certain spaces of elliptic boundary value problems', *Comm. Math. Phys.* **121**, 1-9.
11. —, —: 1993, *Elliptic Boundary Problems for Dirac Operators*, Birkhäuser, Boston (in print).
12. Branson, Th.P., and Gilkey, P.B.: 1992, 'Residues of the eta function for an operator of Dirac type', *J. Funct. Anal.* **108**, 47-87.
13. Bunke, U.: 1993, *A Glueing Formula for the eta-Invariant*, Preprint, Humboldt Universität Berlin.
14. Calderón, A.P.: 1963, 'Boundary value problems for elliptic equations', in: *Outlines of the Joint Soviet-American Symposium on Partial Differential Equations*, Novosibirsk, pp. 303-304.
15. Douglas, R.G., and Wojciechowski, K.P.: 1991, 'Adiabatic limits of the η-invariants. The odd-dimensional Atiyah-Patodi-Singer problem', *Comm. Math. Phys.* **142**, 139-168.
16. Gilkey, P.B.: 1984, *Invariance Theory, the Heat Equation, and the Atiyah-Singer Index Theorem*, Publish or Perish, Wilmington. New revised edition in preparation.
17. —: 1989, 'The geometrical index theorem for Clifford modules', in: Rassias, T.M. (ed.), *Topics in Mathematical Analysis*, World Scientific Press, Singapore, pp. 315-327.
18. —: 1991, *On the Index of Geometrical Operators for Riemannian Manifolds with Boundary*, Preprint, University of Oregon, Eugene.
19. Gilkey, P.B., and Smith, L.: 1983a, 'The eta invariant for a class of elliptic boundary value problems', *Comm. Pure Appl. Math.* **36**, 85-131.
20. —, —: 1983b, 'The twisted index problem for manifolds with boundary', *J. Differential Geom.* **18**, 393-444.
21. Hitchin, N.: 1974, 'Harmonic spinors', *Adv. Math.* **14**, 1-55.
22. Karoubi, M.: 1978, *K-Theory*, Springer, Berlin.
23. Kirk, P.A., and Klassen, E.P.: 1993, *Computing Spectral Flow via Cup Products*, Preprint, Indiana University, Bloomington.
24. Klimek, S., and Wojciechowski, K.P.: 1992, *Adiabatic Cobordism Theorems for Analytic Torsion and η-Invariant*, Preprint, Purdue University, IUPUI, Indianapolis.
25. —, —: 1993, 'η-Ivariants on manifolds with cylindrical end', *Differ. Geom. Appl.* **3** (1993), to appear.
26. Lawson, H.B., and Michelsohn, M.-L.: 1989, *Spin Geometry*, Princeton University Press, Princeton.
27. Lesch, M., and Wojciechowski, K.P.: 1993, *On the η-invariant of generalized Atiyah-Patodi-Singer boundary value problems*, Report No. 278, Universität Augsburg.
28. Lück, W.: 1993, 'Analytic and topological torsion for manifolds with boundary and symmetry', *J. Differential Geom.* **37**, 263-322.
29. Mazzeo, R.R, and Melrose, R.B.: 1992, *Analytic Surgery and the Eta Invariant*, Preprint, M.I.T.
30. Müller, W.: 1993, *Eta-Invariants and Manifolds with Boundary*, Preprint, Max-Planck-Inst. f. Math., Bonn.
31. Nicolaescu, L.: 1993, *The Maslov Index, the Spectral Flow, and Splittings of Manifolds*, Preprint, Michigan State University, East Lansing.
32. Palais, R.S. (ed.): 1965, *Seminar on the Atiyah-Singer Index Theorem*, Ann. of Math. Studies 57, Princeton University Press, Princeton.
33. Roe, J.: 1988, 'Partitioning non-compact manifolds and the dual Toeplitz problem', in: *Operator algebras and application. Vol. 1*, Lond. Math. Soc. Lect. Note Ser. 135, pp. 187-228.
34. Seeley, R.T.: 1966, 'Singular integrals and boundary value problems', *Amer. J. Math.* **88**, 781-809.
35. —: 1969, 'Topics in pseudo-differential operators', in: *CIME Conference on Pseudo-Differential Operators (Stresa 1968)*. Ed. Cremonese, Rome, 1969, pp. 167-305.
36. Singer, I.M.: 1988, 'The η-invariant and the index', in: Yau, S.-T. (ed.), *Mathematical Aspects of String Theory*, World Scientific Press, Singapore, pp. 239-258.
37. Taubes, C.H.: 1990, 'Casson's invariant and gauge theory', *J. Differential Geom.* **31**, 547-599.
38. Witten, E.: 1989, 'Quantum field theory and the Jones polynomial', *Comm. Math. Phys.* **121**, 351-400.
39. Wojciechowski, K.P.: 1985, 'Elliptic operators and relative K-homology groups on manifolds with boundary', *C.R. Math. Rep. Acad. Sci. Canada* **7**, 149-154.

SPIN STRUCTURES AND HARMONIC SPINORS ON RIEMANN SURFACES

JAROLÍM BUREŠ
Mathematical Institute of Charles University
Sokolovská 83, 18600 Prague
Czech republic.

Abstract. The aim of the paper is a description of some relations between spin-structures, the Dirac operators and harmonic spinors on the one side and the complex geometry of Riemann surfaces on the other side. The case of hyperelliptic surfaces is studied in more details.

Key words: Dirac operator, Harmonic spinors,compact Riemann surface, hyperelliptic surface

1. Introduction

The Dirac operator on a riemannian spin-manifold M belongs to the intensively studied operators in geometry and mathematical physics. Its kernel is called the space of harmonic spinors. In general the dimension of the space of harmonic spinors depends on the metric and spin-structure used to define the Dirac operator. In the Hitchin's paper [1] the space of harmonic spinors on compact Riemann surfaces is studied. In this paper we give complete description of the space of harmonic spinors on hyperelliptic surfaces and some results on harmonic spinors for the nonhyperelliptic surfaces. More details and proofs will be published in [3]. The similar results vas obtained independently by Ch.Bar and P.Schmutz in [6].

2. Harmonic spinors on Riemannian manifolds

Let (M,g) be an oriented riemannian manifold, let B_g be the principal fibre bundle of oriented orthonormal frames with respect to g on M.

A spin-structure on (M,g) is a principal fibre bundle $\tilde{B}$ on M with the group $Spin(n)$ and a projection $\lambda : \tilde{B} \to B_g$ which is 2-1 covering and equivariant map. There exists a spin structure on M iff the second Stiefel-Whitney class vanishes (it is a topological condition).

Let us denote S (M,g) the set of all unequivalent spin-structures on (M,g). Then card(S $(M,g)) = $ card$(H^1(M,Z_2))$.

Let $\xi_0 : \widetilde{Gl^+}(n,R) \to Gl^+(n,R)$ be the unique nontrivial double covering (universal for $n \geq 3$). Then $\xi_0^{-1}(SO(n)) = Spin(n)$.

Let F^+ be the principal fibre bundle of oriented frames on M (with the structural group $Gl^+(n,R)$), let $\mathcal{F}$ (M) be the set of unequivalent double coverings $\xi : \tilde{F} \to F^+$, where $\tilde{F}$ is a principal fibre bundle with the group $\widetilde{Gl^+(n,R)}$ and and ξ is an equivariant map.

F. Brackx et al. (eds.), Clifford Algebras and their Applications in Mathematical Physics, 67–74.

For an arbitrary riemannian metric g on M we have the natural bijection ι_g : $\mathcal{F}(M) \to \mathrm{S}\,(M,g)$ defined in the following way:

$$\iota_g(\tilde{F}) = \xi^{-1}(B_g).$$

and for any two metric g, g' on M the bijection $\eta_{g,g'} : \mathrm{S}\,(M,g) \to \mathrm{S}\,(M,g')$ defined as a composition $\eta_{g,g'} = \iota_{g'}^{-1} \circ \iota_g$.

2.1. DIRAC OPERATOR.

To any spin-structure $s \in \mathrm{S}\,(M,g)$ it corresponds the fundamental spinor bundle S_s which is an associated vector bundle to $\tilde{B}_g^s$ with respect to the representation of the group Spin(n) which is restriction of irreducible representation of Clifford algebra C_n to Spin(n). (We always suppose $Spin(n) \subset C_n$.)

The classical Dirac operator D^s associated to s is defined using the connection ∇ on S_s which is inducedfrom the Levi-Civita connection on (M,g).

The operator

$$D^s : \Gamma(M, S_s) \to \Gamma(M, S_s)$$

is composition of maps:

$$\Gamma(M,S_s) \xrightarrow{\nabla} \Gamma(M, S_s \otimes T^*) \xrightarrow{\hat{g}} \Gamma(M, S_s \otimes T) \xrightarrow{\mu} \Gamma(M, S_s)$$

where $\hat{g} : S_s \otimes T^* \to S_s \otimes T$ is identity times duality induced from riemannian metric g, μ is given by the action of C_n on S_s.

A local expression of D^s using a local orthonormal frame $\{e_1, \ldots, e_n\}$ is

$$D^s = \sum_{i=1}^{n} e_i.\nabla_{e_i}.$$

The space of harmonic spinors for the spin structure s is

$$\mathbf{H}^s = \{\xi \in \Gamma(M, S_s); D^s\xi = 0\}.$$

If n = 2m even, we have a decomposition of spinor bundle $S_s = S_s^+ \oplus S_s^-$, the Dirac operator D^s maps $\Gamma(M, S_s^+)$ into $\Gamma(M, S_s^-)$ and $\Gamma(M, S_s^-)$ into $\Gamma(M, S_s^+)$, and the space of harmonic spinors decomposes into spaces of positive harmonic spinor $\mathbf{H}^{s+}$ and negative harmonic spinors $\mathbf{H}^{s-}$.

2.2. HARMONIC SPINORS.

Let M be an oriented manifold which admits spin-structures (i.e $w_2(M) = 0$). Fix some (let say standard) riemannian metric g_0 on M and spin-structure $s_0 \in \mathrm{S}(M, g_0)$. Then we can to identify

$$H^1(M, Z_2) \equiv \mathrm{S}(M, g_0) \overset{\alpha_{g_0,g}}{\to} \mathrm{S}(M,g)$$

for an arbitrary riemannian metric g on M. If g is a riemannian metric on M and $s \in H^1(M, Z_2)$ spin-structure on (M,g) we shall consider the corresponding spinor bundle $S_{g,s}$, the corresponding Dirac operator D^s_g, its space of harmonic spinors $\mathrm{H}^s_g = KerD^s_g$ and denote

$$h^s_g = dim\mathrm{H}^s_g$$

We shall say that dimension of harmonic spinors on M depends on metric iff there exists $s \in H^1(M, Z_2)$ such that h^s_g depends on g. It is well-known fact that the dimension of harmonic spinor is conformal invariant, namely if g is conformal to g' then $h^s_g = h^s_{g'}$.

For any even dimension n = 2m we have the decomposition of spinor spaces into two parts given above and we can denote further $h^{s+}_g = dimH^{s+}_g and h^{s-}_g = dimH^{s-}_g$.

In the concrete cases the notions defined in this and previous section will be shortened in a natural way (omitting some letters).

3. Spin-structures on complex manifold

Let (M,g,J) be a complex Kaehler manifold of complex dimension m,
let $K_M = \Lambda^{(m,0)}$ be the canonical holomorphic line bundle of M. If moreover (M,g) is a spin-manifold, then we have the theorem.

THEOREM 3.1 *[1] Let (M,g,J) be a Kaehler spin-manifold ,then*

A) Spin structures on (M,g) are in one-to-one correspondence with holomorphic line bundles L on (M,J) which are square-roots of the canonical line bundle K_M. (i.e L satisties $L \otimes L = K$.)

B) For a holomorphic line bundle L^s which is holomorphic square root of K and corresponds to the spin structure $s \in S(M, g)$ we have the following identifications:

a) for the spinor bundle S_s

$$S_s = \Lambda^{(0,*)}(M) \otimes L^s$$

b) for the Dirac operator D^s

$$D_s = \bar{\partial}_{L^s} + \bar{\partial}^*_{L^s} : \Lambda^{0,*} \otimes L^s \to \Lambda^{0,*} \otimes L^s$$

(It is the standard operator on (0,)-forms with values in the holomorphic line bundle L^s.)*

c) the splitting of the spinor bundle

$$S_s = S^+_s \oplus S^-_s$$

corresponds to the decomposition

$$\Lambda^{0,*} \otimes L^s = \Lambda^{0,even} \otimes L^s \oplus \Lambda^{0,odd} \otimes L^s$$

and the Dirac operator can be (after restriction) identified with two operators $D^+_s (D^-_s)$ from even (odd) forms into odd (even) forms with values in L^s.

c) Let us demote by $\mathcal{O}(L)$ the sheaf of germs of holomorphic sections of the holomorphic bundle L.

For the corresponding spaces of (odd and even) harmonic spinors there are isomorphisms coming from the Hodge theory.

$$H^{s+} \simeq H^{even}(M, \mathcal{O}(L))$$

$$H^{s-} \simeq H^{odd}(M, \mathcal{O}(L))$$

4. Spin structures on Riemann surfaces

A Riemann surface M is a one dimensional complex manifold. We restrict ourselves on compact Riemann surfaces, so in this paper we shall use the notion Riemann surface for compact Riemann surface only. There is only one topological invariant of M called genus of M.

Let g be a riemannian metric on M compactible with complex structure J on M. Then (M, J, g) is Kaehler manifold. Every two metrics compatible with J are conformally equivalent.

There is one-to-one correspondence between the set of complex structures on a two dimensional manifold M and the set $C(\mathrm{M})$ of conformal classes of metrics on M.

Every oriented 2-dimensional manifold M is a spin manifold, so that Riemann surface is also spin manifold.

LEMMA 4.1 : *Let M be of genus g. There are*

$$N(g) = cardH^1(M, Z_2) = 2^{2g}$$

different spin structures on M.

Let K be the canonical line bundle of Riemann surface M. From the Theorem 3.1 it follows that there are isomorphisms:

$$\mathrm{H}^{s+} \simeq H^0(M, \mathcal{O}(L))$$

$$\mathrm{H}^{s-} \simeq H^1(M, \mathcal{O}(L))$$

Let us denote $h^i = dimH^i(M, \mathcal{O}(L))$, then from the Serre duality follows that

$$H^0(M, \mathcal{O}(L)) \simeq H^1(M, \mathcal{O}(L))$$

and $h^0 = h^1$.

For any $g \in \mathbf{Z}^+$ we fix some riemannian metric g_0 on a Riemann surface M_g, and some spin-structure $s_0 \in \mathrm{S}\ (M_g, g)$.

Then we can associate to every element $s \in H^1(M, Z_2)$

a) spin structure s on (M_g, g) (a principal Spin(2)-bundle $\widetilde{B}_s$ which is 2-1 covering of B_g.

b) holomorphic line bundle L^s on (M_g, g) satisfying $L^s \otimes L^s = K_M$.

Let us denote $h^0_{s,g} = dimH^0(M, \mathcal{O}(L^s)$, then

$$h^s_g = 2.h^0_{s,g}$$

THEOREM 4.1 *For any metric g and $\alpha \in H^1(M, Z_2)$ on a surface of genus g we have an inequality:*

$$h^0_{\alpha,g} \leq [\frac{1}{2}(g+1)]$$

THEOREM 4.2 ([2]) *Let M be a Riemann surface of genus g. Then for any metric g there are precisely $2^{g-1}(2^g+1)$ spin structures α on M for which $h^0_{\alpha,g}$ is an even number. These spin structures are called even spin structures on M.*

For the other $2^{g-1}(2^g-1)$ spin structures α on M, $h^0_{\alpha,g}$ is an odd number. These spin structures are called odd spin structures.

From these two theorems immediately follows :

THEOREM 4.3 ([1]) *If the genus g of M is less then 3 then the dimension of the space of harmonic spinors on M is independent on the metric.*

We shall study the situation for different these genera separately.

0) For g= 0 , $M = P^1(C) = S^2$ which is simple connected. There exists just one spin structure on M which does not admits nonzero harmonic spinors. So $h^0 = h^1 = 0$.

1) For g = 1, there exist altogether 4 spin-structures on M, namely 3 even spin-structures which does not admit nonzero harmonic spinors ($h^0 = 0$), one odd spin structure (the trivial one) has space of positive harmonic spinors one dimensional.

2) For g = 2 there exist altogether 16 spin structures on M, there are 10 even spin structures with no nonzero harmonic spinors ($h^0 = 0$), and also 6 odd spin-structures with one-dimensional space of positive harmonic spinors ($h^0 = h^1 = 1$).

Let us present here a rewiev of some results. For the definition and properties of hyperelliptic Riemann surfaces see next section.

THEOREM 4.4 ([1]) *The dimension of the space of harmonic spinors on a Riemann surface of genus $g \geq 3$ varies with the choice of metric.*

THEOREM 4.5 ([1]) *If (M,J) is hyperelliptic, there exists spin structure α on (M,J) such that $h^0_{\alpha,g} = [\frac{1}{2}(g+1)]$. Moreover if g is even, there are just 2(g +1) such structures.*

THEOREM 4.6 ([4]) *If there exist on Riemann surface spin structures α such that $h^0_{\alpha,g} = [\frac{1}{2}(g+1)]$ then (M,J) is one of the following types*

(a) hyperelliptic

(b) $g = 4$

(c) $g = 6$

In the nonhyperelliptic case of $g = 4$ and 6 , there is only one spin structure having $h^0_{\alpha,g} = [\frac{1}{2}(g+1)]$.

REMARK 4.1 *From the theory of Riemann surfaces we have on Riemann surfaces of M of genus g$\geq$ 3 the following identifications:*

a)Harmonic spinors $\leftrightarrow$ holomorphic sections of line bundles associated to the divisors D of order g- 1 and satisfying the equality 2.D = K $\leftrightarrow$ meromorphic functions on M having divisors greater or equal to D.

b)Spin-structures on M are in 1-1 correspondence with θ-characteristics

REMARK 4.2 *There is a closed connections between compact Riemann surfaces and projective algebraic curves. Compact Riemann surfaces (curves) are divided from certain point of view into two classes, the first class consists of so called hyperelliptic surfaces (curves), the others are called nonhyperelliptic (see next section). There exist as hyperelliptic as well as nonhyperelliptic surfaces (curves) for all genera g, with $g \geq 3$.*

4.1. HYPERELLIPTIC RIEMANN SURFACES.

From the results of a complex projective algebraic geometry and theory of theta functions we can get the complete description of spin-structures and harmonic spinors for all hyperelliptic cases (see e.g. [5]).

DEFINITION 4.1 : *A hyperelliptic curve C is a complex projective curve, which admits a rational surjective map π onto the projective line $P^1(C)$ which is 2-1 up to finite set of point, which are called branching points of π.*

Let $\mathcal{B}$be the set of branching points of π. The genus gof C is

$$g = \frac{1}{2}(card\mathcal{B}) - 1 = k - 1.$$

There is an involution $\iota : C \to C$ with $\pi(P) = \pi(\iota(P)$ for any $P \in C$, having the set of fixpoints equal to the set of branching points $\mathcal{B}$and divisor class

$$L_C := \{P + \iota(P); P \in C\}.$$

Let K_C be the canonical divisor class of C (i.e class of divisors of abelian differentials on C).

Let

$$\sum = \{D; 2.D = K_C\}$$

be set of divisor classes on C, called set of θ- characteristics.

For any divisor class D (or divisor D) we denote by L(D) the line bundle on C which corresponds to this divisor class D (or to the divisor D) and $\mathcal{L}$(D) the space of all holomorphic sections of L(D). We use the standard language of the theory of Riemann surfaces without comments (e.g the relation between divisor and divisor class for an induced notions).

THEOREM 4.7 *[5]:*

(i) $K_C = (g-1)L_C$,

(ii) Any θ-characteristics has a form

$$f_T := \sum_{P \in T} P + \frac{1}{2}(g - 1 - cardT).L_C$$

where $T \subset \mathcal{B}$, with $cardT \equiv g + 1$ mod 2

(iii) We have

$$f_{T_1} \equiv f_{T_2} \Leftrightarrow T_1 = T_2 or\ T_1 = cT_2$$

$$\sum \equiv \{T \subset \mathcal{B}; cardT \equiv (g+1) mod 2\}/T \sim cT$$

(iv) For all such T there exist points $\tilde{P}_1, ..., \tilde{P}_{g-1}$ of C with $f_T \equiv \sum_{i=1}^{i=g-1} \tilde{P}_i$ iff $cardT \neq g+1$ and if $cardT \leq g$, then

$$dim\mathcal{L}(f_T) = \frac{1}{2}(g+1-cardT)$$

(if card $T \geq (g+2)$ we change cT for T itself).

(v) $\mathcal{L}(f_T) \simeq \{$ The set of polynoms in t of degree $\leq \frac{1}{2}(g-1+cardT)$ vanishing in all points $P \in T\}$.

From this theorem we get the following results:

(1) Spin-structures on hyperelliptic curve C correspond in one-to-one way to θ-characteristics. Let us denote $L(f_T)$ the spin structure corresponding to the θ-characteristics f_T, where T is a subset of $\mathcal{B}$with card T $\leq$ g+ 1 (under an equivalence relation $T \sim cT$).

(2) The dimension of positive harmonic spinors for spin-structure L(f_T) is :

$$h^0_{L(f_T)} = \frac{1}{2}(\mathrm{g}+1-cardT)$$

(3) The set of positive harmonic spinors for a given spin structure L(f_T) is $\mathrm{H}^+_{L(f_T)}$ = The set of all polynoms in t of degree less or equal $\frac{1}{2}(\mathrm{g}-1+cardT)$ vanishing at all points $P \in T$.

EXAMPLE 4.1 *Let us describe the cases of genera $g = 3, 4and6$ more detaily.*

Let g= 3. Then N(3) = 64, $N_{odd} = 28, N_{ev} = 36, h^0 \leq 2$.

$$\sum = \{T \subset \{1, \ldots, 8\}, cardT = 0, 2, 4, 6, 8\}/T \sim cT$$

a) odd spin structures:

$N_{odd} = 28$ spin structures correspond to $T = \{i, j\}, 1 \leq i < j \leq 8\}$ with $h^0 = 1$.

b) even spin structures:

one spin structure, corresponding to $T = \emptyset$ with $h^0 = 2$.

35 spin-structures corresponding to $T = \{1, i, j, k\}; 2 \leq i < j < k \leq 8\}$ with $h^0 = 0$.

Let g= 4. Then N(4) = 64, $N_{odd} = 28, N_{ev} = 36, h^0 \leq 2$.

$$\sum = \{T \subset \{1, \ldots, 10\}, cardT = 1, 3, 5, 7, 9\}/T \sim cT$$

a) odd spin structures:

$N_{odd} = 120$ spin structures correspond to $T = \{i, j, k\}, 1 \leq i < j < k \leq 10\}$ with $h^0 = 1$.

b) even spin structures:

10 spin-structures, corresponding to $T = \{i\}; 1 \leq i \leq 10$ with $h^0 = 2$.

126 spin-structures corresponding to $T = \{1, i, j, k, l\}; 2 \leq i < j < k < l \leq 10\}$ with $h^0 = 0$.

Let g= 6. Then N(6) =2^{12} , $N_{odd} = 2^5(2^6-1), N_{ev} = 2^5(2^6+1), h^0 \leq 3$.

$$\sum = \{T \subset \{1, \ldots, 14\}, cardT = 1, 3, 5, 7, 9, 11, 13\}/T \sim cT$$

a) odd spin structures:
14 spin-structures corresponding to $T = \{i\}, 1 \leq i \leq 14$, *with* $h^0 = 3$. C_{14}^5*spin-structures corresponding to* $T = \{i, j, k, l, m\}, 1 \leq i < j < k < l < m \leq 14\}$ *with* $h^0 = 1$.

b) even spin structures:
C_{14}^3 *spin-structures, corresponding to* $T = \{i, j, k\}; 1 \leq i < j < k \leq 14$ *with* $h^0 = 2$. C_{13}^6 *spin-structures corresponding to* $T = \{1, i, j, k, l, m, n\}; 2 \leq i < j < k < l < m < n \leq 14\}$ *with* $h^0 = 0$.

4.2. NONHYPERELLIPTIC RIEMANN SURFACES.

Only some remarks on nonhyperelliptic Riemann surfaces of genera 3,4 and 6 will be presented here. More details and other examples will be published in the paper [3].

If g= 3, then N(3) = 64, $N_{odd} = 28, N_{ev} = 36, h^0 \leq 2$. We have for all odd spin-structures $h^0 = 1$, and for all even spin-structures $h^0 = 0$.

REMARK 4.3 *a) Nonhyperelliptic Riemann surfaces of genus 3 are just regular quartics in* $P^2(C)$.

Let g= 4. There is unique even spin-structure (corresponding to the Weierstrass point) with $h^0 = 2$. Other 135 even spin-structures have $h^0 = 0$. All odd spin-structures have $h^0 = 1$.

Let g= 6. There is unique odd spin-structure with $h^0 = 3$. Other odd spin-structures have $h^0 = 1$.

REMARK 4.4 *There exists a system of nonhyperelliptic curves C of genus 6, namely regular quintics in* $P^2(C)$ *with the canonical line bundle* $K = O(2)/C$, *and spin structure* $L = O(1)/C$ *with* $h_L^0 = 3$. *(O(k) is the k-th power of the dual of tautological bundle on* $P^2(C)$*).*

References

1. Hitchin N.: Harmonic spinors, Advances in Mathematics 14, (1974), 1-55.
2. Atiyah M.F.: Riemann surfaces and spin structures, Ann.scient. Ec. Norm. Sup. t.4, (1971), 47-62.
3. Bureš J.: Dirac operator and Harmonic spinors on Riemann surfaces, Proc.Winter School "Geometry and Physics" Srni 1993,Suppl.Rend.Circ.Math.Palermo will be published.
4. Farkas H.M.: Special divisors and analytic subloci of Teichmueler space, Amer.J.Math 88, (1966), 881-901
5. Mumford D.: Tata lectures on Theta II, Progress in Mathematics 43, Birkhauser 1984.
6. Bar Ch.,Schmutz P.: Harmonic spinors on Riemann surfaces, Annals of Global Analysis and Geometry 10, (1992), 263-273.

SPHERICAL GEOMETRY AND MÖBIUS TRANSFORMATIONS

J. CNOPS
Dept. of Mathematical Analysis, State University of Ghent, Galglaan 2, B-9000 Gent, Belgium

1. Introduction

Clifford algebras: notations. The Clifford algebra over the space R^{pq} with bilinear form $\mathcal{B}(\vec{x}, \vec{y})$ and an orthonormal basis e_i, $i = 1, \cdots, n = p+q$, is generated by R^{pq} with the relations

$$e_i^2 = \mathcal{B}(e_i, e_i) \qquad e_i e_j + e_j e_i = 0 \quad i \neq j,$$

where $\mathcal{B}(e_i, e_i) = 1$ for $1 \leq i \leq p$ and $\mathcal{B}(e_i, e_i) = -1$ for $p < i \leq n$. Except for the basis elements e_i and sometimes the zero vector in this article always a vector arrow is used for elements of the underlying vector space of the Clifford algebra. An automorphism $a \to a'$ ($(ab)' = a'b'$) is defined by $\vec{x}' = -\vec{x}$, a first antiautomorphism $a \to \overline{a}$ ($\overline{ab} = \overline{b}\overline{a}$) is defined by $\overline{\vec{x}} = -\vec{x}$ and a second antiautomorphism by $a^* = \overline{a'}$. For a vector $\vec{x}^2 = \mathcal{B}(\vec{x}, \vec{x})$ is real and $\vec{x}$ is invertible if and only if $\vec{x}$ is not isotropic. For two different vectors $\mathcal{B}(\vec{x}, \vec{y}) = \frac{1}{2}(\vec{x}\vec{y} + \vec{y}\vec{x})$ which is sometimes denoted as $\vec{x} \cdot \vec{y}$. As the Clifford algebra for $n > 1$ is not commutative it must be remarked that in $\frac{a}{b}$ the inverse is put on the right hand side i. e. $\frac{a}{b} = a/b = ab^{-1}$. The space generated (as a linear space over R) by the elements $e_{i_1} \cdots e_{i_k}$ of lenght k ($i_1 < \cdots i_k$) is called the space of k-vectors. the projection of $a \in R_{pq}$ on this space is denoted as $[a]_k$.

Three subgroups of the Clifford algebra are needed:

- The Clifford group $\Gamma(p,q)$ of products of invertible vectors. For $\in \Gamma(p,q)$ $aa^* = \pm a\overline{a}$ is real and non-zero.
- the Pin group $\text{Pin}(p,q)$ which is the subgroup of $\Gamma(p,q)$ of elements such that $aa^* = \pm 1$.
- the Spin group $\text{Spin}(p,q)$ which is the subgroup of $\text{Pin}(p,q)$ of products of an even number of vectors.

For s in the Clifford group the mapping $\text{Pin}(s) : \vec{x} \to Pin(s)\vec{x} = s\vec{x}s'^{-1}$ is an orthogonal transformation. This way $\text{Pin}(p,q)$ becomes a double covering of $O(p,q)$ ($\text{Pin}(-1)$ is the identity) and $\text{Spin}(p,q)$ of $SO(p,q)$.

Moreover a periodicity theorem is needed: let $R_{p+1,q+1}$ be the algebra over $R^{p+1,q+1}$ with the orthonormal basis $e_1, \cdots, e_n, e_+, e_-$, with $e_\pm^2 = \pm 1$. Then $R_{p+1,q+1}$ is isomorphic to $R_{pq}^{2\times 2}$ and an isomorphism is given by

$$a + be_+ + ce_- + de_+e_- \longrightarrow \begin{pmatrix} a+d & -b+c \\ b'+c' & a'-d' \end{pmatrix},$$

F. Brackx et al. (eds.), Clifford Algebras and their Applications in Mathematical Physics, 75–84.

or, in the other direction

$$\begin{pmatrix} a & b \\ c & d \end{pmatrix} \longrightarrow \frac{1}{2}\left((a+d') + (a-d')e_+e_- + (b+c')e_+ + (-b+c')e_-\right).$$

For this theorem we refer to Porteous ([4]). In the sequel both the matrix notation and the classical notation for an element of $R_{p+1,q+1}$ will be used. An explicit calculation gives the explicit formulae for the automorphism and the antiautomorphisms, e.g.

$$\begin{pmatrix} a & b \\ c & d \end{pmatrix}' = \begin{pmatrix} a' & -b' \\ -c' & d' \end{pmatrix},$$

$$\overline{\begin{pmatrix} a & b \\ c & d \end{pmatrix}} = \begin{pmatrix} d^* & -b^* \\ -c^* & a^* \end{pmatrix}.$$

The matrix form of a vector $\vec{Y} \in R^{p+1,q+1}$ is $\begin{pmatrix} \vec{y} & \mu \\ \nu & -\vec{y} \end{pmatrix}$ where $\vec{y}$ is a vector in R^{pq} and μ and ν are real.

The *pseudodeterminant* $\Delta(A)$ of the matrix $A = \begin{pmatrix} a & b \\ c & d \end{pmatrix}$ is given by $ad^* - bc^*$. This is the first entry of the product $A\overline{A}$ and so, if $A\overline{A}$ is real, then $\Delta(A) = A\overline{A}$. Moreover in this case $\Delta(BA) = \Delta(B)\Delta(A)$ for arbitrary B, since $BA\overline{(BA)} = A\overline{A}B\overline{B} = B\overline{B}\Delta(A)$.

For the Pin-group representation we need A'^{-1} for $A \in \Gamma(p+1, q+1)$. For such A the equality $A'^{-1} = cA^*$, where $c = \pm 1$. As we shall work projectively, this is sufficient for our purposes without explicitating c.

Möbius transformations. A Möbius transformation on R^{pq} is by definition a bijection from the compactified space $\overline{R^{pq}}$ (to be defined strictly later on) into itself which (a) maps spheres to spheres and (b) is conformal. For $p+q > 2$ condition (a) is superfluous, as in this case every conformal mapping is a Möbius transformation (see Haantjes [2]). The group of Möbius transformations will be denoted as $\mathcal{M}(p,q)$. The group of even Möbius transformations which are sense preserving is denoted as $\mathcal{M}^+(p,q)$.

Spheres. The equation for a sphere in R^{pq} with centre $\vec{m}$ and radius r (where r^2 is real, but not necessarily positive, we call a sphere positive or negative according to wether r^2 is positive or negative, if $r^2 = 0$ we have a zero sphere) can be written as

$$(\vec{y} - \vec{m})^2 = r^2$$

or

$$y^2 - 2B(\vec{y}, \vec{m}) + \left(\vec{m}^2 - r^2\right) = 0. \tag{1}$$

We introduce projective coordinates for such a sphere:

$$\rho(\vec{m}, 1, \vec{m}^2 - r^2) = (\vec{\mu}, \nu_1, \nu_2)$$

where ρ is an arbitrary non zero factor. A point $\vec{y}$ can be identified with the sphere of radius 0 and centre $\vec{y}$ having coordinates

$$\rho(\vec{y}, 1, \vec{y}^2) = (\vec{\xi}, \eta_1, \eta_2)$$

while a hyperplane with equation $\mathcal{B}(\vec{m},\vec{x}) - b = 0$ has coordinates $\rho(\vec{m}, 0, 2b)$. This way spheres are identified with rays in $R^{p+1,q+1}$ and points with rays in the light cone of $R^{p+1,q+1}$. It can be proved that this way every orthogonal transformation on $R^{p+1,q+1}$ induces a Möbius transformation on $\overline{R^{pq}}$, the set of rays of the light cone. This way a double covering is obtained, because obviously the mapping $\vec{X} \to -\vec{X}$ induces the identity. This way also a four-fold covering of $\mathcal{M}(p,q)$ by Spin($(p+1, q+1)$) is obtained, and using the representation of $R_{p+1,q+1}$ by 2×2 matrices with entries in R_{pq}, with each Möbius transformation 4 matrices can be associated, the identity being represented by

$$\pm\begin{pmatrix} 1 & 0 \\ 0 & 1 \end{pmatrix} \text{ and } \pm\begin{pmatrix} e_1 \dots e_n & 0 \\ 0 & -(e_1 \dots e_n)' \end{pmatrix}$$

For an element of the Spin group $\begin{pmatrix} a & b \\ c & d \end{pmatrix}$ the associated Möbius transformation sends $\vec{x}$ to $\frac{a\vec{x}+b}{c\vec{x}+d}$. A nice characterisation of the matrices in the Spin group was given in [1]. Notice that all such matrices have pseudodeterminant ± 1.

2. Transform of the metric

From now on we suppose that g is the Möbius transformation associated with the element $A = \begin{pmatrix} a & b \\ c & d \end{pmatrix}$ of Pin$(p+1, q+1)$. The image of a vector $\vec{x}$ is written as $g\vec{x}$. We can rewrite the expression $g\vec{u} - g\vec{v}$ (since $g\vec{v} = (g\vec{v})^*$) as

$$\begin{aligned}
&((a\vec{u}+b)(c\vec{u}+d)^{-1} - (c\vec{v}+d)^{*-1}(a\vec{v}+b)^* \\
&= (c\vec{v}+d)^{*-1}\left[(c\vec{v}+d)^*(a\vec{u}+b) - (a\vec{v}+b)^*(c\vec{u}+d)\right](c\vec{u}+d)^{-1} \\
&= (c\vec{v}+d)^{*-1}\left[\vec{v}(c^*b - a^*d) + (d^*a - b^*c)\vec{u} + \vec{v}(c^*a - a^*c)\vec{u} + d^*b - b^*d\right] \\
&\qquad\qquad\qquad\qquad\qquad\qquad\qquad\qquad \times(c\vec{u}+d)^{-1} \\
&= \pm(c\vec{v}+d)^{*-1}\left[\vec{u} - \vec{v}\right](c\vec{u}+d)^{-1}
\end{aligned}$$

where the $\pm$ sign agrees with the $\pm$ sign in $\Delta(A) = \pm 1$. Taking the limit for $\vec{v} \to \vec{u}$ gives $\mu_g(\vec{u})|d\vec{u}| = |dg\vec{u}|$, where $\mu_g(\vec{u}) = |c\vec{u}+d|^{-2}$ is the local *contraction factor*. When there is no confusion possible, we shall omit the subindex g. Obviously $\mu_g(\vec{x}) = \left(\mu_{g^{-1}}(g\vec{x})\right)^{-1}$. As the inverse transformation is given by the matrix $\overline{A} = \begin{pmatrix} d^* & -b^* \\ -c^* & a^* \end{pmatrix}$ we have, with $\vec{y} = g\vec{x}$, that

$$\mu_g(\vec{x}) = |c\vec{x}+d|^{-2} = \mu_{g^{-1}}(\vec{y}) = |-c^*\vec{y}+a^*|^2 = |\vec{y}a - c|^2.$$

When also the direction of $d\vec{x}$ and $d\vec{y}$ is taken into account we have the relations $d\vec{y} = \pm(c\vec{x}+d)^{*-1}d\vec{x}(c\vec{x}+d)^{-1}$ and $d\vec{x} = \pm(-c^*\vec{y}+a^*)^{*-1}d\vec{y}(-c^*\vec{y}+a^*)^{-1}$ where the $\pm$ signs agree since $\Delta(A) = \Delta(\overline{A})$. As a result $(-c^*\vec{y}+a^*) = \pm(c\vec{x}+d)^{-1}$.

In order to link the contraction factor to the orthogonal group representant of the Möbius transformation a normalisation of the representants of spheres is needed. For spheres with radius different from zero one of the representants with spinorial norm 1, i. e. a representant s with $ss^* = \pm 1$ (one could introduce the notion of oriented sphere to distinguish between the two representants) is taken. Because A is in the Spin group AsA^* is a normalised representant of the image of the sphere.

For spheres with radius zero the representant with third entry equal to 1 is taken, i. e. the vector $\vec{x}$ of R^{pq} is represented by $\vec{X} = \begin{pmatrix} \vec{x} & -\vec{x}^2 \\ 1 & -\vec{x} \end{pmatrix}$. It is easily seen by explicit calculation that, if $\vec{Y}$ is the normalised representant of $g\vec{x}$, then $A\vec{X}A^* = \pm\mu^{-1}(\vec{x})\vec{Y}$. This leads to the following theorem:

Theorem 2.1 *Let $\mathcal{A}$ be a subgroup of $Pin(p+1, q+1)$ containing $e_1 \ldots e_+ e_-$ and -1, and let G be the group of Möbius transformations associated to $\mathcal{A}$. Let moreover f be a scalar function defined on R^{pq}. Then the following statements are equivalent:*
(i) The metric $|d\vec{x}|/f(\vec{x})$ is invariant under G.
(ii) The function F defined on the isotropic cone K of $R^{p+1,q+1}$ by $F(\lambda\vec{X}) = |\lambda| f(\vec{x})$ for any real λ, where $\vec{X}$ is the standard representant of $\vec{x}$, is invariant under $\mathcal{A}$.

Proof.
Let g be an arbitrary element of G, and A be one of the four associated elements in $\text{Pin}(p+1, q+1)$ (that all four are in $\mathcal{A}$ follows from the condition imposed on $\mathcal{A}$).

If (i) holds then, for $\vec{y} = g\vec{x}$ we have that $f(\vec{x})\mu(\vec{x}) = f(\vec{y})$ and as a result that $F(A(\lambda\vec{X})A'^{-1}) = F(\pm\lambda(\mu(\vec{x}))^{-1}\vec{Y}) = |\lambda| f(\vec{x}) = F(\lambda\vec{X})$, which proves that F is invariant under $\mathcal{A}$. The converse is easily seen reading the proof backwards. ∎

The most classical class of these invariant metrics is obtained by taking for F the absolute value of the inner product with a vector in $R^{p+1,q+1}$. For an interpretation in R^{pq} we introduce the 'product distance' between a point $\vec{x}$ and a sphere S, denoted by $(\vec{x}, S)$ as the absolute value of the inner product of the representants, i. e. $(\vec{x}, S) = |B(\vec{X}, s)|$. Notice that $(\vec{x}, S)$ is zero if and only if $\vec{x} \in S$. The transform of the metric can now be described as follows: Let S be a sphere not going through $\vec{x}$, (so $(\vec{x}, S) \neq 0$) with image $gS = K$, with normalised representants s and k. Then we have that $(\vec{x}, S) = |B(\vec{X}, s)| = |B(A\vec{X}A^*, AsA^*)| = |\mu^{-1}(\vec{x})B(\vec{Y}, k)| = \mu^{-1}(\vec{x})(\vec{y}, K)$, or

$$\mu(\vec{x}) = \frac{(g\vec{x}, gS)}{(\vec{x}, S)}.$$

If S is a zero radius sphere, then this formula still holds up to a constant independent of $\vec{x}$. If the centre of this sphere is $\vec{v}$, we indeed have that $AsA^* = \pm\mu^{-1}(\vec{v})k$ and so

$$\mu(\vec{x}) = \frac{(g\vec{x}, gS)}{(\vec{x}, S)}\mu(\vec{v}).$$

Denote by G_S the group of Möbius transformations associated with elements A such that $AsA^* = \pm s$.For a non-null sphere S one has that G_S is the fixgroup in $\mathcal{M}(p, q)$ of S, while for a null sphere S with centre $\vec{v}$, G_S is the group of Möbius transformations keeping S fixed and satisfying $\mu(\vec{v}) = 1$. Taking the viewpoint of $\text{Pin}(p+1, q+1)$, G_S is generated by the non-isotropic vectors orthogonal to s, which means in R^{pq} that G_S is generated by the symmetries round spheres orthogonal to S. Then we have the following result:

Corollary 2.1 *The metric $d\vec{x}(\vec{x}, S)^{-1}$ is invariant under Möbius transformations in G_S.*

This leads to 6 invariant metrics, depending on wether the sphere is positive, negative or null, and wether it is a finite sphere or an infinite one. For the Euclidean case ($q = 0$) there is no negative sphere at infinity (this being a hyperplane with normal vector $\vec{m}$ with $\vec{m}^2 < 0$, geometrically speaking).

kind of sphere	s	s^2	metric	invariance group
hyperplane	$\begin{pmatrix} e_m & 0 \\ 0 & -e_m \end{pmatrix}$	1	$\frac{dx}{\lvert x_m \rvert}$	Poincaré group
unit sphere	$\begin{pmatrix} 0 & 1 \\ 1 & 0 \end{pmatrix}$	1	$\frac{dx}{\lvert 1 - \vec{x}^2 \rvert}$	$R\left\{\begin{pmatrix} 1 & \vec{u} \\ -\vec{u} & 1 \end{pmatrix}, \vec{u}^2 \neq 1\right\}$
imaginary unit sphere	$\begin{pmatrix} 0 & -1 \\ 1 & 0 \end{pmatrix}$	-1	$\frac{dx}{\lvert 1 + \vec{x}^2 \rvert}$	$R\left\{\begin{pmatrix} 1 & \vec{u} \\ \vec{u} & 1 \end{pmatrix}, \vec{u}^2 \neq 1\right\}$
point at infinity	$\begin{pmatrix} 0 & 1 \\ 0 & 0 \end{pmatrix}$	0	dx	$O(m)T$
origin	$\begin{pmatrix} 0 & 0 \\ 1 & 0 \end{pmatrix}$	0	$\frac{dx}{\lvert \vec{x}^2 \rvert}$	$O(m)K$

Here R is the group of (anti)rotations, T is the translation group, and K the conjugate of T under the inversion $\vec{x} \to 1/\vec{x}$. This is by no means the most general kind of invariant metric which can be associated with subgroups of $\mathcal{M}(p,q)$. Two other examples are

(i) Fixgroups of a certain number of spheres. If we take two spheres S_1 and S_2 and take the intersection of G_{S_1} and G_{S_2} we can write

$$\mu(\vec{x}) = \frac{f\left((g\vec{x}, S_1), (g\vec{x}, S_2)\right)}{f\left((\vec{x}, S_1), (\vec{x}, S_2)\right)},$$

where f has to satisfy $f(\lambda a, \lambda b) = \lambda f(a, b)$ for $\lambda \geq 0$. Similar constructions can be made for a greater number of spheres.

(ii) Let P be a hyperplane in $R^{p+1,q+1}$ and define $(\vec{x}, P)$ to be the absolute value of the square of the projection of $\vec{X}$ onto P. Suppose for ease of thought that P is spanned by two mutually orthogonal non-isotropic vectors, and let G_P be the obvious fixgroup generated by unit vectors orthogonal to P and the unit vectors of P itself (The imposed condition makes the latter part a classical orthogonal group). Then again we can express $\mu(\vec{x})$ as

$$\mu(\vec{x}) = \frac{(g\vec{x}, P)}{(\vec{x}, P)}$$

and we have an invariant metric. This case is essentially different from what we had before, since G_P does not leave any sphere invariant. All spheres in P intersect in the same 'circle' (i. e. m-2 dimensional sphere) and this sphere is invariant under P.

Not only the maximal invariance group for a metric can be calcuted, but also all the invariant metrics for a given group can be decided, as is shown in the following theorem:

Theorem 2.2 *Let A be a subgroup of $Pin(p+1, q+1)$ and let G be the group of associated Möbius transformations. Then each metric of the form $|d\vec{x}|/f(\vec{x})$ can be described as follows: Let S be the set of singular points for the group G defined by*

$$S = \{\vec{x} \in \overline{R^{pq}}\} : \exists g \in G : g\vec{x} = \vec{x} \text{ and } \mu_g(\vec{x}) \neq 1\}$$

(it is clear that for each $\vec{x}$ in S f must be zero or infinity). Let $\mathcal{P} = \overline{R^{pq}}/G$ be the set of orbits of G (notice that $\mathcal{P}$ partitions both S and $\overline{R^{pq}} \setminus S$) and let P be a set containing one element of each orbit outside S (it is possible that one needs the axiom of choice to do this). Denote for $\vec{x}$ in P the equivalence class in $\mathcal{P}$ by $[\vec{x}]$. For a function $F : P \to R^+$ the function f is defined as follows: for $\vec{y} \in [\vec{x}]$ there exists a $g \in G$ such that $g\vec{x} = \vec{y}$. Put then $f(\vec{y}) = F(\vec{x})\mu_g(\vec{x})$.

Corollary 2.2 *A sphere S is the set of singular points of the fixgroup G_S. The metric $|d\vec{x}|/(\vec{x}, S)$ is up to a constant the unique invariant metric for this group.*

It is possible to calculate the image of the Laplacian and the Dirac operator, and the preceding results can be used to obtain an invariant Laplacian for the fixgroups of the theorem above. First we define for a geometrical transformation g the operator $\hat{g}$ acting on functions by $\hat{g}F(\vec{y}) = F(g^{-1}\vec{y})$. As an example we have that $\hat{g}\mu(\vec{y}) = |\vec{y}c - a|^2$. For an operator $\mathcal{A}$ the conjugate of $\mathcal{A}$ with respect to g is defined as being $\hat{g}\mathcal{A}\hat{g}^{-1}$. An operator thus is invariant under (the action of) a group G if $\hat{g}\mathcal{A}\hat{g}^{-1} = \mathcal{A}$ for all g in G. In the sequel we work with a definite metric for ease of notation, but generalisation to R^{pq} is straightforward. In accordance with general use, which admittedly is rather awkward here, we use anti-Euclidean space, and to avoid confusion we shall speak of real and imaginary spheres instead of negative and positive ones.

Theorem 2.3 *(i) The conjugate of the Laplacian with respect to an arbitrary Möbius transformation g is given by*

$$\hat{g}\Delta\hat{g}^{-1} = (\hat{g}\mu)^{1+m/2}\Delta(\hat{g}\mu)^{1-m/2}. \quad (2)$$

(ii) The operator

$$f^{1+m/2}\Delta f^{1-m/2} \tag{3}$$

is invariant under the group G if and only if the metric $d\vec{x}/f(\vec{x})$ is invariant under G.

Proof.
As the matrix A is the product of unit vectors in $R^{1,m+1}$, it is sufficient to prove (i) for such vectors, as the general case follows by composition. Essentially there are two possibilities for such vectors: they induce either the reflection round a hyperplane or the symmetry round a sphere with finite centre. As the Laplacian is invariant under reflections the formula holds for the first case. For the second case we assume without loss of generality that we have a real sphere, e.g. the unit sphere. Indeed, if we have $-r^2 < 0$, the inversion round the sphere can be written as a product of inversions round real spheres: if we take e.g. e_+ we can use $e_1 \ldots e_n e_-$, since $e_1 \ldots e_n e_+ e_-$ induces the identity Möbius transform. In this case $\vec{y} = g\vec{x} = -\vec{x}^{-1}$ and $\mu(\vec{x}) = |\vec{x}^{-2}|$. Introducing the spherical decomposition for the Laplacian:

$$\Delta = \partial_r^2 + \frac{m-1}{r}\partial_r + \frac{1}{r^2}\Delta_{\vec{\xi}}$$

(i) is easily proved by explicit calculation.

For (ii) the starting point is that $\mu(\vec{x})f(\vec{x}) = f(g\vec{x})$, in other symbols $\hat{g}f = f/\hat{g}\mu$. Hence

$$\begin{aligned}\hat{g}\left(f^{1+m/2}\Delta f^{1-m/2}\right)\hat{g}^{-1} &= f^{1+m/2}(\hat{g}\mu)^{-1-m/2}\hat{g}\Delta\hat{g}^{-1}(\hat{g}\mu)^{-1+m/2}f^{1-m/2} \\ &= f^{1+m/2}\Delta f^{1-m/2}\end{aligned}$$

∎

Likewise the transform of the Dirac operator D can be calculated.

Theorem 2.4 *The conjugate of the Dirac operator with respect to an arbitrary Möbius transformation is given by*

$$\begin{aligned}\hat{g}D\hat{g}^{-1}F(\vec{y}) &= (cg^{-1}\vec{y}+d)^*(\hat{g}\mu(\vec{y}))^{1+m/2}D(cg^{-1}\vec{y}+d)(\hat{g}\mu(\vec{y}))^{1-m/2}F(\vec{y}) \\ &= \pm(-c^*\vec{y}+a^*)(\hat{g}\mu(\vec{y}))^{m/2}D(\vec{y}c-a)\hat{g}\mu(\vec{y}))^{-m/2}F(\vec{y}).\end{aligned}$$

where the $\pm$ sign agrees with $\Delta(A)$.

In the case that F is a scalar valued function we can take into account that $(\vec{y}c - a)\hat{g}\mu(\vec{y}))^{-m/2}$ is a monogenic function so $D\left((\vec{y}c-a)\hat{g}\mu(\vec{y}))^{-m/2}F(\vec{y})\right) = (DF(\vec{y}))(\vec{y}c-a)\hat{g}\mu(\vec{y}))^{m/2}$ and we have that

$$\hat{g}D\hat{g}^{-1}F(\vec{y}) = (-c^*\vec{y}+a^*)(DF(\vec{y}))(\vec{y}c-a). \tag{4}$$

3. Transforms of harmonic and monogenic functions

The explicit formulae for the transform of the Laplacian and the Dirac operator immediately give the following result:

Theorem 3.1 - *Let h be harmonic. then*

$$(\hat{g}\mu)^{1-m/2}\hat{g}h$$

is harmonic.

- *Let $u(\vec{x})$ be monogenic. Then the function $(cg^{-1}\vec{y}+d)\mu(g\vec{y})^{1-m/2}\hat{g}u(\vec{y})$ is monogenic in $\vec{y}$. This function (up to a $\pm$ sign) can also be written as*

$$\gamma u(\vec{y}) = \frac{(\vec{y}c-a)}{|\vec{y}c-a|^m}\hat{g}u(\vec{y}). \tag{5}$$

The first result however leads to another transformation for vector valued monogenic functions. Indeed, take h harmonic and scalar valued. Then $u = Dh$ is monogenic and vector valued. Conversely, if u is monogenic and vector valued we can, at least locally, find a real valued harmonic function such that $u = Dh$ holds. According to the formula for the transform of the Laplacian we have that

$$\hat{g}u(\vec{y}) = (-c^*\vec{y}+a^*)(D\hat{g}h(\vec{y}))(\vec{y}c-a).$$

Transforming h on the other hand we see that $|c|^{m-2}(\hat{g}\mu)^{1-m/2}\hat{g}h = |\vec{y}-ac^{-1}|^{2-m}\hat{g}h$ (we assume $c \neq 0$) is harmonic and so

$$D|\vec{y}-ac^{-1}|^{2-m}\hat{g}h = (2-m)\frac{\vec{y}-ac^{-1}}{|\vec{y}-ac^{-1}|^m}\hat{g}h + |\vec{y}-ac^{-1}|^{2-m}D\hat{g}h$$

is monogenic. Applying D to this function gives

$$(4-2m)\frac{\vec{y}-ac^{-1}}{|\vec{y}-ac^{-1}|^m}D\hat{g}h + |\vec{y}-ac^{-1}|^{2-m}D^2\hat{g}h = 0.$$

Applying D to the function $|\vec{y}-ac^{-1}|^{4-2m}D\hat{g}h$ gives

$$(4-2m)\frac{\vec{y}-ac^{-1}}{|\vec{y}-ac^{-1}|^{2m-2}}D\hat{g}h + |\vec{y}-ac^{-1}|^{4-2m}D^2\hat{g}h$$

which also must be zero. Hence the function $|\vec{y}-ac^{-1}|^{4-2m}D\hat{g}h$ is monogenic and comparison with the expression for $\hat{g}u$ gives up to a constant the function

$$\frac{(\vec{y}c-a)\hat{g}u(\vec{y})(c^*\vec{y}-a^*)}{|\vec{y}c-a|^{2m}}. \tag{6}$$

An application was given by Leutwiler in [3] for the case of the Poincaré metric. He called a function hyperbolic harmonic (h.h.) if it satisfies

$$\left((x_n)^{1+m/2}\Delta(x_n)^{1-m/2} - \lambda\right)h = 0$$

for a certain scalar constant λ, which will be given later. Obviously, if g is in the Poincaré group, then $\hat{g}h$ is also h.h. He then studied functions which locally have the form $u = Dh$, (call them hyperbolic monogenic, h.m.). If u is h.m., then $\hat{g}u(\vec{y}) = (-c^*\vec{y} + a^*)(D\hat{g}h(\vec{y}))(\vec{y}c - a)$, and so we obtain the h.m. function

$$\frac{(\vec{y}c - a)\hat{g}u(\vec{y})(-c^*\vec{y} + a)}{|\vec{y}c - a|^2}.$$

Now $\Delta\left((x_n)^{1-m/2}h\right) = (\Delta(x_n)^{1-m/2})h - 2(D(x_n)^{1-m/2})\cdot(Dh) + (x_n)^{1-m/2}\Delta h$ $= (1-m/2)(-m/2)(x_n)^{-1-m/2}h - (2-m)(x_n)^{-m/2}e_n\cdot(Dh) + (x_n)^{1-m/2}\Delta h$ so we can write the equation for h as

$$x_n\Delta h - (2-m)e_n\cdot(Dh) + (1-m/2)(-m/2)h - \lambda h = 0.$$

Leutwiler put $\lambda = (1-m/2)(-m/2)$, and we can write the equation for hyperbolically monogenic functions as

$$-Du - (1-m/2)e_n\cdot Du = 0.$$

This can be generalised: Take a group G with invariant metric $|d\vec{x}|/f(\vec{x})$ which is moreover transitive on the part of $\overline{R^m}$ where the metric is not singular. The associated invariant Laplacian is $f^{1+m/2}\Delta f^{1-m/2}$. We define invariant harmonic functions (i.h.) h as satisfying

$$\left(f^{1+m/2}\Delta f^{1-m/2} - \lambda\right)h = 0$$

where we calculate λ later, and invariant monogenic functions (i.m.) as functions u which can locally be written as Dh. Again we have that $\Delta\left(f^{1-m/2}h\right) = (\Delta f^{1-m/2})h - 2(Df^{1-m/2})\cdot(Dh) + f^{1-m/2}\Delta h$ and the equation for h becomes

$$f\left(f\Delta h - (2-m)(Df)\cdot Dh\right) + (f^{1+m/2}\Delta f^{1-m/2} - \lambda)h = 0.$$

It is easy to see that the function $f^{1+m/2}\Delta f^{1-m/2}$ must be a real constant. Indeed, take the function $w(\vec{x}) = 1$. As $\hat{g}w = w$ for all Möbius transformations we have that $f^{1+m/2}\Delta f^{1-m/2} = f^{1+m/2}\Delta f^{1-m/2}w$ is invariant under G. Since G is transitive, this means that $f^{1+m/2}\Delta f^{1-m/2}$ is independent of $\vec{x}$, in other words, a constant. Thus, if we take λ equal to this constant, we have the defining relations for i.h. and of i.m. functions:

$$f\Delta h - (2-m)(Df)\cdot Dh = 0 \qquad (7)$$

$$fDu + (2-m)(Df)\cdot u = 0 \qquad (8)$$

together with the transformation relation that if u is i.m., then

$$\frac{(\vec{y}c - a)\hat{g}u(\vec{y})(-c^*\vec{y} + a)}{|\vec{y}c - a|^2}$$

is also, and this for any g in the group G.

We can relate the theory for invariant harmonic and monogenic functions for different groups. A first theorem deals with conjugate groups.

Theorem 3.2 *Let G be a group of Möbius transformations with invariant metric $d\vec{x}/f(\vec{x})$. Let g be a Möbius transformation and let H be the conjugate group gGg^{-1}. Then H has invariant metric $d\vec{y}/\hat{g}f(\vec{y})$.*

Proof.
Take F and A as in theorem 1.1. Let g be one of the elements of $Pin(p+1, q+1)$ represented by g. Then obviously gAg^{-1} is represented by gGg^{-1} and $\hat{g}F$ is invariant under gAg^{-1}. As $\hat{g}F$ is associated with $\hat{g}f$ this proves the theorem. ∎

Corollary 3.1 *Let Δ_G be the invariant Laplacian for the group G. Then*

$$\hat{g}\Delta_G\hat{g}^{-1} = \Delta_{gGg^{-1}}.$$

As an example we can describe functions in a ball as hypermonogenic functions. If we take any Möbius transformation g mapping the m-dimensional upper half space onto the unit ball we get the spherical model for hyperbolic space with metric $d\vec{x}/(1-r^2)$ and with invariant Laplacian $(1-r^2)^{1+m/2}\Delta(1-r^2)^{1-m/2}$. If we take any real valued hyperharmonic function h in the sense of Leutwiler, then $\hat{g}h$ satisfies the equation

$$(1-r^2)\Delta\hat{g}h + (4-2m)\vec{x}.D(\hat{g}h) = 0.$$

An invariant monogenic function u for the unit ball hence satisfies

$$(1-r^2)Du - (4-2m)\vec{x}.u = 0.$$

The expression for an invariant monogenic function u for the unit ball derived from an i.m. function for the half space can be derived from (4), putting $F = \hat{g}h$, which gives $\hat{g}v(\vec{y}) = (-c^*\vec{y} + a^*)u(\vec{y})(\vec{y}c - a)$. This of course depends on the Möbius transformation used to map the half space to the ball. A logical choice is

$$g\vec{x} = \frac{\vec{x} - e_m}{-e_m\vec{x} + 1}$$

with Spin matrix $\frac{1}{\sqrt{2}}\begin{pmatrix} 1 & -e_m \\ -e_m & 1 \end{pmatrix}$, which gives the transition formula

$$u(\vec{y}) = -\frac{(\vec{y}e_m + 1)\hat{g}v(\vec{y})(e_m\vec{y} + 1)}{|e_m\vec{y} + 1|^2}.$$

REFERENCES

[1] **J. Fillmore and A. Springer** 'Möbius groups over general fields using Clifford algebras associated with spheres', *Int. J. Theo. Phys.*, **29**(1990) pp. 225-246.

[2] **J. Haantjes:** 'Conformal representations of an n-dimensional euclidean space with a non-definite fundamental form on itself', *Proc. Kon. Nederl. Acad. Wetensch.*, 40(1937), pp. 700-705.

[3] **H. Leutwiler:** 'Modified Clifford Analysis', *Complex Variables*, **17**(1992), pp. 153-171.

[4] **I. Porteous:** *Topological geometry*, Van Rostrand Reinhold Company, London, 1969.

CAUCHY TRANSFORMS AND BI-AXIAL MONOGENIC POWER FUNCTIONS

A.K. COMMON
Institute of Mathematics and Statistics
University of Kent, Canterbury, Kent CT2 7NF, England.

and

F. SOMMEN
Seminar of Algebra and Functional Analysis
State University of Ghent, Galglaan 2. B-9000 Ghent, Belgium.

Abstract. We consider the transform of suitable Clifford valued functions or distributions $f(\vec{x}_1, \vec{x}_2)$ defined on bi-axially symmetric domains in R^{p+q} with $p+q = m$. Such functions have the expansion

$$f(\vec{x}_1, \vec{x}_2) = \sum_{k,l} f_{k,l}(|\vec{x}_1|, |\vec{x}_2|) P_{k,l}(\vec{x}_1, \vec{x}_2),$$

where $P_{k,l}(\vec{x}_1, \vec{x}_2)$ are bi-axial spherical monogenics of degree k in $\vec{x}_1$ and degree l in $\vec{x}_2$. We define generalised Cauchy transforms for the case when the $f_{k,l}$ are independent of $|\vec{x}_2|$:

$$\Lambda^{(1)}_{k,l}(f)(\vec{x}) = -\frac{1}{\omega_m} \int_{R^p} \frac{[\vec{x}_1 + \vec{x}_2 - \vec{u}]}{|\vec{x}_1 + \vec{x}_2 - \vec{u}|^{m+2l}} P_{k,l}(\vec{\eta}, \vec{x}_2) f_{k,l}(\lambda) d^p\vec{u}$$

$$\Lambda^{(2)}(f)(\vec{x}) = -\frac{1}{\omega_m} \int_{R^p} \frac{[\vec{x}_1 + \vec{x}_2 - \vec{u}]}{|\vec{x}_1 + \vec{x}_2 - \vec{u}|^{m+2l}} \vec{\eta} P_{k,l}(\vec{\eta}, \vec{x}_2) f_{k,l}(\lambda) d^p\vec{u}$$

where $\vec{u} = \lambda\vec{\eta}$, $|\vec{\eta}| = 1$.

The angular integrations are explicitly carried out and conditions for the existence of $\Lambda^{(j)}_{k,l\alpha} \equiv \Lambda^{(j)}_{k,l}(\lambda^\alpha)$ are derived. Finally INNER and OUTER bi-axial power functions are defined in analogy to the axial case and are related to the Cauchy transforms $\Lambda^{(j)}_{k,l,\alpha}$.

1. Introduction

An important class of monogenic functions is that defined over bi-axially symmetric domains (Sommen, 1984) and we have studied special functions of this type in a recent work (Common and Sommen, 1992). The approach is to consider the splitting $R^m = R^p + R^q$ and to denote a general element $\vec{x}$ of R^m by $\vec{x} = \vec{x}_1 + \vec{x}_2 = \rho_1\vec{\omega}_1 + \rho_2\vec{\omega}_2$, where $\rho_1 = |\vec{x}_1|, \rho_2 = |\vec{x}_2|$ and $\vec{x}_1 \in R^p, \vec{x}_2 \in R^q$. Inner spherical monogenics $P_{k,l}(\vec{x}_1, \vec{x}_2)$ may be defined (Jank and Sommen, 1990). They have the property of being polynomials which are homogeneous of degrees k in $\vec{x}_1$ and l in $\vec{x}_2$ and satisfy

$$\partial_{\vec{x}_1} P_{k,l}(\vec{x}_1, \vec{x}_2) = \partial_{\vec{x}_2} P_{k,l}(\vec{x}_1, \vec{x}_2) = 0 \tag{1}$$

F. Brackx et al. (eds.), Clifford Algebras and their Applications in Mathematical Physics, 85–90.

We may use these spherical monogenics to define generalised Cauchy transforms of suitable Clifford valued functions or distributions f defined on R^p by,

$$\Lambda^{(1)}_{k,l}(f)(\vec{x}) = -\frac{1}{\omega_m}\int_{R^p}\frac{[\vec{x}_1+\vec{x}_2-\vec{u}]P_{k,l}(\vec{\eta},\vec{x}_2)f(\lambda)d^p\vec{u}}{|\vec{x}_1+\vec{x}_2-\vec{u}|^{m+2l}} \tag{2}$$

$$\Lambda^{(2)}_{k,l}(f)(\vec{x}) = -\frac{1}{\omega_m}\int_{R^p}\frac{[\vec{x}_1+\vec{x}_2-\vec{u}]\vec{\eta}P_{k,l}(\vec{\eta},\vec{x}_2)f(\lambda)d^p\vec{u}}{|\vec{x}_1+\vec{x}_2-\vec{u}|^{m+2l}} \tag{3}$$

where $\vec{u}=\lambda\vec{\eta}$ and ω_m is the area of the unit sphere in m dimensions. It is straightforward to show that these transforms are left monogenic in $R^m\backslash R^p$ since in the latter case one may assume $\vec{\eta}$ commutes or anti-commutes with $P_{k,l}(\vec{\eta},\vec{x}_2)$. The corresponding transforms in the axial monogenic case were introduced by Sommen (1988) and have been generalised by Ryan (1991).

In §2 we will use the monogenicity of $\Lambda^{(j)}_{k,l}(f)$ to derive relations between the transforms of f and f'. The integrations over solid angle in R^p for the right-hand sides of (2) and (3) will also be done explicitly giving expressions for $\Lambda^{(j)}_{k,l}(f)$ as integrals over λ. These properties will be used in §3 to study the transforms $\Lambda^{(j)}_{k,l}(\lambda^\alpha)\equiv\Lambda^{(j)}_{k,l,\alpha}$. It will be demonstrated that $\Lambda^{(1)}_{k,l,\alpha}$ exists for $-(p+k)<Re\alpha<k+2l+q$ and may be extended to $Re\alpha<-(p+k)$ for $Re\alpha\neq-(p+k)-2n$, $n=1,2,\ldots$. Similarly $\Lambda^{(2)}_{k,l,\alpha}$ exists for $-(p+k+1)<Re\alpha<k+2l+q-1$ and may be extended to $Re\alpha<-(p+k+1)$ for $Re\alpha\neq-(p+k+1)-2n$, $n=1,2,\ldots$. The behaviour of $\Lambda^{(j)}_{k,l,\alpha}(\vec{x})$ will be studied in the limit as $|\vec{x}_1|\to 0$ with $|\vec{x}_2|\neq 0$ fixed. It will thus be demonstrated that the $\Lambda^{(j)}_{k,l,\alpha}(\vec{x})$ are representations for the INNER and OUTER bi-axial power functions introduced by us recently (Common and Sommen, 1992).

2. Generalised Cauchy Transforms

Before considering the Cauchy transforms (2), (3) for $f(\lambda)=\lambda^\alpha$, we will derive some of their properties in the general case including a representation where the integration over solid angle in R^p has been performed. Since $\Lambda^{(j)}_{k,l}(f)$ are monogenic,

$$\partial_{\vec{x}_2}\Lambda^{(1)}_{k,l}(f)(\vec{x}) = \frac{1}{\omega_m}\int_{R^p}\partial_{\vec{x}_1}\left[\frac{\vec{x}_1+\vec{x}_2-\vec{u}}{|\vec{x}_1+\vec{x}_2-\vec{u}|^{m+2l}}\right]P_{k,l}(\vec{\eta},\vec{x}_2)f(\lambda)d^p\vec{u} \tag{4}$$

$$= -\frac{1}{\omega_m}\int_{R^p}\partial_{\vec{u}}\left[\frac{\vec{x}_1+\vec{x}_2-\vec{u}}{|\vec{x}_1+\vec{x}_2-\vec{u}|^{m+2l}}\right]P_{k,l}(\vec{\eta},\vec{x}_2)f(\lambda)d^p\vec{u}. \tag{5}$$

$$= -\frac{1}{\omega_m}\int_{R^p}\left\{\left[\frac{\vec{x}_1+\vec{x}_2-\vec{u}}{|\vec{x}_1+\vec{x}_2-\vec{u}|^{m+2l}}\right]\partial_{\overleftarrow{u}}\right\}P_{k,l}(\vec{\eta},\vec{x}_2)f(\lambda)d^p\vec{u}$$
$$-\frac{(m+2l)}{\omega_m}\int_{R^p}\frac{[\vec{x}_2(\vec{u}-\vec{x}_1)-(\vec{u}-\vec{x}_1)\vec{x}_2]}{|\vec{x}_1+\vec{x}_2-\vec{u}|^{m+2l+2}}P_{k,l}(\vec{\eta},\vec{x}_2)f(\lambda)d^p\vec{u}\,. \tag{6}$$

On integrating by parts in the first integral,

$$\partial_{\vec{x}_2}\Lambda^{(1)}_{k,l}(f)(\vec{x}) = -\Lambda^{(2)}_{k,l}(f'-\frac{k}{\lambda}f)(\vec{x})+\frac{1}{\omega_m}\partial_{\vec{x}_1}\left[\int_{R^p}\frac{2\vec{x}_2}{|\vec{x}_1+\vec{x}_2-\vec{u}|^{m+2l}}P_{k,l}(\vec{\eta},\vec{x}_2)f(\lambda)d^p\vec{u}\right] \tag{7}$$

$$= -\Lambda^{(2)}_{k,l}(f' - \frac{k}{\lambda}f)(\vec{x}) - \partial_{\vec{x}_1}\left[\Lambda^{(1)}_{k,l}(f)(\vec{x}_1+\vec{x}_2) - (-)^l\Lambda^{(1)}_{k,l}(f)(\vec{x}_1-\vec{x}_2)\right] \tag{8}$$

where we have used

$$\partial_{\vec{u}} = \vec{\eta}\left[\frac{\partial}{\partial\lambda} + \frac{1}{\lambda}\Gamma_\eta\right], \Gamma_\eta P_{k,l}(\vec{\eta},\vec{x}_2) = -kP_{k,l}(\vec{\eta},\vec{x}_2) \tag{9}$$

and

$$P_{k,l}(\vec{\eta},-\vec{x}_2) = (-)^l P_{k,l}(\vec{\eta},\vec{x}_2)\,. \tag{10}$$

Similarly

$$\partial_{\vec{x}_2}\Lambda^{(2)}_{k,l}(f)(\vec{x}) = \Lambda^{(1)}_{k,l}\left(f' + \frac{k+p-1}{\lambda}f\right)(\vec{x})$$

$$- \partial_{\vec{x}_1}\left[\Lambda^{(2)}_{k,l}(f)(\vec{x}_1+\vec{x}_2) - (-)^l\Lambda^{(2)}_{k,l}(f)(\vec{x}_1-\vec{x}_2)\right]\,. \tag{11}$$

To do the angular integrations in (2), (3) we write

$$\Lambda^{(j)}_{k,l}(f)(\vec{x}) = [(\vec{x}_1+\vec{x}_2)A^{(j)}(\vec{x}) + B^{(j)}(\vec{x})]\,; \quad j=1,2, \tag{12}$$

where

$$A^{(1)}(\vec{x}) = -\frac{1}{\omega_m}\int_0^\infty I_1(\vec{x}_1,\vec{x}_2,\lambda)f(\lambda)\lambda^{p-1}d\lambda \tag{13}$$

$$B^{(1)}(\vec{x}) = \frac{1}{\omega_m}\int_0^\infty I_2(\vec{x}_1,\vec{x}_2,\lambda)f(\lambda)\lambda^{p}d\lambda \tag{14}$$

$$A^{(2)}(\vec{x}) = -\frac{1}{\omega_m}\int_0^\infty I_2(\vec{x}_1,\vec{x}_2,\lambda)f(\lambda)\lambda^{p-1}d\lambda \tag{15}$$

$$B^{(2)}(\vec{x}) = -\frac{1}{\omega_m}\int_0^\infty I_1(\vec{x}_1,\vec{x}_2,\lambda)f(\lambda)\lambda^{p}d\lambda \tag{16}$$

and

$$I_1 = \int_{S^{p-1}}\frac{P_{k,l}(\vec{\eta},\vec{x}_2)d\vec{\eta}}{[\rho_1^2+\rho_2^2-2\lambda<\vec{x}_1,\vec{\eta}>+\lambda^2]^{l+m/2}} \tag{17}$$

$$I_2 = \int_{S^{p-1}}\frac{\vec{\eta}P_{k,l}(\vec{\eta},\vec{x}_2)d\vec{\eta}}{[\rho_1^2+\rho_2^2-2\lambda<\vec{x}_1,\vec{\eta}>+\lambda^2]^{l+m/2}}\,. \tag{18}$$

Using the Funk-Hecke Theorem(Hochstadt, 1970)

$$I_1(\vec{x}_1,\vec{x}_2,\lambda) = T_{l,k,m}\left(\sqrt{\rho_1^2+\rho_2^2+\lambda^2}\,,\ \lambda\rho_1\right)P_{k,l}(\vec{\omega}_1,\vec{x}_2) \tag{19}$$

$$I_2(\vec{x}_1,\vec{x}_2,\lambda) = T_{l,k+1,m}\left(\sqrt{\rho_1^2+\rho_2^2+\lambda^2}\,,\ \lambda\rho_1\right)\vec{\omega}_1P_{k,l}(\vec{\omega}_1,\vec{x}_2) \tag{20}$$

where

$$T_{l,k,m}(a,b) = \omega_{p-1}\int_{-1}^1\frac{P_{k,p}(s)(1-s^2)^{(p-3)/2}ds}{[a^2-2bs]^{l+m/2}}\,. \tag{21}$$

Here $P_{k,p}(s)$ is the Legendre polynomial of order k in p-dimensions and is given by the Rodrigues' formula

$$P_{k,p}(s) = \frac{(-1/2)^k\Gamma[(p-1)/2]}{\Gamma[k+(p-1)/2]}(1-s^2)^{(3-p)/2}\frac{d^k}{ds^k}(1-s^2)^{k+(p-3)/2}\,. \tag{22}$$

Substituting in (21) and integrating by parts

$$T_{l,k,m}(a,b) = \frac{2\pi^{(p-1)/2}b^k\Gamma(k+l+m/2)}{\Gamma(l+m/2)\Gamma[k+(p-1)/2]}\int_{-1}^{1}\frac{(1-s^2)^{k+(p-3)/2}ds}{(a^2-2bs)^{k+l+m/2}} \tag{23}$$

since

$$\omega_p = 2\pi^{p/2}/\Gamma(p/2)\,. \tag{24}$$

By making the substitution $s = 2t-1$ in the integral we obtain the standard representation of a hypergeometric function and so finally,

$$T_{l,k,m}(a,b) = \frac{2^{p-1}\pi^{(p-1)/2}\Gamma(k+l+m/2)\Gamma(k+(p-1)/2)}{\Gamma(l+m/2)\Gamma(2k+p-1)}\left[\frac{(4b)^k}{(a^2+2b)^{k+l+m/2}}\right]$$

$$\times\, {}_2F_1\left(k+(p-1)/2\,,\; k+l+m/2\,;\; 2k+p-1\,;\; \frac{4b}{a^2+2b}\right)\,. \tag{25}$$

Substituting this expression for $T_{l,k,m}$ in (19), (20) gives the angular integrals $I_j(\vec{x}_1, \vec{x}_2,$ which may then be substituted in (12) to (16) to give the $\Lambda^{(j)}_{k,l}(f)$ as integrals over λ.

3. Generalised Cauchy Transforms of Power Functions

We consider now the special case when $f(\lambda) \equiv \lambda^\alpha,\;\; \alpha \in C$. From (19), (20) with the $T_{l,k,m}$ given by (25) it follows that as $\lambda \to \infty$

$$I_1(\vec{x}_1,\vec{x}_2,\lambda) = \mathcal{O}(\lambda^{-(k+2l+m)})\,,\;\; I_2(\vec{x}_1,\vec{x}_2,\lambda) = \mathcal{O}(\lambda^{-(k+2l+m+1)})\,, \tag{26}$$

and as $\lambda \to 0$

$$I_1(\vec{x}_1,\vec{x}_2,\lambda) = \mathcal{O}(\lambda^k) \qquad , \; I_2(\vec{x}_1,\vec{x}_2,\lambda) = \mathcal{O}(\lambda^{k+1})\,. \tag{27}$$

The integrals in (13) to (16) then have the following existence requirements:

(a) For $A^{(1)}$ to exist,

$$-(p+k) < Re\alpha < k+2l+m-p \tag{28}$$

(b) For $B^{(1)}$ to exist,

$$-(p+k+2) < Re\alpha < k+2l+m-p \tag{29}$$

(c) For $A^{(2)}$ to exist,

$$-(p+k+1) < Re\alpha < k+2l+m-p+1 \tag{30}$$

(d) For $B^{(2)}$ to exist,

$$-(p+k+1) < Re\alpha < k+2l+m-p-1 \tag{31}$$

Combining these results, $\Lambda^{(1)}_{k,l,\alpha}$ exists for $-(p+k) < Re\alpha < k+2l+m-p$ and $\Lambda^{(2)}_{k,l,\alpha}$ exists for $-(p+k+1) < Re\alpha < k+2l+m-p-1$.

However from (8) and (11) with $f = \lambda^{\alpha}$,

$$\partial_{\vec{x}_2}\Lambda^{(1)}_{k,l,\alpha}(\vec{x}) + \partial_{\vec{x}_1}\left[\Lambda^{(1)}_{k,l,\alpha}(\vec{x}_1+\vec{x}_2) - (-)^l\Lambda^{(1)}_{k,l,\alpha}(\vec{x}_1-\vec{x}_2)\right]$$

$$= -(\alpha-k)\Lambda^{(2)}_{k,l,\alpha-1}(\vec{x}) \tag{32}$$

$$\partial_{\vec{x}_2}\Lambda^{(2)}_{k,l,\alpha}(\vec{x}) + \partial_{\vec{x}_1}\left[\Lambda^{(2)}_{k,l,\alpha}(\vec{x}_1+\vec{x}_2) - (-)^l\Lambda^{(2)}_{k,l,\alpha}(\vec{x}_1-\vec{x}_2)\right]$$

$$= (\alpha+k+p-1)\Lambda^{(1)}_{k,l,\alpha-1}(\vec{x})\,. \tag{33}$$

These relations may be used to define $\Lambda^{(1)}_{k,l\alpha}(\vec{x})$ for $Re\alpha < -(p+k)$, $Re\alpha \neq -(p+k)-2n,\ n=1,2,\ldots$ and to define $\Lambda^{(2)}_{k,l\alpha}(\vec{x})$ for $Re\alpha < -(p+k+1)$, $Re\alpha \neq -(p+k+1)-2n,\ n=1,2,\ldots$.

It has been shown by one of us (Sommen, 1992) that there are unique monogenic functions $p_{\alpha;k,l;p,q}(\vec{x})P_{k,l}(\vec{x}_1,\vec{\omega}_2)$ and $q_{\alpha;k,l;p,q}(\vec{x})P_{k,l}(\vec{x}_1,\vec{\omega}_2)$ such that as $\rho_1 \to 0$,

$$p_{\alpha;k,l;p,q}(\vec{x}) \approx \rho_2^{\alpha}\ ,\quad q_{\alpha;k,l;p,q}(\vec{x}) \approx \vec{\omega}_2\rho_2^{\alpha} \tag{34}$$

In analogy with the axial case (Sommen, 1988) we call them respectively INNER and OUTER Bi-axial power functions. We will now investigate how the transforms $\Lambda^{(j)}_{k,l,\alpha}(\vec{x})$ are related to these functions.

In the limit $\vec{x}_1 \to 0$ and $\vec{x}_2$ fixed non-zero,

$$T_{k,l,m}\left(\sqrt{\rho_1^2+\rho_2^2+\lambda^2}\ ,\ \lambda\rho_1\right) \approx \frac{2\pi^{p/2}\Gamma(k+l+m/2)(\lambda\rho_1)^k}{\Gamma(l+m/2)\Gamma(k+p/2)[\rho_2^2+\lambda^2]^{l+k+m/2}} \tag{35}$$

so that

$$A^{(1)} \approx -\frac{1}{\omega_m}\int_0^{\infty}\frac{\lambda^{\alpha+p-1}2\pi^{p/2}\Gamma(k+l+m/2)(\lambda\rho_1)^kP_{k,l}(\vec{\omega}_1,\vec{x}_2)d\lambda}{\Gamma(l+m/2)\Gamma(k+p/2)[\rho_2^2+\lambda^2]^{l+k+m/2}} \tag{36}$$

$$= -\frac{2\pi^{p/2}\Gamma(k+l+m/2)\sigma_{m-1,l+k}(\alpha+p+k-1)}{\Gamma(l+m/2)\Gamma(k+p/2)}\rho_2^{\alpha-k-2l-q}P_{k,l}(\vec{x}_1,\vec{x}_2) \tag{37}$$

where

$$\sigma_{m,j}(\beta) = \frac{1}{\omega_{m+1}}\int_0^{\infty}t^{\beta}(1+t^2)^{-[j+(m+1)/2]}dt = \frac{1}{2\omega_{m+1}}B[(m+2j-\beta)/2,\ (\beta+1)/2]\,. \tag{38}$$

Similarly in this limit

$$B^{(1)} \approx \frac{2\pi^{p/2}\Gamma(k+l+1+m/2)\sigma_{m-1,l+k+1}(\alpha+p+k+1)}{\Gamma(l+m/2)\Gamma(k+1+p/2)}\rho_2^{\alpha-k-2l-q}\vec{x}_1 P_{k,l}(\vec{x}_1,\vec{x}_2) \tag{39}$$

and we note the extra power of ρ_1 compared with $A^{(1)}$. Therefore as $\rho_1 \to 0$,

$$\Lambda^{(1)}_{k,l,\alpha}(\vec{x}) \approx \vec{x}_2 A^{(1)}$$

$$\approx -\frac{2\pi^{p/2}\Gamma(k+l+m/2)\sigma_{m-1,l+k}(\alpha+p+k-1)\vec{\omega}_2}{\Gamma(l+m/2)\Gamma(k+p/2)}\rho_2^{\alpha+1-l-k-q}P_{k,l}(\vec{x}_1,\vec{\omega}_2)\,. \tag{40}$$

Comparing the right hand sides of (37) and the second of (34) and because of the above stated uniqueness we see that for $\vec{x} \in R^m \backslash R^p$,

$$\Lambda^{(1)}_{k,l,\alpha}(\vec{x}) = -\frac{2\pi^{p/2}\Gamma(k+l+m/2)\sigma_{m-1,l+k}(\alpha+p+k-1)}{\Gamma(l+m/2)\Gamma(k+p/2)}q_{\alpha+1-l-k-q;k,l;p,q}(\vec{x}) \times P_{k,l}(\vec{x}_1,\vec{\omega}_2)\,. \tag{41}$$

Similarly it may be shown that

$$\Lambda^{(2)}_{k,l,\alpha}(\vec{x}) = -\frac{2\pi^{p/2}\Gamma(k+l+m/2)\sigma_{m-1,l+k}(\alpha+p+k)}{\Gamma(l+m/2)\Gamma(k+p/2)}p_{\alpha+1-l-k-q;k,l;p,q}(\vec{x}) \times P_{k,l}(\vec{x}_1,\vec{\omega}_2)\,. \tag{42}$$

The relations (41), (42) may be inverted to give expressions for the INNER and OUTER bi-axial power function as generalised Cauchy transforms.

References

Common, A.K. and Sommen, F.: 'Special Bi-Axial Monogenic Functions', University of Kent preprint IMS/A92/9, submitted to *Journ.Math.Anal.App.* for publication.

Hochstadt, H.: 1971, 'The Functions of Mathematical Physics' *Pure and Applied Math.* **23**, Wiley-Intersciences, New York.

Jank, G. and Sommen F.: 1990, 'Clifford Analysis, Bi-Axial Symmetry and Pseudo-analytic Functions', *Complex Variable* **13**, pp. 195-212.

Ryan, J.: 1991, 'Applications of Clifford Analysis to Axially Symmetric Partial Differential Equations', *Complex Variables: Theory and Applications* **16**, pp. 137-51.

Sommen, F.: 1984, 'Plane Elliptic Systems and Monogenic Functions in Symmetric Domains', *Supp.Rend.Circ.Mat.Palermo* (2) **6**, pp. 259-269.

Sommen, F.: 1988, 'Special Functions in Clifford Analysis and Axial Symmetry', *Journ.Math.Anal.App.* **130**, pp. 100-133.

Sommen, F.: 1992, Lecture Notes, University of Ghent.

NOTE ON THE USE OF SPHERICAL VECTORFIELDS IN CLIFFORD ANALYSIS

J. DE GRAAF
Eindhoven University of Technology

Summary and Introduction

In this note I discuss some properties and (anti)-commutation relations of Clifford differentiation operators on $\mathbb{R}^n$ which are geometrical invariants. The results are a happy and accidental combination of earlier work on spherical vector fields [G] and inspiration drawn from Van Acker's PhD-Thesis [A] on Clifford differential operators.

A bit of polemic is unavoidable here because in [A] Van Acker introduces 'spherical' coordinates r, $\underline{\omega}$ on $\mathbb{R}^q$ by $\underline{x} = r\underline{\omega}$, $r = |\underline{x}|$, $\underline{\omega} = \sum_{j=1}^{q} \omega^j \underline{e}_j$. So she introduces $q+1$ variables, $\omega^k = \dfrac{x^k}{|\underline{x}|}$, $r = |\underline{x}|$, instead of q. Of course, there is the relation $(\omega^1)^2 + (\omega^2)^2 + \ldots + (\omega^q)^2 = 1$. But then the range of the variables ω^k is no longer a linear space, the ω^k are no longer independent and one gets worried about too reckless applications of the chain rule. One expects that some differential geometry is in place here.

In our approach we try to explain the differentiations in [A], such as $\dfrac{\partial}{\partial_{\omega_k}} = \partial_{\omega_k}$, as directional derivatives with respect to vector fields. Doing this we mostly arrive at the same algebraic relations as Van Acker. However we find that, in our notation, $\sum_{k=1}^{q} \omega^k \Omega_k = 0$, which should correspond to $\sum_{k=1}^{q} \omega_k \partial_{\omega_k} = 0$. In [A] this observation has not been made and, although no apert contradictions are visible, this discrepancy is a bit of a mystery which questions the formal application of the chain rule.

It looks as if the approach of this note is also useful for finding Clifford geometrical differential operators corresponding to other fibrations (than spheres) for $\mathbb{R}^q\backslash\{\cdots\}$. The Appendix can be read independently. Also several concepts in [CF] can be better understood by looking at them in our 'tensorial' way.

F. Brackx et al. (eds.), Clifford Algebras and their Applications in Mathematical Physics, 91–100.

Notations

In expressions like

$$\underline{x} = \sum_{j=1}^{q} x^j \underline{e}_j \in \mathbb{R}^q$$

we ommit the summation sign (Einstein convention). If an index occurs twice in any one term, once as an *upper* and once as a *lower* index, summation has to be effected, cf. [S].
The usual (Euclidean) inner product of vectors $\underline{x}$ and $\underline{y}$ is denoted by $(\underline{x}, \underline{y})$.
Following [A] the Clifford algebra $\mathbb{R}_q$ is based on the geometric product

$$\underline{e}_j\underline{e}_k + \underline{e}_k\underline{e}_j = -2\delta_{jk}.$$

Here $(\underline{e}_j)$ is an orthonormal basis in the usual Euclidean sense, so $(\underline{e}_j, \underline{e}_k) = \delta_{jk}$. with respect to the 'negative' inner product $-(\cdot,\cdot)$ on $\mathbb{R}^q$ we introduce the reciprocal basis $(\underline{e}^k)$ of a basis $(\underline{e}_j)$ by $-(\underline{e}^k, \underline{e}_j) = \delta_j^k$. In this definition $(\underline{e}_j)$ need not be orthonormal, however *if* it is orthonormal, raising and lowering of indices corresponds to multiplication by -1. So

$$\underline{e}_j = -\underline{e}^j \;, \quad x_j = -x^j \;, \quad \underline{x} = x^i\underline{e}_i = x_k\underline{e}^k \in \mathbb{R}^q$$

and

$$\underline{e}^j\underline{e}_k + \underline{e}_k\underline{e}^j = 2\delta_k^j.$$

The vector field approach

If

$$\underline{v}(\underline{x}) = v^1(\underline{x})\underline{e}_1 + \ldots + v^q(\underline{x})\underline{e}_q = v^i(\underline{x})\underline{e}_i$$

is a vector field on an open subset of $\mathbb{R}^q$ then the *directional derivative* $\mathcal{L}_{\underline{v}}f$ of a function f with respect to $\underline{v}$ is defined by

$$\mathcal{L}_{\underline{v}}f = \langle df, v\rangle = (\underline{v}, \nabla f) = v^i(\underline{x})\partial_i f(\underline{x}).$$

Here ∂_i denotes $\dfrac{\partial}{\partial x^i}$, i.e. partial differentiation with respect to a Cartesian coordinate system $x^1, \ldots, x^q$.
On $\mathbb{R}^q\backslash\{\underline{0}\}$ we define the functions r and ω^i, $1 \le i \le q$, by

$$r(\underline{x}) = |\underline{x}| = (\delta_{ij}x^ix^j)^{\frac{1}{2}} = (-x_ix^i)^{\frac{1}{2}}$$

$$\omega^i(\underline{x}) = x^i|\underline{x}|^{-1}.$$

Next we introduce the vector fields

$$\underline{x} = x^i\underline{e}_i \qquad \underline{b}_j = r\underline{e}_j - (\underline{e}_j, \underline{x})\frac{1}{r}\,\underline{x} = r\underline{x}_j + x_j\,\frac{1}{r}\,\underline{x} = r\underline{e}_j + \omega_j\underline{x}$$

and their corresponding directional derivatives $\mathcal{E}$ and Ω_j, $1 \le j \le q$:

$$\mathcal{E}f = (\underline{x}, \nabla f) = x^k \partial_k f,$$

$$\Omega_j f = (\underline{b}_j, \nabla f) = r\partial_j f + x_j \frac{x^k}{r} \partial_k f = r\partial_j f + \omega_j \mathcal{E} f.$$

we consider Ω_j and $\mathcal{E}$ as differentiation operators. They play a prominent role in our exegesis of the operators $\frac{\partial}{\partial \omega_j}$, $1 \leq j \leq q$, in [A].

Note that at each point $\underline{x}$ in space the directional derivative Ω_j is tangent to the sphere with radius $r = |\underline{x}|$, whereas $\mathcal{E}$ is normal to that sphere. Already at this stage we find

Proposition 1.

$$\sum_{j=1}^{q} -x_j \Omega_j = x^j \Omega_j = 0.$$

Proof.

$$x^j \Omega_j = r x^j \partial_j + x^j \omega_j \mathcal{E} = r\mathcal{E} + r\omega^j \omega_j \mathcal{E} = r\mathcal{E} - r\mathcal{E} = 0. \qquad \square$$

For future reference we now list the actions of the operators Ω_j and $\mathcal{E}$ on some 'special' functions.

Table I. (Elementary actions).

$\mathcal{E} r^\alpha = \alpha r^\alpha, \quad \alpha \in \mathbb{R}$	$\Omega_k r^\alpha = 0, \quad \alpha \in \mathbb{R}$
$\mathcal{E} x^j = x^j$	$\Omega_k x^j = r\delta_k^j + \frac{x_k x^j}{r}$
$\mathcal{E} \omega^j = \mathcal{E} \frac{x^j}{r} = 0$	$\Omega_k \omega^j = \delta_k^j + \omega_k \omega^j$

$$x^j \Omega_j = 0$$

The operators of multiplication by x^j, ω^j and r are denoted again by the same symbols. We introduce, corresponding to [A], the operator $L_{ij} = -(x_i \partial_j - x_j \partial_i)$. The following table is a matter of straightforward calculation.

Table II. (Elementary commutators).

$[\partial_j, x^k] = \delta_j^k$	$[\mathcal{E}, \partial_j] = -\partial_j$
$[\partial_j, r] = -\frac{1}{r}\, x_j$	$[\mathcal{E}, x^j] = x^j$
$[\partial_j, \omega^k] = \frac{1}{r}\, \delta_j^k + \frac{1}{r}\, \omega_j \omega^k$	$[\mathcal{E}, r] = r$
	$[\mathcal{E}, \omega^j] = 0$
	$[\mathcal{E}, L_{ij}] = 0$
$[\Omega_k, x^j] = r\delta_k^j + \omega_k x^j$	
$[\Omega_k, r] = 0$	
$[\Omega_k, \omega^j] = \delta_k^j + \omega_k \omega^j$	
$[\Omega_k, \mathcal{E}] = 0$	
$[\Omega_i, \Omega_j] = L_{ij}$	
$[\Omega_k, \partial_j] = \frac{1}{r}\, L_{kj} + \frac{1}{r}(\delta_{kj} - \omega_k \omega_j)\mathcal{E}$	
$[\Omega_k, L_{ij}] + [\Omega_j, L_{ki}] + [\Omega_i, L_{jk}] = 0$	

Proof (specimen).

- $$\Omega_k \partial_j f = r\partial_k \partial_j f + \omega_k \mathcal{E}(\partial_j f) = \partial_j r \partial_k f + \omega_j \partial_k f + \omega_k(-\partial_j f + \partial_j \mathcal{E} f)$$

- $$\partial_j \Omega_k f = \partial_j r \partial_k f + \partial_j(\omega_k \mathcal{E} f) = \partial_j r \partial_k f - \frac{1}{r}\, \delta_{jk} \mathcal{E} f + \frac{1}{r}\, \omega_j \omega_k \mathcal{E} f + \omega_k \partial_j \mathcal{E} f$$

- $$[\mathcal{E}, L_{ij}] = [\mathcal{E}, [\Omega_i, \Omega_j]] = -[\Omega_j, [, E, \Omega_i]] - [\Omega_i, [\Omega_j, \mathcal{E}]] = 0.$$ □

From now on all above mentioned operators are supposed to act on Clifford valued functions $f : \mathbb{R}^q \backslash \{\underline{0}\} \to \mathbb{R}_q$. They can be looked upon as 'scalar' operators. In addition to those scalar operators we introduce the Clifford multiplication operators

$$\underline{e}_i\,, \qquad \underline{x} = x^i \underline{e}_i\,, \qquad \underline{\omega} = \omega^i \underline{e}_i\,,$$

and the Clifford differentiation operators

$$\partial_{\underline{x}} = -\underline{e}^j \partial_j\,, \qquad \partial_{\underline{\omega}} = -\underline{e}^i \Omega_i\,.$$

They are all supposed to act from the left! In the next three propositions some elementary relations are presented. Note that $\partial_{\underline{x}}$ and $\partial_{\underline{\omega}}$ are geometrical objects!

Proposition 2.

$$-(\underline{x}\partial_{\underline{x}} + \partial_{\underline{x}}\underline{x}) = 2\mathcal{E} + q \ .$$

Proof.

$$\underline{x}\partial_{\underline{x}}f + \partial_{\underline{x}}\underline{x}f = -x^i\underline{e}_i\underline{e}^j\partial_j f - \underline{e}^j\partial_j(x^i\underline{e}_i f) = -x^i\underline{e}_i\underline{e}^j\partial_j f - \underline{e}^j\delta^i_j\underline{e}_i f - \underline{e}^j x^i\underline{e}_i\partial_j f$$

$$= -x^i(\underline{e}_i\underline{e}^j + \underline{e}^j\underline{e}_i)\partial_j f - \underline{e}^j\underline{e}_j f = -x^i(2\delta^j_i)\partial_j f - qf = -2\mathcal{E}f - qf \ . \qquad \square$$

Proposition 3.

$$\partial_{\underline{\omega}} = r\partial_{\underline{x}} - \underline{\omega}\mathcal{E} \ .$$

Proof. From the definition

$$\Omega_j = r\partial_j + \frac{1}{r}\, x_j\mathcal{E}$$

it follows

$$\underline{e}^j\partial_j = \frac{1}{r}\,\underline{e}^j\Omega_j - \frac{1}{r^2}\,\underline{e}^j x_j\mathcal{E}$$

$$\partial_{\underline{x}} = \frac{1}{r}\,\partial_{\underline{\omega}} + \frac{1}{r^2}\,\underline{x}\mathcal{E} \ .$$

From which the wanted result follows. $\square$

Proposition 4.

- $$\underline{\omega}\partial_{\underline{\omega}} = \underline{x}\partial_{\underline{x}} + \mathcal{E}$$
- $$-\underline{x}\partial_{\underline{x}} = \Gamma + \mathcal{E} \ ,$$

with $\Gamma = -\underline{\omega}\partial_{\underline{\omega}}$, unlike [A], p. 9, cf. also the discrepancy with [A], p. 25, line ↑ 6.

Proof. From proposition 3

$$\underline{\omega}\partial_{\underline{\omega}} = \underline{\omega}r\partial_{\underline{x}} - \underline{\omega}\ \underline{\omega}\mathcal{E} = \underline{x}\partial_{\underline{x}} + \mathcal{E} \ . \qquad \square$$

In addition to the commutation relations for the scalar operators in Table II we now gather some commutation relations for Clifford operators in Table III.

Table III (Commutators of scalar- and Clifford–operators)

$[\partial_i, \underline{e}_j] = 0$	$[x^k, \partial_{\underline{x}}] = \underline{e}^k$
$[\partial_i, \underline{x}] = \underline{e}_i$	$[x^k, \partial_{\underline{\omega}}] = r\underline{e}^k + x^k\underline{\omega}$
$[\partial_i, \underline{\omega}] = -\frac{1}{r}(\underline{e}_i + \underline{\omega})$	$[\omega^k, \partial_{\underline{x}}] = \frac{1}{r}\ \underline{e}^k + \frac{1}{r}\ \omega^k\underline{\omega}$
$[\partial_i, \partial_{\underline{x}}] = 0$	$[\omega^k, \partial_{\underline{\omega}}] = \underline{e}^k + \omega^k\underline{\omega}$
$[\partial_i, \partial_{\underline{\omega}}] = \frac{1}{r}\ L_{ki}\underline{e}^k + \frac{1}{r}(\underline{e}_i + \omega_i\underline{\omega})\mathcal{E}$	$[r, \partial_{\underline{\omega}}] = 0$
$[\mathcal{E}, \underline{e}_j] = 0$	$[\Omega_k, \underline{e}_j] = 0$
$[\mathcal{E}, \underline{x}] = \underline{x}$	$[\Omega_k, \underline{x}] = r\underline{e}_k + \omega_k\underline{x}$
$[\mathcal{E}, \underline{\omega}] = 0$	$[\Omega_k, \underline{\omega}] = \underline{e}_k + \omega_k\underline{\omega}$
$[\mathcal{E}, \partial_{\underline{x}}] = -\partial_{\underline{x}}$	$[\Omega_k, \partial_{\underline{x}}] = \frac{1}{r}\ L_{jk}\underline{e}^j + \frac{1}{r}\{\underline{e}_k + \omega_k\underline{\omega}\}\mathcal{E}$
$[\mathcal{E}, \partial_{\underline{\omega}}] = 0$	$[\Omega_k, \partial_{\underline{\omega}}] = L_{jk}\underline{e}^j = -(x_k\partial_{\underline{x}} + \underline{x}\partial_k)$
$[\mathcal{E}, \underline{\omega}\partial_{\underline{\omega}}] = 0$	$[\Omega_k, \underline{\omega}\partial_{\underline{\omega}}] = \omega_k\underline{\omega}\partial_{\underline{\omega}} + \underline{e}_k\partial_{\underline{\omega}} + \underline{\omega}L_{jk}\underline{e}^j$
$[\mathcal{E}, \underline{x}\partial_{\underline{x}}] = 0$	

Proof (specimen).

- $$\mathcal{E}\underline{x}f = x^k\partial_k(x^\ell\underline{e}_\ell f) = x^k\delta_k^\ell\underline{e}_\ell f + x^k x^\ell\underline{e}_\ell\partial_k f = \underline{x}f + \underline{x}\mathcal{E}f$$
- $$\mathcal{E}\partial_{\underline{x}}f = -x^k\partial_k\underline{e}^m\partial_m f = -\underline{e}^m\partial_m(x^k\partial_k f) + \underline{e}^m\delta_m^k\partial_k f = \partial_{\underline{x}}\mathcal{E}f - \partial_{\underline{x}}f$$
- $$\mathcal{E}_{\underline{x}}\partial_{\underline{x}}f = \underline{x}\mathcal{E}\partial_{\underline{x}}f + \underline{x}\partial_{\underline{x}}f = \underline{x}\partial_{\underline{x}}\mathcal{E}f - \underline{x}\partial_{\underline{x}}f + \underline{x}\partial_{\underline{x}}f = \underline{x}\partial_{\underline{x}}\mathcal{E}f$$
- $$\Omega_k\partial_{\underline{x}}f = -\underline{e}^j\Omega_k\partial_j f = -\underline{e}^j\{\frac{1}{r}\ L_{kj} + \frac{1}{r}(\delta_{kj} - \omega_k\omega_j)\mathcal{E}\}f - \underline{e}^j(\partial_j\Omega_k f)$$

$$= \frac{1}{r}\ L_{jk}\underline{e}^j f + \frac{1}{r}\{\underline{e}_k + \omega_k\underline{\omega}\}\mathcal{E}f - \partial_{\underline{x}}\Omega_k f\ .$$ □

Proposition 5. [A], p. 10.

$$\Gamma = -\underline{\omega}\partial_{\underline{\omega}} = -\sum_{i<j}\underline{e}_i\underline{e}_j L_{ij} = -\sum_{i<j}\underline{e}_i\underline{e}_j[\Omega_i, \Omega_j]\ .$$

Proof.

$$\Gamma = -\mathcal{E} - \underline{x}\partial_{\underline{x}} = -x^i\partial_i + x^j\underline{e}_j\underline{e}^i\partial_i = \sum_{i\neq j} x^j\underline{e}_j\underline{e}^i\partial_i$$

$$= \sum_{i<j} x^j\underline{e}_j\underline{e}^i\partial_i + \sum_{i>j} x^j\underline{e}_j\underline{e}^i\partial_i = \sum_{i<j} x^j\underline{e}_j\underline{e}^i\partial_i + \sum_{j>i} x^i\underline{e}_i\underline{e}^j\partial_j$$

$$= \sum_{i<j} x^j\underline{e}_j\underline{e}^i\partial_i - \sum_{i<j} x^i\underline{e}^j\underline{e}_i\partial_j = -\sum_{i<j} \underline{e}_i\underline{e}_j(x^i\partial_j - x^j\partial_i)$$

$$= \sum_{i<j} \underline{e}^i\underline{e}^j(x_i\partial_j - x_j\partial_i)\ . \qquad \square$$

In the next proposition we calculate the action of the operator $\underline{\omega}\cdot\partial_{\underline{\omega}}$ on the Clifford function $G = \underline{\omega}$.

Proposition 6. [A], p. 11, ↓ 3.

$$-\Gamma\underline{\omega} = \underline{\omega}\partial_{\underline{\omega}}(\underline{\omega}) = -(q-1)\underline{\omega}\ .$$

Proof.

$$\underline{\omega}\partial_{\underline{\omega}}(\underline{\omega}) = -\omega^i\underline{e}_i\underline{e}^j\Omega_j(\omega^k\underline{e}_k) = -\omega^i\underline{e}_i\underline{e}^j\{\delta_j^k + \omega_j\omega^k\}\underline{e}_k$$

$$= -\omega^i\underline{e}_i\underline{e}^k\underline{e}_k - \omega^i\underline{e}_i\omega_j\underline{e}^j\omega^k\underline{e}_k = -q\underline{\omega} + \underline{\omega}\ . \qquad \square$$

Table IV (Anti–commutators of Clifford–operators).

$\partial_{\underline{x}}\underline{e}_i + \underline{e}_i\partial_{\underline{x}} = -2\partial_i$	$\partial_{\underline{\omega}}\underline{x} + \underline{x}\partial_{\underline{\omega}} = -(q-1)r$
$\partial_{\underline{x}}\underline{x} + \underline{x}\partial_{\underline{x}} = -(qI + 2\mathcal{E})$	$\partial_{\underline{\omega}}\underline{\omega} + \underline{\omega}\partial_{\underline{\omega}} = -(q-1)$
$\partial_{\underline{x}}\underline{\omega} + \underline{\omega}\partial_{\underline{x}} = -\frac{1}{r}(q-1+2\mathcal{E})$	$\underline{\omega}\partial_{\underline{\omega}}\underline{\omega} + \underline{\omega}\ \underline{\omega}\partial_{\underline{\omega}} = -(q-1)\underline{\omega}$
$\partial_{\underline{x}}\partial_{\underline{x}} + \partial_{\underline{\omega}}\partial_{\underline{x}} = \frac{1}{r}(2\partial_{\underline{\omega}}\partial_{\underline{\omega}} - (q-1)\mathcal{E})$	$\underline{\omega}\partial_{\underline{\omega}}\underline{\omega} + \underline{x}\ \underline{\omega}\partial_{\underline{\omega}} = -(q-1)\underline{x}\ .$

Proof. (specimen)

• Let $G : \mathbb{R}^q \to \mathbb{R}_q$. Then

•
$$\partial_{\underline{\omega}}\underline{\omega}G = -\underline{e}^j\Omega_j(\omega^i\underline{e}_iG) = -\underline{e}^j(\Omega_j(\omega^i))\underline{e}_iG - \underline{e}^j\omega^i\underline{e}_i\Omega_jG$$

$$= -\underline{e}^j\{\delta_j^i + \omega_j\omega^i\}\underline{e}_iG + \omega^i\underline{e}_i\underline{e}^j\Omega_jG - 2\delta_i^j\omega^i\Omega_jG$$

$$= -\underline{e}^i\underline{e}_iG - \underline{\omega}\ \underline{\omega}G - \underline{\omega}\partial_{\underline{\omega}}G - 2\omega^j\Omega_jG = -qG + G - \underline{\omega}\partial_{\underline{\omega}}G$$

- $$\partial_{\underline{x}}\partial_{\underline{\omega}} + \partial_{\underline{\omega}}\partial_{\underline{x}} = \underline{e}^j\partial_j\underline{e}^k\Omega_k + \underline{e}^k\Omega_k\underline{e}^j\partial_j$$

$$= \underline{e}^j\underline{e}^k[\partial_j,\Omega_k] - 2r\Delta + 2\mathcal{E}\,\frac{1}{r}\,\mathcal{E},\ \textit{with}\ \Delta = \sum_{j=1}^{q}\partial_j\partial_j\ ,$$

$$= \underline{e}^j\underline{e}^k\{-\frac{1}{r}\,L_{kj} - \frac{1}{r}\,\delta_{kj}\mathcal{E} + \frac{1}{r}\,\omega_k\omega_j\mathcal{E}\} - 2r\Delta + 2\mathcal{E}\,\frac{1}{r}\,\mathcal{E}$$

$$= \frac{1}{r}\{2\underline{\omega}\partial_{\underline{\omega}} + (q-1)\mathcal{E} - 2r^2\Delta + 2r\mathcal{E}\,\frac{1}{r}\,\mathcal{E}\} = \frac{1}{r}(2\partial_{\underline{\omega}}\partial_{\underline{\omega}} - (q-1)\mathcal{E})$$

(use the expression for Δ in Proposition 7). □

Proposition 7 (Factorization of the Laplacian), [A], p. 12.

$$\Delta_{\underline{x}} = -\partial_{\underline{x}}\partial_{\underline{x}} = \frac{1}{r^2}\{\mathcal{E}^2 + (q-2)\mathcal{E} + \Delta^*_{\underline{\omega}}\}$$

with $\Delta^*_\omega = -(q-2+\underline{\omega}\partial_{\underline{\omega}})\underline{\omega}\partial_{\underline{\omega}}$.

Proof. From proposition 3

$$\partial_{\underline{x}}\partial_{\underline{x}} = \frac{1}{r}(\underline{\omega}\mathcal{E} + \partial_{\underline{\omega}})\frac{1}{r}(\underline{\omega}\mathcal{E} + \partial_{\underline{\omega}})\ .$$

With Tables I and II

$$\partial_{\underline{x}}\partial_{\underline{x}} = \frac{1}{r^2}(-\underline{\omega} + \underline{\omega}\mathcal{E} + \partial_{\underline{\omega}})(\underline{\omega}\mathcal{E} + \partial_{\underline{\omega}})\ .$$

Because of $[\underline{\omega},\mathcal{E}] = 0$, $[\mathcal{E},\partial_{\underline{\omega}}] = 0$ and Table IV

$$\partial_{\underline{x}}\partial_{\underline{x}} = \frac{1}{r^2}(\mathcal{E} - \underline{\omega}\partial_{\underline{\omega}} + \underline{\omega}\mathcal{E}\underline{\omega}\mathcal{E} + \underline{\omega}\mathcal{E}\partial_{\underline{\omega}} + \partial_{\underline{\omega}}\underline{\omega}\mathcal{E} + \partial_{\underline{\omega}}\partial_{\underline{\omega}})$$

$$= \frac{1}{r^2}(\mathcal{E} - \underline{\omega}\partial_{\underline{\omega}} - \mathcal{E}^2 + (\underline{\omega}\partial_{\underline{\omega}} + \partial_{\underline{\omega}}\underline{\omega})\mathcal{E} + \partial_{\underline{\omega}}\partial_{\underline{\omega}})$$

$$= -\frac{1}{r^2}(\mathcal{E}^2 + (q-2)\mathcal{E} + \underline{\omega}\partial_{\underline{\omega}} - \partial_{\underline{\omega}}\partial_{\underline{\omega}})\ .$$

Now, put $\Delta^*_{\underline{\omega}} = \underline{\omega}\partial_{\underline{\omega}} - \partial_{\underline{\omega}}\partial_{\underline{\omega}}$, and observe (cf. Prop. 6)

$$\partial_{\underline{\omega}}\partial_{\underline{\omega}} = -\underline{\omega}\ \underline{\omega}\partial_{\underline{\omega}}\partial_{\underline{\omega}} = -\underline{\omega}(-\partial_{\underline{\omega}}\underline{\omega} - q + 1)\partial_{\underline{\omega}} = \underline{\omega}\partial_{\underline{\omega}}\underline{\omega}\partial_{\underline{\omega}} + (q-1)\underline{\omega}\partial_{\underline{\omega}}\ .$$

Therefore

$$\Delta^*_{\underline{\omega}} = -(\underline{\omega}\partial_{\underline{\omega}})^2 - (q-2)\underline{\omega}\partial_{\underline{\omega}} = -(q-2+\underline{\omega}\partial_{\underline{\omega}})\underline{\omega}\partial_{\underline{\omega}}\ .$$

With $\Gamma = -\underline{\omega}\partial_{\underline{\omega}}$ we still find, cf. [A], p. 12,

$$\Delta^*_{\underline{\omega}} = (q-2-\Gamma)\Gamma\ .$$ □

Corresponding to [A]. p. 12.

Appendix

This Appendix is completely independent from the spherical vectorfield approach. We make sure that the variables $\underline{\omega}$ and r range in the linear space $\mathbb{R}^{q+1}$. Then playing the game of the Chain Rule in well established circumstances leads to an interesting result for

$$\sum_{j=1}^{q} \omega_j \frac{\partial}{\partial \omega_j} .$$

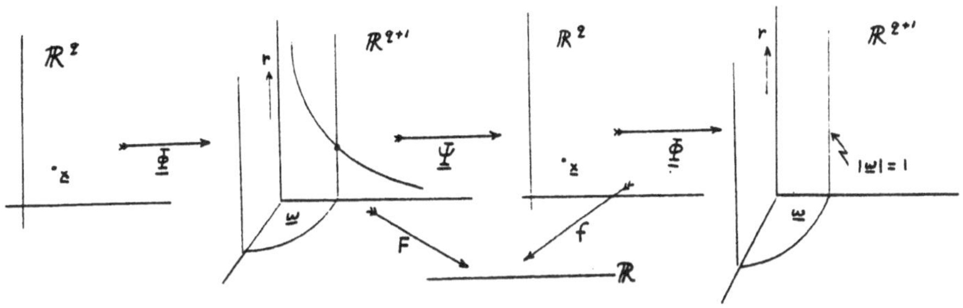

Define $\underline{\Phi} : \mathbb{R}^q \backslash \{\underline{0}\} \to \mathbb{R}^{q+1}$ by $\quad \underline{y} = \begin{pmatrix} \underline{\omega} \\ r \end{pmatrix} = \underline{\Phi}(\underline{x}) = \begin{pmatrix} \dfrac{\underline{x}}{|\underline{x}|} \\ \\ |\underline{x}| \end{pmatrix}$

$$\underline{\Psi} : \mathbb{R}^{q+1} \to \mathbb{R}^q \quad \text{by} \quad \underline{x} = \underline{\Psi}(\underline{y}) = \underline{\Psi}(\underline{\omega}, r) = r\underline{\omega}.$$

Note that

$$\underline{\Psi}^{\leftarrow}(\underline{a}) = \left\{ \begin{pmatrix} \dfrac{1}{t}\,\underline{a} \\ \\ t \end{pmatrix} \in \mathbb{R}^{q+1} \;\middle|\; t \in \mathbb{R} \right\} .$$

Calculate the following derivatives and compositions

- $$D\underline{\Phi}(\underline{x}) = \begin{bmatrix} \dfrac{1}{|\underline{x}|}\left(I - \dfrac{\underline{x}\,\underline{x}^T}{|\underline{x}|^2}\right) \\ \\ \dfrac{1}{|\underline{x}|}\,\underline{x}^T \end{bmatrix} \in \mathbb{R}^{(q+1)\times q} \text{ , with } \underline{x}^T = row(x^1, \ldots, x^q)$$
- $$D\underline{\Psi}(\underline{\omega}, r) = [rI \quad \underline{\omega}] \in \mathbb{R}^{q\times(q+1)}$$
- $$(\underline{\Psi} \circ \underline{\Phi})(\underline{x}) = |\underline{x}|\, \frac{\underline{x}}{|\underline{x}|} = \underline{x}$$
- $$(\underline{\Phi} \circ \underline{\Psi})(\underline{\omega}, r) = (sgn\ r\ \frac{\underline{\omega}}{|\underline{\omega}|}\ ,\ |r\underline{\omega}|)$$

- $D\underline{\Psi} \cdot D\underline{\Phi} = I$

- $$(D\underline{\Phi} \cdot D\underline{\Psi})(\underline{\omega}, r) = \frac{sgn\ r}{|\underline{\omega}|} \begin{bmatrix} I - \dfrac{\underline{\omega}\ \underline{\omega}^T}{|\underline{\omega}|^2} & \underline{0} \\ r\underline{\omega}^T & |\underline{\omega}|^2 \end{bmatrix} \in \mathbb{R}^{q+1)\times(q+1)} .$$

If the function $f : \mathbb{R}^q \to \mathbb{R}$ is given, then the function $F : \mathbb{R}^{q+1} \to \mathbb{R}$, defined by the pull–back $F(\underline{\omega}, r) = f(r\underline{\omega})$, is constant on the hyperbolic curves $t \mapsto (\frac{1}{t}\ \underline{a}, t)$, $\underline{a} \in \mathbb{R}^q$. Such functions F are determined by their values on the cylinder $|\underline{\omega}| = 1$.
They have to satisfy

$$F \circ \underline{\Phi} \circ \underline{\Psi} = F \quad \text{for } r \neq 0, |\underline{\omega}| \neq 0 .$$

Taking the derivative on both sides

$$(DF)((\underline{\Phi} \circ \underline{\Psi})(\underline{\omega}, r)) \cdot D((\underline{\Phi} \circ \underline{\Psi}))(\underline{\omega}, r) = DF(\underline{\omega}, r)$$

leads, after transposition, to

$$\frac{sgn\ r}{|\underline{\omega}|} \begin{bmatrix} I - \dfrac{\underline{\omega}\ \underline{\omega}^T}{|\underline{\omega}|^T} & r\underline{\omega} \\ \underline{0}^T & |\underline{\omega}|^2 \end{bmatrix} \begin{pmatrix} \dfrac{\partial F}{\partial \underline{\omega}} \left(\dfrac{r\underline{\omega}}{|r\underline{\omega}|}\ ,\ |r\underline{\omega}| \right) \\ \dfrac{\partial F}{\partial r} \left(\dfrac{r\underline{\omega}}{|r\underline{\omega}|}\ ,\ |r\underline{\omega}| \right) \end{pmatrix} = \begin{pmatrix} \dfrac{\partial F}{\partial \underline{\omega}}(\underline{\omega}, r) \\ \dfrac{\partial F}{\partial r}(\underline{\omega}, r) \end{pmatrix} .$$

Evaluation for $r > 0$ and $|\underline{\omega}| = 1$ leads to

$$\frac{\partial F}{\partial \underline{\omega}}\ \underline{\omega} = r\ \frac{\partial F}{\partial r} .$$

To put is more provocative:
For functions $F : \mathbb{R}^{q+1} \to \mathbb{R}$ which are the pull–back $f \circ \underline{\Phi}$ of a function $f : \mathbb{R}^q \to \mathbb{R}$ one has

$$\sum_{j=1}^{q} \omega_j\ \frac{\partial}{\partial \omega_j} F = \mathcal{E} F$$

References

[A] Acker, N. van, Clifford differentiaaloperatoren en randwaardetheorie van de nuloplossingen ervan op de sfeer en de Lie–sfeer. Proefschrift, Rijksuniversiteit Gent, 1991-1992.

[CF] Chisholm, J.S.R. and R.S. Farwell, CLifford approach to metric manifolds. Proc. Winterschool on Geometry and Physics. SRNI, 6-13 January, 1990, pp. 123-133.

[G] Graaf, J. de, Skew Hermitean Representations of Lie Algebras of Vector Fields on the Unit–Sphere. Mitteilungen der Math. Gesellschaft Hamburg. Band XII, Heft 3, 1991, 705-711.

[S] Schouten, J.A., Tensor analysis for physicists. Dover Edition, New York, 1989.

QUATERNIONIC ANALYSIS AND TRANSMISSION PROBLEMS

K. GÜRLEBECK
Technische Universität Chemnitz
Fachbereich Mathematik
PSF 964
D-O-9010 Chemnitz
Germany

Abstract. It will be considered a transmission problem for the Laplace equation in a bounded domain $G \subset R^3$ with a smooth boundary and k inclusions G_i. We find an integral representation of the solution using quaternionic analysis. The unknown densities in this integral representation are determined by a system of integral equations on the boundaries of the inclusions. We show that this system is equivalent to a generalized Riemann-Hilbert transmission problem for a function which is H-regular outside of all boundaries of the domain and of the inclusions. This problem will be solved. Using these results an explicit representation formula for the solution of the transmission problem will be given. At the end there is a short discussion of a transmission problem for the Lame' system.

Key words: quaternionic analysis, transmission problems, integral representation

1. Introduction

It will be considered a transmission problem in a bounded domain $G \subset R^3$ with a smooth boundary and k inclusions G_i

$$\Delta^* u^{(i)} = 0 \quad \text{in } G_i, \quad i = 0, ..., k \tag{1}$$

$$u^{(i)} = u^{(0)} \quad \text{on } \Gamma_i, \quad i = 1, ...k \tag{2}$$

$$u^{(0)} = 0 \quad \text{on } \Gamma_0 \tag{3}$$

$$\tau_n u^{(i)} - \tau_n u^{(0)} = h_i \quad \text{on } \Gamma_i, \quad i = 1, ..., k \tag{4}$$

where $\bar{G}_i \subset G, G_0 = G \setminus \cup_{i=1}^k \bar{G}_i, \bar{G}_i \cap \bar{G}_j = \emptyset$ for $i \neq j, \Gamma_i = \partial G_i, \Gamma_0 = \partial G$ and the operator Δ^* stands for the Lame' operator $\Delta + \frac{m}{m-2} grad\ div$ using the number $m = 1/\nu$, where ν is the Poisson ratio. The operator τ_n is the operator of the normal stresses on Γ_i.

The class of such transmission problems is well studied. In the plane case it is usual to solve the problem using complex function theory (see e.g. [2], [4] and further references included there). It is the aim of this paper to show how quaternionic analysis can be applied to find the solution of (1)-(4). For simplicity we start with the case $m = 0$ ($\Delta^* = \Delta$ and $\tau_n = \frac{\partial u}{\partial n}$). We give a description of the orthogonal complement of the subspace of all H-regular functions in $L_2(G_0)$ and then we can prove criterions which allow to answer the question if a given set of functions $u_0, ..., u_k$ defined on $\Gamma_0, ..., \Gamma_k$ has an H-regular extension into the domain G_0. Using these results an explicit representation formula for the solution of the transmission

F. Brackx et al. (eds.), *Clifford Algebras and their Applications in Mathematical Physics*, 101–108.

problem will be given. Considering the boundary values and the jump relations on Γ_i we obtain a system of integral equations for the unknown density $(u_1, ..., u_k)$ on $(\Gamma_1 \times \ldots \times \Gamma_k)$. We discuss uniqueness and existence of the solution of the integral equation system in a constructive way. At the end we will give some information for the general case $m \neq 0$.

2. Preliminaries

Let Ω be a sufficiently smooth bounded domain in R^3 with the boundary γ . Let $e_0 = 1, e_1, e_2, e_3$ be a basis in the algebra H of quaternions. Each H-valued function $u(x)$ may be represented by $u(x) = u_0(x)e_0 + u_1(x)e_1 + u_2(x)e_2 + u_3(x)e_3$.
Furthermore let $\underline{u}(x) = u_1(x)e_1 + u_2(x)e_2 + u_3(x)e_3$ and $\overline{u}(x) = u_0(x)e_0 - \underline{u}(x)$. Sometimes we use the notations $\underline{u} = Im\ u$ and $u_0 = Re\ u$. The spaces of H-valued functions $C_H^{0,\alpha}, W_{2,H}^k, L_{2,H}$ and $\overset{0}{W}{}_{2,H}^1$ are defined componentwise. Let D_i, $i = 1,2,3$ the i-th partial derivative. Then $D = e_1D_1 + e_2D_2 + e_3D_3$ denotes a generalized Cauchy-Riemann operator, the so-called Moisil-Teodorescu operator. Functions u with $Du = 0$ are called H-regular. It is necessary to define the integral operators

$$(F_\gamma u)(x) = \frac{1}{4\pi}\int_\gamma \frac{y-x}{|y-x|^3}n(y)u(y)d\gamma_y,\ x \notin \gamma$$

and

$$(T_\Omega u)(x) = \frac{1}{4\pi}\int_\Omega \frac{x-y}{|x-y|^3}u(y)d\Omega_y,\ x \in R^3.$$

Here $n(y) = n_1(y)e_1 + n_2(y)e_2 + n_3(y)e_3$ defines the outward directed normal on the boundary γ at the point y . We mention the known relations:

(i) $DT_\Omega u = u$ in Ω, if $u \in L_{2,H}(\Omega)$

(ii) Let $u \in W_{2,H}^1(\Omega)$, then
$F_\gamma u + T_\Omega Du = u$ in Ω (Borel-Pompeiu's formula)

(iii) Let $l = 0,1,\ldots$ We have
$T_\Omega : W_{2,H}^l(\Omega) \longrightarrow W_{2,H}^{l+1}(\Omega)$,
$F_\gamma : W_{2,H}^{l-1/2}(\gamma) \longrightarrow W_{2,H}^l(\Omega) \cap ker\, D$.

(iv) Let $u \in C_H^{0,\alpha}(\gamma), 0 < \alpha < 1$
$\lim_{\substack{x \to x_0 \\ x \in \Omega^\pm}} (F_\gamma u)(x) = \frac{1}{2}[\pm u(x_0) + (S_\gamma u)(x_0)]$ (Plemelj's formula)
where $\Omega^+ = \Omega, \Omega^- = R^3 \setminus \overline{\Omega}$, with

$$(S_\gamma u)(x) = \frac{1}{2\pi}\int_\gamma \frac{x_0-y}{|x_0-y|^3}n(y)u(y)d\gamma_y.$$

Here S_γ has to be understood in the sense of Cauchy's principal value. Notice that this result may be generalized to the space $L_{2,H}(\gamma)$. Denote with $P_\gamma = \frac{1}{2}(I + S_\gamma)$

the projection onto the space of all H-valued functions defined on γ which have a H-regular extension into Ω^+. The operator $Q_\gamma = \frac{1}{2}(I - S_\gamma)$ denotes the projection onto the space of all H-valued functions which are H-regular extendable into Ω^- and vanish at infinity.

By the help of the above introduced operators we can write $\Delta^* = DMD$ in G_i with a multiplication operator M. The multiplication operator M is defined by $Mf = \frac{2m-2}{m-2} Ref + Imf$.

3. Boundary Value Problems in Multiply Connected Domains

It is clear that all cited results can be applied to operators which are connected with G_0, G_i, Γ_0 and Γ_i. The generalization of Borel-Pompeiu's formula to the case of the multiply connected domain G_0 reads as follows:

$$u = F_{\Gamma_0} u - \sum_{i=1}^{k} F_{\Gamma_i} u + T_{G_0} Du \qquad \forall u \in W_2^1(G_0) \tag{5}$$

We denote by tr_j the trace operator on Γ_j, where Γ_j is considered as part of ∂G_0. Applying tr_j we get from (5)

$$u_j := tr_j u = tr_j F_{\Gamma_0} u_0 - \sum_{i=1}^{k} tr_j F_{\Gamma_i} u_i + tr_j T_{G_0} Du$$

and for functions $u \in kerD(G_0)$ we have

$$u_j = tr_j F_{\Gamma_0} u_0 - \sum_{i=1}^{k} tr_j F_{\Gamma_i} u_i \qquad j = 0, ...k.$$

If we collect these equations we get the following system

$$\begin{pmatrix} u_0 \\ u_1 \\ \vdots \\ u_k \end{pmatrix} = \begin{pmatrix} P_{\Gamma_0} & -tr_0 F_{\Gamma_1} & \dots & -tr_0 F_{\Gamma_k} \\ tr_1 F_{\Gamma_0} & Q_{\Gamma_1} & \dots & -tr_1 F_{\Gamma_k} \\ \vdots & \vdots & \ddots & \vdots \\ tr_k F_{\Gamma_0} & -tr_k F_{\Gamma_1} & \dots & Q_{\Gamma_k} \end{pmatrix} \begin{pmatrix} u_0 \\ u_1 \\ \vdots \\ u_k \end{pmatrix} \tag{6}$$

We denote this matrix operator by P_Γ and we can prove now the following theorems using the same ideas as in case of a simply connected domain [1].

Theorem 1 *The operator P_Γ is a projection.*

Theorem 2 *It holds*

$(u_0, \dots, u_k)^T \in imP_\Gamma \cap W_2^{1/2}(\partial G_0) \Longleftrightarrow$

$\exists u \in kerD \cap \overset{0}{W}{}_2^1(G_0) : tr_j u = u_j \quad j = 1, ...k$

Theorem 3 *$L_2(G_0)$ admits the following orthogonal decomposition*

$L_2(G_0) = L_2(G_0) \cap kerD(G_0) \oplus D(\overset{0}{W}{}_2^1(G_0))$

The corresponding orthoprojections onto $ker D(G_0)$ and $\overset{0}{W}{}_2^1(G_0)$ are denoted by P_{G_0} and Q_{G_0}

$P_{G_0} : L_2(G_0) \longrightarrow ker D(G_0)$

$Q_{G_0} : L_2(G_0) \longrightarrow D(\overset{0}{W}{}_2^1(G_0))$.

We want to investigate now the matrix operator

$$tr\ T_{G_0}\vec{F_\Gamma} := \begin{pmatrix} tr_0 \\ tr_1 \\ \vdots \\ tr_k \end{pmatrix} T_{G_0}(F_{\Gamma_0}, -F_{\Gamma_1}, \ldots, -F_k). \tag{7}$$

In general this operator is not invertible but if we use some suitable restrictions we get the following result.

Theorem 4 *The operator*

$tr\ T_{G_0}\vec{F_\Gamma} : imP_\Gamma \cap W_2^{1/2}(\partial G_0) \longrightarrow imQ_\Gamma \cap W_2^{3/2}(\partial G_0)$

is invertible.

Proof: Using the orthoprojections P_{G_0} and Q_{G_0} we can show that the boundary value problem $\{\Delta u = 0\ in\ G_0, tr_i u = u_i\ on\ \Gamma_i\ i = 1, ..., k\}$ has a unique solution for arbitrarily given functions $u_i, i = 1, ..., k$. The function

$$u = F_{\Gamma_0}u_0 - \sum_{i=1}^{k} F_{\Gamma_i}u_i + T_{G_0}P_{G_0}Dh$$

is a solution for an arbitrary extension (with correct regularity) h of the boundary values u_i into G_0. We get the uniqueness by the help of Borel-Pompeiu's formula for multiply connected domains.

If we have now $TrT_{G_0}\vec{F_\Gamma}\vec{g} = 0$ with $\vec{g} \in imP_\Gamma$ then it follows $T_{G_0}\vec{F_\Gamma}\vec{g} = 0$, $\vec{F_\Gamma}\vec{g} = 0$ and $\vec{g} = 0$.

Furthermore, let $\vec{g} \in imQ_\Gamma$. We solve the problem $\{\Delta u = 0, tru = \vec{g}\}$ and then we obtain

$$u = T_{G_0}P_{G_0}Dh = T_{G_0}\vec{F_\Gamma}\Phi,\ tr\Phi \in imP_\Gamma$$

Therefore $tr\ u = \vec{g} = trT_{G_0}\vec{F_\Gamma}tr\Phi$ and the proof is finished.

Using Theorem 4 we can describe the orthoprojection P_{G_0} explicitly.

Theorem 5 $P_{G_0} = \vec{F_\Gamma}\{trT_{G_0}\vec{F_\Gamma}\}^{-1}trT_{G_0}$

In this way we obtain a representation for the solution of the Dirichlet problem in the multiply connected domain G_0.

$$u^{(0)} = \vec{F_\Gamma}\vec{g} + T_{G_0}\vec{F_\Gamma}\{trT_{G_0}\vec{F_\Gamma}\}^{-1}trT_{G_0}Dh$$

In this formula $\vec{g}$ stands for the vector $(g_0, g_1, \ldots, g_n)^T$ and the function h is an arbitrarily choosen extension of $\vec{g}$ into the domain G_0. By the help of Borel-Pompeiu's formula we have $trT_{G_0}Dh = Q_\Gamma\vec{g}$ and we can prove Theorem 6.

Theorem 6 $u^{(0)} = \vec{F_\Gamma}\vec{g} + T_{G_0}\vec{F_\Gamma}\{trT_{G_0}\vec{F_\Gamma}\}^{-1}Q_\Gamma\vec{g}$

4. Solution of the Transmission Problem

In each subdomain G_i we can use the known representation formula [1]

$u^{(i)} = F_{\Gamma_i}u_i + T_{G_i}F_{\Gamma_i}(tr_iT_{G_i}F_{\Gamma_i})^{-1}Q_{\Gamma_i}u_i$

From this we have a formal solution in G

$$u = \begin{cases} \vec{F_\Gamma}\vec{u} + T_{G_0}\vec{F_\Gamma}\{trT_{G_0}\vec{F_\Gamma}\}^{-1}Q_\Gamma\vec{u} & in \quad G_0 \\ F_{\Gamma_i}u_i + T_{G_i}F_{\Gamma_i}(tr_iT_{G_i}F_{\Gamma_i})^{-1}Q_{\Gamma_i}u_i & in \quad G_i, \quad i = 1, ..., k \end{cases} \tag{8}$$

where we used $\vec{u} = (0, u_1, \ldots, u_n)^T$. This representation formula can be simplified. From the representation in G_i we can see that

$T_{G_i}F_{\Gamma_i}(trT_{G_i}F_{\Gamma_i})^{-1}Q_{\Gamma_i}u_i \in kerD(co\ \overline{G_i})$.

Furthermore it holds $tr_iF_{\Gamma_i}u_i = -Q_{\Gamma_i}u_i$ and $-F_{\Gamma_i}u_i \in kerD(co\ \overline{G_i})$. Using the identity theorem for H-regular functions [3] we get

$$-F_{\Gamma_i}u_i = T_{G_i}F_{\Gamma_i}(trT_{G_i}F_{\Gamma_i})^{-1}Q_{\Gamma_i}u_i \quad in \quad G_0 \qquad \forall i \neq 0. \tag{9}$$

In a similar way we can show that

$$\sum_{s=1}^{k} F_{\Gamma_s}u_s = T_{G_0}\vec{F_\Gamma}\{trT_{G_0}\vec{F_\Gamma}\}^{-1}Q_\Gamma\vec{u} \tag{10}$$

By the help of the equations (9) and (10) we get a unique representation of the solution in the whole domain G.

$$u = T_{G_0}\vec{F_\Gamma}\{trT_{G_0}\vec{F_\Gamma}\}^{-1}Q_\Gamma\vec{u} + \sum_{s=1}^{k} T_{G_s}F_{\Gamma_s}(trT_{G_s}F_{\Gamma_s})^{-1}Q_{\Gamma_s}u_s \tag{11}$$

This function u fulfils the equations (1)-(3). The open problem is to satisfy also the equation (4) by controlling the unknown densities u_i on the boundaries Γ_i. For this purpose we introduce the operator V_γ defined by

$V_\gamma f = \frac{1}{4\pi}\int\limits_\gamma \frac{n(y)}{|x-y|}f(y)d\gamma_y$

It is known $V_\gamma f = K_\Omega DF_\gamma + T_\Omega f$ with the volume potential K_Ω. In the case of $f \in kerD(\Omega)$ we have $V_\gamma trf = T_\Omega f$ in Ω. In our situation it follows that

$$T_{G_i}F_{\Gamma_i}(trT_{G_i}F_{\Gamma_i})^{-1}Q_{\Gamma_i}u_i = V_{\Gamma_i}(trT_{G_i}F_{\Gamma_i})^{-1}Q_{\Gamma_i}u_i \quad in \quad R^3 \tag{12}$$

using again the identity theorem for H-regular functions. We abbreviate as follows

$\vec{\psi} = (\psi_0, \ldots, \psi_n)^T := \{trT_{G_0}\vec{F_\Gamma}\}^{-1}Q_\Gamma\vec{u} \ni imP_\Gamma$

and start the investigation of $T_{G_0}\vec{F_\Gamma}\vec{\psi}$. We find that $T_{G_0}\vec{F_\Gamma}\vec{\psi} = V_{\Gamma_0}\psi_0 - \sum\limits_{i=1}^{k} V_{\Gamma_i}\psi_i$.

Now we recall into the mind the jump formula

$[\frac{\partial}{\partial n_{ext}} - \frac{\partial}{\partial n_{int}}]V_{\Gamma_i}\phi = -n_i(y)\phi_i(y)$,

where n_i denotes the unit normal on Γ_i directed into G_0. If this formula will be

applied to equation (12) we obtain
$n_i\psi_i - n_i(trT_{G_i}F_{\Gamma_i})^{-1}Q_{\Gamma_i}u_i = h_i$ or

$$(trT_{G_i}F_{\Gamma_i})^{-1}Q_{\Gamma_i}u_i - \psi_i = n_i h_i \qquad i = 1, ..., k \tag{13}$$

We have to prove now that this system of equations determines the unknown functions u_i. First we are looking for the uniqueness and we assume that there are two different solutions. Then we get for the difference which we will denote by ψ again the following properties.
$\psi_i \in imP_{\Gamma_i} \quad i = 1, ..., k \quad \Rightarrow \psi_0 \in P_{\Gamma_0}$.
From (13), (12) and the definition of $\vec{\psi}$ we have $u = V_{\Gamma_0}\psi_0$ and it holds
$tr_0 u = tr_0 V_{\Gamma_0}\psi_0 = 0 \Rightarrow V_{\Gamma_0}\psi_0 = 0 \Rightarrow DV_{\Gamma_0}\psi_0 = F_{\Gamma_0}\psi_0 = 0 \Rightarrow \psi_0 = 0$.
From $\psi_i \in imP_{\Gamma_i}$ and $\psi_i = -\sum_{s=1, s\neq i}^{k} tr_i F_{\Gamma_s}\psi_s + Q_{\Gamma_i}\psi_i = 0$ it follows that $\vec{\psi} = 0$.
$\Rightarrow Q_{\Gamma_i}u_i = 0 \quad i = 1, ...k \Rightarrow u_i \in imP_{\Gamma_i}, u_0 = 0$
$\Rightarrow u = T_{G_0}\vec{F_\Gamma}\{trT_{G_0}\vec{F_\Gamma}\}^{-1}Q_\Gamma\vec{u} \Rightarrow tru = Q_\Gamma\vec{u} \Rightarrow P_\Gamma\vec{u} = 0$. On the other hand we have that with $\vec{\psi} = 0$ it is also valid $\{trT_{G_0}\vec{F_\Gamma}\}^{-1}Q_\Gamma\vec{u} = 0$ and this means $Q_\Gamma\vec{u} = 0 \Rightarrow u_i = 0 \quad \forall i$.

Theorem 7 *If the system*

$$(trT_{G_i}F_{\Gamma_i})^{-1}Q_{\Gamma_i}u_i - \psi_i = n_i h_i \qquad i = 1, ..., k \tag{14}$$

has a solution it is unique.

If we now differentiate our representation (8) of the solution by applicating the operator D we get

$$Du = \begin{cases} \vec{F_\Gamma}\{trT_{G_0}\vec{F_\Gamma}\}^{-1}Q_\Gamma\vec{u} & in \quad G_0 \\ F_{\Gamma_j}(tr_jT_{G_j}F_{\Gamma_j})^{-1}Q_{\Gamma_j}u_j & in \quad G_j, \quad j = 1, ..., k \end{cases}$$

A consideration of the boundary values of Du leads to

$tr_{j,ext}Du = \psi_j$ and $tr_{j,int}Du = (tr_jT_{G_j}F_{\Gamma_j})^{-1}Q_{\Gamma_j}u_j$

It follows

$$(tr_{j,ext} - tr_{j,int})(n_jDu) = n_j\psi_j - n_j(tr_jT_{G_j}F_{\Gamma_j})^{-1}Q_{\Gamma_j}u_j = h_j$$

and that means that the functions n_jDu and $\frac{\partial u}{\partial n_j}$ have the same jump if we are crossing the boundary Γ_j. Obviously the equation $tr_0Du = \psi_0$ holds and our problem (1)-(4) can be transformed into the following generalized Riemann-Hilbert transmission problem for the function $v = Du$.

$$Dv = 0 \qquad in \quad G_i \tag{15}$$

$$(tr_{j,ext} - tr_{j,int})v = -n_jh_j \quad on \quad \Gamma_j, j = 1, ..., k \tag{16}$$

$$tr_0v = \psi_0 \qquad on \quad \Gamma_0 \tag{17}$$

We try to find a solution of this problem in the following way

$v = \sum_{s=1}^{k} F_{\Gamma_s}(n_s h_s) - F_{\Gamma_0}\theta_0$

Using this expression for computing $tr_0 v$ and $tr_{i,ext} v$ we come to

$$\begin{pmatrix} \psi_0 \\ \vdots \\ \psi_j \\ \vdots \end{pmatrix} = \begin{pmatrix} -P_{\Gamma_0}\theta_0 + \sum_{i=1}^{k} tr_0 F_{\Gamma_i}(n_i h_i) \\ \vdots \\ -tr_j F_{\Gamma_0}\theta_0 + \sum_{i=1, i\neq j}^{k} tr_j F_{\Gamma_i}(n_i h_i) - Q_{\Gamma_j}(n_j h_j) \\ \vdots \end{pmatrix}$$

By the help of Borel-Pompeiu's formula for $u^{(i)}$ in G_i and $v = Du$ we obtain

$Q_{\Gamma_i} u_i = \sum_{s=1}^{k} tr_i T_{G_i} F_{\Gamma_s}(n_s h_s) - tr_i T_{G_i} F_{\Gamma_0}\theta_0 \qquad i = 1, ..., k$

From (10) we have

$P_{\Gamma_i} u_i = tr_i T_{G_0}(F_{\Gamma_0}\psi_0 - \sum_{s=1}^{k} F_{\Gamma_s}\psi_s) + tr_i \sum_{s=1, s\neq i}^{k} F_{\Gamma_s} Q_{\Gamma_s} u_s$

and by inserting the already known functions $\psi_i, Q_{\Gamma_i} u_i$ we obtain after some computation the following result.

$u_i = -tr_i T_G F_{\Gamma_0}\theta_0 + tr_i T_G \sum_{s=1}^{k} F_{\Gamma_s}(n_s h_s)$

Finally, we get by inserting in equation (11)

$tr_0 T_G F_{\Gamma_0}\theta_0 = tr_0 T_G N(h) \quad \text{or} \quad \theta_0 = (tr_0 T_G F_{\Gamma_0})^{-1} tr_0 T_G N(h),$

where we used the abbreviation $N(h) = \sum_{s=1}^{k} F_{\Gamma_s}(n_s h_s)$. The above considerations prove our main theorem.

Theorem 8 *The transmission problem (1)-(4) has the solution*

$$u = T_G(I - F_{\Gamma_0}(tr_0 T_G F_{\Gamma_0})^{-1} tr_0 T_G) N(h) \tag{18}$$

for arbitrarily given functions $h_i \in W_2^{1/2}(\Gamma_i)$.

5. Transmission Problems for the Lame' system

We will give now some information what we have to do in the case of the Lame' system ($m > 2$). For the representation of a solution u of (1)-(4) in G_i we can use the following result

$u^{(i)} = F_{\Gamma_i} u_i + T_{G_i} M^{-1} F_{\Gamma_i}(tr_i T_{G_i} M^{-1} F_{\Gamma_i})^{-1} Q_{\Gamma_i} u_i$

Similar as in the previous section we have to consider the matrix operator $\{tr T_{G_0} M^{-1} \vec{F_\Gamma}\}$ and we obtain the following representation formula

$$u = T_{G_0} M^{-1} \vec{F_\Gamma} \{tr T_{G_0} M^{-1} \vec{F_\Gamma}\}^{-1} Q_\Gamma \vec{u} + \sum_{s=1}^{k} F_{\Gamma_s} u_s \tag{19}$$

Now we have to investigate the operator $K_\Omega D$, where K_Ω is the volume potential operator of the linear elasticity and we get the following result.

$V_\gamma^{elast} f = \frac{m-2}{2m-2} M^{-1} T M^{-1} f \qquad \forall f \in ker D$

where the operator V_γ^{elast} is of single layer type defined by

$$V_\gamma^{elast} f(x) = -\frac{1}{4\pi} \int_\gamma \sum_{i=1}^{4} H_i(x-y)[n\, f]_i(y) d\gamma_y$$

with $H_0(x) = \frac{1}{|x|}$, $H_i(x) = \frac{3m-4}{4m-4} \frac{1}{|x|} e_i - \frac{m}{4m-4} \sum_{j=1}^{3} \frac{x_i x_j}{|x|^3} e_j$.

We can prove the jump formula

$(\tau_{n,ext} - t_{n,int}) V_\gamma^{elast} f = n\, f$

and then our further consideration can be done in the same way as above.
At the end we remark that the representation formulas in the form of equation (18) allow to study the analytical properties of the solution of the transmission problem in a convenient way. We can look for stability and regularity results but it is also possible to study the behaviour of the solution if we disturb the boundaries Γ_i. On the other hand the presented method is open for numerical applications. If we use ideas of discrete quaternionic analysis it should be possible to repeat our considerations for discrete functions on a lattice and the corresponding operators acting on lattice functions.

References

1. Gürlebeck, K. and Sprößig, W.: Quaternionic Analysis and Elliptic Boundary Value Problems, ISNM 89, Birkhäuser Verlag Basel 1990
2. Muschelischvili, N.I.: Grundaufgaben zur mathematischen Elastizitätstheorie, Leipzig 1966
3. Sprößig, W.: Über eine mehrdimensionale Operatorenrechnung über beschränkten Gebieten des E_n, Thesis, TH Karl-Marx-Stadt 1979
4. Tutschke, W.: Classical and Modern Methods of Complex Analysis, in: Lanckau, E. and Tutschke, W. (Ed.): Complex Analysis, Akademie-Verlag Berlin 1983, pp. 15–58

C*-ALGEBRAS OF NONLOCAL QUATERNIONIC CONVOLUTION TYPE OPERATORS

YURI I. KARLOVICH*
Department of Mathematics
Odessa State University
Petra Velikogo str. 2
Odessa 270057, UKRAINE.

Abstract. The author studies the Fredholm property for some C*-algebras of quaternionic convolution type operators with discrete amenable groups of shifts on spaces $L_2(\Gamma, \mathbb{H}(\mathbb{C}))$, where Γ is the unit sphere $S^m (m = 1, 2, \ldots)$ and $\mathbb{H}(\mathbb{C})$) is the algebra of complex quaternions. The investigation is based on the local-trajectory method for studying the invertibility of bounded linear operators with shifts in a Hilbert space.

Key words: quaternion, convolution type operator, shift, C*-algebra, local-trajectory method, invertibility, semi-Fredholm property

1. Introduction

Clifford algebras and Clifford analysis play the important role in contemporary multidimensional complex analysis, the theory of representations of groups and algebras, operator theory and their applications in mathematical physics (see [1, 2, 3]). The skew-field of quaternions $\mathbb{H}$ is the most convenient and investigated model of non-commutative Clifford algebras. It is well known the significance of quaternionic analogues of singular integral operators in the theory of hyperholomorphic functions and hyperholomorphic Toeplitz operators. Some C*- algebras of quaternionic singular integral operators without shift and with involution shift have been studied in [4, 5] respectively. Algebras of such operators with infinite groups of shifts were not consider yet. On the other hand the investigations of nonlocal pseudodifferential operators of zero order with infinite groups of shifts with respect to direct and dual variables in the sence of Fourier are continuously extended now (see [6, 7, 8, 9, 10, 11, 12, 13]).

This report devoted to the construction of symbol algebras and studying the Fredholm property for some C*-algebras of quaternionic convolution type operators with discrete amenable groups of shifts and continuous or piecewise continuous coefficients on space $L_2(\Gamma, \mathbb{H}(\mathbb{C}))$, where Γ is sphere S^m $(m = 1, 2, \ldots)$ and $\mathbb{H}(\mathbb{C})$) is the algebra of complex quaternions. The investigation is based on the local-trajectory method [11, 13] of studying the invertibility of bounded linear operators with shifts in a Hilbert space. This method is closely connected with the investigation of crossed products of C*-algebras and groups of their automorphisms. The character of results essentially depends on the structure of set of fixed points of shifts.

* I am grateful to *University of Gent* (België) and the *Ukrainian Renaissance foundation* offered grants me for participation at the Conference.

F. Brackx et al. (eds.), Clifford Algebras and their Applications in Mathematical Physics, 109–118.

2. Local-trajectory method

Let $\mathfrak{A}$ be a C*-subalgebra (with identity) of the algebra $\mathfrak{L}(H)$ of bounded linear operators in a Hilbert space H, let Z be a central C*-subalgebra of $\mathfrak{A}$ with identity, let G be a discrete group, $U : g \to U_g$ a unitary reprezentation of G in H, and let $\mathfrak{B} = C^*(\mathfrak{A}, U_G)$ be the C*-subalgebra of $\mathfrak{L}(H)$ generated by all operators $A \in \mathfrak{A}$ and U_g $(g \in G)$. The essence of the local-trajectory method [11, 13] consists in obtaining criteria for the one- and two-sided invertibility of operators $B \in \mathfrak{B}$ in terms of the same invertibility of local reprezentatives corresponding to the orbits of points $t \in M$, generated by G, where M is the compact space of maximal ideals of the central C*-subalgebra $Z \subset \mathfrak{A}$. This method is a natural generalization of the local method [14, 15] to nonlocal type operators. As a result, we obtain a universal method of studying the (semi-) Fredholm property in an L_2-space for various algebras of one-dimensional and multidimensional convolution type operators with discrete groups of shifts with respect to direct and dual variables in the sense of Fourier, in the presense of discontinuities of the coefficients and presymbols.

We introduce starting assumptions.

(A1) For all $g \in G$ the mappings $\alpha_g : A \to U_g A U_g^*$ are *-automorphisms of the C*-algebras $\mathfrak{A}$ and Z.

In view of (A1) the C*-algebra $\mathfrak{B}$ is closure of set of operators $B = \sum A_g U_g$, where $A_g \in \mathfrak{A}$ and g runs through finite subsets of G. Due to the bijection of the nonzero complex homomorphisms $\tilde{t}$ of Z and the maximal ideals $t = Ker\tilde{t} \in M$, under assumption (A1) the group G can be realized as a homeomorphism group of M according to the rule:

$$[\widetilde{g(t)}](z) = \tilde{t}[\alpha_g(z)], z \in Z, t \in M, g \in G.$$

(A2) G is an amenable discrete group.

As is known [16], discrete group G is called amenable if on the space $l_\infty(G)$ of all bounded complex-valued functions $f : G \to \mathbb{C}$ there exists an invariant mean, i.e. a state μ satisfying the condition: $\mu(f) = \mu({}_h f) = \mu(f_h)$ for all $f \in l_\infty(G)$ and $h \in G$, where $({}_h f)(g) := f(h^{-1}g)$ and $(f_h)(g) := f(gh)$ for $g \in G$. In particular, finite groups, commutative groups, subexponential groups and solvable groups are amenable.

Let $P_{\mathfrak{A}}$ denote the set of pure states of the C*-algebra $\mathfrak{A}$ and e denote the unit in the group G.

(A3) For every finite set $G_0 \subset G$ and every open set $W \subset P_{\mathfrak{A}}$ there exists a point $\nu \in W$ such that $g(t_\nu) \neq t_\nu$ for all $g \in G_0 \setminus \{e\}$, where $t_\nu = Z \cap Ker\nu \in M$.

We remark that (A3) is fulfilled also in the case $G = \{e\}$.

Let J_t be the closed two-sided ideal of $\mathfrak{A}$ generated by the ideal $t \in M$, let $\Omega(X)$ be the set of G-orbits of the points $t \in X(\subset M)$, $\Omega = \Omega(M)$, let t_ω be an arbitrary fixed point of an orbit $\omega \in \Omega$, let H_ω be the space of an isometric representation $\tilde{\pi}_\omega$ of the quotient algebra $\mathfrak{A}/J_{t_\omega}$, and let ρ_ω be the natural homomorphism $\mathfrak{A} \to \mathfrak{A}/J_{t_\omega}$, $\pi'_\omega = \tilde{\pi}_\omega \circ \rho_\omega$. We consider the representation $\pi_\omega : \mathfrak{B} \to \mathfrak{L}(l_2(G, H_\omega))$ defined by

$$(\pi_\omega(A)f)(g) = \pi'_\omega(\alpha_g(A))f(g), (\pi_\omega(U_h)f)(g) = f(gh), A \in \mathfrak{A}; g, h \in G.$$

Theorem 1. If assumptions (A1)-(A3) are satisfied, then an operator $B \in \mathfrak{B}$ is invertible (left invertible, right invertible) in H if and only if for every orbit $\omega \in \Omega$ the operator $\pi_\omega(B)$ is invertible (left invertible, right invertible) in $l_2(G, H_\omega)$.

We shall obtain an analogy of Theorem 1 when (A3) is violated.

Let $P(\cdot)$ be the spectral measure of the commutative algebra Z, let $\mathfrak{R}(M)$ be the σ-algebra of Borel subsets of M and let $\mathfrak{R}_G(M) := \{\Delta \in \mathfrak{R}(M) : g(\Delta) = \Delta, \forall g \in G\}$.

If $\Delta \in \mathfrak{R}(M)$ and $P(\Delta) \neq 0$, then the commutative C*-algebra $Z_\Delta := P(\Delta)ZP(\Delta)$ is isometrically isomorphic to the C*-algebra $C(\tilde{\Delta})$, where $\tilde{\Delta}$ is the closed set of points $t \in M$ for any open neighborhoods $u_t \subset M$ of which, $P(u_t \cap \Delta) \neq 0$. It is obvious that $AP(\Delta) = P(\Delta)A$ for all $A \in \mathfrak{A}$ and $\Delta \in \mathfrak{R}(M)$. But if $\Delta \in \mathfrak{R}_G(M)$ and (A1) is satisfied, then according to [11] $BP(\Delta) = P(\Delta)B$ already for all $B \in \mathfrak{B}$.

Let $\Delta \in \mathfrak{R}_G(M)$, $P(\Delta) \neq 0$ and $G_\Delta := \{g \in G : P(\Delta)U_gP(\Delta) = P(\Delta)\}$. It is obvious that G_Δ is a normal subgroup of G. We consider the unitary representation $U_\Delta : h \to U_{h,\Delta}$ of the factor group G/G_Δ in the Hilbert space $P(\Delta)H$ defined by $U_{h,\Delta} := P(\Delta)U_gP(\Delta)$, where g is an arbitrary element from the coset $h \in G/G_\Delta$. By (A1), for every $h \in G/G_\Delta$ the mapping $A \to U_{h,\Delta}AU_{h,\Delta}^*$ is a *-automorphism of the C*-algebra $Z_\Delta \cong C(\tilde{\Delta})$, and consequently the group G/G_Δ can be realized as a group of homeomorphisms of the compact space $\tilde{\Delta}$: $h(x) := g(x)$ for all $x \in \tilde{\Delta}$ and $g \in h$. In G/G_Δ we choose the normal subgroup $\tilde{G}_\Delta := \{h \in G/G_\Delta : h(x) = x, \forall x \in \tilde{\Delta}\}$. It is clear that Z_0 is a central subalgebra of the C*-algebra $\mathfrak{A}_\Delta := P(\Delta)C^*(\mathfrak{A}, U_{G^0_\Delta})P(\Delta)$, where $G^0_\Delta := \{g \in G : g(x) = x, \forall x \in \tilde{\Delta}\}$.

Assume the following.

(A4) The set M can be partitioned into finitely many sets $M_j \in \mathfrak{R}_G(M)$ such that, for every j,

1) $P(M_j) \neq 0$ and the factor group G/G_{M_j} can be decomposed in a semidirect product $\tilde{G}_j \rtimes G_j$ of its subgroups $\tilde{G}_j = \tilde{G}_{M_j}$ and G_j, i.e. for every $h \in G_j$ the mapping $s \to hsh^{-1}$ is an automorphism of the group $\tilde{G}_j$, and

2) (A3) is satisfied if we replace $\mathfrak{A}, G$ and M with $\mathfrak{A}_j = \mathfrak{A}_{M_j}$, G_j and $\tilde{M}_j$.

Let $U^{(j)}$ denote the restriction of the representation U_{M_j} onto G_j. Due to (A4) each C*-algebra $\mathfrak{B}_j := P(M_j)\mathfrak{B}P(M_j)$ can be written in the form $C^*(\mathfrak{A}_j, U^{(j)}_{G_j})$, and assumptions (A1)-(A3) are satisfied for one.

Let $\Omega_j = \Omega(\tilde{M}_j)$. For every $\omega \in \Omega_j$ in analogy with $l_2(G, H_\omega)$ and π_ω, we construct a space $l_2(G_j, H^{(j)}_\omega)$ and a representation $\pi^{(j)}_\omega$ of $\mathfrak{B}_j$ in it, replacing $\mathfrak{A}, G, U$ and M with $\mathfrak{A}_j, G_j, U^{(j)}$ and $\tilde{M}_j$.

Theorem 2. If assumptions (A1),(A2) and (A4) are fulfilled, then an operator $B \in \mathfrak{B}$ is invertible (left invertible, right invertible) in H if and only if for every j and every orbit $\omega \in \Omega_j$ the operator $\pi^{(j)}_\omega(B)$ is invertible (left invertible, right invertible) in $l_2(G_j, H^{(j)}_\omega)$.

Remark 1. Theorems 1-2 allow to formulate the criteria for the (semi-) Fredholm property for an operator $B \in \mathfrak{B}$ acting in the Hilbert space H, because in this case Fredholmness (n-normality, d-normality) of B is equivalent to the invertibility (left invertibility, right invertibility) of the corresponding coset $\hat{B}$ in Calkin algebra $\mathfrak{L}(H)/\mathfrak{L}_0(H)$, where $\mathfrak{L}_0(H)$ is the ideal of compact operators in H.

Remark 2. The function $\mathcal{B}$ defined at the disjoint union $\sqcup\Omega_j$ by the rule $\mathcal{B}(j, \omega) := \pi^{(j)}_\omega(B)$ for all $\omega \in \Omega_j$ is called a symbol of an operator $B \in \mathfrak{B}$. But if $\mathfrak{B} \cap \mathfrak{L}_0(H) \neq \emptyset$

then according to Remark 1 a symbol of an operator $B \in \mathfrak{B}$ is usually definded as the symbol of the coset $\hat{B} \in \mathfrak{L}(H)/\mathfrak{L}_0(H)$.

3. C*-algebras of abstract nonlocal quaternionic operators

As is well known, the non-commutative skew-field $\mathbb{H}$ of the real quaternions consists of elements

$$y = y_0 + y_1 i_1 + y_2 i_2 + y_3 i_3, \tag{1}$$

where $y_n \in \mathbb{R}(n = 0, 1, 2, 3)$ and $i_1^2 = i_2^2 = i_3^2 = i_1 i_2 i_3 = -1$. We shall consider further the corresponding algebra $\mathbb{H}(\mathbb{C})$ of the complex quaternions having the form (1), where $y_n \in \mathbb{C}(n = 0, 1, 2, 3)$ and the imaginary unit $i \in \mathbb{C}$ commutes with each imaginary unit $i_k(k = 1, 2, 3)$.

Let Γ be the unit sphere S^m in $\mathbb{R}^{m+1}(m = 1, 2, \ldots)$ and let $H = L_2(\Gamma, \mathbb{H}(\mathbb{C}))$ be the complex Hilbert space of functions $f : \Gamma \to \mathbb{H}(\mathbb{C})$ with the norm $\| f \| = (\int_\Gamma \sum_{n=0}^{3} | f_n(x) |^2 \, dx)^{1/2}$, where dx is the m-dimensional Lebesgue measure on Γ.

Let $\mathcal{N}$ denote a C*-subalgebra of $\mathfrak{L}(L_2(\Gamma))$ with identity. For $A_n \in \mathcal{N}(n = 0, 1, 2, 3)$ in $\mathfrak{L}(H)$ we consider the operators

$$A^l = (A_0 + \sum_{k=1}^{3} A_k i_k)^l, A^r = (A_0 + \sum_{k=1}^{3} A_k i_k)^r \tag{2}$$

defined by the rule

$$A^l(f_0 + f_1 i_1 + f_2 i_2 + f_3 i_3) = (A_0 f_0 - A_1 f_1 - A_2 f_2 - A_3 f_3) + (A_1 f_0 + A_0 f_1 - A_3 f_2 + A_2 f_3) i_1$$

$$+ (A_2 f_0 + A_3 f_1 + A_0 f_2 - A_1 f_3) i_2 + (A_3 f_0 - A_2 f_1 + A_1 f_2 + A_0 f_3) i_3,$$

$$A^r(f_0 + f_1 i_1 + f_2 i_2 + f_3 i_3) = (A_0 f_0 - A_1 f_1 - A_2 f_2 - A_3 f_3) + (A_1 f_0 + A_0 f_1 + A_3 f_2 - A_2 f_3) i_1$$

$$+ (A_2 f_0 - A_3 f_1 + A_0 f_2 + A_1 f_3) i_2 + (A_3 f_0 + A_2 f_1 - A_1 f_2 + A_0 f_3) i_3$$

and named the left quaternionic operator and the right quaternionic operator respectively. Let $\mathfrak{A}$ be a C*-subalgebra of $\mathfrak{L}(H)$ generated by operators A^l and A^r with $A_n \in \mathcal{N}(n = 0, 1, 2, 3)$.

Let Z_0 be a central subalgebra of $\mathcal{N}$ with identity and let Z be the set of all operators (2) with $A_0 \in Z_0$ and $A_k = 0(k = 1, 2, 3)$. It is obvious that Z is a central subalgebra of $\mathfrak{A}$ with identity.

Let G be an amenable discrete group, let $U : g \to U_g$ be a unitary representation of G in H and let $\mathfrak{B} = C^*(\mathfrak{A}, U_G)$. We assume also that (A1), (A3) (or (A1), (A4)) are satisfied for the algebra $\mathfrak{B}$.

For convenience we consider the operator of a dequaternionization $\mathbb{D} : \mathbb{H}(\mathbb{C}) \to \mathbb{C}^4$ defined by the rule

$$\mathbb{D} : y = y_0 + \sum_{k=1}^{3} y_k i_k \to \tilde{y} = \{y_n\}_{n=0}^{3}.$$

Then $\tilde{\mathfrak{A}} := \mathbb{D}\mathfrak{A}\mathbb{D}^{-1}$ is the algebra of (4×4)-matrices with elements from $\mathcal{N}$, $\tilde{Z} := \mathbb{D}Z\mathbb{D}^{-1}$ is the algebra of diagonal matrices with equal elements from Z_0, and

$$\mathbb{D}A^l\mathbb{D}^{-1} = \begin{pmatrix} A_0 & -A_1 & -A_2 & -A_3 \\ A_1 & A_0 & -A_3 & A_2 \\ A_2 & A_3 & A_0 & -A_1 \\ A_3 & -A_2 & A_1 & A_0 \end{pmatrix}, \mathbb{D}A^r\mathbb{D}^{-1} = \begin{pmatrix} A_0 & -A_1 & -A_2 & -A_3 \\ A_1 & A_0 & A_3 & -A_2 \\ A_2 & -A_3 & A_0 & A_1 \\ A_3 & A_2 & -A_1 & A_0 \end{pmatrix}.$$

According to (A1) the unitary operators $\tilde{U}_g := \mathbb{D}U_g\mathbb{D}^{-1}$ generate *-automorphisms of C*-algebras $\tilde{U}$ and $\tilde{Z}$. Consequently, the algebra $\tilde{\mathfrak{B}} := \mathbb{D}\mathfrak{B}\mathbb{D}^{-1}$ can be represented in the form $\tilde{\mathfrak{B}} = C^*(\tilde{\mathfrak{A}}, \tilde{U}_G)$, and assumptions (A1)-(A3) (or (A1), (A2), (A4)) are fulfilled for one.

As a result, we can apply Theorem 1 (or Theorem 2) for studying the invertibility of nonlocal quaternionic operators $B \in \mathfrak{B}$ or equivalently operators $\tilde{B} \in \tilde{\mathfrak{B}}$.

4. C*-algebras of one-dimensional nonlocal quaternionic convolution type operators

Now we consider the case $m = 1$.

Let Γ be the unit circle S^1, let G be an amenable discrete group of the orientation-preserving homeomorphisms of Γ onto itself with piecewise-continuous derivatives, acting topologically freely on Γ, i.e. for any subset $F \subset G$ and any arc $\gamma \subset \Gamma$ there exists a point $t \in \gamma$ with pairwise disjointed points $g(t)(g \in F)$.

In the space $L_2(\Gamma)$ we consider the C*-algebra $\mathcal{N}$ of singular integral operators (SIO) generated by operators $aI(a \in PC_\Gamma)$ and S , where I is the identity operator, PC_Γ is the set of ruled functions (i.e. uniform limits of step functions) at Γ and

$$(S\phi)(t) = (\pi i)^{-1}\int_\Gamma (\tau - t)^{-1}\phi(\tau)d\tau, t \in \Gamma.$$

Let $\mathfrak{A}$ denote the C*-algebra of quaternionic SIO with ruled coefficients in the space $H = L_2(\Gamma, \mathbb{H}(\mathbb{C}))$ generated by all operators (2) with components $A_n \in \mathcal{N}(n = 0, 1, 2, 3)$. We consider also the unitary shift operators $U_g \in \mathfrak{L}(H), g \in G$, defined by

$$[U_g(f_0 + \sum_{k=1}^{3} f_k i_k)](t) = \mid g'(t) \mid^{1/2} (f_0[g(t)] + \sum_{k=1}^{3} f_k[g(t)]i_k), t \in \Gamma. \tag{3}$$

Then $\mathfrak{B} = C^*(\mathfrak{A}, U_g)$ is the C*-algebra of quaternionic SIO with shifts consisting of uniform limits of operators of the form $B = \sum A_g U_g$, where $A_g \in \mathfrak{A}$.

Let $\bar{\mathbb{R}} = [-\infty, +\infty]$. As is known [17], a symbol of an operator $A \in \mathcal{N}$ is the matrix-function $A(\cdot,\cdot) : \Gamma \times \bar{\mathbb{R}} \to \mathbb{C}^{2\times 2}$ defined for generators of the algebra $\mathcal{N}$ by the formulas

$$(aI)(t,x) = \begin{pmatrix} a(t+0) & 0 \\ 0 & a(t-0) \end{pmatrix}, S(t,x) = \begin{pmatrix} \tanh \pi x & (\cosh \pi x)^{-1} \\ (\cosh \pi x)^{-1} & -\tanh \pi x \end{pmatrix},$$

$(t, x) \in \Gamma \times \bar{\mathbb{R}}$, and extended by the homomorphism $A \to A(\cdot,\cdot)$ onto all operators $A \in \mathcal{N}$. Respectively, symbols of operators $A \in \mathfrak{A}$ are 8×8 matrix-functions $\tilde{A}(\cdot,\cdot)$

given for generators (2) of the algebra $\mathfrak{A}$ by the rule

$$\tilde{A}^{l,r}(t,x) = \begin{pmatrix} A_0(t,x) & -A_1(t,x) & -A_2(t,x) & -A_3(t,x) \\ A_1(t,x) & A_0(t,x) & \epsilon_l A_3(t,x) & \epsilon_r A_2(t,x) \\ A_2(t,x) & \epsilon_r A_3(t,x) & A_0(t,x) & \epsilon_l A_1(t,x) \\ A_3(t,x) & \epsilon_l A_2(t,x) & \epsilon_r A_1(t,x) & A_0(t,x) \end{pmatrix}, (t,x) \in \Gamma \times \bar{\mathbb{R}}, \tag{4}$$

where $-\epsilon_l = \epsilon_r = 1$ for A^l and $\epsilon_l = -\epsilon_r = 1$ for A^r.

To each point $t \in \Gamma$ and each operator $B = \sum A_g U_g \in \mathfrak{B}$ with $A_g \in \mathfrak{A}$ there correspond discrete operators $B_t^{\pm}$ acting in $l_2(G, \mathbb{C}^8)$ according to the rule

$$(B_t^{\pm} f)(h) = \sum_g \tilde{A}_g(h(t), \pm\infty) f(hg), h \in G,$$

where $(hg)(\cdot) = g[h(\cdot)]$.

Since group G acts topologically freely on Γ then we can obtain the following.

Lemma 1. For every $t \in \Gamma$ the mappings $B \to B_t^{\pm}$ can be extended to homomorphisms of $\mathfrak{B}$ into $\mathfrak{L}(l_2(G, \mathbb{C}^8))$ and

$$\| B_t^{\pm} \| \leq | B | := inf\{\| B + K \| : K \in \mathfrak{L}_0(H)\}.$$

Let Λ be the set of all G-orbits $G(t) = \{g(t) : g \in G\}$ of points $t \in \Gamma$ and let I_k stand for the kth-order identity matrix. We consider the homomorphisms $\psi_{\lambda,x} : \mathfrak{B} \to \mathfrak{L}(l_2(\lambda, \mathbb{C}^8)), \lambda \in \Lambda, x \in \mathbb{R}$, defined on generators $A \in \mathfrak{A}$ and $U_g (g \in G)$ of the algebra $\mathfrak{B}$ by

$$\psi_{\lambda,x}(A) = diag\{\tilde{A}(t,x)\}_{t \in \lambda}, \psi_{\lambda,x}(U_g) = diag\{I_4 \otimes u_{t,\tau}\}_{t,\tau \in \lambda},$$

$$u_{t,\tau} = \begin{cases} diag\{| g'(t+0) |^{ix}, | g'(t-0) |^{ix}\}, & \text{if } g(t) = \tau, \\ 0, & \text{if } g(t) \neq \tau. \end{cases}$$

Theorem 3. If G is an amenable discrete group of shifts acting topologically freely on Γ then an operator $B \in \mathfrak{B}$ is Fredholm (n-normal, d-normal) in H if and only if for all $t \in \Gamma$ the operators $B_t^{\pm}$ are invertible (left invertible, right invertible) in $l_2(G, \mathbb{C}^8)$ and for all $\lambda \in \Lambda$ and $x \in \mathbb{R}$ the operators $\psi_{\lambda,x}(B)$ are invertible (left invertible, right invertible) in $l_2(\lambda, \mathbb{C}^8)$.

Further, let $\tilde{\mathcal{N}}$ be the minimal extension of the C*-algebra $\mathcal{N}$ of SIO in $L_2(\Gamma)$ with ruled coefficients, invariant under transformations

$$A \to u_{h,\lambda} A u_{h,\lambda}^{-1} (h \in \mathbb{R}, \lambda \in \Gamma),$$

where $u_{h,\lambda} = exp[h(t+\lambda)/(t-\lambda)]$ is continuous on $\Gamma \backslash \{\lambda\}$ functions admitting discontinuity of periodic type at the point λ. Let $\mathfrak{M}$ be the set of orientation-preserving diffeomorphisms $g : \Gamma \to \Gamma$, possessing second Taylor derivatives, i.e. there exist the limits

$$2lim_{t \to \tau}[g(t) - g(\tau) - g'(\tau)(t-\tau)](t-\tau)^{-2}, \tau \in \Gamma.$$

A set of generators of the C*-algebra $\tilde{\mathcal{N}}$ is formed by the following operators

$$aI (a \in PC_\Gamma), S_{h,\lambda} = u_{h,\lambda} S u_{h,\lambda}^{-1} (h \in \mathbb{R}, \lambda \in \Gamma).$$

Besides SIO and convolution type operators with kernels of class L_1, transplanted from $\mathbb{R}$ to Γ with the help of a linear-fractional transformation, the C*-algebra $\tilde{\mathcal{N}}$ contains operators $aU_{\tilde{g}}^{-1}F^{-1}bFU_{\tilde{g}}$ (where $a \in PC_\Gamma, b \in PC_{\bar{\mathbb{R}}}, \tilde{g}(x) = g[(x-i)/(x+i)]$ and

$$(F\phi)(x) = (2\pi)^{-1/2}\int_{\mathbb{R}} e^{-ix\tau}\phi(\tau)d\tau, (U_{\tilde{g}}\phi)(x) = \mid \tilde{g}'(x) \mid^{1/2} \phi[\tilde{g}(x)], x \in \mathbb{R}, g \in \mathfrak{M},$$

sums of products of these operators and their uniform limits.

Let $\dot{\mathbb{R}} = \mathbb{R} \cup \{\infty\}, Q = (\dot{\mathbb{R}} \times \bar{\mathbb{R}})\backslash(\{\infty\} \times \mathbb{R})$ and $Y = \Gamma \times Q$. To each generator A of the C*-algebra $\tilde{\mathcal{N}}$ we assign the symbol $A(\cdot,\cdot,\cdot) : Y \to \mathbb{C}^{2\times 2}$ by the rule

$$(aI)(t,x,\xi) = diag\{a(t-0), a(t+0)\},$$

$$S_{h,\lambda}(t,x,\xi) = \begin{cases} I_2, & \text{if } x > h(1 - sign \mid t - \lambda \mid), \\ \begin{pmatrix} \tanh \pi\xi & (\cosh \pi\xi)^{-1} \\ (\cosh \pi\xi)^{-1} & -\tanh \pi\xi \end{pmatrix}, & \text{if } x = h(1 - sign \mid t - \lambda \mid), \\ -I_2, & \text{if } x < h(1 - sign \mid t - \lambda \mid). \end{cases}$$

According to [12] the mapping $A \to A(\cdot,\cdot,\cdot)$ can be extended to *-homomorphism of the C*-algebra $\tilde{\mathcal{N}}$ onto some C*-algebra of 2×2 matrix-functions.

Further we consider the C*-subalgebra $\mathfrak{A} \subset \mathfrak{L}(H)$ generated by quaternionic operators (2) with $A_n \in \tilde{\mathcal{N}} (n = 0,1,2,3)$. By analogy with (4) we construct symbols $\tilde{A} : Y \to \mathbb{C}^{8\times 8}$ of operators $A \in \mathfrak{A}$. Let $\mathfrak{B}, \mathfrak{C}, \mathfrak{D}$ be a C*-algebras of $\mathfrak{L}(H)$ generated respectively by nonlocal quaternionic operators

$$B = \sum_g A_g U_g, C = \sum_{h,\lambda} A_{h,\lambda} u_{h,\lambda} I, D = \sum_g C_g U_g,$$

where $A_g, A_{h,\lambda} \in \mathfrak{A}, C_g \in \mathfrak{C}; g, h$ and λ run through finite subsets of $G, \mathbb{R}$ and Γ, G is an amenable discrete group of shifts $g \in \mathfrak{M}$, and U_g defined by (3). It is obvious that C*-algebras $\mathfrak{C}$ and $\mathfrak{D}$ coincide respectively with an algebra of SIO with coefficients admitting discontinuities of semi-almost-periodic type and with an algebra of such SIO with shifts.

For each $(t,x,\xi) \in Y$ let $\pi_{t,x,\xi}$ be the representation of $\mathfrak{B}$ in $l_2(G,\mathbb{C}^8)$ defined by

$$[\pi_{t,x,\xi}(\sum_s A_s U_s)f](g) = \sum_s \tilde{A}_s(g(t), \mid g'(t) \mid x, \xi) f(gs), g \in G.$$

Theorem 4. If G is an amenable discrete group of shifts of class $\mathfrak{M}$, acting freely on Γ (i.e. each shift $g \in G\backslash\{e\}$ hasn't fixed points on Γ), then an operator $B \in \mathfrak{B}$ is Fredholm (n-normal, d-normal) in H if and only if the operators $\pi_{t,x,\xi}(B)$ are invertible (left invertible, right invertible) in $l_2(G,\mathbb{C}^8)$.

The conditions of Theorem 4 are satisfied, for example, if G is a group of rotations of Γ.

If shifts $g \in G$ have fixed points on Γ then situation is complicated: the space of representations becomes inhomogeneous.

Let for simplicity G be an amenable group of shifts of class $\mathfrak{M}$ with a finite set $\Lambda \subset \Gamma$ of common fixed points, acting freely on $\Gamma\backslash\Lambda$. For $\lambda \in \Lambda$ the set $G_\lambda = \{g'(\lambda) :$

$g \in G\}$ is group with respect to multiplication, acting freely on $\mathbb{R}_+ = (0, +\infty)$ by the rule $x \to kx (k \in G_\lambda)$. Let $\Lambda_0 = \Lambda \backslash \Lambda_1, \Lambda_1 = \{\lambda \in \Lambda : g'(\lambda) = 1, \forall g \in G\}$ and

$$h_g(\lambda) = \frac{-i}{2\lambda g'(\lambda)} lim_{t \to \lambda} \left(g'(\lambda) \frac{g(t) + \lambda}{t + \lambda} - \frac{g(t) - \lambda}{t - \lambda} \right) (g(t) - \lambda)^{-1}, \lambda \in \Lambda.$$

Theorem 5. Under above mentioned conditions, an operator $B \in \mathfrak{B}$ is Fredholm (n-normal, d-normal) in H if and only if the following operators are invertible (left invertible, right invertible): $\pi_{t,x,\xi}(B)$ for all $(t, x, \xi) \in (\Gamma \backslash \Lambda) \times Q$, $\pi^0_{\lambda,x,\xi}(B)$ for all $\lambda \in \Lambda_0$ and $(\pm x, \xi) \in \mathbb{R}_+ \times \bar{\mathbb{R}}$, and $\pi^1_{\lambda,x,\xi}(B)$ for all $(\lambda, x, \xi) \in (\Lambda_0 \times \{0\} \times \mathbb{R}) \cup (\Lambda_1 \times \mathbb{R} \times \bar{\mathbb{R}})$, where $\pi^0_{\lambda,x,\xi}$ and $\pi^1_{\lambda,x,\xi}$ are representations of $\mathfrak{B}$ in $l_2(G_\lambda, \mathbb{C}^8)$ and $\mathbb{C}^8$, respectively, defined by

$$[\pi^0_{\lambda,x,\xi}(\sum_s A_s U_s) f](k) = \sum_s \tilde{A}_s(\lambda, kx, \xi) exp[iks'(\lambda) h_s(\lambda) x] f(ks'(\lambda)), k \in G_\lambda,$$

$$\pi^1_{\lambda,x,\xi}(\sum_s A_s U_s) f = \sum_s \tilde{A}_s(\lambda, x, \xi) exp[i(-\xi lns'(\lambda) + x h_s(\lambda))] f$$

(we put $\xi lns = 0$ for $\xi = \pm\infty, s = 1$).

Theorem 6. An operator $C \in \mathfrak{C}$ is Fredholm (n-normal, d-normal) in H if and only if for all $(t, \xi) \in \Gamma \times \bar{\mathbb{R}}$ the operators $\tilde{\pi}_{t,\xi}(C)$ are invertible (left invertible, right invertible) in $l_2(\mathbb{R}, \mathbb{C}^8)$, where $\tilde{\pi}_{t,\xi}$ are the representations of $\mathfrak{C}$ in $l_2(\mathbb{R}, \mathbb{C}^8)$, defined by

$$[\tilde{\pi}_{t,\xi}(\sum_{h,\lambda} A_{h,\lambda} u_{h,\lambda} I) f](x) = \sum_h \tilde{\tilde{A}}_h(t, x, \xi) f(x - h), x \in \mathbb{R},$$

where

$$\tilde{\tilde{A}}_0(t, x, \xi) = \tilde{A}_{0,t}(t, x, \xi) + \sum_h \sum_{\lambda \neq t} \tilde{A}_{h,\lambda}(t, x, \xi) u_{h,\lambda}(t), \tilde{\tilde{A}}_h(t, x, \xi) = \tilde{A}_{h,t}(t, x, \xi), h \neq 0. \tag{5}$$

We recall that the C*-algebra $\mathfrak{D}$ is generated by operators

$$K = \sum_{h,\lambda,s} A_{h,\lambda,s} u_{h,\lambda} U_s (A_{h,\lambda} \in \mathfrak{A}, h \in \mathbb{R}, \lambda \in \Gamma, s \in G).$$

Theorem 7. If the hypotheses of Theorem 4 are satisfied then an operator $D \in \mathfrak{D}$ is Fredholm (n-normal, d-normal) in H if and only if for all $(t, \xi) \in \Gamma \times \bar{\mathbb{R}}$ the operators $\pi_{t,\xi}(D)$ are invertible (left invertible, right invertible) in $l_2(G \times \mathbb{R}, \mathbb{C}^8)$, where $\pi_{t,\xi}$ are representations of $\mathfrak{D}$ in $l_2(G \times \mathbb{R}, \mathbb{C}^8)$ having the form

$$[\pi_{t,\xi}(K) f](g, x) = \sum_{h,s} \tilde{\tilde{A}}_{h,s}(g(t), \mid g'(t) \mid x, \xi) e^{ih_s(t)} f(gs, x - \mid g'(t) \mid^{-1} h),$$

$(g, x) \in G \times \mathbb{R}$, and the $\tilde{\tilde{A}}_{h,s}$ are obtained from $\tilde{\tilde{A}}_h$ by replacing $\tilde{A}_{h,\lambda}$ by $\tilde{A}_{h,\lambda,s}$ in (5).

5. C*-algebras of multidimensional nonlocal quaternionic convolution type operators

Let $m \geq 2$. We consider the C*-algebra $\mathcal{N}$ of multidimensional SIO of Michlin-Calderon-Zygmund type along the unit sphere $\Gamma = S^m$ in the Hilbert space $L_2(\Gamma)$.

Let $\eta = w(x)$ be the stereographic projection, i.e. the transformation of $\dot{\mathbb{R}}^m$ onto unit sphere $S^m \subset \mathbb{R}^{m+1}$ defined by the formulas

$$\eta_k = \frac{2x_k}{|x|^2+1}(k=1,\ldots,m), \eta_{m+1} = \frac{|x|^2-1}{|x|^2+1},$$

where $|x|^2 = x_1^2 + \ldots + x_m^2$. Then the operator $V : L_2(S^m) \to L_2(\mathbb{R}^m)$ defined by

$$(V\phi)(x) = \left(\frac{2}{|x|^2+1}\right)^{m/2} \phi[w(x)], x \in \mathbb{R}^m,$$

provides a unitary equivalence of an operator $A \in \mathcal{N}$ with some m-dimensional SIO $VAV^{-1} \in \mathfrak{L}(L_2(\mathbb{R}^m))$.

Let $\dot{\mathbb{R}}^m = \mathbb{R}^m \cup \{\infty\}$ be the one-point compactification of $\mathbb{R}^m$, let $H(C(S^{m-1}))$ be the algebra of zero degree homogeneous functions on $\mathbb{R}^m$ with continuous restrictions on S^{m-1} and

$$(F\phi)(x) = (2\pi)^{-m/2} \int_{\mathbb{R}^m} e^{-ix\cdot\tau}\phi(\tau)d\tau, x \in \mathbb{R}^m.$$

Further we assume that $\mathcal{N} = V^{-1}\Psi(\mathcal{A},\mathcal{B})V$, where $\Psi(\mathcal{A},\mathcal{B})$ is the C*-subalgebra of $\mathfrak{L}(L_2(\mathbb{R}^m))$ generated by operators $aF^{-1}bF(a \in \mathcal{A}, b \in \mathcal{B})$ and $\mathcal{A} = C(\dot{\mathbb{R}}^m), \mathcal{B} = H(C(S^{m-1}))$. Then according to [18] the C*-algebra of symbols of operators $A \in \mathcal{N}$ coincides with $C(\dot{\mathbb{R}}^m \times S^{m-1})$, where a generator $A = V^{-1}aF^{-1}bFV$ of the algebra $\mathcal{N}$ has the symbol $A(x,\eta) = a(x)b(\eta), (x,\eta) \in \dot{\mathbb{R}}^m \times S^{m-1}$.

Let $H = L_2(\Gamma, \mathbb{H}(\mathbb{C}))$. As usually, we consider the C*-algebra $\mathfrak{A} \subset \mathfrak{L}(H)$ generated by quaternionic operators (2) with components $A_n \in \mathcal{N}(n = 0,1,2,3)$. Then 4×4 matrix-functions $\tilde{A} \in C(\dot{\mathbb{R}}^m \times S^{m-1}, \mathbb{C}^{4\times4})$ defined for generators (2) by the formulas (4) are the symbols of operators $A \in \mathfrak{A}$.

Let G be an amenable discrete group of diffeomorphisms of $\Gamma = S^m$ onto itself with Hőlder derivatives and fixed point (0,...,0,1), in which every diffeomorphism $g \in G$ has the identity Jacobi matrix. For each $g \in G$ we consider the shift operator $U_g \in \mathfrak{L}(H)$ defined by

$$[U_g(f_0 + \sum_{k=1}^{3} f_k i_k)](t) = |J_g(t)|^{1/2} (f_0[g(t)] + \sum_{k=1}^{3} f_k[g(t)]i_k), t \in \Gamma,$$

where $J_g(t)$ is a Jacobi matrix of the mapping g at the point t, $|J_g(t)| := |detJ_g(t)|$. According to [18] for every $g \in G$ and every $A \in \mathfrak{A}$ the operator $N = U_gAU_g^{-1} \in \mathfrak{A}$ and its symbol has the form

$$\tilde{N}(x,\eta) = \tilde{A}(\tilde{g}(x), ({}^TJ_{\tilde{g}}^{-1}(x))\eta), (x,\eta) \in \dot{\mathbb{R}}^m \times S^{m-1},$$

where $\tilde{g} = w^{-1} \circ g \circ w : \dot{\mathbb{R}}^m \to \dot{\mathbb{R}}^m$ and ${}^TJ_{\tilde{g}}^{-1}(x)$ is the inverse matrix to transposed Jacobi matrix $J_{\tilde{g}}(x)$. Consequently, the group G induces the action of one on $\dot{\mathbb{R}}^m \times$

S^{m-1} by the rule $g:(x,\eta)\to(\tilde{g}(x),({}^TJ_{\tilde{g}}^{-1}(x))\eta/\parallel({}^TJ_{\tilde{g}}^{-1}(x))\eta\parallel)$. Further we assume that G acts topologically freely on $\dot{\mathbb{R}}^m\times S^{m-1}$. Let $\pi_{x,\eta}$ be a representation of the C*-algebra $\mathfrak{B}=C^*(\mathfrak{A},U_G)$ in $l_2(G,\mathbb{C}^4)$ defined by

$$[\pi_{x,\eta}(\sum_s A_sU_s)f](g)=\sum_s\tilde{A}_s(\tilde{g}(x),({}^TJ_{\tilde{g}}^{-1}(x))\eta)f(gs),g\in G.$$

Theorem 8. Under above mentioned conditions an operator $B\in\mathfrak{B}$ is Fredholm (n-normal, d-normal) in H if and only if the operators $\pi_{x,\eta}(B)$ are invertible (left invertible, right invertible) in $l_2(G,\mathbb{C}^4)$ for all $x,\eta\in\dot{\mathbb{R}}^m\times S^{m-1}$.

References

1. F.Brackx, R.Delanghe, F.Sommen, *Clifford Analysis*, Boston, Pitman, 1982.
2. R.Delange, F.Sommen, V.Souček, *Clifford Analysis and Spinor Valued Functions: Function Theory for the Dirac Operator*, Dordrecht, Klewer Acad. Pub., 1992.
3. K.Gürlebeck, W.Sprössig, *Quaternionic Analysis and Elliptic Boundary Value Problem*, Berlin, Akademie-Verlag, 1989.
4. M.V.Shapiro, N.L.Vasilevski, *Quaternionionic ψ- hyperholomorphic Functions, Singular Operators with Quaternionic Caushy Kernel and Analogues of the Riemann Boundary Value Problem*, Reporto Interno #102, Departamento de Matematicas, CINVESTAV del I.P.N., Mexico, 1992.
5. N.L.Vasilevski, *On an Algebra Generated by Abstract Singular Operators and Shift Operator*, Reporto Interno #105, Departamento de Matematicas, CINVESTAV del I.P.N., Mexico, 1992.
6. A.B.Antonevich, *Linear Functional Equations: The Operator Approach*, Minsk, Univ. Press, 1988 (Russian).
7. A.P.Soldatov, *One-dimensional Singular Operators and Boundary Value Problems of Function Theory*, Moscow, Vyssh. Shkola, 1991 (Russian).
8. Yu.I.Karlovich, V.G.Kravchenko, *Systems of Singular Integral Equations with a Shift*, Math. USSR Sbornik, 44(1983), 75- 95.
9. Yu.I.Karlovich, V.G.Kravchenko, *An Algebra of Singular Integral Operators with Piecewise-continuous Coefficients and a Piecewise-smooth shift on a composite contour*, Math. USSR Izvestiya, **23**(1984), 307-352.
10. Yu.I.Karlovich, I.M.Spitkovskii, *Factorization of Almost Periodic Matrix-valued Functions and the Noether Theory for Certain Classes of Equations of Convolution Type*, Math. USSR Izvestiya, **34** (1990), 281-316.
11. Yu.I.Karlovich, *The Local-trajectory Method of Studying Invertibility in C*-algebras of Operators with Discrete Groups of Shifts*, Soviet Math. Dokl., **37** (1988), 407-412.
12. Yu.I.Karlovich, *C*-algebras of Operators of Convolution Type with Discrete Groups of Shifts and Oscillating Coefficients*, Soviet Math. Dokl., **38** (1989), 301-307.
13. Yu.I.Karlovich, V.G.Kravchenko, G.S.Litvinchuk, *On Noethericity and Mikhlin Symbols of Operators of the Type of Singular Integral Operators with Shift*, Z. Analysis Anw., **9** (1990), 15-32.
14. I.B.Simonenko, *A new general method of studying linear operator equations of the type of singular integral equations*,I,II, Izv. Akad. Nauk SSSR. Ser. Mat., **29** (1965), 567-586, 757-782 (Russian).
15. R.G.Douglas, *Banach Algebra Technique in Operator Theory*, New York and London, Academic Press, 1972.
16. F.Greenleaf, *Invariant Means on Topological Groups and Their Applications*, Van Nostr. Reinhold Co., New York, 1969.
17. I.Ts.Gokhberg, N.Ya.Krupnik, *On the algebra generated by the one-dimensional singular integral operators with piecewise continuous coefficients*, Funktsional. Anal. i Prilozhen.4 (1970), no. 3, 26-36 (Russian).
18. S.G.Michlin, S.Prössdorf, *Singuläre Integraloperatoren*, Akademie-Verlag, Berlin, 1980.

ON THE SOLUTONS OF $D^N \hat{D}^M F = 0$

B. KLEIN OBBINK
Eindhoven University of Technology

Abstract

We show that the general solution of the equation $D^n \hat{D}^m f = 0$ on a domain $\Omega \subset R^3$, with D the Cauchy–Riemann operator, $n, m \in N$ and f a $C\ell_{0,2}$-valued function, can be expressed as a sum of products of polynomials and left and anti–left monogenic functions. Thereby, our key assumption is that Ω is special x-normal. As an application of this result, we show how the general solution of Stokes' equations in three dimensions can be represented by two left monogenic functions.

1. Introduction and Summary

It is known from classical function theory that for each holomorphic function on a simply connected domain $\Omega \subset R^2 \simeq C$, there exists a primitive; i.e. given a function f satisfying $\partial_{\bar{z}} f = 0$, there is a function F satisfying:

$$\partial_z F = f \ , \quad \partial_{\bar{z}} F = 0 \ . \tag{1a,b}$$

In the attempt to generalize this result for Clifford analysis, one may pose the following problem: Find all domains $\Omega \subset R^3$ such that every left monogenic function on it has a left monogenic primitive; i.e. all those domains such that there exists a function F satisfying:

$$\hat{D} F = f \ , \quad DF = 0 \ , \tag{2a,b}$$

given a function f on Ω satisfying $Df = 0$. In section 2, we will give both a necessary condition and a sufficient condition on Ω for it to posses this property. It is remarked that de Rham's theorem can not be used as primitives in the sense of relation (2a) correspond to line integrals while Cauchy's theorem corresponds to surface integrals. The conditions on Ω are not purely homological. In the literature ([1],[2]), one usually considers equation (2a) on its own.
From the classical result for the existence of a primitive of an analytical function, one can show that the solution of the equation:

$$\partial_{\bar{z}}^n \partial_z^m f = 0 \ , \qquad n, m \in N \times N / \{(0,0)\} \tag{3}$$

on simply connected domains can be expressed in terms of holomorphic and anti–holomorphic functions ([3],[6]). In section 3, we generalize this result, solving:

F. Brackx et al. (eds.), Clifford Algebras and their Applications in Mathematical Physics, 119–127.

$$D^n\hat{D}^m f = 0\ , \qquad n,m \in N \times N/\{(0,0)\}\ . \tag{4}$$

Here ^ denotes the grade involution (Notice that although the underlying algebra $C\ell_{0,2}$ is isomorphic to H, this operation has not an analogue in H).
In section 4, it will be shown that Stokes' equations on certain domains $\Omega \subset R^3$ can be rewritten as a special case of equation (4). This will lead to a representation of the general solution of Stokes' equations in terms of left monogenic functions, thus generalizing the Kolosov–Muskhelishvili method ([4],[5]).

2. Left monogenic primitives

Definitions. A domain $\Omega \subset R^3$ is called x-normal if there is a plane $V : \{(x,y,z) \in R^3 \mid x = a\}$ such that every line segment connecting a point in Ω with its perpendicular projection on V lies entirely in V. A domain $\Omega \subset R^3$ is called special x-normal if it is x-normal with respect to a plane V such that $V \cap \Omega$ is simply connected.

Proposition 1. Let f be a left monogenic function on a special x-normal domain $\Omega \subset R^3$. Then there exists a left monogenic function F such that:

$$\hat{D}F = f\ . \tag{5}$$

Proof. Let f be decomposed as follows:

$$f = p + q_1 e_1 + q_2 e_2 + r\iota \tag{6}$$

where ι denotes $e_1 e_2 \in C\ell^2_{0,2}$. It is known that the functions p, q_1, q_2 and r are very smooth (real analytical). We prove the existence of functions P, Q_1, Q_2 and R such that:

$$F = \tfrac{1}{2}(P + Q_1 e_1 + Q_2 e_2 + R\iota) \tag{7}$$

satisfies:

$$DF = 0\ , \quad \hat{D}F = f\ . \tag{8a,b}$$

Let $V : x = a$ be the plane as mentioned in the definitions. Consider the following Poisson equation for a function f_2 on $V \cap \Omega$:

$$\Delta f_2 = f_{2,yy}(y,z) + f_{2,zz}(y,z) = (p,z - r,y)|_{(a,y,z)}\ . \tag{9}$$

As the right hand side is smooth, there does exist a smooth solution f_2. Next, consider the following set of partial differential equations for a function f_1 on $V \cap \Omega$:

$$f_{1,y}(y,z) = p(a,y,z) - f_{2,z}(y,z) \tag{10a}$$

$$f_{1,z}(y,z) = r(a,y,z) - f_{2,y}(y,z)\ . \tag{10b}$$

By the method of characteristics one proves that there does exist a smooth solution f_1 as $V \cap \Omega$ is simply connected and as the compatibility condition is satisfied.
As Ω is x-normal, the following functions are well defined:

$$Q_i(x,y,z) := f_i(y,z) + \int_a^x q_i(\xi, y, z) d\xi \,, \qquad i = 1,2 \,. \tag{11}$$

Using the generalized Cauchy–Riemann equations, one checks:

$$Q_{1,x} = q_1 \,, \quad Q_{2,x} = q_2 \tag{12a,b}$$

$$Q_{1,y} + Q_{2,z} = p \,, \quad -Q_{2,y} + Q_{1,z} = r \,. \tag{12c,d}$$

By the same method, one constructs functions P and R satisfying:

$$P,x = p \,, \quad R,x = r \tag{13a,b}$$

$$P,y + R,z = -q_1 \,, \quad -R,y + P,z = -q_2 \,. \tag{13c,d}$$

The relations (6), (7), (12) and (13) imply the relations (8a,b). □

One easily proves the following assertion: If a x-normal domain $\Omega \subset R^3$ is such that every left monogenic function on it has a left monogenic primitive, then Ω is special x-normal. The next proposition shows why the domain Ω is required to be x-normal.

Definition. A domain $\Omega \subset R^3$ is called x-convex if every line segment connecting two points in Ω with equal y- and z-coordinates lies entirely in Ω.

Proposition 2. Let $\Omega \subset R^3$ be a domain such that every left monogenic funcion on Ω has a left monogenic primitive. Then Ω is x-convex.

Proof. We will prove the proposition by showing that the assumption that Ω is not x-convex but does have the stated property leads to a contradiction. Without any loss of generality, we may assume that $(x_1,0,0), (x_2,0,0) \in \Omega$, with $x_1 > 0$, $x_2 < 0$, while $O = (0,0,0) \notin \Omega$.
The function

$$e(x,y,z) = -2\,\frac{x - ye_1 - ze_2}{r^3} \,, \qquad (x,y,z) \in \Omega \tag{14}$$

with $r = (x^2+y^2+z^2)^{\frac{1}{2}}$, is a properly defined, left monogenic function on Ω. By assumption, there exists a function

$$E = P + Q_1 e_1 + Q_2 e_2 + R\iota \tag{15}$$

such that:

$$\hat{D}E = e \,, \quad DE = 0 \,. \tag{16a,b}$$

Among other things, this implies the existence of a function $Q = Q_1(x,y,z)$ such that:

$$Q,x = \frac{y}{r^3} \,, \quad \Delta Q = 0 \,. \tag{17}$$

On Ω minus the x-axis, Q can be written as:

$$Q(x,y,z) = \frac{xy}{(y^2+z^2)r} + f(x,y,z) \ . \tag{18}$$

Substituting this equation into the equation (17) leads to:

$$f,_x = 0 \ , \quad \Delta f = 0 \ . \tag{19}$$

One argues that f is a function of the variables y and z only, using the fact that the harmonic extention of a harmonic function is uniquely determined. Therefore we write $f = f(x,y,z) = f(y,z)$.
As the function Q is harmonic on the whole domain Ω, it is continuous in some neighbourhoods of the points $(x_1, 0, 0)$ and $(x_2, 0, 0)$. Therefore, the following limits converge:

$$\lim_{y\to 0} Q(x_1, y, 0) = \lim_{y\to 0} \ 1/y + f(y,0) \tag{20a}$$

$$\lim_{y\to 0} Q(x_2, y, 0) = \lim_{y\to 0} \ -1/y + f(y,0) \tag{20b}$$

As the existence of one limit contradicts the existence of the other, the assumption that Ω is not x-convex must be false. □

Exercise: Find out why a similar reasoning does not hold in the two dimensional case.

3. The equation $D^n \hat{D}^m f = 0$

Using proposition 1, it is possible to express the general solution of the equation $D^n \hat{D}^m f = 0$ in terms of left monogenic and anti–left monogenic functions. Thereby, an anti–left monogenic function f is defined as a solution of the equation:

$$\hat{D} f = 0 \ . \tag{21}$$

Notice that the grade involution of an anti–left monogenic function is left monogenic.

Proposition 3. Let n and m be two integers not both equal to zero, let $\Omega \subset R^3$ be a special x-normal domain and let f be a $C\ell_{0,2}$-valued $(n+m)$-continuously differentiable function on Ω satisfying:

$$D^n \hat{D}^m f = 0 \ . \tag{22}$$

Then there exist left monogenic functions g_k, $k = 1, \dots, n$ and anti–left monogenic functions h_k, $k = 1, \dots, m$ such that:

$$f = \sum_{k=1}^{n} x^{k-1} g_k + \sum_{k=1}^{m} x^{k-1} h_k \ . \tag{23}$$

Proof. The statements for $n=1$, $m=0$ and $n=0$, $m=1$ are trivial. The statement for an arbitrary pair n, m can then be proven by two inductions. Only the following step is proven: if the statements holds for a certain pair n, m then it holds also for the pair $n, m+1$, the other induction step goes similarly.
Consider the equation:

$$D^n \hat{D}^{m+1} f = (D^n \hat{D}^m)(\hat{D} f) = 0 . \tag{24}$$

There exist left monogenic function g_k, $k = 1, \ldots, n$ and anti-left monogenic functions h_k, $k = 1, \ldots, m$ such that:

$$\hat{D} f = \sum_{k=1}^{n} x^{k-1} g_k + \sum_{k=1}^{m} x^{k-1} h_k . \tag{25}$$

Next, consider the following n coupled partial differential equations:

$$\hat{D} G_n = g_n \tag{26a}$$

$$\hat{D} G_i + i G_{i+1} = g_i , \qquad i = 1, \ldots, n-1 . \tag{26b}$$

With the aid of proposition 1, one argues that there is a set of left monogenic functions G_k, $k = 1, \ldots, n$ satisfying the equations (26). One then checks that the well–defined function

$$F := \sum_{k=1}^{n} x^{k-1} G_k + \sum_{k=2}^{m+1} \frac{x^{k-1}}{k-1} h_{k-1} \tag{27}$$

is a particular solution of equation (25):

$$\hat{D} F = \sum_{k=1}^{n} \{(\hat{D} x^{k-1}) G_k + x^{k-1} \hat{D} G_k\} + \sum_{k=2}^{m+1} \left\{ \left(\hat{D} \frac{x^{k-1}}{k-1}\right) h_{k-1} + \frac{x^{k-1}}{k-1} \hat{D} h_{k-1} \right\} =$$

$$\sum_{k=1}^{n} \{(k-1) x^{k-2} G_k + x^{k-1} \hat{D} G_k\} + \sum_{k=2}^{m+1} x^{k-2} h_{k-1} =$$

$$\sum_{k=1}^{n-1} \{k x^{k-1} G_{k+1} + x^{k-1} \hat{D} G_k\} + x^{n-1} \hat{D} G_n + \sum_{k=2}^{m+1} x^{k-2} h_{k-1} =$$

$$\sum_{k=1}^{n} x^{k-1} g_k + \sum_{k=1}^{m} x^{k-1} h_k . \tag{28}$$

Hence:

$$\hat{D}(f - F) = 0 \tag{29}$$

and there exists an anti–left monogenic function h_0 on Ω such that:

$$f = F + h_0 \ . \tag{30}$$

From the indentities (27) and (30), the relation (23) follows for the pair $n, m+1$. □

Remarks.

- The reverse of this proposition also holds: if f is a function which can be written in the form of relation (24), then it satisfies equation (23). This is easily proved by a straightforward calculation, using the fact that every (anti-)left monogenic function on a domain Ω is in $C^\infty(\Omega)$.

- Proposition 3 states for the case $n=m=1$ that a $C\ell_{0,2}$-valued harmonic function f can be written as the sum of a left monogenic function g and an anti–left monogenic function h. If f is real (i.e. $C\ell_{0,2}^0$-valued), it is the scalar part of a left monogenic function:

$$f = (g + h)_0 = (g + \hat{h})_0 \ . \tag{31}$$

  Compare this to proposition 8.6 of [1], where it is shown that every real harmonic function on a star–shaped domain can be written as the scalar part of a left monogenic function.

- The proposition does not state to what extent the functions h_k, $k = 1, \dots, n$ and g_k, $k = 1, \dots, m$ are determined, given the function f. To find out, one uses the next proposition which proof can be found in [6].

Proposition 4. Let m be a positive integer. Let $\Omega \subset R^3$ be a x-normal domain and let f be a left monogenic function on Ω satisfying:

$$\hat{D}^m f = 0 \ . \tag{32}$$

Then there exist $C\ell_{0,2}^+$-valued functions s_j and p_j, $j = 1, \dots, m$, anti–holomorphic with respect to the variable $w = y + \iota z$ such that:

$$f = \sum_{j=1}^{m} \sum_{k=1}^{j} \frac{x^{j-k}}{(j-k)!} \left(\frac{e_1}{2}\right)^{k-1} \frac{w^{[(k-1)/2]}}{[(k-1)/2]!} \left\{ s_j^{([j/2]-[k/2])} + t_j^{([j/2]-[k/2])} e_1 \right\} , \tag{33}$$

where $[\ell]$ denotes the entier of a number ℓ and where $q^{(n)}$, $n \in N$ denotes $\partial_{\overline{w}}^n q$ for an anti–holomorphic function q.

4. Stokes' equations

Let p and $\underline{v} = (v_1, v_2, v_3)$ satisfy Stokes' equations on a domain $\Omega \subset R^3$:

$$\Delta \underline{v} = \nabla p \tag{34}$$

$$\nabla \cdot \underline{v} = 0 \ . \tag{35}$$

If Ω is assumed to be 2-connected (i.e. $H_2(\Omega)$ is trivial), there exists a vector field $\underline{\psi}$ such that:

$$\underline{v} = \nabla \wedge \underline{\psi} . \tag{36}$$

If $\underline{\psi}$ satisfies this relation, then also does $\underline{\tilde{\psi}} = \underline{\psi} + \nabla f$, and one may put:

$$\nabla \cdot \underline{\psi} = 0 \tag{37}$$

by solving $\Delta f = -\nabla \cdot \underline{\psi}$ and omitting the tilde. As p is harmonic, it is smooth and there exists a function ϕ such that:

$$p = -\Delta \phi . \tag{38}$$

Now, let a vector $\underline{w} = (w_1, w_2, w_3)$ be identified with the Clifford number

$$\underline{w} = w_1 + w_2 e_1 + w_3 e_2 . \tag{39}$$

Defining

$$F := \phi + \iota \underline{\psi} , \tag{40}$$

the relations (36), (38) and (34) then become respectively:

$$\underline{v} = D \iota \underline{\psi} \tag{41a}$$

$$p = -D\hat{D}\psi \tag{41b}$$

$$DD\hat{D}F = 0 . \tag{41c}$$

If Ω is assumed to be special x-normal (and hence 2-connected), proposition 3 states that there exist a left monogenic function θ and a $C\ell_{0,2}$ valued harmonic function τ such that:

$$F = x\theta + \tau . \tag{42}$$

If we perform what physicists call a gauge transformation:

$$\phi \rightarrow \tilde{\phi} = \phi + g \tag{43a}$$

$$\underline{\psi} \rightarrow \underline{\tilde{\psi}} = \underline{\psi} + Df \tag{43b}$$

with f and g arbitrary real harmonic functions on Ω, the function F transforms as:

$$F \rightarrow \tilde{F} = \tilde{\phi} + \iota \underline{\tilde{\psi}} = x\theta + \tau + g + \iota D f . \tag{44}$$

The next proposition states that f and g can be "gauged" such that

$$\chi = \tau + g + \iota D f \tag{45}$$

is an anti–left monogenic function.

Proposition 5. Let τ be a $C\ell_{0,2}$-valued harmonic function on a special x-normal domain $\Omega \subset R^3$. Then there exist real harmonic functions f and g such that:

$$\hat{D}(\tau + g + \iota Df) = 0 \,. \tag{46}$$

Proof. Decompose τ as follows:

$$\tau = \tau_0 + t_1 e_1 + \tau_2 e_2 + \tau_{12}\iota \,. \tag{47}$$

Let $V : x = a$ be a plane as before (that is, Ω is special x-normal with respect to V) and consider the following Poisson equations for g_1 and F_1 on $\Omega \cap V$:

$$\Delta g_1(y,z) = (\tau_{0,xx} + \tau_{1,yx} + \tau_{2,zx})|_{(a,y,z)} \tag{48a}$$

$$\Delta F_1(y,z) = (\tau_{12,xx} - \tau_{2,yx} + \tau_{1,zx})|_{(a,y,z)} \,. \tag{48b}$$

As the terms at the right hand sides of the equations (48) are smooth, solutions g_1 and F_1 do exist.
Next define:

$$\sigma(y,z) := \sigma_1(y,z) + \iota\sigma_2(y,z) \tag{49a}$$

with:

$$\sigma_1(y,z) := (\tau_{1,x} - \tau_{0,y} - \tau_{12,z})|_{(a,y,z)} - g_{1,y}(y,z) - F_{1,z}(y,z) \tag{49b}$$

$$\sigma_2(y,z) := (\tau_{2,x} + \tau_{12,y} - \tau_{0,z})|_{(a,y,z)} - g_{1,z}(y,z) + F_{1,y}(y,z) \,. \tag{49c}$$

Using the harmonicity of τ and the relations (48), one checks by a straightforward computation that σ is an anti–holomoprhic function with respect to the variable $w = y + \iota z$:

$$\partial_w \sigma = 0 \,. \tag{50}$$

As $V \cap \Omega$ is simply connected, there exists an anti–holomorphic function

$$h = 2(g_0 - \iota F_0) \tag{51}$$

such that:

$$\partial_{\overline{w}} h = \sigma \,.$$

Then it is checked that the functiuons f and g properly defined on Ω by:

$$g(x,y,z) = -\int_a^x (\tau_{0,\xi} + \tau_{1,y} + \tau_{2,z})|_{(\xi,y,z)} d\xi + g_1(y,z) + g_0(y,z) \tag{52a}$$

$$F(x,y,z) = \int_a^x (-\tau_{12,\xi} + \tau_{2,y} - \tau_{1,z})|_{(\xi,y,z)} d\xi + F_1(y,z) + F_0(y,z) \tag{52b}$$

satisfy:

$$\Delta g = 0 \,, \quad \Delta F = 0 \tag{53a,b}$$

$$g_{,x} = -(\tau_{0,x} + \tau_{1,y} + \tau_{2,z}) \tag{53c}$$

$$F_{,x} = -\tau_{12,x} + \tau_{2,y} - \tau_{1,z} \tag{53d}$$

$$g_{,y} + F_{,z} = \tau_{1,x} - \tau_{0,y} - \tau_{12,z} \tag{53e}$$

$$g_{,z} - F_{,y} = \tau_{2,x} + \tau_{12,y} - \tau_{0,z} \,. \tag{53f}$$

Finally, one constructs a harmonic function f on Ω satisfying:

$$f_{,x} = \tfrac{1}{2} F \tag{54}$$

and checks that the identities (53) and (54) are equivalent with identity (46). □

So we conclude that every solution of Stokes' equations on a special x-normal domain $\Omega \subset R^3$ can be represented by a function

$$F = x\theta + \chi \tag{55}$$

with θ left monogenic and χ anti–left monogenic (or $\hat{\chi}$ monogenic). By a straightforward calculation, one checks easily that such a function F indeed represents a solution of Stokes' equation if:

$$(\theta + \chi, x)_2 = 0 \tag{56}$$

where $(\ )_2$ denotes the bivectorial part.

The author likes to thank J. de Graaf for critically reading the manuscript.

References

1. Brackx, F., Delanghe, R., Sommen, F.; Clifford analysis, Research Notes in Math. Vol. 76, Pitman, London, 1982.

2. Gürlebeck, K., Sprössig, W.; Quaternionic Analysis and Elliptic Boundary Value Problems, Math. Research Vol. 56, Akademic–Verlag, Berlin, 1989.

3. Goursat, E.; Cours d'analyse mathématique Vol. III, 3rd ed. Gauthier–Villars, Paris, 1927.

4. Muskhelishvili, N.I.; Some basic problems in the math theory of elasticity (Eng. transl.; Noordhof, Groningen, 1963.

5. Pimenov, A.A., Pushkarev, V.I.; The use of quaternions to generalize the Kolosov–Muskhelishvili method... (Eng. transl.), J. Appl. Math. Mechs, Vol. 55, no. 3, pp. 343-347, 1991.

6. Klein Obbink, B.; On the solutions of $D^n \hat{D}^m f = 0$, Rana 93–07, Eindhoven (to be published).

BIREGULAR QUATERNIONIC FUNCTIONS

WIESŁAW KRÓLIKOWSKI
Institute of Mathematics, Polish Academy of Sciences
Łódź Branch, Narutowicza 56, PL-90-136, Łódź, Poland

and

R. MICHAEL PORTER
Department of Mathematics, Centro de Investigación y
de Estudios Avanzados del IPN
Apartado Postal 14-740, 07000 México, D.F., Mexico

Abstract. We examine regular functions of a quaternionic variable whose inverse is also regular. The most interesting case is when the inverse of a left regular function is right regular.

1. Introduction

The Cauchy-Riemann equations characterizing holomorphicity for a single complex variable may be easily generalized to quaternionic functions, as in [3], [9]; the corresponding functions may be referred to as "hyperholomorphic" or simply "regular". Regular functions express in many ways the spirit of complex analysis in the quaternionic context, as many classical results (e.g., Cauchy's integral formula, Morera's theorem, the Laurent expansion) carry over in a more or less natural way [11],[12].

In this note we present some observations on biregular functions, that is, invertible regular functions with regular inverse. We find that a more interesting concept is that of functions which are, for example, left regular with a right regular inverse; we call them LR-biregular mappings. Regular and LR-biregular functions can be on affine quaternionic manifolds; the manifold point of view also suggests that the notion of LR-biregularity is more natural than biregularity.

2. Regular Functions and Chain Rules

Since we will be looking at inverse functions, it will be relevant to make some comments about chain rules for quaternionic functions, which are interesting in their own right. We cite first a few well known facts, see [11],[12].

Consider a quaternion $q = x_0+ix_1+jx_2+kx_3 \in H$ where $x_\alpha \in R$, $\alpha = 0,1,2,3$, and where the standard quaternionic units satisfy $i^2 = j^2 = k^2 = ijk = -1$. Write $\partial_\alpha = \partial/\partial x_\alpha$ and form the following symbolic differential expressions:

$$D^+ = \frac{1}{4}(\partial_0 + i\partial_1 + j\partial_2 + k\partial_3)\ , \quad D = \frac{1}{4}(\partial_0 - i\partial_1 - j\partial_2 - k\partial_3)\ .$$

We say that a differentiable function $f = f^0+if^1+jf^2+kf^3 : \Omega \to H$ is *left regular* (or simply, *regular*) in the domain $\Omega \subseteq H$ if $D^+ \cdot f = 0$ in Ω; it is *left antiregular* if

F. Brackx et al. (eds.), Clifford Algebras and their Applications in Mathematical Physics, 129–135.

$D \cdot f = 0$. Similarly, f is ***right regular*** (resp. ***right antiregular***) if $f \cdot D^+ = 0$ (resp. $f \cdot D = 0$). If f is invertible and both f and f^{-1} are regular, then f is said to be ***biregular***.

The following "complex notation" is well known. The quaternion q can be expressed uniquely as $q = u + vj$ where $u = x_0 + ix_1$, $v = x_2 + ix_3 \in R + iR$. Then for differentiable $f : \Omega \to H$ we have $D^+ f = 0$ if and only if the following system is satisfied:

$$\phi_v = -\overline{\psi}_u , \quad \psi_v = \overline{\phi}_u . \tag{1}$$

Similarly, $Df = 0$ if and only if $\phi_v = \overline{\psi}_{\overline{u}}$, $\psi_v = -\overline{\phi}_{\overline{u}}$.

The following 4×4 matrix notation will also prove convenient for our purposes. For f as above define

$$J_C(f) = \begin{pmatrix} \phi_u & \phi_{\overline{u}} & \phi_v & \phi_{\overline{v}} \\ \overline{\phi}_u & \overline{\phi}_{\overline{u}} & \overline{\phi}_v & \overline{\phi}_{\overline{v}} \\ \psi_u & \psi_{\overline{u}} & \psi_v & \psi_{\overline{v}} \\ \overline{\psi}_u & \overline{\psi}_{\overline{u}} & \overline{\psi}_v & \overline{\psi}_{\overline{v}} \end{pmatrix} \in \mathcal{M} ,$$

where $\mathcal{M} \subseteq C^{4\times 4}$ is the collection of matrices, written in 2×2 block form $B = (B_{00}, B_{01};\ B_{10}, B_{11})$, where $B_{\alpha\beta}$ is of the form $(a, \overline{b};\ b, \overline{a})$, $a, b \in C$. The second and fourth columns of matrices in $\mathcal{M}$ are redundant; this notation permits expressing the Chain Rule as $J_C(f \circ g) = J_C(f) \circ J_C(g)$, $J_C(f^{-1}) = (J_C f)^{-1}$. The complex equations (1) say that $D^+ f = 0$ is equivalent to $J_C(f) \in \mathcal{F}_C$, where

$$\mathcal{F}_C = \{(b_{\alpha\beta}) \in \mathcal{M} :\ b_{02} = -b_{30}, b_{22} = b_{10}\} . \tag{2}$$

Similarly, $Df = 0$ corresponds to the conditions $b_{02} = \overline{b_{20}}$, $b_{22} = -\overline{b_{00}}$, while $fD^+ = 0$ corresponds to $b_{12} = -\overline{b_{20}}$, $b_{22} = \overline{b_{10}}$ and $fD = 0$ corresponds to $b_{12} = b_{30}$, $b_{22} = -b_{00}$.

We describe one further way of expressing regularity. Introduce the following "conjugations,"

$$\begin{aligned} \overline{q} &= x_0 - ix_1 - jx_2 - kx_3 , \\ \overline{q}^{(1)} &= x_0 - ix_1 + jx_2 + kx_3 , \\ \overline{q}^{(2)} &= x_0 + ix_1 - jx_2 + kx_3 , \\ \overline{q}^{(3)} &= x_0 + ix_1 + jx_2 - kx_3 , \end{aligned} \tag{3}$$

in terms of which may solve formally for the coordinates x_α in terms of the conjugates $\overline{q}, \overline{q}^{(1)}, \ldots$ To define symbolic partial derivatives with respect to the $\overline{q}^{(\alpha)}$, consider more generally any four R-linear isomorphisms $\sigma_0, \sigma_1, \sigma_2, \sigma_3 : H \to H$ with the property that one may solve for the x_α uniquely in terms of the $\sigma_\alpha(q)$; that is, the expressions $\partial x_\alpha / \partial \sigma_\beta(q)$ are defined uniquely by the condition

$$x_\alpha = \sum_\beta \sigma_\beta(q) \cdot \frac{\partial x_\alpha}{\partial \sigma_\beta(q)} \tag{4}$$

for $0 \le \alpha \le 3$. For any quaternionic function f define then $\partial f / \partial \sigma_\alpha(q)$ by

$$\frac{\partial f}{\partial \sigma_\alpha(q)} := \sum_\beta \frac{\partial x_\beta}{\partial \sigma_\alpha(q)} \cdot \frac{\partial f}{\partial x_\beta} . \tag{5}$$

From this it is not difficult to verify the following.

Proposition 2.1 *Let $p \mapsto q \mapsto f$ be smooth mappings. Then*

$$\frac{\partial f}{\partial \sigma_\alpha(p)} = \sum_\beta \frac{\partial \sigma_\beta(q)}{\partial \sigma_\alpha(p)} \cdot \frac{\partial f}{\partial \sigma_\beta(q)} .$$

Returning to the specific conjugations (3), we see that $\partial x_\alpha / \partial \overline{q} = \partial x_\alpha / \sigma_0(q)$ takes the values $1/4, i/4, j/4, k/4$ for $\alpha = 0, 1, 2, 3$. Therefore according to (5) we have defined $\partial f / \partial \overline{q} = \sum_\alpha (\partial x_\alpha / \partial \overline{q}) \partial_\alpha f = D^+ f$.

If we take instead $\sigma_0(q)$,$\sigma_1(q)$,$\sigma_2(q)$,$\sigma_3(q)$ equal to

$$\begin{aligned} q &= x_0 + ix_1 + jx_2 + kx_3 , \\ q^{(1)} &= x_0 + ix_1 - jx_2 - kx_3 , \\ q^{(2)} &= x_0 - ix_1 + jx_2 - kx_3 , \\ q^{(3)} &= x_0 - ix_1 - jx_2 + kx_3 , \end{aligned} \tag{6}$$

respectively, (thus $q^{(1)} = iqi^{-1}$, etc.) then it will be found easily that $\partial f / \partial q = D \cdot f$.

It must be stressed that $\partial / \partial \overline{q}$, $\partial / \partial q$ are not intrinsically determined by the definitions of $\overline{q}$,q but depend on the complete system of four conjugations used. It is not difficult to find systems other than (3) for which $\partial / \partial q$ is still equal to the left operator D, but with the remaining symbolic partial derivatives $\partial / \partial \overline{q}^{(1)}$,$\partial / \partial \overline{q}^{(2)}$,$\partial / \partial \overline{q}^{(3)}$ turning out quite different. From Proposition 2.1 one may derive various "Chain Rules" such as the following.

Proposition 2.2 *Let $f : \Omega_2 \to H$, $g : \Omega_1 \to \Omega_2$ be differentiable mappings of domains in H. Let $\partial / \partial \overline{q}^{(\alpha)}$, $\partial / \partial q^{(\alpha)}$ be determined by the systems (3),(6) respectively. Then*

$$\begin{aligned} D^+(f \circ g) &= (D^+ \overline{g}) \cdot ((D^+ f) \circ g) + \sum_{\alpha=1}^{3} (D^+ \overline{g}^{(\alpha)}) \cdot (\frac{\partial f}{\partial \overline{q}^{(\alpha)}} \circ g) , \\ D(f \circ g) &= (Dg) \cdot ((Df) \circ g) + \sum_{\alpha=1}^{3} (Dg^{(\alpha)}) \cdot (\frac{\partial f}{\partial q^{(\alpha)}} \circ g) , \\ D^+(f \circ g) &= \sum_{\alpha=0}^{3} (D^+ g^\alpha) \cdot (\partial_\alpha f) \circ g , \\ D(f \circ g) &= \sum_{\alpha=0}^{3} (D g^\alpha) \cdot (\partial_\alpha f) \circ g , \end{aligned}$$

where in the latter two formulas, $g = g^0 + ig^1 + jg^2 + kg^3$.

As an application of the Chain Rules we have the following elementary formulas, which may also be verified directly.

Corollary 2.3 *Let $g(q) = aq + b$, $h(q) = qc + d$ be left and right affine quaternionic mappings, $a, b, c, d \in H$. Then*

$$\begin{aligned} D^+(f \circ g) &= \bar{a}(D^+ f) \circ g\,, \\ D(f \circ g) &= (a_0 Df + ia_1 \frac{\partial f}{\partial q^{(1)}} + ja_2 \frac{\partial f}{\partial q^{(2)}} + ka_3 \frac{\partial f}{\partial q^{(3)}}) \circ g\,, \\ D^+(f \circ h) &= (c_0 D^+ f - ic_1 \frac{\partial f}{\partial \overline{q}^{(1)}} - jc_2 \frac{\partial f}{\partial \overline{q}^{(2)}} - kc_3 \frac{\partial f}{\partial \overline{q}^{(3)}}) \circ h\,, \\ D(f \circ h) &= c(Df) \circ h\,, \end{aligned}$$

where $a = a_0 + ia_1 + ja_2 + ka_3$, $c = c_0 + ic_1 + jc_2 + kc_3$. Further, $D^+(h \circ f) = (D^+ f)c$, $D(h \circ f) = (Df)c$.

With the aid of these formulas it is not difficult to show the following.

Theorem 2.4 *Let $A : H \to H$ be R-linear. If $\partial A/\partial \overline{q}^{(1)} = \partial A/\partial \overline{q}^{(2)} = \partial A/\partial \overline{q}^{(3)} = 0$, then $\overline{A}$ is a left multiplication mapping $q \mapsto aq$. If $\partial \overline{A}/\partial q^{(1)} = \partial \overline{A}/\partial q^{(2)} = \partial \overline{A}/\partial q^{(3)} = 0$, then $\overline{A}$ is a right multiplication mapping $q \mapsto qc$. Similarly, if $D^+ \overline{A}^{(1)} = D^+ \overline{A}^{(2)} = D^+ \overline{A}^{(3)} = 0$, then A is a left multiplication; if $DA^{(1)} = DA^{(2)} = DA^{(3)} = 0$, then A is a right multiplication.*

3. Biregular and LR-Biregular Mappings

Define $\mathcal{F}_C^* = \{B \in \mathcal{F}_C : \det B \neq 0,\ B^{-1} \in \mathcal{F}_C\}$. By (2), an invertible $B \in \mathcal{F}_C$ is in $\mathcal{F}_C^*$ precisely when $(B^{-1})_{0,2} = -(B^{-1})_{3,0}$ and $(B^{-1})_{2,2} = (B^{-1})_{1,0}$. These elements of the inverse matrix can be easily obtained via expansion by minors, and we have the following characterization of biregular matrices.

Theorem 3.1 *Let $B = (b_{\alpha\beta}) \in \mathcal{M}$. Then $B \in \mathcal{F}_C^*$ if and only if $\det B \neq 0$, $b_{02} = -b_{30}$, $b_{22} = b_{10}$, and*

$$\begin{vmatrix} \overline{b_{10}} & -b_{30} & \overline{b_{12}} \\ \overline{b_{00}} & b_{12} & -\overline{b_{30}} \\ \overline{b_{20}} & b_{32} & \overline{b_{10}} \end{vmatrix} = \begin{vmatrix} b_{10} & \overline{b_{00}} & b_{12} \\ b_{20} & \overline{b_{30}} & b_{10} \\ b_{30} & \overline{b_{20}} & b_{32} \end{vmatrix}, \tag{7}$$

$$\begin{vmatrix} b_{00} & \overline{b_{10}} & \overline{b_{12}} \\ b_{10} & \overline{b_{00}} & -\overline{b_{30}} \\ b_{30} & \overline{b_{20}} & \overline{b_{10}} \end{vmatrix} = -\begin{vmatrix} b_{10} & b_{12} & -\overline{b_{30}} \\ b_{20} & b_{10} & \overline{b_{32}} \\ b_{30} & b_{32} & \overline{b_{10}} \end{vmatrix}.$$

LR-biregular mappings. We will say that the invertible mapping f is *left-right biregular* (*LR-biregular*) when f is (left) regular and f^{-1} is right regular. Let $\mathcal{F}_C^{\mathrm{LR}} \subseteq \mathcal{M}$ denote the set of matrices corresponding to LR-biregular mappings. By the same reasoning as above, one finds that if $B \in \mathcal{F}_C^*$ is invertible, then B^{-1} is right regular if and only if $(B^{-1})_{1,2} = -(\overline{B}^{-1})_{2,0}$, $(B^{-1})_{2,2} = (\overline{B}^{-1})_{1,0}$; that is,

$$-\begin{vmatrix} b_{00} & -b_{30} & \overline{b_{12}} \\ b_{10} & b_{12} & -\overline{b_{30}} \\ b_{30} & b_{32} & \overline{b_{10}} \end{vmatrix} = -\begin{vmatrix} \overline{b_{10}} & b_{00} & -b_{30} \\ \overline{b_{20}} & b_{30} & b_{32} \\ \overline{b_{30}} & b_{20} & b_{10} \end{vmatrix}, \tag{8}$$

$$\begin{vmatrix} b_{00} & \overline{b_{10}} & \overline{b_{12}} \\ b_{10} & \overline{b_{00}} & -\overline{b_{30}} \\ b_{30} & \overline{b_{20}} & \overline{b_{10}} \end{vmatrix} = -\begin{vmatrix} \overline{b_{10}} & \overline{b_{12}} & -b_{30} \\ \overline{b_{20}} & \overline{b_{10}} & b_{32} \\ \overline{b_{30}} & \overline{b_{32}} & b_{10} \end{vmatrix} .$$

Equations (8) turn out to be considerably simpler than (7). After expansion of these determinants and some rearranging, we have the following characterization of LR-biregular matrices.

Theorem 3.2 *Let* $B = (b_{\alpha\beta}) \in \mathcal{M}$. *Then* $B \in \mathcal{F}_C^{LR}$ *if and only if* $\det B \neq 0$, $b_{22} = b_{10}$, $b_{02} = -b_{30}$, *and*

$$\begin{aligned} b_{12}(b_{00}\overline{b_{10}} - \overline{b_{12}}b_{30}) + b_{10}(b_{00}\overline{b_{20}} + \overline{b_{12}}b_{32}) + b_{20}(\overline{b_{10}}b_{32} + \overline{b_{20}}b_{30}) &= 0 , \quad (9) \\ \overline{b_{00}}(b_{00}\overline{b_{10}} - \overline{b_{12}}b_{30}) + \overline{b_{30}}(b_{00}\overline{b_{20}} + \overline{b_{12}}b_{32}) - \overline{b_{32}}(\overline{b_{10}}b_{32} + \overline{b_{20}}b_{30}) &= 0 . \end{aligned}$$

It follows from this that if $t_1, t_2, t_3, t_4 \in C$ and $t_1 t_2 + t_3^2 \neq 0$, then the following matrix

$$B = \begin{pmatrix} t_1 & t_3 t_4 & -t_3 & t_1 t_4 \\ \overline{t_3 t_4} & \overline{t_1} & \overline{t_1 t_4} & -\overline{t_3} \\ -\overline{t_2 t_4} & \overline{t_3} & \overline{t_3 t_4} & \overline{t_2} \\ t_3 & -t_2 t_4 & t_2 & t_3 t_4 \end{pmatrix} \qquad (10)$$

is an element of $\mathcal{F}_C^{\mathrm{LR}}$.

This is by no means the most general LR-biregular linear mapping. We give here a few examples of non affine LR-biregular mappings in this family. Suppose ϕ and $\overline{\psi}$ are holomorphic functions of the complex variables u and v; that is, $\phi_{\overline{u}} = \psi_u = 0$, $\phi_{\overline{v}} = \psi_v = 0$. Suppose further that $\underline{\overline{\psi}_u} = -\phi_v$. (As a simple illustration, consider $\phi(u,v) = \lambda(u) + \mu(v)$, $\psi(u,v) = -\overline{u\mu'(v)}$ for λ,μ holomorphic). Then for $f = \phi + \psi j$ we have that $J_C f$ is of the form (10). The nonsingularity condition $\phi_u \overline{\psi}_v + (\overline{\psi}_u)^2 \neq 0$ is easily achieved.

One may see from these examples that the class $\mathcal{F}_C^{\mathrm{LR}}$ is not preserved under composition.

4. Regular Functions on Manifolds

Let M be a real differentiable manifold of dimension 4. Let $x = (x_0, x_1, x_2, x_3) : U \to R^4$ be a smooth local coordinate system in an open set U in M. Let $q = x_0 + ix_1 + jx_2 + kx_3$. Then a function $F : M \to H$ is *regular* with respect to this coordinate system if $f = F \circ x^{-1}$ is a regular function of q. The following fact limits the manifolds which can admit regular functions in this sense.

Proposition 4.1 *Let* $g : \Omega_2 \to \Omega_1$ *be a diffeomorphism of domains in* H. *Suppose* $f \circ g$ *is regular in* Ω_2 *for every regular* f *defined in* Ω_1. *Then* g *is a left affine map* $g(q) = aq + b$, $a, b \in H$. *Conversely, for any left affine* g, $f \circ g$ *is regular whenever* f *is.*

Proof. According to Proposition 2.2, we have

$$D^+\overline{g}^{(1)}\frac{\partial f}{\partial \overline{q}^{(1)}} + D^+\overline{g}^{(2)}\frac{\partial f}{\partial \overline{q}^{(2)}} + D^+\overline{g}^{(3)}\frac{\partial f}{\partial \overline{q}^{(3)}} = 0$$

whenever $D^+f = 0$. For this to hold for for all regular f, we must have

$$D^+\overline{g}^{(1)} = D^+\overline{g}^{(2)} = D^+\overline{g}^{(3)} = 0 .$$

By Theorem 2.4 the differential $dg \in R^{4\times 4}$ has the form of a left multiplication mapping. It is well known [12] that any quaternionic function whose differential is of this form at each point is left affine. Conversely, by Corollary 2.3, pre-composition by such functions conserves regularity.

There are, of course, contexts in which mappings more general than affine can conserve regularity in a weaker sense. (For instance, let $g(q) = (aq+b)(cq+d)^{-1}$ $(a,b,c,d \in H,\ a^{-1}b - c^{-1}d \neq 0)$ be a quaternionic Möbius transformation. Then f is regular if and only if $\rho \cdot (f \circ g)$ is regular [12], where the left factor ρ is given by $\rho(q) = (|b - ac^{-1}d|^{-2}|cq+d|^{-2})(cq+d)^{-1}$.) However, the definition we have given for regular F on M implies that if x^α, x^β are compatible local coordinates, and if $f^\alpha = F \circ (x^\alpha)^{-1}$, $f^\beta = F \circ (x^\beta)^{-1}$, then $f^\beta = f^\alpha \circ g$ where $g = x^\alpha \circ (x^\beta)^{-1}$ must satisfy the condtion of Proposition 4.1. This may be summarized as follows.

Corollary 4.2 *The only manifolds M modelled locally on H and admitting a well-defined notion of "regular function" $F : M \to H$ are the left affine manifolds.*

Similarly, in order for regular functions $F : M \to N$ to be defined between two manifolds, M must be left affine and N right affine. Since the natural functions defined on N are the right regular ones, the natural notion of invertibly regular function in this context is that of LR-biregular. Corollary 2.3 gives the following.

Theorem 4.3 *Let M, M' be left affine quaternionic manifolds, and let N, N' be right affine. Let $G : M \to M'$, $H : N \to N'$ be left and right affine mappings respectively. Let $F : M \to N$ be LR-biregular. Then $H \circ F \circ G$ is LR-biregular.*

Similar statements hold when "regular" is replaced by "antiregular" and "right" is exchanged with "left."

Acknowledgment is made for use of the computing facilities at the CINVESTAV-IPN, donated in part by CONACyT, in the realization of this work.

References

1. Bonan, E., "Sur les G-structures de type quaternionien," *Cahiers Top. et Geom. Diff.* 9 (1967) 389-461.
2. Curtis, M.L., *Matrix Groups*, Springer-Verlag (1984).
3. Fueter, R., "Die Funktionentheorie der Differentialgleichungen $\Delta u = 0$ und $\Delta\Delta u = 0$ mit vier rellen Variablen," *Comment. Math. Helv.* 7 (1935) 307-330.
4. Fueter, R., "Über die analytische Darsetllung der regulären Funktionen einer quaternionen-variablen," *Comment. Math. Helv.* 8 (1936) 371-378.
5. Królikowski, W., "Quaterniconic mappings and harmonicity", preprint .
6. Królikowski, W. and Ramírez de Arellano, E., "Fueter-Hurwitz regular mappings and an integral representation," *Clifford Algebras and their Applications in Mathematical Physics*, Proceedings, Montpellier 1989 ed. A. Micali, Kluwer Academic Publishers, Dordrecht-Boston-London (1992) 221-237.
7. Królikowski, W. and Ramírez de Arellano, E., "On polynomial solutions of the Fueter-Hurwitz equation," *Proc. Intl. Conf. on Complex Analysis*, Univ. of Wisconsin, 1991, ed. A. Nagel and E. L. Stout, A.M.S. series *"Contemporary Mathematics"* 137 (1992) 297-305.

8. Narasimhan, R., *Analysis on real and complex manifolds*, North-Holland Publishing Company, Amsterdam (1968).
9. Shapiro, M. V. and Vasilevski, N., "Quaternionic ψ-hyperholomorphic functions, singular integral operators and boundary value problems" (I and II), to appear in *Complex Variables: Theory and Applications*.
10. Sommese, A. J., "Quaternionic manifolds," *Math. Ann.* 212 (1975)191-214.
11. Souček, V., "Holomorphicity in quaternionic analysis," Seminari di Geometria 1982-1983, Università di Bologna, Istituto de Geometria, Dipartimento de Matematica (1984).
12. Sudbery, A., "Quaternionic Analysis," *Math. Proc. Camb. Phil. Soc.* 85 (1979) 199-225.

MONOGENIC AND HOLOMORPHIC FUNCTIONS

G.LAVILLE
Caen-France

Abstract. The theory of holomorphic functions of several variables and the theory of monogenic functions (Clifford-analytic functions) are similar in the following sense. We write $C^m = R^m \oplus iR^m$ and $R^{m+1} = R^m \oplus R$. Let f be a real-valued, real analytic function defined in an open subset of R^m; f may be extended to a holomorphic function on an open subset of C^m or to a monogenic function on an open set of R^{m+1}; roughly speaking the knowledge of the latter is equivalent of the knowledge of the former. This motivates an explicit change from a holomorphic function to a monogenic function.

Notation

$A = R_{0,m}$, universal Clifford algebra of the anti-euclidean space R^m. For $x \in R^{m+1}$, put $x = (x_0, \underline{x})$ and also by abuse of notation

$$x = x_0 + \sum_{i=1}^{m} e_i x_i = x_0 + \underline{x} \in A.$$

A_C is the complexification of A

$$\omega'(\zeta) = \sum_{j=1}^{m} (-1)^{j-1} \zeta_j d\zeta_1 \wedge \ldots \wedge \widehat{d\zeta_j} \wedge \ldots \wedge d\zeta_m,$$

$$\sum = \{t \in R^m; t_j \geq 0 \quad j = 1, \ldots, m, \sum_{j=1}^{m} t_j = 1\}.$$

For $a_1, \ldots, a_m \in A_C$ and $\zeta_1, \ldots, \zeta_m \in C$, we write

$$\langle \zeta, a \rangle = \sum_{i=1}^{m} \zeta_i a_i.$$

1. Monogenic kernel

The starting point is the following (see [3], [4], [5], [6]).
Let $a_1, \ldots, a_m$, be m invertible elements of the algebra A_C. Then there exist a simplex $\Lambda \subset C^m$, diffeomorphic to Σ, such that
(i) $\zeta \in \Lambda \Rightarrow \langle \zeta, a \rangle$ is invertible in A_C.

F. Brackx et al. (eds.), Clifford Algebras and their Applications in Mathematical Physics, 137–139.

(ii) $\int_\Lambda \omega'(\zeta)/\langle\zeta, a\rangle^m$ does not depend on the particular choice of Λ.
(iii) if $a_1, \ldots, a_m \in R$, then

$$\int_\Lambda \omega'(\zeta)/\langle\zeta, a\rangle^m = \frac{(-1)^{m-1}}{(m-1)!}\,\frac{1}{a_1 \ldots a_m}$$

For $z \in C^m$, $x \in R^{m+1}$, let

$$K(z, x) = (-1)^{m-1}(m-1)! \int_\Lambda \frac{\omega'(\zeta)}{\langle\zeta, z - (x - ex_0)\rangle^m}$$

This kernel does not depend on Λ. It is well defined as soon as, for all j, $z_j - x_j + e_j x_0$ is invertible, that is, $K(z, x)$ is well defined for

$$(z_j - x_j)^2 + x_0^2 \neq 0 \quad \text{for} \quad j = 1, \ldots, m.$$

Let $\Omega_K = \{(z, x) \in C^m \times R^{m+1};\ \forall j,\ z_j \neq x_j \pm i x_0\}$.

PROPERTIES OF K :

(1) K is an A_C-valued function defined on Ω_K.
(2) K does not depend on λ.
(3) K is left and right monogenic in x and holomorphic in z.
(4) $K(z, 0, \underline{x}) = \dfrac{1}{(z_1 - x_1)\ldots(z_n - x_n)}$ and $K(z, x)$ is the Cauchy-Kovalevska extension of that function.
(5) For all permutations σ of $\{1, \ldots, m\}$

$$K(z_{\sigma(1)}, \ldots, z_{\sigma(m)}, x_0, x_{\sigma(1)}, \ldots, x_{\sigma(m)}) = K(z_1, \ldots, z_m, x_0, x_1, \ldots, x_m)$$

Proof.
We have already proved 1) and 2).
3) follows by differentiating

$$(\sum_{j=1}^m \zeta_j(z_j - x_j) + \sum_{j=1}^m \zeta_j e_j x_0)^{-m}$$

where only scalars and $\sum_{j=1}^m \zeta_j e_j$ appear in the computation.
4) Apply (iii) with $x_0 = 0$. The Cauchy-Kovalevska extension is unique.
5) Invariance by permutation of $\zeta_1, \ldots, \zeta_m$. □

PROPOSITION (SERIES EXPANSION OF K)

We have

$$K(z, x) = \sum_{l=0}^{\infty} \sum_{|L|=l} \frac{v_L(x)}{z_1^{l_1+1} \ldots z_m^{l_m+1}}$$

where $L = (l_1, \ldots, l_m)$, $|L| = l_1 + \ldots + l_m$ and $v_L(x)$ are themonogenic homogeneous polynomials defined (up to a constant) in [1] or [2]

$$v_L(0, \underline{x}) = x_1^{l_1} \ldots x_m^{l_m}$$

This series converges for $|z_j| > |x_j - e_j x_0|$, $j = 1, \ldots, m$

$$K(z,x) = \sum_{l=0}^{\infty} \sum_{|L|=l} z_1^{l_1} \ldots z_m^{l_m} \frac{(-1)^{m-1}(l! - 1)}{(l_1-1)! \ldots (l_m - 1)!} \int_{\Lambda} \frac{\zeta_1^{l_1} \ldots \zeta_m^{l_m} \omega'(\zeta)}{\langle \zeta, x - e x_0 \rangle^{|L|}}$$

This series converges for $|z_j| < |x_j - e_j x_0|$, $j = 1, \ldots, m$

2. From holomorphic to monogenic function

Let Ω be an open subset in R^{m+1}. We define $\chi(\Omega) \subset C^m$ such that

$$\chi(\Omega) = \{z = (x_1 + iy_1, \ldots, x_m + iy_m); (\pm y_1, x) \in \Omega, \ldots, (\pm y_m, \underline{x}) \in \Omega\}$$

with $\underline{x} = (x_1, \ldots, x_m)$.
Note that if $(x_0, \underline{x}) \in \Omega$ then $(x_1 \pm ix_0, \ldots, x_m \pm ix_0) \in \Omega$.

Theorem 2.1 *Let f be an A_C-valued holomorphic function defined on $\chi(\Omega)$. Let*

$$\Lambda(f)(x) = \frac{1}{(2i\pi)^m} \oint_{C_1} \ldots \oint_{C_m} f(z) K(z,x) dz_1 \ldots dz_m$$

where C_j is a closed path in C such that $C_1 \times \ldots \times C_m \subset \chi(\Omega)$ and $x_j \pm ix_0$ is inside the domain defined by the corresponding path.
Then $\Lambda(f)$ is a monogenic function defined an Ω, with values in A_C.

Proof.
Immediate consequence of the properties of the kernel K. □

PROPERTIES OF Λ

(1) If $p(z) = z_1^{l_1} \ldots z_m^{l_m}$ then $\Lambda(p)(x) = v_L(x)$.
(2) Λ is an algebra homomorphism from the algebra of A_C-valued holomorphicfunctions defined on $\chi(\Omega)$ to the A_C-valued monogenic functions defined onΩ.

REFERENCES

[1] F. Brackx, R. Delanghe and F. Sommen. "Clifford Analysis", Research Notes in Math. 76, Pitman, London (1982).

[2] R. Delanghe, F. Sommen and V. Souček. "Clifford Algebra and Spinor-valued functions: a function theory for the Dirac-operator", Mathematics and Its Applications 53, Kluwer Academic Publishers, Dordrecht (1992).

[3] L. Fantappié. "L'indicatrice proiettiva del funzionali linearie e i prodotti funzionali proiettivi", Ann. Math. Pura Appl. (4)(22) pp. 181-289 (1943).

[4] R.P. Feynmann. "Theory of positions", Phys. Rev. 76, 749 (1949).

[5] G. Henkin. "Integral representations of differential forms on C-R manifolds and the theory of C-R functions", Uspekhi Mat. Nauk 39, 3, pp. 39-106.

[6] G. Laville. Jour. of functional Analysis; vol. 101 n 1, pp. 25-37(1991).

HYPERCOMPLEX DIFFERENTIABILTY AND ITS APPLICATIONS

HELMUTH R. MALONEK
Universidade de Aveiro
Departamento de Matemática
P-3800 Aveiro, Portugal

Abstract.
We give a survey about the development of an elementary concept of hypercomplex differentiable $\mathcal{A}$-valued functions defined in open subsets Ω of $\mathbf{R}^{m+1}$, whereby $\mathcal{A}$ is a Clifford algebra over the field of real numbers. Using a different from the usual one hypercomplex structure of $\mathbf{R}^{m+1}$ we get by this way a natural generalization of the Cauchy approach to monogenic functions which seems to be not possible so far. Exemplary this concept applies to transfer important properties of holomorphic functions in the plane. The results are of wide formal uniformity with the theory of functions of several complex variables.

Key words: Monogenic functions, generalized Cauchy approach, hypercomplex differentiability

1. Introduction

About ten years ago, R. DELANGHE, F. BRACKX AND F. SOMMEN published the book *CLIFFORD ANALYSIS* [1] which was mainly devoted to a function theory of the so called generalized Cauchy-Riemann operator

$$D = \frac{\partial}{\partial x_0} + \frac{\partial}{\partial x_1}e_1 + \ldots + \frac{\partial}{\partial x_m}e_m \tag{1}$$

where $e_1, \ldots, e_m$ are the orthonormal base of the vector space $\mathbf{R}^m$ and a subset of $e_1, \ldots, e_n$, $n \geq m$, generating the 2^n-dimensional universal Clifford algebra $\mathcal{A}$=C($V_{n,0}$)=R_{0+n} over $\mathbf{R}$ according to the multiplication rules $e_k e_l + e_l e_k = -2\delta_{kl}$, $k, l = 1, \ldots, m$. Classes of regular $\mathcal{A}$-valued functions of the form

$$f(x) = \sum_A f_A(x)e_A, \quad f_A(x) \in \mathbf{R}, \quad x \in \mathbf{R}^{m+1}$$

were delimited by the conditions $Df = 0$ (left monogenic functions) resp. $fD = 0$ (right monogenic functions) and where $\{e_A : A \subseteq \{1, \ldots, n\}\}$ form the basis of $\mathcal{A}$ with $e_A = e_{h_1}e_{h_2}\ldots e_{h_r}, 1 \leq h_1 < \ldots < h_r \leq n$, $e_\emptyset = e_0 = 1$. Thereby (1) generalizes the Cauchy-Riemann operator for the case $m = 1$ in the Wirtinger form

$$\frac{\partial}{\partial \bar{z}} = \frac{1}{2}(\frac{\partial}{\partial x_0} + i\frac{\partial}{\partial x_1}) \tag{2}$$

and the monogenic functions are generalizations of holomorphic functions $f(z) = f(x_0 + ix_1)$ for $m = n = 1$ being solutions of $\frac{\partial f(z)}{\partial \bar{z}} = 0$.

F. Brackx et al. (eds.), Clifford Algebras and their Applications in Mathematical Physics, 141–150.

The theory of monogenic functions described in [1] (and more recently in [2]) uses for the hypercomplex structure of the pre-image set the identification of $x = (x_0, x_1, \ldots, x_m) = (x_0, \vec{x})$ with $z = x_0 + x_1 e_1 + \ldots + x_m e_m$. Thereby

$$\mathbb{R}^{m+1} \cong \tilde{\mathcal{A}} = span\{1, e_1, \ldots, e_m\}$$

and there are considered functions $f : \tilde{\mathcal{A}} \longrightarrow \mathcal{A}$, whereby classes of regular functions are defined as null-solutions of D (acting from the left resp. right) according to the Riemann approach for holomorphic functions.

As was demonstrated by SUDBERY [13] in the special case of functions of a quaternion variable this generalization of the Riemann approach (proposed already by FUETER [3]) is the only one to get a generalized class of holomorphic functions in an appropriate way. (cf.[1]). But there is a way to come out of this situation which will be described in Section **3**. Section **2** is devoted to some simple background and Section **4** build a bridge to the direct generalization of the Weierstrass approach. In Section **5** are briefly listed some applications - more or less only as examples for the elementary treatment of monogenic function theory using the derived concept. For complete proofs we refer to the original papers.

2. Two Remarkable Observations

The idea of a new approach via hypercomplex differentiabilty can be illustrated by two very suggestive formal observations.

The first is related to the relation of real and complex dimensions between $\mathbb{R}^2$ and $\mathbb{C}$. The complex dimension of $\mathbb{C}$ is equal to 1 and so $\mathbb{R}^2 \cong \mathbb{C}$ reduces the real dimension *to 1*. The same procedure is made considering $\mathbb{R}^{m+1} \cong \tilde{\mathcal{A}}$.

But we can also remark that this could be a reduction of dimensions *by the half.* Clearly this is the way to the theory of several complex variables: $\mathbb{R}^{2m} \cong \mathbb{C}^m$.

Finally there exists a third possibility - the reduction *by 1* dimension. We will see that this is the proposed ***second way*** to monogenic functions closely connected with the usual one and formally even with the case of several complex variables, besides them thereby giving the natural generalizations of the Cauchy and Weierstrass approaches.

A similar suggestive observation could be done by using the property of harmonicity of monogenic functions which results from $D\bar{D} = \bar{D}D = \Delta_{m+1}$, where $\bar{D} = \frac{\partial}{\partial x_0} - \sum_{k=1}^{m} \frac{\partial}{\partial x_k} e_k$ is the conjugated Cauchy-Riemann operator. Using $\bar{z} = x_0 - x_1 e_1 - \ldots - x_m e_m$ we obtain easily

$$x_0 = \frac{1}{2}(\bar{z} + z), \quad x_k = \frac{1}{2}(\bar{z} e_k - e_k z) = \frac{1}{2}(e_k \bar{z} - z e_k) \tag{3}$$

which for $m = 1$ turns out to $x_0 = \frac{1}{2}(\bar{z} + z)$ and $x_1 = \frac{i}{2}(\bar{z} - z)$. The transformation (3) allows us to change the real variables in the formal Taylor series (written down for simplicity in the neighborhood of the origin)

$$f(z) = \sum_{k=0}^{\infty} \frac{1}{k!} \left\{ x_0 \frac{\partial}{\partial x_0} + x_1 \frac{\partial}{\partial x_1} + \cdots + x_m \frac{\partial}{\partial x_m} \right\}^k f(0) =$$

$$= \sum_{k=0}^{\infty} \frac{1}{k!} \left\{ \frac{z+\bar{z}}{2} \frac{\partial}{\partial x_0} + \frac{1}{2}(\bar{z}e_1 - e_1 z) \frac{\partial}{\partial x_1} + \cdots + \frac{1}{2}(\bar{z}e_m - e_m z) \frac{\partial}{\partial x_m} \right\}^k f(0)$$

$$= \sum_{k=0}^{\infty} \frac{1}{k!} \left\{ \frac{\bar{z}}{2}(\frac{\partial}{\partial x_0} + e_1 \frac{\partial}{\partial x_1} + \cdots + e_m \frac{\partial}{\partial x_m}) + \right. \tag{4}$$

$$\left. + \frac{1}{2}(z\frac{\partial}{\partial x_0} - e_1 z \frac{\partial}{\partial x_1} - \cdots - e_m z \frac{\partial}{\partial x_m}) \right\}^k f(0).$$

Its obvious to see that as a consequence of $Df = 0$ the dependence on $\bar{z}$ disappears similar to the relations between real and complex analytic functions. Using in the second summand of the right hand side again $\frac{\partial f}{\partial x_0} = -\frac{\partial f}{\partial x_1} e_1 - \cdots - \frac{\partial f}{\partial x_m} e_m$ we get from (4)

$$f(z) = \sum_{k=0}^{\infty} \frac{1}{k!} \left\{ z_1 \frac{\partial}{\partial x_1} + z_2 \frac{\partial}{\partial x_2} + \cdots + z_m \frac{\partial}{\partial x_m} \right\}^k f(0), \tag{5}$$

which is an important hint for the special role of the monogenic variables

$$z_k = x_k - x_0 e_k = -\frac{z e_k + e_k z}{2},; \quad k = 1, \ldots, m, \tag{6}$$

in the monogenic function theory.

Whereas in papers about power series analogons (cf.[1]) the z_k are arising in a natural way the nonmonogenic basic variable $z \in \tilde{\mathcal{A}}$ never appeared.

In this way it was more or less cogent to ask for the reasons of such a situation and to try to investigate the special role of z_k. At first we can see that there exists an isomorphism between $\mathbb{R}^{m+1}$ and the set $\mathbb{H}^m$ of all such z_k and we obtain

$$\mathbb{R}^{m+1} \cong \mathbb{H}^m = \{\vec{z} : \vec{z} = (z_1, \ldots, z_m), z_k = x_k - x_0 e_k; x_0, x_k \in \mathbb{R}\}$$

Therefore we consider now $f \in C_1(\Omega, \mathcal{A})$ as a mapping from $\mathbb{H}^m$ into the Clifford algebra $\mathcal{A}$.

3. Hypercomplex Differentiability

Keeping in mind the problem of an adequate Cauchy approach, i.e. the problem to define the property of differentiability as characteristic for a proper subset of $\mathcal{A}$-valued functions we have now to consider the pre-image set as an m-dimensional set. This leads immediatly to another understanding of the main problem: differentiability has to be discussed as a *general property of local approximation by linearization.* Obviously, the situation is more delicate as for vector spaces. It's necessary to take in account that the related linear mappings must be $\mathcal{A}$-linear and that the underlying algebra generates only a module over a noncommutative ring apart from the case $m = 1$.

On this line we are able to point out that $\mathbb{H}^m$ is a special subset of $\mathcal{A}^m = \mathcal{A} \times \mathcal{A} \times \cdots \times \mathcal{A}$ but not an $\mathcal{A}$-submodule since $\lambda\vec{z}$, $\vec{z}\lambda$ belong to $\mathbb{H}$ if and only if $\lambda \in \mathbb{R}$. Nevertheless the imbedding of $\mathbb{H}^m$ in the module $\mathcal{A}^m$ enables us to use the properties of $\mathcal{A}^m$ to describe $\mathcal{A}$-linear mappings from $\mathbb{H}^m$ to $\mathcal{A}$ in the following way (cf.[6]):

Theorem 1 *Every $\mathcal{A}$-linear from the left (right) mapping $l_L(l_R) \in \mathcal{L}(\mathsf{H}^m; \mathcal{A})$ may be represented in a unique way in the form*

$$l_L(\vec{z}) = z_1 A_1 + \cdots + z_m A_m \quad resp. \quad l_R(\vec{z}) = A_1 z_1 + \cdots + A_m z_m \quad .$$

where $A_k \in \mathcal{A}$, $(k = 1, \ldots, m)$.

Formally there are no differences to the situation in vector spaces apart from the distinction of left- and right-linear mappings (which relates in a straightforward manner to the distinction of left- and right-monogenic functions).But it should be remarked that on account of $z_k \vec{s_k} \notin \mathsf{H}^m$ only the imbedding of H^m in $\mathcal{A}^m$ enables us to use for the proof of Theorem 1 (similar to the case for linear mappings between vector spaces) the canonical basis-representation

$$\vec{z} = z_1 \vec{s_1} + \cdots + z_m \vec{s_m} \quad \text{resp.} \quad \vec{z} = \vec{s_1} z_1 + \cdots + \vec{s_m} z_m$$

where $\vec{s_k} = (0, \ldots, 0, e_0, 0, \ldots, 0)$, $(k = 1, \ldots, m)$ are the elements of the canonical base of $\mathcal{A}^m$. As a corollary it follows that the components of $\vec{z}$ form a base for the algebraic dual $\mathcal{L}_L(\mathsf{H}^m; \mathcal{A})$ resp. $\mathcal{L}_R(\mathsf{H}^m; \mathcal{A})$ of H^m.

Provided with the knowledge about $\mathcal{A}$-linear mappings from H^m into $\mathcal{A}$ we still have to introduce with the help of the following inner product its induced norm

$$(\vec{z}, \vec{\zeta}) = \overline{(\vec{\zeta}, \vec{z})} = \sum_{k=1}^{m} \overline{z_k} \zeta_k$$

$$\|\vec{z}\| = (\vec{z}, \vec{z})^{1/2} = \left(\sum_{k=1}^{m} \overline{z_k} z_k \right)^{1/2} = (m x_0^2 + x_1^2 + \cdots + x_m^2)^{1/2}. \tag{7}$$

It's obvious that we got in general $(m \geq 1)$ thereby not an isometric relation between $\mathbf{R}^{m+1}$ and H^m like between $\mathbf{R}^2$ and $\mathbf{C}$, but for $x \in \mathbf{R}^{m+1}, z \in \tilde{\mathcal{A}}$ and $\vec{z} \in \mathsf{H}^m$ holds

$$m^{-1/2} \|\vec{z}\| \leq |z| = |x| \leq \|\vec{z}\|. \tag{8}$$

This leads (cf.[6]) to the following natural definition of *hypercomplex differentiability:*

Definition 1 *Let f be a continuous mapping of a neighborhood of a point $\vec{z} \in \mathsf{H}^m$ into the algebra $\mathcal{A}$. The function $f = f(\vec{z})$ is called hypercomplex differentiable from the left (right) at $\vec{z}$ if there exists a left (right) $\mathcal{A}$-linear mapping $l \in \mathcal{L}(\mathsf{H}^m; \mathcal{A})$ so that*

$$\lim_{\Delta \vec{z} \to 0} \frac{|f(\vec{z} + \Delta \vec{z}) - f(\vec{z}) - l(\Delta \vec{z})|}{\|\Delta \vec{z}\|} = 0 \tag{9}$$

We say that the function is hypercomplex differentiable in $\Omega \subset \mathbf{R}^{m+1} \cong \mathsf{H}^m$ if the function is hypercomplex differentiable at all points of Ω. The linear mapping $l_L(l_R)$ is called the left (right) hypercomplex derivative $f'_L(f'_R)$ of the function f.

In the same way as for vector spaces it's easy to verify (cf.[6])

Theorem 2 *If $f(\vec{z})$ is hypercomplex L-(R-) differentiable then the corresponding linear mapping (the L-(R-) derivative) is determined in a unique way ([6]).*

In view of Theorem 1 the relation (9) is equivalent to

$$f(\vec{z}+\Delta\vec{z})-f(\vec{z})=A_1\Delta z_1+\cdots+A_m\Delta z_m+o(||\Delta\vec{z}||) \tag{10}$$

in the case of hypercomplex R-differentiability where $A_k\in\mathcal{A}$ and

$$\lim_{\Delta\vec{z}\to 0}\frac{o(||\Delta\vec{z}||)}{||\Delta\vec{z}||}=0$$

(analogously for the case of L-differentiability). We remember also that for the concept considering f as a mapping from $\tilde{\mathcal{A}}$ into $\mathcal{A}$ we get by an analogue procedure only $f(z+\Delta z)-f(z)=A\Delta z+o(|\Delta z|)$ as a demand of differentiability. This is the reason for the negative results in [13]. To see that the proposed concept of hypercomplex differentiability represents the Cauchy approach in the theory of monogenic functions, we prove the following:

Theorem 3 *Let $f=f(\vec{z})$ be continuously real differentiable at a point $\vec{z}\in\mathrm{H}^m$. The function is hypercomplex L- (R-) diffferentiable at $\vec{z}$, iff f is L- (R-) monogenic at $\vec{z}$.*

Proof: We will only consider hypercomplex R- differentiable functions. The proof for L- differentiable functions is entirely analogue. From the assumption of real differentiability it follows that the increment of $f(\vec{z})$ has the form

$$f(\vec{z}+\Delta\vec{z})-f(\vec{z})=\frac{\partial f}{\partial x_0}\Delta x_0+\frac{\partial f}{\partial x_1}\Delta x_1+\cdots+\frac{\partial f}{\partial x_m}\Delta x_m+o(|\Delta x|) \tag{11}$$

with

$$\lim_{\Delta x\to 0}\frac{o(|\Delta x|)}{|\Delta x|}=0.$$

By means of the transformation

$$x_0=z_0 \text{ and } x_k=e_kz_0+z_k,$$
$$\text{invers to}\quad z_0=x_0,\ \text{and}\ z_k=x_k-x_0e_k,\ \ k=1,\ldots,m. \tag{12}$$

equation (11) together with (8) leads to the following

$$\Delta f(\vec{z})\ =\ fD\Delta z_0+\frac{\partial f}{\partial x_1}\Delta z_1+\cdots+\frac{\partial f}{\partial x_m}\Delta z_m+o(||\Delta\vec{z}||). \tag{13}$$

The necessary condition of R-monogenicity, i.e. $fD=0$ follows from Theorem 2 about the unique determination of f'_R, the $\mathcal{A}$-linear independence of $z_0,z_1,\ldots,z_m$ and formula (10). The sufficiency of monogenicity follows from Definition 1.

Remarks:

1. For the proof of the case of L-differentiability (12) must be replaced by the "left" transformation $x_0 = z_0$ and $x_k = z_0 e_k + z_k, \quad k = 1, \ldots m$.
2. We get the same result also by application of

$$\begin{aligned} x_0 &= \frac{1}{m+1}(\bar{z} + e_1 z_1 + \cdots + e_m z_m) \\ x_k &= \frac{1}{m+1}(e_k \bar{z} + e_k e_1 z_1 + \cdots + m z_k + \cdots + e_k e_m z_m) \end{aligned} \tag{14}$$

$(k = 1, \ldots, m)$ a transformation which is inverse to

$$\bar{z} = x_0 - e_1 x_1 - \cdots - e_m x_m, \quad z_k = -e_k x_0 + x_k \quad k = 1, \ldots, m, \tag{15}$$

and generalizes the "Wirtinger calculus". Instead of (13) $\Delta f(\vec{z})$ becomes equal to

$$\begin{aligned} \Delta f(\vec{z}) = {} & \frac{1}{m+1} f D \Delta \bar{z} + \left(\frac{\partial f}{\partial x_1} + \frac{1}{m+1} f D e_1 \right) \Delta z_1 + \cdots + \\ & + \left(\frac{\partial f}{\partial x_m} + \frac{1}{m+1} f D e_m \right) \Delta z_m + o(\|\Delta \vec{z}\|). \end{aligned} \tag{16}$$

and D could be changed by $\frac{1}{m+1}D$, which illustrates the role of the factor $\frac{1}{2}$ in the complex partial derivative (2) (the case $m = 1$).
3. Contrary, from the viewpoint of the generalized Cauchy approach the plane case $(m = 1)$ should be considered not only using the "complex variable" $z = x_0 + e_1 x_1$ and its conjugate $z = x_0 - e_1 x_1$, but should be also considered in relation to $z_0 = x_0$ and $z_1 = x_1 - x_0 e_1 = -e_1 z$. The specification of (13) to the case $m = 1$

$$\begin{aligned} \Delta f(z_1) &= \left(\frac{\partial f}{\partial x_0} + \frac{\partial f}{\partial x_1} e_1 \right) \Delta z_0 + + \frac{\partial f}{\partial x_1} \Delta z_1 + o(\|\Delta z_1\|) \\ &= f D \Delta z_0 + \frac{\partial f}{\partial x_1} \Delta z_1 + o(\|\Delta z_1\|) \end{aligned} \tag{17}$$

shows even more. Comparing with the usual complex form in "Wirtinger derivatives" (which would be more closely connected with the $\tilde{\mathcal{A}}$-approach)

$$\Delta f(z) = \frac{1}{2} f D \Delta \bar{z} + \frac{1}{2} f \bar{D} \Delta z + o(\|\Delta z\|) \tag{18}$$

we see that instead of $\bar{D}$ the gradient $(\frac{\partial}{\partial x_1}, \ldots, \frac{\partial}{\partial x_m})$ has to play a more essential role. The following explanation is widly a demonstration of this fact.

4. Elementary Relations between the Elements of $\tilde{\mathcal{A}}$ and H^m. Definition of a Symmetric Product

Looking at formula (6) we can recognize that the components z_k of $\vec{z} \in \mathrm{H}^m$ are some kind of projection of $z \in \tilde{\mathcal{A}}$ in direction of the canonical base vector $\vec{s_k} \in \mathrm{H}^m$, given by symmetric multiplication with $-e_k \in \tilde{\mathcal{A}}$. In a symbolic form we could write

$$z_k = -z \times e_k = -\frac{z e_k + e_k z}{2}, \quad k = 1, \ldots, m. \tag{19}$$

Taking in account this simple observation we can consider the special cases $z = e_j, j = 1, \ldots, m$, and also $z = e_0 = 1$. We get

$$-e_j \times e_k = \delta_{jk}, \quad -e_0 \times e_k = -e_k \quad k, l = 1, \ldots, m$$

and consequently

$$\vec{e_k} = (0, \ldots, 0, e_0, \ldots, 0) = \vec{s_k}, \quad \vec{e_0} = (-e_1, \ldots, -e_m) = \hat{i} \in \mathsf{H}^m. \tag{20}$$

¿From (20) now follows $\vec{z} = x_0\vec{e_0} + x_1\vec{e_1} + \cdots x_m\vec{e_m}$ and with the help of (7) $(\vec{e_k}, \vec{z}) = z_k, \quad (\hat{i}, \vec{z}) = mx_0 + x_1e_1 + \cdots + x_me_m$. The different relations of $z \in \tilde{A}$ and $\vec{z} \in \mathsf{H}^m$ to $x \in \mathbb{R}^{m+1}$ are illustrated by

$$z = x_0e_0 + (\hat{i}, \vec{x}) \in \tilde{A}, \quad \text{resp.} \quad \vec{z} = \vec{x}e_0 + \hat{i}x_0 \in \mathsf{H}^m \tag{21}$$

(the inverse relations are easy to verify using the conjugated elements). As a consequence of (21) we have now to consider $\vec{x}$ as the *real part* of $\vec{z}$ and x_0 as the *imaginary part* (in contrast to the situation in $\tilde{A}$). But this coincides very well with the process of hypercomplex extension of multiple power series in $\vec{x} \in \mathbb{R}^m$ being described as so-called Cauchy-Kowalewsky extension in [1] (see Section 5).

Comparing with the Weierstrass approach in the theory of functions of m complex variables the application of m hypercomplex variables leads in a natural way to the question, whether all products of z_k are also monogenic or not.(Thereby remembering that the main difficulty for developing the theory in the sense of Weierstrass in $\tilde{A}$ was that neither z nor any power of z are monogenic for $m > 1$). Unfortunately the answer is negative, e.g. $f(z_1, z_2) = z_1z_2$, is not monogenic. Again the way-out comes from a symmetric product which generalizes the relation (19)(cf.[7]).

Definition 2 *Let $V_{+,\cdot}$ be some commutative or non-commutative ring, $a_k \in V$ $(k = 1, \ldots, n)$, then the " $\times$ "-product is defined by*

$$a_1 \times a_2 \times \cdots \times a_n = \frac{1}{n!} \sum_{\pi(i_1, \ldots, i_n)} a_{i_1}a_{i_2}\cdots a_{i_n} \tag{22}$$

where the sum runs over all *permutations of all $(i_1, \ldots, i_n)$.*

For the further application of our approach to a natural generalization of the direct Weierstrass approach and a suggestiv definition of generalized power series the following *convention* plays an essential role:

Convention If the factor a_j occurs μ_j-times in (22), we briefly write

$$\underbrace{a_1 \times \cdots \times a_1}_{\mu_1} \times \cdots \times \underbrace{a_n \times \cdots \times a_n}_{\mu_n} = a_1^{\mu_1} \times a_2^{\mu_2} \times \cdots \times a_n^{\mu_n} = \vec{a}^{\mu} \tag{23}$$

and set parentheses if the powers are understood in the ordinary way, $\vec{a} = (a_1, a_2, \ldots, a_n)$ and $\mu = (\mu_1, \mu_2, \ldots, \mu_n)$ as a multiindex.

(For properties - including a recursion formula - and examples concerning the symmetric n-ary product defined by (22) compare e.g. [7].)

5. Some Applications of Hypercomplex Differentiability

The basic ideas explained in the preceding sections can be applied to transfer important properties of holomorphic functions in the plane ($m = 1$) to the higherdimensional case with the help of an instrumentarium which is easy to handle. Along this line it could be possible to enrich the theory of monogenic functions, which sometimes has seemed to be too modest as a self-contained theory like it was mentioned by DAVID HESTENES in his talk at this conference.

Certainly, further investigations also have to show that besides generalizations of well known facts in the plane case (which are of course also of great own interest and perhaps the only one way to make progress - apart from a chance!) there exist some important qualitativ new facts. But it seems to be not hopeless that for instance the uncovered suggestive conformity with some tools in the theory of functions of several complex variables gives new clues in such a direction. The results presented by GUY LAVILLE (cf.[5]) at this conference seem to be directed just to the same goal. Still open until now is also the question of a proper approach along a more geometric line (in analogy to conformal mappings) like it was already mentioned some time ago by KLAUS HABETHA([4]).

1. The generalized Weierstrass approach Using the representation by spherical monogenics [1] or Gegenbauer polynomials [4] of the following monogenic function (which is nothing else as the generalized Cauchy kernel of an appropriate m-dimensional Cauchy-integral cf.[1])

$$E(\zeta, z) = \frac{\overline{\zeta - z}}{|\zeta - z|^{m+1}}, \tag{24}$$

the monogenic analogons of z^n for the generalized Weierstrass approach of convergent power series are recognized in the form of so-called FUETER-polynomials $P_\nu(x)$ given in our terminology by $P_\nu(x) = \frac{1}{\nu!}\vec{z}^\nu$ where ν is a multiindex and (23) is used.

Like it was already mentioned on the end of section 4 the idea of hypercomplex differentiability (i.e. the use of $\mathbf{H}^m$ instead of $\tilde{A}$) together with the symmetric product (22) lead directly to the following

Definition 3 *Let $\nu = (\nu_1, \ldots, \nu_m)$ be a multiindex, then*

$$P(\vec{a}, \vec{z}) = \sum_\nu c_\nu(\vec{z} - \vec{a})^\nu \quad resp. \quad P(\vec{z}, \vec{a}) = \sum_\nu (\vec{z} - \vec{a})^\nu c_\nu \tag{25}$$

where $c_\nu \in \mathcal{A}, \vec{a} = \vec{\alpha} + \alpha_0\hat{i} \in \Omega \subset \mathbf{H}^m, (\alpha_0, \vec{\alpha} \in \mathbf{R}^{m+1})$ are called right (resp. left) (multiple) hypercomplex power series.

An elementary proof for the convergence to a monogenic function in a polycylindric domain of the form $\mathcal{U}(\vec{r}) = \{\vec{z} \in \mathbf{H}^m : |z_k| = (x_0{}^2 + x_k{}^2)^{1/2} < r_k,\ k = 1, \ldots, m\}$, the coincidence with the Taylor series representation (we recognize in (25) the form (5) written in full and not ordered by homogeneous polynomials - which could be also done (cf.[7])) all this and other properties could be find in [7].

We still would like to mention that (25) coincides with the so-called right (resp. left) Cauchy-Kowalewsky extension of the real multiple power series

$$P(\vec{\alpha},\vec{x}) = \sum_{\nu} c_\nu(\vec{x}-\vec{a})^\nu$$

where $\vec{x},\vec{\alpha} \in \mathbf{R}^m$ and the Cauchy-Kowalewsky product of two hypercomplex power series P and Q with coefficients c_ν resp. d_μ can be represented as a hypercomplex power series with coefficients b_σ where $\sigma = \nu + \mu$ (like a generalized Cauchy product). Briefly we refer also that defining hypercomplex differentiabilty of higher order in an inductive way (cf.[7]) we get immediatly

Theorem 4 *Every R- (L-) monogenic function defined in an open domain $\Omega \subset \mathsf{H}^m$ is infinitely R- (L-) hypercomplex differentiable.*

2.Related differential forms It is very well known that real differential forms in connection with the outer product are an powerful instrument to derive in a more or less easy formal way deep results like the Stokes' or other integral theorems in euclidean spaces of any dimension. Under the assumption that the occuring functions are real differentiable in the considered domain the complex form of the Stokes' theorem leads directly to Cauchy's integral theorem and the Cauchy integral formula as the main theorem of complex analysis. Therefore, the interest to hypercomplex differential forms was quite natural. In [12] SOMMEN has developed the complete de Rham - theory of monogenic differential forms. For quaternionic valued functions SUDBERY considered in [13] also some special differential forms. But in both cases the outer product of two considered basic 1-forms was not alternative and the structure of p-forms ($1 < p \leq m+1$) not so transparent for an easy formalism. But using the transformations (12) we have

Definition 4 *The outer product of two basic hypercomplex 1-forms, given by $dz_0 = dx_0$ resp. $dz_k = -e_k dx_0 + dx_k$, $k = 1,\ldots,m$ is defined by*

$$\begin{aligned} dz_k \wedge dz_l &= (-e_k dx_0 + dx_k) \wedge (-e_l dx_0 + dx_l) \\ &= dx_k dx_l - e_k dx_0 \wedge dx_l - e_l dx_k dx_0 \wedge dx_0 \\ dz_k \wedge dz_0 &= (-e_k dx_0 + dx_k) \wedge dx_0 = dx_k \wedge dx_0 \quad (k,l = 1,\ldots,m) \end{aligned} \tag{26}$$

A basic hypercomplex p-form is the outer product of p different basic hypercomplex 1-forms.

The advantage of such basic differential 1-forms can briefly described by the following corollaries of Def.4 easy to verify:

1. The outer product is alternative: $dz_k \wedge dz_l = -dz_l \wedge dz_k$, $k \neq l$, $dz_k \wedge dz_k = 0$.
2. The ordered m-form $d\vec{z} = dz_1 \wedge \cdots \wedge dz_m$ is the hypercomplex surface-element and $dz_0 \wedge d\vec{z} = dx_0 \wedge dx_1 \wedge \cdots \wedge dz_m$ is the volume-element which are occuring in the hypercomplex form of Stokes' theorem.
3. If we consider the m-form $\omega = fd\vec{z}g$ both facts, together with the differential of f following from (13) as $df = fDdz_0 + \frac{\partial f}{\partial x_1}dz_1 + \cdots + \frac{\partial f}{\partial x_m}dz_m = dz_0 Df + dz_1\frac{\partial f}{\partial x_1} + \cdots + dz_m\frac{\partial f}{\partial x_m}$ lead to $d\omega = d(f\vec{z}g) = (fD.g + f.Dg)dz_0 \wedge d\vec{z}$.

Using Stokes'theorem for differential forms we get immediatly the hypercomplex Cauchy theorem for monogenic functions. In [10] it has been shown that using the notion of hypercomplex differentiability the validity of the Cauchy theorem can be proved without the assumption of real differentabilty. Therefor Goursat's proof for the plane case was generalized.

3. Singular integral operators The notion of hypercomplex differentiability can be used to investigate the existence and the behaviour of solution of a non-linear nonhomogeneous hypercomplex differential equation for $\mathcal{A}$-valued functions w defined in $\mathbf{H}^m$ of the form $Dw = F(z, w, \nabla_{\bar{z}} w)$. Therefor a system of equivalent integro-differential equations can be considered. (For the plane case cf.e.g.[14]). Besides the well usual inverse integral operator (cf.e.g.[1]) T of D with the Cauchy kernel $E(\zeta, z)$ from (24), in [9] was defined and discussed a singular integral operator of the form

$$\vec{\Pi}_\Omega f = -\frac{1}{\omega}\int_\Omega \vec{K}(\zeta, z) f(\zeta) d\Omega_\zeta, \quad \text{with } \vec{K}(\zeta, z) = \nabla_{\bar{z}} E(\zeta, z)$$

acting e.g. on Hoelder-continuous functions. The relation between T and $\vec{\Pi}$ which is appropriate for the proof of fix-point theorems for the derived integral system is given by (cf. also [11] about Bergmann kernel methods)

$$\nabla_{\bar{z}} Tf = \vec{\Pi} f + \frac{1}{m+1} f\hat{i}.$$

References

1. Brackx, F., Delanghe, R. and Sommen, F., *Clifford Analysis*, Pitman research notes in mathematics **76**, Pitman 1982
2. Delanghe, R., Sommen, F. and Souček, V., *Clifford algebra and spinor-valued functions* Kluwer Academic Publishers, Dordrecht 1992
3. Fueter, R., Über die analytische Darstellung der regulären Funktionen einer Quaternionen-variablen, *Comment. Math. Helv.*, 1934/36, Vol. **8**, pp. 371-378
4. Habetha, K. Function theory in algebras,in *Complex Analysis, Methods, Trends and Applications*, ed. by E. Lanckau and W. Tutschke, Akademie-Verlag, Berlin, **1983**
5. Laville, G., On Cauchy-Kowalewsky Extension, *Journal of Functional Analysis*, 1991, Vol. **101**,No. 1, pp. 25-37
6. Malonek, H., A new hypercomplex structure of the Euclidean space $\mathbf{R}^{m+1}$ and the concept of hypercomplex differentiability, *Complex Variables Theory Appl.*, 1990, Vol. 14, pp. 25-33
7. Malonek, H., Power series representation for monogenic functions in $\mathbf{R}^{m+1}$ based on a permutational product, *Complex Variables Theory Appl.*, 1990, Vol. 15, pp. 181-191
8. Malonek, H., The concept of hypercomplex differentiability and related differential forms, in *Studies in complex analysis and its applications to partial differential equations 1* ed. by R. Kühnau and W. Tutschke, Pitman **256**, Longman 1991, pp.193-202
9. Malonek, H. and Müller, B., Definition and properties of a hypercomplex singular integral operator, *Results in Mathematics*, 1992, Vol. **22** pp. 713-724
10. Malonek, H., and Wirthgen, B. Zur Übertragung des Goursatschen Beweises des Cauchyschen Integralsatzes auf hyperkomplex differenzierbare Funktionen im $\mathbf{R}^{m+1}$, *Wiss. Zeitschr. PH Halle-Köthen*, 1990, Vol. XXVIII,4, pp. 34-38
11. Shapiro,M. V. and Vasilevski, N. L., On the Bergmann Kernel Function in Clifford Analysis, *see these Proceedings*
12. Sommen, F., Monogenic differential forms and homology theory, *Proc. R. Irish Acad.*, 1984, Vol. **84A**,No. **2**, pp. 87-109
13. Sudbery, A., Quaternionic analysis, *Math. Proc. Camb. Phil. Soc.*, 1979, Vol. **85**, pp. 199-225
14. Vekua, I. N., *Generalized analytic functions*, Reading 1962

CLIFFORD ALGEBRAS AND BOUNDARY ESTIMATES FOR HARMONIC FUNCTIONS

MARIUS MITREA
*Institute of Mathematics of the Romanian Academy, P.O.Box 1-764 RO-70700 Bucharest, Romania**

Abstract. A Clifford algebra technique approach for proving boundary estimates for harmonic functions in nonsmooth domains is presented. In particular these estimates are used for studying the Dirichlet and Neumann problems for the Laplace operator on Lipschitz domains in a unified manner.

1. Introduction

There is a deep interdependence between the theory of harmonic functions of two real variables and that of analytic functions of one complex variable (we do not attempt to provide an account of the vast literature on this subject; however see [20]).

The key fact whatsoever is that, at least locally, any harmonic function is the real part of an analytic one. Nevertheless, for $n \geq 3$ real variables the connection between harmonic and analytic functions ceases to be so effective.

In this note we present some instances where it is possible to use Clifford algebras as a substitute for the field of complex numbers in higher dimensions in proving certain results of the theory of harmonic functions of several real variables (cf. also [14]).

The point of view we adopt here is fundamentally very simple. Every harmonic function suitably small at infinity is the real part of a Clifford-analytic function preserving some of its characteristics. Some harmonic and Clifford-analytic Hardy type spaces arise as the most natural setting within which our results can be stated.

In this framework, some Riesz type transforms are introduced and studied. Essentially, they are the vehicle for passing from harmonic functions to Clifford-analytic functions. In particular, they are used to produce boundary estimates and deduce an integral representation formula for functions in these Hardy spaces.

We conclude by briefly indicating how these estimates can be used in the treatment of the classical boundary value problems for harmonic functions in Lipschitz domains. Although part of them can also be obtained by other means as e.g. the techniques developed in [4], [5], [6], [19], the present approach is particularly simple and gives a new insight to the problem.

* *Current address :* Department of Mathematics, University of South Carolina, Columbia, SC 29208, USA.

F. Brackx et al. (eds.), *Clifford Algebras and their Applications in Mathematical Physics*, 151–158.

2. Preliminaries

First recall the definition of the real 2^n–dimensional Clifford algebra $\mathbb{R}_{(n)}$ as the real algebra freely generated by the standard basis $e_0, e_1, ..., e_n$ in $\mathbb{R}^{n+1}$ subject to the conditions $e_0 = 1$ and $e_j e_k + e_k e_j = -2\delta_{jk}$ for $1 \leq j, k \leq n$ (we refer to e.g. [1], [9] for the very basic facts about $\mathbb{R}_{(n)}$). Note that e.g. $\mathbb{R}_{(0)}$ is the field of real numbers, $\mathbb{R}_{(1)}$ is the field of complex numbers and $\mathbb{R}_{(2)} = \mathbb{H}$ the quaternionic skew field, respectively.

Canonically we embed $\mathbb{R}^{n+1}$ in $\mathbb{R}_{(n)}$. For $x \in \mathbb{R}_{(n)}$, $\operatorname{Re} x$, *the real part of* x, will stand for the e_0–component of x and $\operatorname{Im} x := x - (Re\, x)e_0$.

We also equip $\mathbb{R}_{(n)}$ with the Euclidean norm $|x|^2 := Re\,(x\overline{x})$, where the conjugation $-$ is the unique linear morphism of $\mathbb{R}_{(n)}$ with $\overline{e_0} = e_0$, $\overline{e_j} = -e_j$ for $1 \leq j \leq n$ and $\overline{xy} = \overline{y}\,\overline{x}$ for all $x, y \in \mathbb{R}_{(n)}$.

Suggested by the case $n = 1$, call a $\mathbb{R}_{(n)}$–valued function F in $\mathbb{R}^{n+1}$ *Clifford-analytic*, provided it is annihilated by *the Dirac operator* $D := \sum e_j \partial_j$, i.e.

$$DF := \sum_{j=0}^{n} e_j \partial_j F = 0.$$

Introducing $\overline{D} := \sum \overline{e_j} \partial_j$ it is easy to see that $D\overline{D} = \overline{D}D = \Delta :=$ the Laplacian in $\mathbb{R}^{n+1}$ so that, as in the classical case, the components of a Clifford-analytic function are harmonic functions.

We conclude this section by stating a result whose verification is straightforward.

Proposition 1 *Given the* $(n+1)$–*tuple of real valued functions* $U = (u_j)_{j=0}^{n}$, *then* $F := u_0 e_0 - u_1 e_1 - ... - u_n e_n$ *is Clifford-analytic if and only if* U *is a system of conjugate harmonic functions in the sense of Moisil-Teodorescu [15], [16], i.e.* U *satisfies the generalized Cauchy-Riemann equations* $div\, U = 0$, $curl\, U = 0$.

3. Hardy Spaces in Lipschitz Domains

In this section we record some facts concerning Hardy spaces of Clifford-analytic functions that will be relevant for us in the sequel.

Recall that *an unbounded Lipschitz domain* Ω in $\mathbb{R}^{n+1}$ is the domain located above the graph of a Lipschitz function $\varphi : \mathbb{R}^{n+1} \longrightarrow \mathbb{R}$, i.e. $\Omega := \{(t, x)\,;\, \varphi(x) < t\}$.

Next, for a $\mathbb{R}_{(n)}$–valued function F in Ω, define *the nontangential maximal function* $F^*(X) := \|F\|_{\infty, \Gamma + X}$, $X \in \partial\Omega$, where Γ denotes a fixed sufficiently sharp upright cone in $\mathbb{R}^{n+1}$ centered at the origin of the system.

Also, *the Lusin area-function* is

$$\mathcal{A}(F)(X) := \left(\iint_{\Gamma + X} |\partial_0 F(Y)|^2 |Y - X|^{1-n} dY \right)^{1/2}, \quad X \in \partial\Omega,$$

and its radial analogue, *the Littlewood-Paley g-function*

$$g(F)(X) := \left(\int_0^{+\infty} |\partial_0 F(X + te_0)|^2 t dt \right)^{1/2}, \quad X \in \partial\Omega.$$

Next recall that *the higher dimensional Cauchy operator* $\mathcal{C}$ acting functions f defined on $\partial\Omega$ which are $I\!R_{(n)}$–valued by

$$\mathcal{C}f(X) := \frac{1}{\sigma_n}\int_{\partial\Omega}\frac{\overline{Y-X}}{|Y-X|^{n+1}}N(Y)f(Y)dS(Y), \quad X\in\Omega.$$

Here $N(Y)$ is the outward unit normal of Ω defined for a.e. $Y \in \partial\Omega$, σ_n is the area of the unit sphere in $I\!R^{n+1}$ and the integrand must be interpreted in the sense of pointwise multiplication in $I\!R_{(n)}$ (see also [1]). As in the classical case it reproduces Clifford-analytic functions in a neighborhood of $\overline{\Omega}$ having a suitable decay at infinity. Another useful property is that for real-valued functions on $\partial\Omega$ one has $Re\,\mathcal{C} = \mathcal{D}$ =*the double layer potential operator* (see e.g. [13]).

The following version of a result from [14] will be of importance for us.

Theorem 1 *Consider* $1 < p < +\infty$ *and* $\omega \in A_p$, *the class of Muckenhoupt. Then, for a Clifford-analytic function* F *in* Ω, *the following are equivalent:*

(1) $F^* \in L^p(\partial\Omega, \omega dS)$;

(2) $sup_{\delta>0}||F(\cdot+\delta e_0)||_{L^p(\partial\Omega,\omega dS)} < +\infty$;

(3) $\lim_{t\to\infty} F(X+te_0) = 0$ *for some* $X \in \Sigma$ *and* $\mathcal{A}(F) \in L^p(\partial\Omega, \omega dS)$;

(4) $\lim_{t\to\infty} F(X+te_0) = 0$ *for some* $X \in \Sigma$ *and* $g(F) \in L^p(\partial\Omega, \omega dS)$;

(5) F *has a nontangential boundary limit (i.e. the the approach to the boundary takes place within a vertical cone) at a.e. point of* $\partial\Omega$. *The limit function, denoted by the same symbol* F, *belongs to* $L^p(\partial\Omega, \omega dS)$, *and* $F = \mathcal{C}F$.

In addition, in the above conditions,

$$||F^*||_{L^p_\omega} \approx sup_{\delta>0}||F(\cdot+\delta e_0)||_{L^p_\omega} \approx ||\mathcal{A}(F)||_{L^p_\omega} \approx ||g(F)||_{L^p_\omega} \approx ||F||_{L^p_\omega}. \quad (1)$$

It is natural then to introduce *the Hardy spaces* $\mathcal{H}^p_\omega(\Omega)$ as the collection of all Clifford-analytic functions satisfying one of the conditions (1)-(5) listed above. Also, we let $||F||_{\mathcal{H}^p_\omega}$ denote any of the equivalent quantities from (1).

4. Harmonic Hardy Spaces and Riesz Transforms

The main aim of this section is to investigate the spaces

$$H^p_\omega(\Omega) := \{u \ harmonic in\Omega\,;\, u^* \in L^p(\partial\Omega, \omega dS)\},$$

which we endow with the norm $||u||_{H^p_\omega} := ||u^*||_{L^p(\partial\Omega,\omega dS)}$. Throughout the section we shall assume that $1 < p < +\infty$ and that $\omega \in A_p$, the class of Muckenhoupt.

In the light of Proposition 1, for a real-valued harmonic function u in Ω, set $\mathcal{R}_j u$ for the j–th harmonic conjugate of u, i.e. $\mathcal{R}_j u := u_j$ provided there exists $U = (u_j)_{j=0}^n$ system of conjugate harmonic functions in Ω with $u_0 = u$ and $\lim_{t\to\infty} U(X_0+te_0) = 0$ for some $X_0 \in \Omega$.

In the sequel, $\mathcal{R}_j$ will be referred to as *the jth Riesz transform on* Ω. The connection with the classical jth Riesz transform R_j on $I\!R^n$ (see [17]), comes from the fact that for the upper-half space case, any element u in $H^p_\omega(I\!R^{n+1}_+)$ has the form $u = P_t * f$ for some function f in $L^p(I\!R^n, \omega dx)$ (P_t being the Poisson kernel in $I\!R^{n+1}_+$; see [11] p.177) so that $\mathcal{R}_j u = P_t * (R_j f)$. Nevertheless, as in the upper-half space, we could have defined these Riesz transforms in terms of the boundary trace of u but generally this requires specific restrictions for p and ω; see [4].

The basic properties of these operators are collected in the next theorem.

Theorem 2 *The following hold true.*

(1) The operators $(\mathcal{R}_j)_j$ *are well-defined and bounded on* $H^p_\omega(\Omega)$.

(2) $\mathcal{R}_j\mathcal{R}_k = \mathcal{R}_k\mathcal{R}_j$ *and* $\sum_{j=1}^n \mathcal{R}_j{}^2 = -I$.

(3) The mappings $\Phi : \mathcal{H}^p_\omega(\Omega) \longrightarrow H^p_\omega(\Omega)$, $\Phi(F) := \mathrm{Re}\, F$, *and* $\Psi : H^p_\omega(\Omega) \longrightarrow \mathcal{H}^p_\omega(\Omega)$, $\Psi(u) := u - \sum_j e_j\mathcal{R}_j u$, *are well-defined, bounded and inverse to each other.*

Proof Let u be arbitrarily chosen in $H^p_\omega(\Omega)$. Defining $U(X) := -\int_0^\infty \nabla u(X + te_0)dt$ and with the same arguments as in [8] it is easy to see that $U =: (u_j)_{j=0}^n$ exists and is a system of conjugate harmonic functions in Ω. In fact, as $\partial_0 U = \nabla u_0$, this is the unique system of conjugate harmonic functions in Ω with $u_0 = u$ and which vanishes at infinity.

Then using the results of [2] and arguing on the same lines as in [3] we obtain that $u_j \in H^p_\omega(\Omega)$ and $\|u_j^*\|_{L^p(\partial\Omega,\omega dS)} \le C\|u^*\|_{L^p(\partial\Omega,\omega dS)}$ for all j, i.e. the operator $\mathcal{R}_j u = u_j$ is well-defined and continuous for all j.

Finally, for any $X \in \Omega$,

$$\sum_{j=1}^n (\mathcal{R}_j)^2 u(X) = \sum_{j=1}^n - \int_0^{+\infty} \partial_j(\mathcal{R}_j u)(X+te_0)dt = \int_0^{+\infty} (\partial_0 u)(X+te_0)dt = -u(X).$$

The commutation relations follow similarly, Q.E.D.

We note some immediate corollaries. The first one provides boundary estimates for the Riesz transforms in terms of the nontangential maximal operator.

Corollary 1 *Any harmonic function u from $H^p_\omega(\Omega)$ is the real part of a Clifford-analytic function F from $\mathcal{H}^p_\omega(\Omega)$. In fact, we can take $F := u - \sum_j e_j\mathcal{R}_j u$ so that u and $\mathcal{R}_j u$ have nontangential boundary limits in $L^p(\partial\Omega, \omega d\sigma)$. Also,*

$$\|u\|_{L^p(\partial\Omega,\omega dS)} + \sum_{j=1}^n \|\mathcal{R}_j u\|_{L^p(\partial\Omega,\omega dS)} \approx \|u^*\|_{L^p(\partial\Omega,\omega dS)},$$

uniformly for $u \in H^p_\omega(\Omega)$.

Corollary 2 *Let* $1 < p, q < +\infty$ *and* $\omega \in A_p$, $\mu \in A_q$. *Then for* $u \in H^p_\omega(\Omega)$ *one has that* $u \in H^q_\mu(\Omega)$ *if and only if the boundary trace of* $\mathcal{R}_j u$ *is in* $L^q(\partial\Omega, \mu dS)$ *for* $j = 0, 1, ..., n$.

Our last application concerns an integral representation for functions in $H^p_\omega(\Omega)$ in terms of the boundary traces of the Riesz transforms.

To this end, we recall that note the reproducing formula $CF = F$, for any $F \in \mathcal{H}^p_\omega(\Omega)$, which we shall use for $F := u - \sum_j e_j \mathcal{R}_j u$, u in $H^p_\omega(\Omega)$. In order to take the real parts of both sides we need the following simple identity

$$Re\,(xyz) = -2(Re\,x)(Re\,y)(Re\,z) + \sum Re\,x\,\langle y, \overline{z}\rangle,$$

valid for any $x, y, z \in I\!R^{n+1} \subseteq I\!R_{(n)}$. Here $\langle\cdot,\cdot\rangle$ stands for the usual inner product in $I\!R^{n+1}$. Using this for $x = \overline{(Y-X)}/|Y-X|^{n+1}$, $y = N$ and $z = u - \sum_j e_j \mathcal{R}_j u$, we finally obtain the following.

Corollary 3 *For any u in $H^p_\omega(\Omega)$ one has*

$$u(X) = \frac{1}{2}\mathcal{D}u(X) + \frac{1}{\sigma_n}\int_{\partial\Omega}\Big(\sum_{j=1}^{n} N_j \mathcal{R}_j u\Big)(Y)\frac{Re\,(Y-X)}{|Y-X|^{n+1}}dS(Y)$$

$$-\frac{1}{\sigma_n}\int_{\partial\Omega}\frac{Re\,N(Y)}{|Y-X|^{n+1}}\sum_{j=1}^{n}(\mathcal{R}_j u)(Y)(Y-X)_j\,dS(Y),$$

$X \in \Omega$, where $\mathcal{D}$ stands for the usual double layer potential operator in Ω and the subscript j denotes the e_j–coordinate of the corresponding Clifford number.

5. L^2–Boundary Estimates and Applications

In this section we shall specialize some of our results for the case $p = 2$ and $\omega \equiv 1$ (we shall therefore omit to write ω as a subscript).

First we note a boundary cancellation property for functions in $\mathcal{H}^2(\Omega)$ (see [14] for a proof).

Proposition 2 *For any $I\!R^{n+1}$–valued function F in $\mathcal{H}^2(\Omega)$ one has*

$$\int_{\partial\Omega} F\,N\,F dS = 0. \tag{2}$$

For $n = 1$ this statement corresponds precisely to the Cauchy theorem for the analytic function F^2.

The main result of this section is the following.

Theorem 3 *For any $I\!R^{n+1}$–valued function F in $\mathcal{H}^2(\Omega)$ one has*

$$||F||_{L^2(\partial\Omega)} \approx ||Re\,F||_{L^2(\partial\Omega)} \approx ||Re\,(FN)||_{L^2(\partial\Omega)} \approx ||FN||_{L^2(\partial\Omega)} \approx ||Im\,(FN)||_{L^2(\partial\Omega)}.$$

The proof is an immediate consequence of Schwarz inequality, the fact that $\mathrm{Re}\,N \leq C < 0$ a.e. on $\partial\Omega$ and the identities contained in the following lemma.

Lemma 1 *In the above hypotheses, one has*

$$2\int_{\partial\Omega} Re\,(FN)F dS = -2\int_{\partial\Omega} Im\,(FN)F dS = \int_{\partial\Omega}\overline{N}|F|^2 dS. \tag{3}$$

Proof Everything can be seen from (2) and $2\,Re\,(FN) = FN + \overline{FN} = FN + \overline{NF}$, $2\,Im(FN) = FN - \overline{FN} = FN - \overline{NF}$, respectively. Q.E.D.

Next, we deduce the corresponding estimates at the boundary in terms of harmonic functions.

Theorem 4 *If u is harmonic in Ω and $(\nabla u)^* \in L^2(\partial\Omega)$, then*

$$\|(\nabla u)^*\|_{L^2(\partial\Omega)} \approx \|\nabla u\|_{L^2(\partial\Omega)} \approx \left\|\frac{\partial u}{\partial N}\right\|_{L^2(\partial\Omega)} \approx \|\nabla_T u\|_{L^2(\partial\Omega)}, \tag{4}$$

where ∇_T is the usual tangential gradient; see [19]. As a corollary, the L^2-Neumann problem for Laplace's operator in Ω has a unique solution modulo an additive constant.

Proof Considering $F := \overline{D}u$, the hypotheses imply that F is a $I\!R^{n+1}-$valued function in $\mathcal{H}^2(\Omega)$. Furthermore, if $\langle \cdot, \cdot \rangle$ stands for the usual inner product in $I\!R^{n+1}$ (or $I\!R_{(n)}$), then

$$\frac{\partial u}{\partial N} = \langle \overline{D}u, \overline{N}\rangle = \langle F, \overline{N}\rangle = Re\,(FN),$$

while $|\nabla_T u| = |Im\,(FN)|$, and since $|F| = |\nabla u|$, everything is a consequence of Theorem 1 and Theorem 3, Q.E.D.

It is a well-known fact that the last equivalence in (4) is the key element that Verchota [19] used for proving the invertibility of the double layer potential on Lipschitz graphs.

Before stating the next result, recall that for a harmonic function in Ω, area- and g-functions are defined as before except the fact that ∂_0 is replaced by the usual gradient.

Theorem 5 *If $u \in H^2(\Omega)$, then*

$$\|u^*\|_{L^2(\partial\Omega)} \approx \|g(u)\|_{L^2(\partial\Omega)} \approx \|\mathcal{A}(u)\|_{L^2(\partial\Omega)} \approx \|u\|_{L^2(\partial\Omega)}.$$

In particular, the L^2-Dirichlet problem for Laplace's operator in Ω (see [4]) has a unique solution.

Proof Using Corollary 1, we realize u as the real part of a $I\!R^{n+1}-$valued function F in $\mathcal{H}^2(\Omega)$, i.e. $F = u_0e_0 - u_1e_1 - \ldots - u_ne_n$ where $(u_j)_{j=0}^n$ is a system of conjugate harmonic functions with $u_0 = u$. Hence

$$|\partial_0 F|^2 = \sum_{j=0}^n |\partial_0 u_j|^2 = \sum_{j=0}^n |\partial_j u_0|^2 = |\nabla u|^2,$$

thus $\mathcal{A}(F) = \mathcal{A}(u)$ and $g(F) = g(u)$. Finally, by Theorem 1,

$$\|u^*\|_{L^2(\partial\Omega)} \geq \|u\|_{L^2(\partial\Omega)} = \|Re\,F\|_{L^2(\partial\Omega)} \approx \|F\|_{L^2(\partial\Omega)} \approx \|F^*\|_{L^2(\partial\Omega)} \geq \|u^*\|_{L^2(\partial\Omega)}$$

and the conclusion follows. Q.E.D.

We conclude by discussing the following corollary (cf. also [5]), and some of its consequences.

Corollary 4 *For any u in $H^2(\Omega)$,*

$$\int_{\partial\Omega} |u|^2 dS \approx \iint_{\Omega} |\nabla u|^2 dist\{X, \partial\Omega\} dX.$$

Proof Obviously, this is an equivalent formulation of $||u||_{L^2(\partial\Omega)} \approx ||g(u)||_{L^2(\partial\Omega)}$. Q.E.D.

It has been shown in [7] that for any harmonic function u in $H^2(\Omega)$,

$$\iint_{\Omega} |\nabla u|^2 dist\{X, \partial\Omega\} dX \approx \iint_{\Omega} \iint_{\Omega} \frac{|u(X) - u(Y)|^2}{|X - Y|^{n+2}} dX dY.$$

In conjunction with the above corollary, this implies that $H^2(\Omega)$ is a subset of the homogeneous Sobolev-Besov space $W^{1/2}(\Omega)$ (see e.g. [10]) and that, for any $u \in H^2(\Omega)$ we have $||u||_{H^2} \approx ||u||_{W^{1/2}}$.

Finally, it worth mentioning that actually $H^2(\Omega)$ coincide precisely with the subset of harmonic functions from $W^{1/2}(\Omega)$ (cf. [7]).

References

1. Brackx, F., Delanghe, R. and Sommen, F. "Clifford Analysis", Pitman Advanced Pub. Program (1982).
2. Banuolos, R. and Moore, C., N. "Sharp estimates for the nontangential maximal function and the Lusin area function in Lipschitz domains", Trans. Amer. Math. Soc., 312 (1989), 641–662.
3. Brukholder, D., L. and Gundy, R., F. "Distribution functions inequalities for the area integral", Studia Math., 44 (1972), 527–544.
4. Dahlberg, B., E., J. "On estimates of harmonic measures", Arch. Rational Mech. and Anal. 65 (1977), 272–288.
5. Dahlberg, B., E., J. "Weighted norm inequalities for the Lusin area integral and the nontangential maximal function for functions harmonic in a Lipschitz domain", Studia Math. 67 (1980), 297–314.
6. Dahlberg, B., E., J. and Kenig, C., E. "Hardy spaces and the L^p–Neumann problem for Laplace's equation in a Lipschitz domain", Ann. of Math. 125 (1987), 437–465.
7. Fabes, E., B. "Layer potential methods for boundary value problems on Lipschitz domains", Lecture Notes, Springer Verlag, 1344 (1987), 55-80.
8. Fefferman. C. and Stein, E., M. "H^p spaces of several variables", Acta Math. 129 (1972), 137–193.
9. Gilbert, J. and Murray, M., A. "Clifford Algebras and Dirac Operators in Harmonic Analysis", C.P.U., Cambridge, (1991).
10. Grisvard, P. "Elliptic Problems in Nonsmooth Domains", Pitman Advanced Publishing Program (1985).
11. Gundy, R., F. and Wheeden, R., L. "Weighted integral inequalities for the nontangential maximal function, Lusin area integral and Walsh-Paley series", Studia Math., 49 (1974), 107–124.
12. Jerison, D. and Kenig, C. "The Neumann problem on Lipschitz domains", Bull. A.M.S., 4 (1981), 203–207.
13. McIntosh, A. "Clifford algebras and the higher dimensional Cauchy integral", Approximation theory and Function Spaces, Banach Center Publications, 22 (1989) 253–267.
14. Mitrea, M. "Singular integrals, Hardy spaces and Clifford wavelets", submitted for publication, (1992).
15. Moisil, G., C. "Sur l'équation $\Delta u = 0$", C. R. Acad. Sci. Paris, 191 (1930) p.984.
16. Moisil G., C. and Teodorescu, N. "Fonctions holomorphes dans l'espace", Mathematica Cluj, 5 (1931), 142–150.
17. Stein, E., M. "Singular Integrals and Differentiability Properties of Functions", Princeton University Press, Princeton (1970).

18. Stein, E., M. and Weiss, G. "On the theory of harmonic functions of several variables, I. The theory of H^p spaces", Acta Math., 103 (1960), 25–62.
19. Verchota, G. "Layer potentials and regularity for the Dirichlet problem for Laplace's equation in Lipschitz domains", Journal of Functional Analysis 59 (1984), 572-611.
20. Zygmund, A. "Trigonometric series", Cambridge (1959).

REGULARITY OF FUNCTIONS WITH VALUES IN CLIFFORD ALGEBRA BASED ON A GENERALIZED AXIALLY SYMMETRIC POTENTIAL THEORY OPERATOR

K. NÔNO
Department of Mathematics, Fukuoka University of Education ,
729 Akama, Munakata, Fukuoka 811-41, JAPAN

Abstract. In this paper,we give a regularity of Clifford valued functions based on a generalized axially symmetric potential theory operator and develop a regular function theory. Also, we give a link of regular functions by Riemann approach and regular functions by Weierstrass approach.

Key words: Clifford algebra, Generalized axially symmetric potential theory operator.

1. Introduction

The Clifford algebra was constructed by W.K.Clifford in 1878 as a generalization of the algebra of quaternions. It was studied in 1930s in connection with the theory of spinors([2-5,15-18,25]) and has been studied in Mathematics and Physics([1,6-8,10-14,19-24,26-30]).

The generalized Clifford algebra was constructed as a generalization of the ordinary Clifford algebra and it was well studied ([18-19,23-24]). T.Nôno([23,24]) generalized the concept of the linearization of the wave equation from another point of view. Also, in [22], we give a characterization of linearizations of a second order partial differential equation.

In the holomorphic function theory of one complex variable, there exist three distinct approaches (Riemann approach based on Cauchy-Riemann equation, Cauchy approach based on complex differentiability, Weierstrass approach based on power series). It is well known that these three approaches are equivalent.

In the quaternionic analysis, R.Fueter([8]) has given a regularity of quaternionic functions as smooth solusions of a quaternionic differential operator(generalized Cauchy-Riemann operator) and Fueter and other authors developed a theory of quaternionic regular functions.

Also, in Clifford analysis, R.Delanghe([7]) gave a regularity of functions with valued in Clifford algebra as smooth solusions of generalized Cauchy-Riemann equation which is a linearization of Laplace equation(Riemann approach) and R.Delanghe, F.Brackx and F.Sommen([1]) have developed the function theory (monogenic function theory). If a regularity of functions is given by the differentiability(Cauchy approach),the class of regular functions is too small([30]). If a regularity of functions is given by expansions of power series(Weierstrass approach), the class of regular functions is too large.

In this paper, we consider a second order partial differential equation which is

F. Brackx et al. (eds.), *Clifford Algebras and their Applications in Mathematical Physics* , 159–166.

called the generalized axially symmetric potential theory operator([9]):

$$L_n u = \sum_{i=0}^{n} \frac{\partial^2 u}{\partial x_i^2} - \frac{n-1}{x_n}\frac{\partial u}{\partial x_n} = 0.$$

We give a regularity of functions with values in Clifford algebra as smooth solusions of a linearization of $L_n u$=0 and developed a regular function theory. Also,we give a link of regular functions by Riemann approach and regular functions by Weierstrass approach.

2. Preliminary

Let $A_n(R)$ be a real Clifford algebra over an n-dimensional vector space with basis $\{e_1, e_2, ..., e_n\}$. Then,it is well known that $A_n(R)$ is a real 2^n-dimensional associative,but non-commutative algebra and its basis $\{e_0, e_1, ..., e_n, ..., e_1e_2 \ldots e_n\}$ satisfy the rule: $e_ie_j + e_je_i = -2\delta_{ij}e_0$ $(i, j = 1, 2, ..., n)$, $e_0 = 1$. Since the subspace of $A_n(R)$ spanned by the base elements $\{e_0, e_1, ..., e_n\}$ is identified with R^{n+1}, we denote the subspace by R^{n+1}.

Let $f = \sum_{i=0}^{n} e_i f_i$ be a function with values in R^{n+1}. Then, we see that the following system of first order partial differential equations is a linearization of $L_n f_0 = 0$:

$$Df = -\frac{(n-1)f_j}{x_j} \quad (j = 1, 2, ..., n),$$

where $D = \sum_{i=0}^{n} e_i \frac{\partial}{\partial x_i}$

Let G be a domain in R^{n+1} and $f = \sum_{i=0}^{n} e_i f_i$ be a function defined in G with values in R^{n+1},where $x = \sum_{i=0}^{n} e_i x_i$.

DEFINITION 1. f **is regular in** G iff

(1) f_i is smooth in G $(i = 0, 1, ..., n)$,

(2) $Df = -(n-1)f_j/x_j$ in G $(j = 1, 2, ..., n)$,

where $D = \sum_{i=0}^{n} e_i \frac{\partial}{\partial x_i}$.

REMARK 1. In case $n = 1$, the regular functions are usual holomorphic functions in the complex function theory. Therefore,the regularity of Definition 1 is a natural generalization of the holomorphy in the complex function theory.

REMARK 2. The condition (2) in Definition 1 is equivalent to the system of the following differential equations:

(1) $\frac{\partial f_0}{\partial x_0} - \sum_{i=1}^{n} \frac{\partial f_i}{\partial x_i} = -\frac{(n-1)}{x_k} f_k (k = 1, 2, ..., n)$,

(2) $\frac{\partial f_i}{\partial x_0} = -\frac{\partial f_0}{\partial x_i} (i = 1, 2, ..., n)$,

(3) $\frac{\partial f_j}{\partial x_i} = \frac{\partial f_i}{\partial x_j} (i, j = 1, 2, ..., n)$

DEFINITION 2. Let G be a domain in R^{n+1}, $u(x)$ be a real valued function defined in G. We say that $u(x)$ satisfies the condition GASPTE in G iff

(1) $u(x)$ is twice continuously differentiable in G,

(2) $L_n u(x) = 0$ in G.

REMARK 3. Let G be a domain in R^{n+1}. If $f(x) = \sum_{i=0}^{n} e_i f_i$ is a twice continuously differentiable and regular function in G, then

(1) $\Delta f_0 = \frac{(n-1)}{x_k}\frac{\partial f_0}{\partial x_k} \quad (k = 1, 2, ..., n)$,

(2) $\Delta f_i = \frac{(n-1)}{x_k}\frac{\partial f_i}{\partial x_k} \quad (i \neq k, i, k = 1, 2, ..., n)$,

(3) $\Delta f_i = \frac{(n-1)}{x_i}\left(\frac{\partial f_i}{\partial x_i} - \frac{f_i}{x_i}\right) \quad (i = 1, 2, ..., n)$

in G, where Δ is Laplacian of $(n + 1)$-real variables. Therefore,$f_0(x)$ satisfies GASPTE in G.

3. Regular functions

In this section, we give several properties of regular functions.

The following two results are obtained by direct calculation.

PROPOSITION 1. *Let G be a domain in R^{n+1} and a , b be real number. If f and g are regular in G, then $af + bg$, $\frac{1}{2}(fg + gf)$ and g^{-1} are regular in G.*

PROPOSITION 2. *A composite function $f \circ g$ of regular functions f and g is regular.*

THEOREM 1. *If $f = \sum_{i=0}^{n} e_i f_i$ is regular in G, then*

$$\sum_{i=0}^{n} \frac{\partial f_0}{\partial x_i}\frac{\partial f_j}{\partial x_i} = 0, \quad (j = 1, 2, ..., n)$$

in G.

PROOF. Since f is regular,by Remark 2, we have that

$$\begin{aligned}
\sum_{i=0}^{n} \frac{\partial f_0}{\partial x_i}\frac{\partial f_j}{\partial x_i} &= \left(\sum_{i=1}^{n} \frac{\partial f_i}{\partial x_i} - \frac{n-1}{x_k} f_k\right)\frac{\partial f_j}{\partial x_0} - \frac{\partial f_1}{\partial x_0}\frac{\partial f_j}{\partial x_1} + \sum_{i=2}^{n} \frac{\partial f_0}{\partial x_i}\frac{\partial f_j}{\partial x_i} \\
&= \left(\sum_{i=2}^{n} \frac{\partial f_i}{\partial x_i} - \frac{n-1}{x_k} f_k\right)\frac{\partial f_j}{\partial x_0} + \sum_{i=2}^{n} \frac{\partial f_0}{\partial x_i}\frac{\partial f_j}{\partial x_i} \\
&= \sum_{i=2}^{n} \left(\frac{\partial f_i}{\partial x_i}\frac{\partial f_j}{\partial x_0} + \frac{\partial f_0}{\partial x_i}\frac{\partial f_j}{\partial x_i} - \frac{f_j}{x_j}\frac{\partial f_j}{\partial x_0}\right)
\end{aligned}$$

$$
\begin{aligned}
&= \sum_{i=2}^{n}\left(\frac{\partial f_i}{\partial x_i}\frac{\partial f_j}{\partial x_0} + \frac{\partial f_0}{\partial x_i}\frac{\partial f_j}{\partial x_i} - \frac{f_j}{x_i}\frac{\partial f_i}{\partial x_0}\right) \\
&= \sum_{i=2}^{n}\left(\frac{\partial f_i}{\partial x_i}\frac{\partial f_j}{\partial x_0} + \left(\frac{\partial f_j}{\partial x_i} + \frac{f_j}{x_i}\right)\frac{\partial f_0}{\partial x_i}\right) \\
&= \sum_{i=2}^{n}\left(\frac{\partial f_i}{\partial x_i}\frac{\partial f_j}{\partial x_0} + \frac{x_j}{x_i}\frac{\partial f_i}{\partial x_i}\frac{\partial f_0}{\partial x_i}\right) \\
&= \sum_{i=2}^{n}\frac{\partial f_i}{\partial x_i}\left(\frac{\partial f_j}{\partial x_0} - \frac{x_j}{x_i}\frac{\partial f_i}{\partial x_0}\right) \\
&= 0.
\end{aligned}
$$

THEOREM 2. *Let G be a domain in R^{n+1}. If $f = \sum_{i=0}^{n} e_i f_i$ is regular in G, then there exists a real valued function $K(x)$ in G such that*

$$
J^2 = \left(\frac{\partial(f_0, f_1, \ldots, f_n)}{\partial(x_0, x_1, \ldots, x_n)}\right)^2 = K(x)\left|\frac{\partial f(x)}{\partial x_0}\right|^2
$$

in G.

Proof. Since f is regular in G, by Theorem 1, we have

$$
\sum_{i=0}^{n}\frac{\partial f_0}{\partial x_i}\frac{\partial f_j}{\partial x_i} = 0, \quad (j = 1, 2, \ldots, n) \tag{1}
$$

in G. From (1), we have that

$$
J^2 = \begin{vmatrix}
\sum_{i=0}^{n}(\frac{\partial f_0}{\partial x_i})^2 & 0 & \cdots & 0 \\
0 & \sum_{i=0}^{n}(\frac{\partial f_1}{\partial x_i})^2 & \cdots & \sum_{i=0}^{n}\frac{\partial f_1}{\partial x_i}\frac{\partial f_n}{\partial x_i} \\
\vdots & \vdots & \vdots & \vdots \\
0 & \sum_{i=0}^{n}\frac{\partial f_n}{\partial x_i}\frac{\partial f_1}{\partial x_i} & \cdots & \sum_{i=0}^{n}(\frac{\partial f_n}{\partial x_i})^2
\end{vmatrix}.
$$

Hence, there exists a real valued function $K(x)$ such that

$$
J^2 = \sum_{i=0}^{n}\left(\frac{\partial f_0}{\partial x_i}\right)^2 K(x)
$$

in G.

4. Analytic regular functions

THEOREM 3. *Let $f(x)$ be a regular homogeneous polynomial of degree m with respect to real variables $x_0, x_1, \ldots, x_n$,then*

$$f(x) = \frac{1}{m!}\frac{\partial^m f(x)}{\partial x_0^m} x^m, \quad x \in R^{n+1}. \tag{2}$$

Proof. Since $f = \sum_{i=0}^{n} e_i f_i$ is a regular homogeneous polynomial,

$$\sum_{i=0}^{n} x_i \frac{\partial f}{\partial x_i} = mf, \quad e_k x_k \sum_{i=0}^{n} e_i \frac{\partial f}{\partial x_i} = -(n-1) e_k f_k, \quad (k = 1, 2, \ldots, n).$$

Hence, we have

$$\begin{aligned} mf &= (n-1)\sum_{k=1}^{n} e_k f_k + \sum_{i=0}^{n} x_i \frac{\partial f}{\partial x_i} + \sum_{k=1}^{n} e_k x_k \sum_{i=0}^{n} e_i \frac{\partial f}{\partial x_i} \\ &= (n-1)\sum_{k=1}^{n} e_k f_k + \sum_{k=0}^{n} e_k x_k \frac{\partial f}{\partial x_0} + \sum_{i=1}^{n} x_i \frac{\partial f}{\partial x_i} + \sum_{k=1}^{n}\sum_{i=1}^{n} e_k e_i x_k \frac{\partial f}{\partial x_i}. \end{aligned}$$

Since f is regular, we have

$$\begin{aligned} &(n-1)\sum_{k=1}^{n} e_k f_k + \sum_{i=1}^{n} x_i \frac{\partial f}{\partial x_i} + \sum_{k=1}^{n}\sum_{i=1}^{n} e_k e_i x_k \frac{\partial f}{\partial x_i} \\ &= (n-1)\sum_{k=1}^{n} e_k f_k + \sum_{i<k} e_k e_i \left(x_k \frac{\partial f}{\partial x_i} - x_i \frac{\partial f}{\partial x_k}\right) \\ &= (n-1)\sum_{k=1}^{n} e_k f_k + \sum_{i<k} e_k e_i \left\{\sum_{j=0}^{n} e_j \left(x_k \frac{\partial f_j}{\partial x_i} - x_i \frac{\partial f_j}{\partial x_k}\right)\right\} \\ &= (n-1)\sum_{k=1}^{n} e_k f_k - \sum_{i<k} e_k e_i \left(x_k \frac{\partial f_i}{\partial x_i} - x_i \frac{\partial f_i}{\partial x_k}\right) + \sum_{i<k} e_k e_i \left(x_k \frac{\partial f_0}{\partial x_i} - x_i \frac{\partial f_0}{\partial x_k}\right) \\ &= \sum_{i<k} e_k \left\{(n-1) f_k - x_k \frac{\partial f_i}{\partial x_i} + x_i \frac{\partial f_i}{\partial x_k}\right\} + \sum_{i<k} e_k e_i \left(x_k \frac{\partial f_0}{\partial x_i} - x_i \frac{\partial f_0}{\partial x_k}\right) \\ &= \sum_{k=1}^{n} e_k \sum_{i<k} \left\{(n-1) f_k - x_k \frac{\partial f_i}{\partial x_i} + x_i \frac{\partial f_i}{\partial x_k}\right\} + \sum_{i<k} e_k e_i \left(x_k \frac{\partial f_k}{\partial x_0} - x_i \frac{\partial f_i}{\partial x_0}\right) \\ &= 0. \end{aligned}$$

Hence, $mf = x\frac{\partial f}{\partial x_0}$. Repeating the above argument, We obtain (2).

THEOREM 4. *If $f(x)$ is expanded in a power series of real coefficients:*

$$f(x) = \sum_{n=0}^{\infty} a_n x^n \tag{3}$$

in a neighbourhood U of the point $0 \in R^{n+1}$,then $f(x)$ is regular in U.

Proof. Put $f(x) = \sum_{i=0}^{n} e_i f_i(x)$ and $P_k(x) = a_k x^k (k = 0, 1, ...)$. Since the polynomials $P_k(x)(k = 0, 1, ...)$ are all regular,

$$DP_k(x) = -\frac{P_{kj}(x)}{x_j}.$$

in U. Therefore, by the termwise differentiation of (3), we have that

$$Df = \sum_{k=0}^{\infty} DP_k = -\frac{1}{x_j} \sum_{k=0}^{\infty} P_{kj} = -\frac{f_j}{x_j}, \quad (j = 1, 2, ..., n)$$

in U. Hence, f is regular in U.

THEOREM 5. *If $f(x)$ is an analytic function of real variables $x_0, x_1, ..., x_n$ in a neighbourhood U of the point $0 \in R^{n+1}$ and if $f(x)$ is regular in U, then*

$$f(x) = \sum_{n=0}^{\infty} a_n x^n, \quad a_n = \frac{1}{n!} \frac{\partial^n f}{\partial x_0^n}(0), \quad x \in U.$$

Proof. Since $f(x)$ is analytic with respect to real variables $x_0, x_1, ..., x_n$ in U,$f(x)$ is expanded in a series of homogeneous polynomials $P_k(x)$ of variables $x_0, x_1, ..., x_n$:

$$f(x) = \sum_{k=0}^{\infty} P_k(x), \quad x \in U$$

where, $P_k(x) = \sum_{i=0}^{n} e_i P_{ki}(x)$ is a homogeneous polynomial of degree k. By regularity and analyticity of f, we have that

$$\sum_{k=0}^{\infty} (x_j DP_k - P_{kj}) = x_j Df - f_j = 0 \quad (j = 1, 2, ..., n).$$

Putting $Q_k(x) = x_j DP_k - P_{kj}$, $Q_k(x)$ is a homogemeous polynomial of degree k. Since $\sum_{k=0}^{\infty} Q_k(x) = 0$, we have that

$$Q_k(x) \equiv 0.$$

Hence,$P_k(x)$ is a regular homogeneous polynomial. From Theorem 3, we have

$$f(x) = \sum_{k=0}^{\infty} \frac{1}{k!} \frac{\partial^k P_k}{\partial x_0^k} x^k$$

Since $\frac{\partial^k P_k}{\partial x_0^k} = \frac{\partial^k f}{\partial x_0^k}(0)$, the proof is complete.

REMARK 4. Theorem 4 and 5 say that Riemann approach and Weierstrass approach for Chifford valued analytic functions of real variables $x_0, x_1, ..., x_n$ is equivalent.

5. Elementary functions

We can define the exponential function, trigonometric functions, logarithmic function of a variable x similarly as the complex function theory. These functions are all regular.

For $x = \sum_{i=0}^{n} e_i x_i$, put

$$r = \left(\sum_{i=1}^{n} x_i^2\right)^{\frac{1}{2}}, \quad \omega = \sum_{i=1}^{n} e_i x_i, \quad x^* = x_0 - \omega.$$

[i] Exponential function:

$$e^x = \sum_{n=0}^{\infty} \frac{x^n}{n!}.$$

e^x is represented as follows

$$e^x = e^{x_0}(\cos r + \frac{\omega}{r} \sin r).$$

Also, e^x is regular in R^{n+1} and $\frac{\partial}{\partial x_0} e^x = e^x$.

[ii] Sine and Cosine functions:

$$\sin x = \sum_{n=0}^{\infty} (-1)^{2n+1} \frac{x^n}{(2n+1)!}, \quad \cos x = \sum_{n=0}^{\infty} (-1)^{2n} \frac{x^n}{(2n)!}.$$

$\sin x$ and $\cos x$ are regular in R^{n+1} and $\frac{\partial}{\partial x_0} \sin x = \cos x$, $\frac{\partial}{\partial x_0} \cos x = -\sin x$.

$\sin r$ and $\cos r$ have the following properties:

$$\cos r = \frac{e^x + e^{x^*}}{2e^{x_0}}, \quad \sin r = \frac{e^x - e^{x^*}}{2e^{x_0}} r\omega^{-1}.$$

[iii] Logarithmic function:

$$\log x = \log |x| + \frac{1}{r} \arctan\left(\frac{r}{x_0}\right)\omega.$$

$\log x$ is an infinitely many valued regular function.

References

1. Brackx, F., Delanghe, R. and Sommen, F.: 1982, 'Clifford Analysis', *Research Notes in Mathematics* **76**,1982
2. Brauder, R. and Weyl, H.: 'Spinor in n-dimensions', *Amer. J.Math.* **57**, 1935 ,pp. 425–449
3. Cartan, E.: 'Theory of spinors' *Dover New York*,1966

4. Chevalley, C.: 'The algebraic theory of spinors' *Columbia Univ. Press New York*, 1954
5. Chisholm, J.S.R. and Farwell, R.S.: 'Spin gauge theory of electric magnetic spinors' *Proc. R. Soc. London*, **A337**, 1981, pp. 1–23
6. Deavours, C.A.: 'The quaternion calculas' *Amer.Math.Mon.*, **80**, 1973, pp.138–162
7. Delanghe, R.: 'On regular analytic functions with values in a Clifford algebra' *Math. Ann.*, **185**, 1970, pp. 91–111
8. Fueter, R.: 'Die funktionentheorie der differentialgeleichungen $\Delta u = 0$ und $\Delta\Delta u = 0$ mit vier reellen variablen' *Comment. Math. Helv.*, **7**, 1935, pp. 303–330
9. Gilgert, R.P.: 'Function theoretic methods in the theory of partial differential equations' Mathematics in Siences and Enginering ,**54**,Academic Press, 1969
10. Goldschmit, B.: 'Regularity properties of generalized analytic vectors in R^{m+1}' *Math. Nachr.*, **103**, 1980, pp. 245–254
11. Goldschmit, B.: 'A Cauchy integral formula for a class of elliptic systems of partial differential equations of first order in the space' *Math. Nachr.*, **108**, 1982, pp. 167–178
12. Gürsey, F.and Tze, H.C.: 'Complex and quaternionic analyticity in chiral and gauge theories I' *Ann. Phys.*, **128**, 1980, pp. 29–130
13. Hoshi, S. and Ochiai, H.: 'On analytic functions of quaternion functions' *Mem. Fac. Eng. Miyazaki Univ.*, **4**, 1964, pp. 1–18
14. Imaeda, K.: 'A new formulation of electromagnetism' *Nuovo Cimento*, **326**, 1976, pp.138–162
15. Lounesto, P.: 'Spinor valued regular functions in hypercomplex analysis' *Doctoral thesis, Helsinki Univ. of Tech.*,1979
16. Lounesto, P.: 'Sur les idéaux á gauche des algebres de Clifford et les produits scalaires des spineurs' *Ann. Inst. H. Poincaré*, Sect. A **33**,1980, pp. 53–61
17. Micalli, A. and Villamator, O.E.: 'Sur les algebres de Clifford' *Ecole Normales Sup.*,4é serie 1, 1968, pp. 271-304
18. Morris, A.O.: 'On generalized Clifford algebra' *Q.J.Math.*, **18**, 1967, pp. 7–12
19. Morinaga, K. and Nôno, T.: 'On the linearization of a form of higher degree and its representation' *J. Hiroshima Univ.* (A), **16**, 1952, pp. 13–41
20. Naser, M.: 'Hyperholomorphic functions' *Siberian Math. J.*, **12**, 1971, pp. 959–968
21. Nôno, K.: 'Characterization of domains of holomorphy by the existence of hyper-conjugate harmonic functions' *Rev. Roumaine Math. Pures et Appl.*, **31**, 1986, pp. 159–161
22. Nôno, K.: 'Characterization of linearizations of second order and function theory' *Proceeding on 'International Symposium on Nonassociative algebras and related topics*,Hiroshima, 1990 pp. 175–192
23. Nôno, T.: 'On the linearization of partial differential equation $\nabla^m\psi = c^m\psi$' *Bull. Fukuoka Univ. of Ed. part III*, **21**, 1971, pp. 35–41
24. Nôno, T.: 'Generalized Clifford algebra and linearization of a partial differential equation: $g^{i_1 i_2 \cdots i_m}\nabla_{i_1}\nabla_{i_2}\ldots\nabla_{i_m}\psi = c^m\psi$' *Proc. Conf. on Clifford algebra, it generalization and applications* , MATSIENCE , Madras, 1971 pp. 1–24
25. Riesz, M.: 'Clifford numbers and spinors' Lecture Notes **38**, Institute for Fluid Dynamics and Applied Mathematics , Univ. of Maryland, 1958
26. Ryan, J.: 'Complexified Clifford analysis' *Complex Variables*, 1, 1982, pp. 119–149
27. Ryan, J.: 'Properties of isolated singularities of some functions taking values in real Clifford algebra' *Math. Proc. Camb. Phil. Soc.*, **95**, 1984, pp. 227–298
28. Sommen, F.: 'Some connections between Clifford analysis and complex analysis' *Complex Variables*, 1, 1982, pp. 97–118
29. Stein, E.M. and Weiss, G.: 'Generalizations of the Cauchy-Riemann equations and representations of the rotation group' *Amer. J. Math.*, **90**, 1968, pp. 163–196
30. Sudbery, A.: 'Quaternionic analysis' *Math. Proc. Camb. Phil. Soc.*, **85**, 1979, pp. 199–225

ON THE ANALOGUE OF THE $\overline{\partial}$-PROBLEM IN QUATERNIONIC ANALYSIS*

R. MICHAEL PORTER†, MICHAEL V. SHAPIRO†‡ and NIKOLAI L. VASILEVSKI†‡
† *Departamento de Matemáticas*
Centro de Investigación y de Estudios Avanzados del I.P.N.
A.P. 14-740, 07000 México, D.F., México
and
‡ *Department of Mathematics*
Odessa State University
Odessa, Ukraine

Abstract. The $\overline{\partial}$-problem in complex analysis has stimulated deep and fruitful investigations. Since two-dimensional complex analysis can be embedded into one-dimensional quaternionic analysis, one expects that the study of the analogous problem ${}^{\psi}\!\mathcal{D}[u] = \alpha$ in a domain $\Omega \subset H = \{quaternions\}$, for fixed parameter $\psi \in H^4$, will be useful and promising. We establish some basic facts concerning operators on quaternionic Hilbert modules, and establish connections among the hyperholomorphic Bergman projector, the so-called T-operator, and a special solution of the ${}^{\psi}\!\mathcal{D}$ equation.

Key words: $\overline{\partial}$-problem, hyperholomorphic function, Bergman projector

1. Preliminaries

We study here the analogue of the classical $\overline{\partial}$-problem, that is, ${}^{\psi}\!\mathcal{D}[u] = \alpha$, in one quaternionic variable, where

$$ {}^{\psi}\!\mathcal{D}[f] := \sum_{k=0}^{3} \psi^k \cdot \frac{\partial f}{\partial x_k}, $$

for suitable normalized coefficients ("structural set") $\psi = (\psi^0, \psi^1, \psi^2, \psi^3)$. This continues lines of investigation initiated in [7],[8]. In particular we establish basic properties of various operators related to this problem. Most of the results stated carry over without difficulty to Clifford algebras.

First we fix the basic notation and establish the main definitions related to Hilbert space with quaternionic scalars.

Let H be the set of quaternions $a = \sum_{k=0}^{3} a_k i_k$, where $a_k \in R$, i_0 is the unit, and i_1, i_2, i_3 are the imaginary units.

Let $\mathcal{H} = \mathcal{H}^R$ be a right H-module; that is, there is a multiplication $\mathcal{H} \times H \to \mathcal{H}$ which is additive in both factors and for which

$$ (a\lambda)\mu = a(\lambda\mu) $$

* This work was partially supported by CONACYT project 1821-E9211

F. Brackx et al. (eds.), Clifford Algebras and their Applications in Mathematical Physics, 167–173.

when $a \in H$ and $\lambda, \mu \in H$. A *scalar product* on H^R is a map $\langle \cdot, \cdot \rangle_H : H \times H \to H$ satisfying the following axioms: for any $a, b, c \in H$ and for any $\lambda \in H$, we have

$$\langle a, b \rangle = \overline{\langle b, a \rangle} ,$$
$$\langle a, b \cdot \lambda \rangle = \langle a, b \rangle \cdot \lambda ,$$
$$\langle a + b, c \rangle = \langle a, c \rangle + \langle b, c \rangle , \text{ and}$$
$$\langle a, a \rangle \geq 0 \text{ with } \langle a, a \rangle = 0 \text{ only when } a = 0 .$$

In a left H-module H^L, the second axiom is replaced by $\langle \lambda \cdot a, b \rangle = \lambda \cdot \langle a, b \rangle$. We will sometimes omit the "$\cdot$" indicating multiplication by a quaternion when there is no possibility of confusion.

An H-module $\mathcal{H} = H^R$ or H^L endowed with a scalar product is called a *Hilbert (right or left) module* if it is complete with respect to the norm generated by the scalar product $\|a\| := \sqrt{\langle a, a \rangle}$. A functional $\mathcal{X} : a \in H^R \mapsto \mathcal{X}(a) \in H$ is *linear* when it satisfies $\mathcal{X}(a \cdot \lambda + b) = \mathcal{X}(a) \cdot \lambda + b$; in H^L this takes the form $\mathcal{X}(\lambda \cdot a + b) = \lambda \cdot \mathcal{X}(a) + b$. The Riesz representation theorem holds in a natural way for quaternionic modules.

Let $\mathcal{H} = H^R$, and let the linear operator A be defined on $\operatorname{dom} A \subset \mathcal{H}$ with $\operatorname{dom} A$ a nonclosed dense subspace. The domain of the adjoint is defined as usual by

$$\operatorname{dom} A^* := \{y \in \mathcal{H} \,|\, \text{the functional } \mathcal{X} : x \mapsto \langle y, Ax \rangle \text{ is bounded on } \operatorname{dom} A\} \tag{1}$$

and by means of the Riesz theorem, the operator A^* is defined to satisfy $\langle y, Ax \rangle = \langle A^* y, x \rangle$ for $x \in \operatorname{dom} A, y \in \operatorname{dom} A^*$.

2. Hyperholomorphic Functions and Cauchy Integral-Type Operators

2.1. Hyperholomorphic Functions

Let Ω be a bounded domain in R^4 with its boundary $\partial\Omega$ a compact, sufficiently smooth hypersurface (codimension 1 manifold). We work with H-valued functions $f = \sum_{k=0}^{3} f^k \cdot i_k$ defined in Ω. $L^2(\Omega; H)$ has the structure of an H-bimodule (that is, simultaneously a left and right H-module), and the scalar product

$$\langle f, g \rangle := \int_\Omega \overline{f(t)} \cdot g(t) dx$$

makes $L^2(\Omega; H)$ into a right Hilbert H-module, while $\int_\Omega f(t) \cdot \overline{g(t) dx}$ makes it into a left one. Usually we will take $L^2(\Omega; H)$ to be a right module.

Let $\psi := \{\psi^0, \cdots, \psi^3\} \in H^4$. The operators ${}^{\psi}\mathcal{D}$ and $\mathcal{D}^{\psi}$ are defined on $C^1(\Omega; H)$ by the formulae

$${}^{\psi}\mathcal{D}[f] := \sum_{k=0}^{3} \psi^k \cdot \partial_k f , \quad \mathcal{D}^{\psi}[f] := \sum_{k=0}^{3} \partial_k f \cdot \psi^k , \tag{2}$$

where $\partial_k = \partial / \partial x_k$. If Δ denotes the 4-dimensional Laplace operator, then the equality

$${}^{\psi}\mathcal{D} \cdot {}^{\overline{\psi}}\mathcal{D} = {}^{\overline{\psi}}\mathcal{D} \cdot {}^{\psi}\mathcal{D} = \mathcal{D}^{\psi} \cdot \mathcal{D}^{\overline{\psi}} = \mathcal{D}^{\overline{\psi}} \cdot \mathcal{D}^{\psi} = \Delta \tag{3}$$

is true if and only if for every j, k we have

$$\langle \vec{\psi}^j, \vec{\psi}^k \rangle_{R^4} = \delta_{jk} \cdot \lambda \tag{4}$$

for some real constant λ (see [7], [8]). Any ψ which satisfies this will be called a ***structural set***. We assume further, without any loss of generality, that $\lambda = 1$. The elements of the sets ${}^{\psi}\mathcal{M}(\Omega) := \ker {}^{\psi}\mathcal{D}$ and $\mathcal{M}^{\psi}(\Omega) := \ker \mathcal{D}^{\psi}$ are called left (respectively, right) ψ-hyperhomomorphic H-valued functions in Ω. These spaces are right (respectively, left) Hilbert H modules.

Let $d\hat{x}_k$ denote the differential form obtained from $dx_0 \wedge \ldots \wedge dx_3$ with the factor dx_k omitted, and let $\sigma^{(3)}_{\psi,x} := \sum_{k=0}^{3} (-1)^k \ \psi^k d\hat{x}_k$. Denote by ds_3 the 3 dimensional volume element of a hypersurface in R^4. Then $|\sigma^{(3)}_{\psi,x}| = ds_3$ and we call $\sigma^{(3)}_{\psi,x}$ the H-valued area form associated to ψ.

Denote by $\theta_4(x) = (1/2\pi)|x|^{-2}$, $x \in R^4 - \{0\}$, the fundamental solution of the Laplace operator. Denote by $\mathcal{K}_{\psi}$ the (common) fundamental solution of the operators ${}^{\psi}\mathcal{D}$ and $\mathcal{D}^{\psi}$, namely

$$\mathcal{K}_{\psi}(x) := \frac{1}{2\pi^2|x|^4} \sum_{k=0}^{3} \overline{\psi^k} \cdot x^k \, . \tag{5}$$

The following well known statements can be found in many sources (see, for example, [1], [7]). Here $C^2(\overline{\Omega}; H)$ denotes the subclass of functions which can be extended smoothly to an open set containing the closure $\overline{\Omega}$.

Integral form of the quaternionic Stokes formula. Let Ω be a bounded domain in R^4 with a smooth boundary $\partial\Omega$, and let $f, g \in C^1(\overline{\Omega}; H)$. Then

$$\int_{\partial\Omega} g(x)\sigma_{\psi,x} f(x) = \int_{\Omega} (\mathcal{D}^{\psi}[g](x) f(x) + g(x)\, {}^{\psi}\mathcal{D}[f](x)) dx. \tag{6}$$

Borel-Pompeiu (Cauchy-Green) quaternionic formula. Let Ω be a bounded domain in R^4 with smooth boundary and $f \in C^1(\overline{\Omega}; H)$. Then

$$f(x) = \int_{\partial\Omega} \mathcal{K}_{\psi}(\tau - x)\sigma_{\psi,\tau} f(x) - \int_{\Omega} \mathcal{K}_{\psi}(\tau - \lambda)\, {}^{\psi}\mathcal{D}[f](\tau) d\tau, \quad x \in \Omega. \tag{7}$$

2.2. Quaternionic Distribution Theory

We now consider the notion of H-valued generalized function. Let $C_0^{\infty}(\Omega; H)$ denote the set of all compactly supported functions in $C^{\infty}(\Omega; H)$. It is clear that $C_0^{\infty}(\Omega; H)$ is an H-bimodule, and we consider it at will as a right or as a left module. Let $f \in L^1_{loc}(\Omega; H)$ be a locally integrable function. Then the mappings

$$f^{(R)} : \varphi \longmapsto \int_{\Omega} f(x) \cdot \varphi(x) dx$$

$$f^{(L)} : \varphi \longmapsto \int_{\Omega} \varphi(x) \cdot f(x) dx$$

are respectively right and left linear, continuous functionals on $C_0^\infty(\Omega; H)$ which are called right and left *regular generalized functions*, or right and left *regular distributions*. They satisfy natural linearity properties, which characterize right and left distributions in general. The definitions of ${}^\psi\mathcal{D}$ and $\mathcal{D}^\psi$ extend to distributions in the natural way. As is the case for holomorphic functions of a complex variable, distributions which are annihilated by ${}^\psi\mathcal{D}$ or $\mathcal{D}^\psi$ are regular distributions.

2.3. L^2 Adjoint Operator of ${}^\psi\mathcal{D}$

We want an operator acting on L^2, and thus are led to define

$$\operatorname{dom}{}^\psi\mathcal{D} := \left\{ f | \ f \in L^2 \text{and } \frac{\partial f}{\partial x_k} \in L^2, \text{ all } k \right\}. \tag{8}$$

which is dense in $L^2(\Omega; H)$. The following is easily verified by integration by parts.

Lemma 2.1 *Let $f, h \in C^1(\overline{\Omega}; H)$. Then*

$$\langle h, {}^\psi\mathcal{D}[f] \rangle = \int_{\partial\Omega} \overline{h(x)} \cdot \sigma_{\psi,x} \cdot f(x) - \langle {}^{\overline{\psi}}\mathcal{D}[h], f \rangle .$$

Corollary 2.2 *If $f, h \in C^1(\overline{\Omega}; H)$ and also for each $x \in \partial\Omega$ either $f(x) = 0$ or $h(x) = 0$, then $\langle h, {}^\psi\mathcal{D}[f] \rangle = -\langle {}^{\overline{\psi}}\mathcal{D}[h], f \rangle$.*

We begin now a more formal investigation of the domains of ${}^\psi\mathcal{D}$ and ${}^\psi\mathcal{D}^*$. The following is analogous to a result given by Kohn [3] in a complex analysis setting.

Theorem 2.3 *Let $h \in C^1(\overline{\Omega}; H)$. Then $h \in \operatorname{dom}{}^\psi\mathcal{D}^*$ if and only if $h|\partial\Omega = 0$.*

Proof. First suppose $h \in \operatorname{dom}{}^\psi\mathcal{D}^*$. This means that for all $f \in \operatorname{dom}{}^\psi\mathcal{D}$, $\langle h, {}^\psi\mathcal{D}[f] \rangle = \langle {}^\psi\mathcal{D}^*[h], f \rangle$. By Corollary 2.2, for $\varphi \in C_0^1(\Omega; H)$ and $h \in \operatorname{dom}{}^\psi\mathcal{D}^*$,

$$\langle h, {}^\psi\mathcal{D}[\varphi] \rangle = -\langle {}^{\overline{\psi}}\mathcal{D}[h], \varphi] \rangle, \tag{9}$$

and so, taking into account the fact that $C_0^1(\Omega; H) \subset \operatorname{dom}{}^\psi\mathcal{D}$, we have $\langle {}^\psi\mathcal{D}^*[h], \varphi \rangle = -\langle {}^{\overline{\psi}}\mathcal{D}[h], \varphi \rangle$ for all $\varphi \in C_0^1(\overline{\Omega}; H)$. Since $C_0^1(\Omega; H)$ is dense in $L^2(\Omega; H)$ we obtain ${}^\psi\mathcal{D}^*[h] = -{}^{\overline{\psi}}\mathcal{D}[h]$. Therefore, again from Corollary 2.2 we see

$$\langle {}^\psi\mathcal{D}^*[h], \varphi \rangle = -\langle {}^{\overline{\psi}}\mathcal{D}[h], \varphi \rangle \tag{10}$$

for all $\varphi \in C^1(\overline{\Omega}; H)$, and from Lemma 2.1 we have

$$\int_{\partial\Omega} \overline{h(x)} \cdot \sigma_{\psi,x} \cdot \varphi(x) = 0 \tag{11}$$

for all $\varphi \in C^1(\overline{\Omega}; H)$. It is known that $\sigma_{\psi,x} = n_\psi(x) \cdot ds_3$, where $n_\psi(x) = \sum_{k=0}^{3} n_\psi(x) \cdot \psi^k$ with $n(x) := (n_0(x), \ldots, n_3(x))$ being the outward normal to $\partial\Omega$ at the point x.

Substitute $\varphi(x) := \overline{n_\psi(x)} \cdot h(x)$ in (11) to obtain $\int_{\partial\Omega} |h(x)|^2 \cdot ds = 0$, which implies $h|\partial\Omega = 0$.

For the reverse inclusion, let now $h|\partial\Omega = 0$. Then (11) is true, and thus by Lemma 2.1 the equality (9) holds. Since $C_0^\infty(\Omega; H)$ is dense in $L^2(\Omega; H)$, $\langle h, {}^\psi\mathcal{D}[\varphi]\rangle = -\langle {}^{\bar{\psi}}\mathcal{D}[h], \varphi\rangle$ for all $\varphi \in \operatorname{dom}{}^\psi\mathcal{D}$. Therefore $h \in \operatorname{dom}{}^\psi\mathcal{D}^*$ as claimed.

From (10) we see that $-{}^{\bar{\psi}}\mathcal{D}$, which has the same domain as ${}^\psi\mathcal{D}$, is an extension of ${}^\psi\mathcal{D}^*$. Theorem 2.3 characterizes the elements of the domain of ${}^\psi\mathcal{D}^*$ which are smooth up to the boundary. This is a dense set in the L^2 norm. From a Lemma of Hörmander [5, p. 95] it is not difficult to verify the following.

Theorem 2.4 *Let $\Omega \subset H$ be a bounded domain with smooth (C^1) boundary. (i) The set $C^\infty(\Omega; H) \cap \operatorname{dom}{}^\psi\mathcal{D}$ is dense in $\operatorname{dom}{}^\psi\mathcal{D}$ in the graph norm; (ii) $C^\infty(\Omega; H) \cap \operatorname{dom}{}^\psi\mathcal{D}^*$ is dense in $\operatorname{dom}{}^\psi\mathcal{D}^*$ in the graph norm.*

3. The Quaternionic $\bar{\partial}$-Problem

The $\bar{\partial}$-problem $\bar{\partial}u = \alpha$ is of great importance in the theory of one and several complex variables. Here we will examine the corresponding problem for one quaternionic variable, namely

$$ {}^\psi\mathcal{D}[u] = \alpha. \tag{12} $$

$\Omega \subset H$ will denote throughout a bounded domain for which $\partial\Omega$ is a smooth hypersurface. It is convenient to take into account spaces of functions more general than $u \in L^2(\Omega; H)$ since certain operators such as the T-operator do not necessarily conserve L^2. We limit the discussion to the left operator ${}^\psi\mathcal{D}$ since the theory for $\mathcal{D}^\psi$ is completely analogous.

We begin with the (weakly singular) integral operator ${}^\psi T$. In H, as in the case of C^1, no special "compatibility condition" is necessary for the existence of a solution. The solution given by ${}^\psi T$ is the one described by the normalization ${}^\psi K[u] = 0$. This has certain inconveniences, since ${}^\psi K$ is an unbounded operator, and ${}^\psi K$ cannot be applied to general L^2 functions, since it is defined in terms of their boundary values. We therefore contrast this approach with a second solution of (12) given by means of the quaternionic Bergman kernel; the characterization for this case is orthogonality to the hyperholomorphic L^2 functions.

3.1. T-Operator Solution

In this subsection we will collect some facts about the K and T operators which are essentially the same as for C^1. They are mainly from Goldschmidt [2] and Iftimie [6] (in the more general context of Clifford algebras).

Recall the integral kernel $\mathcal{K}_\psi(x)$ given by (5), and define

$$ {}^\psi K[f](x) = \int_{\partial\Omega} \mathcal{K}_\psi(\tau - x)\sigma_{\psi,\tau} f(\tau) \tag{13} $$

for any f which is integrable on $\partial\Omega$. When $f \in C^0(\overline{\Omega}; H)$ we will write ${}^\psi K[f]$ for

${}^{\psi}K[f|\partial\Omega]$. Define further

$$ {}^{\psi}T[\alpha](x) := -\int_{\Omega} \mathcal{K}_{\psi}(\tau - x)\alpha(\tau)d\tau \tag{14} $$

for any $\alpha \in L^2(\Omega; H)$. It is known that then ${}^{\psi}T[\alpha] \in L^2$, and

$$ {}^{\psi}\mathcal{D} \cdot {}^{\psi}T[\alpha] = \alpha \;, \tag{15} $$

so the solution set of (12) is ${}^{\psi}T[\alpha] + {}^{\psi}\mathfrak{A}_2(\Omega)$. However, for $f \in L^p(\Omega; H)$ the most we can say in general [2] is that ${}^{\psi}T[f] \in L^1_{\text{loc}}(\Omega; H) = \mathrm{L}^1(\Omega; H)$. Therefore we may consider generalized derivatives of ${}^{\psi}T[f]$. With this proviso in mind, we have the following [6].

Proposition 3.1 *Let $p \geq 2$. Then ${}^{\psi}\mathcal{D}{}^{\psi}T[f] = f$ for $f \in L^p(\Omega; H)$.*

It follows that for $\alpha \in L^2(\Omega; H)$ we have ${}^{\psi}T[\alpha] \in \operatorname{dom}{}^{\psi}\mathcal{D}$. The following estimates may be obtained by arguments similar to those in [2].

Proposition 3.2 *Let $p > 4$. If $f \in L^p(\Omega; H)$ and $u := {}^{\psi}T[f] \in L^q(\Omega; H)$, with $1 < q < 4$, then*

$$ |u(z)| \leq M_1 \|f\|_{L^p} \;, \qquad z \in C \;, $$

$$ |u(z_1) - u(z_2)| \leq M_1 \|f\|_{L^p} \cdot |z_1 - z_2|^{(p-4)/p} \;, \quad z_1, z_2 \in \Omega \;, $$

where M_1 is a constant for fixed p and Ω, and M_2 constant for fixed p. Thus ${}^{\psi}T(L^p(\Omega)) \subset C^{0,(p-4)/p}(\Omega; H)$.

From this we deduce that for $f \in L^p(\Omega; H)$, ${}^{\psi}T[f]$ extends continuously to $\partial\Omega$. Therefore it makes sense to speak of ${}^{\psi}K[{}^{\psi}T[f]]$. As a result, The Borel-Pompeiu formula (7), which we will write as

$$ I = {}^{\psi}K + {}^{\psi}T{}^{\psi}\mathcal{D} \;, \tag{16} $$

is valid when applied to functions in the range of ${}^{\psi}T$. For $f \in L^p(\Omega; H)$, $p > 4$, we have that $u = {}^{\psi}T[f]$ is the only solution of (12) with ${}^{\psi}K[u] = 0$.

3.2. Bergman Operator Solution

The hyperholomorphic Bergman kernel, introduced in [8], is given by

$$ {}^{\psi}B[f](x) := \int_{\Omega} {}_{\psi}\mathcal{B}(x, \tau) f(\tau) d\tau \tag{17} $$

where ${}_{\psi}\mathcal{B}(x,\tau) := {}^{\psi}\mathcal{D}_{\tau} \cdot {}^{\psi}\overline{\mathcal{D}}_x[g](x,\tau) = {}^{\psi}\overline{\mathcal{D}}_x \cdot {}^{\psi}\mathcal{D}_{\tau}[g](x,\tau)$, where $g(s,\tau)$ is the classical Green's function for $\Omega \subset R^4$, and $(x,\tau) \in \overline{\Omega} \setminus$ diagonal . The basic facts pertaining to ${}^{\psi}B$ are as follows [8].

Proposition 3.3 *On any $L^p(\Omega; H)$, $(p > 1)$ the relation ${}^{\psi}\mathcal{D}{}^{\psi}B = 0$ holds.*

Proposition 3.4 *${}^{\psi}B : L^2(\Omega; H) \to {}^{\psi}\mathfrak{A}_2(\Omega; H)$ is a self-adjoint orthogonal projection.*

Proposition 3.5 *For functions in the range of ${}^{\psi}T$, ${}^{\psi}B{}^{\psi}K = {}^{\psi}K$.*

3.3. Relations Among the Operators

Now introduce the operator

$$^{\psi}S := (I - {}^{\psi}B)^{\psi}T. \tag{18}$$

which is defined on every $L^p(\Omega; H)$, $p > 1$. In $L^2(\Omega; H)$, clearly ${}^{\psi}S[\alpha]$ is the unique solution of (12) which is orthogonal to ${}^{\psi}\!\mathcal{A}_2(\Omega; H)$. We relate ${}^{\psi}S$ to ${}^{\psi}\mathcal{D}$, ${}^{\psi}K$, and ${}^{\psi}B$ as follows.

Theorem 3.6 *${}^{\psi}B = I - {}^{\psi}S^{\psi}\mathcal{D}$ on dom ${}^{\psi}\mathcal{D}$.*

Proof. By (18) and (16), ${}^{\psi}S^{\psi}\mathcal{D} = (I - {}^{\psi}B)^{\psi}T^{\psi}\mathcal{D} = (I - {}^{\psi}B)(I - {}^{\psi}K) = I - {}^{\psi}B - {}^{\psi}K + {}^{\psi}B^{\psi}K$. By Corollary (3.3) this is equal to $I - {}^{\psi}B$, which proves the result.

Theorem 3.7 *${}^{\psi}K^{\psi}B = {}^{\psi}B$ on the set $\{f \in L^2(\Omega; H) | \ {}^{\psi}B[f] \in C^1(\overline{\Omega}; H)\}$.*

Proof. Using (16) and the previous result we have

$$^{\psi}K^{\psi}B = (I - {}^{\psi}T^{\psi}\mathcal{D})(I - {}^{\psi}S^{\psi}\mathcal{D}) = I - {}^{\psi}T^{\psi}\mathcal{D} - {}^{\psi}S^{\psi}\mathcal{D} + {}^{\psi}T^{\psi}\mathcal{D}^{\psi}S^{\psi}\mathcal{D} .$$

But ${}^{\psi}\mathcal{D}^{\psi}S = {}^{\psi}\mathcal{D}({}^{\psi}T - {}^{\psi}B^{\psi}T) = {}^{\psi}\mathcal{D}^{\psi}T - D^{\psi}B^{\psi}T = I$ by Theorem 3.1 and Corollary 3.3. Therefore ${}^{\psi}K^{\psi}B = I - {}^{\psi}S^{\psi}\mathcal{D} = {}^{\psi}B$ by Theorem 3.6 again.

On $L^2(\Omega; H)$ the operator ${}^{\psi}B$ is bounded. In contrast, ${}^{\psi}K$, with an appropiate natural domain, is not bounded because it annihilates $C_0^{\infty}(\overline{\Omega}; H)$. From the above, it is of interest to contrast the projection $I - {}^{\psi}B$, whose image contains $({}^{\psi}\!\mathcal{A}_2)^{\perp}$, and $I - {}^{\psi}K$, whose image contains the image of ${}^{\psi}T$. It is not known whether the intersection of these two images is trivial. Similarly, it is not known what the kernel of ${}^{\psi}B - {}^{\psi}K$ looks like.

References

1. F. Brackx, R. Delanghe, F. Sommen, *Clifford Analysis*, Pitman (1982)
2. B. Goldschmidt, "Regularity properties of generalized analytic vectors in R^n," *Math. Nachr.* 103 (1981) 245-254
3. J. J. Kohn, "Harmonic integrals on strongly pseudo-convex manifolds," I: *Annals of Math.* 78 (1963) 112-148; II: *Annals of Math.* 79 (1964) 450-472
4. J. J. Kohn, "A survey of the $\bar{\partial}$-Neumann problem,", *Proc. Symp. in Pure Math* 41 (1984) 137-145
5. L. Hörmander, "L^2 estimates and existence theorems for the $\bar{\partial}$ operator," *Acta Math.* 113 (1965) 89-152
6. V. Iftimie, "Fonctions hypercomplexes," *Bull. Math. de la R. S. R.* 9 (1965) 279-332
7. M. V. Shapiro, N. L. Vasilevski, "Quaternionic ψ-hyperholomorphic functions, singular operators with quaternionic Cauchy kernel and analogies of the Riemann boundary problem", Reporte Interno No. 102, CINVESTAV del I.P.N., 75 pp. (1992)
8. M. V. Shapiro, N. L. Vasilevski, "On the Bergman kernel function in hyperholomorphic analysis," Reporte Interno No. 115, CINVESTAV del I.P.N., 35 pp. (1993)
9. I. Vekua, *Generalized Analytic Functions*, Pergamon Press (1962)

HURWITZ PAIRS AND CLIFFORD ALGEBRA REPRESENTATIONS*

ENRIQUE RAMÍREZ DE ARELLANO†, MICHAEL V. SHAPIRO†‡ and NIKOLAI L.VASILEVSKI†‡
†*Department of Mathematics*
Centro de Investigación y de Estudios Avanzados del IPN
Apartado Postal 14-740
México, D.F. 07000 - México
and
‡ *Department of Mathematics*
Odessa State University
Odessa, Ukraine

Abstract. The theory of pseudoeuclidean and symplectic Hurwitz pairs is reviewed. A canonical algorithm for constructing an irreducible representation of a certain Clifford algebra for a given Hurwitz pair is introduced. Conversely, given an irreducible representation of a Clifford algebra, a canonical procedure is given to construct corresponding Hurwitz pairs.

Key words: Hurwitz pairs, Clifford Algebras, irreducible representations.

Introduction

In 1898 A. Hurwitz [H1] posed a famous, still unsolved, problem: Suppose the variables z_ℓ are nondegenerate real bilinear functions of x_j and y_k. Which is the minimum m such that the following polynomial identity holds

$$\left(\sum_{j=1}^{p} x_j^2\right)\left(\sum_{k=1}^{n} y_k^2\right) = \sum_{\ell=1}^{m} z_\ell^2? \tag{0.1}$$

In that paper he gave a solution when $p = n = m$ showing that the only possible cases are $m = 1, 2, 4$, and 8. For a given positive integer p, he gave later a formula for the case $n = m$ in a posthumous paper [H2]. Much has been written since then about the Hurwitz problem and its close relations to several branches of mathematics, including combinatorics, geometry, topology, and differential equations; for an extensive literature see, for instance [Sh] and [SY]. The original problem has been generalized in several directions, e.g. to arbitrary fields of characteristic different from 2, to the pseudoeuclidean and symplectic cases, and so on.

Development of the Hurwitz problem led to the formulation of the notion of Hurwitz multiplication as a bilinear mapping of a pair of vector spaces into a third vector space, these spaces being endowed with arbitrary metrics (pseudoeuclidean and sym-

* This work was partially supported by CONACYT project 1821-E9211

F. Brackx et al. (eds.), Clifford Algebras and their Applications in Mathematical Physics, 175–181.

plectic), and in the polynomial identity (0.1) euclidean metrics are substituted with more general ones.

In this paper we make a detailed study of both euclidean and non–euclidean (pseudoeuclidean and symplectic) Hurwitz pairs. We establish precise relations between Hurwitz pairs and Clifford algebras by constructing explicit algorithms which allow the following: given a Hurwitz pair, to construct an irreducible representation of a certain Clifford algebra, being uniquely defined by this Hurwitz pair; and inversely, given a Clifford algebra and any irreducible representation of it, to construct a certain Hurwitz pair. Any Hurwitz pair can be constructed following our procedure.

Both of these algorithmical procedures are canonical and give a complete description of all possible Hurwitz pairs.

Further, if $(S, V, \circ)$ denotes a Hurwitz pair, we introduce the algebra $Al(S, V, \circ)(\subset \mathrm{Hom}\,(V, V))$ generated by the operators of Hurwitz multiplication by elements of S. The algebra $Al(S, V, \circ)$ which arises naturally in the context of our procedures, is very useful for applications, in particular, in the further development of so–called Hurwitz analysis, see [KR1], [KR2].

1. Hurwitz pairs and their properties

1.1 Let S be a $(p+1)$-dimensional real vector space with basis $\{\varepsilon_\alpha\}, \alpha \in \{0\} \cup N_p, N_p := \{1, \ldots, p\}$, and let the R-bilinear scalar product

$$(\cdot,\cdot)_S : S \times S \to R$$

be defined by the following metric matrix

$$\eta := [\eta_{\alpha\beta}] := [(\varepsilon_\alpha, \varepsilon_\beta)_S] = \mathrm{diag}\ (\underbrace{1, \ldots, 1}_{r+1}; \underbrace{-1, \ldots, -1}_{s}), \tag{1.1}$$

where $p = r + s$.

Introduce also the n-dimensional real vector space V with basis $\{e_j\}, j \in N_n$, provided with an R-bilinear scalar product

$$(\cdot,\cdot)_V : V \times V \to R$$

which is defined by the following nonsingular metric matrix

$$\kappa := [\kappa_{kj}] := [(e_k, e_j)_V]. \tag{1.2}$$

We assume also that the scalar product in V is either symmetric: $\kappa = \kappa^t$, or antisymmetric: $\kappa = -\kappa^t$, where "t" means transposition. Thus for all $\{g, f\} \subset V$,

$$(g, f)_V = \delta(f, g)_V,$$

where $\delta = 1$ for the symmetric case and $\delta = -1$ for the antisymmetric one.

1.2 Let

$$\circ : S \times V \to V$$

be an R-bilinear mapping. As in [LR1–3] we call it a Hurwitz multiplication (of elements from V by elements of S on the left side) if the following axioms are fulfilled:

H.1. For all $\{f, g\} \subset V$ and all $a \in S$,

$$(a,a)_S(f,g)_V = (a \circ f, a \circ g)_V; \tag{1.3}$$

H.2. there exists a unit element ε in S with respect to the mapping "$\circ$", i.e. for all $f \in V, \varepsilon \circ f = f$;

H.3. the mapping "$\circ$" does not leave invariant any proper non singular (with respect to $(\cdot,\cdot)_V$) subspace of V.

The set $(S, V, \circ)$ is usually called a (pseudo–euclidean or symplectic) Hurwitz pair.

1.3 Let us introduce the R-linear isomorphism

$$\nu_S : S \to R^{p+1}$$

by the rule

$$\nu_S : a = \sum_{\alpha=0}^{p} a_\alpha \varepsilon_\alpha \longmapsto \widetilde{a} = \begin{pmatrix} a_0 \\ a_1 \\ \vdots \\ a_p \end{pmatrix} \in R^{p+1}.$$

We always write elements of R^N as column vectors, so $\widetilde{a}^t$ are the corresponding row vectors.

Then for every $a = \sum\limits_{\alpha=0}^{p} a_\alpha \varepsilon_\alpha$ and $b = \sum\limits_{\alpha=0}^{p} b_\alpha \varepsilon_\alpha$ we have

$$(a,b)_S = \widetilde{a}^t \eta \widetilde{b}. \tag{1.4}$$

Define also the R-linear isomorphism

$$\nu_V : V \to R^n$$

by the rule

$$\nu_V : f = \sum_{j=1}^{n} f_j e_j \longmapsto \widetilde{f} = \begin{pmatrix} f_1 \\ \vdots \\ f_n \end{pmatrix} \in R^n.$$

Then for every $f, g \in V$ we have

$$(f,g)_V = \widetilde{f}^t \kappa \widetilde{g}. \tag{1.5}$$

1.4 Given arbitrary basis elements $\varepsilon_\alpha \in S$ and $e_j \in V$, then $\varepsilon_\alpha \circ e_j \in V$. Thus for some real constants $c^k_{\alpha j}$ we have

$$\varepsilon_\alpha \circ e_j = \sum_{k=1}^{n} c^k_{\alpha j} e_k,$$

and for every $f = \sum_{j=1}^{n} f_j e_j \in V$

$$\varepsilon_\alpha \circ f = \sum_{j=1}^{n} f_j \sum_{k=1}^{n} c_{\alpha j}^{k} e_k = \sum_{k=1}^{n} \left(\sum_{j=1}^{n} c_{\alpha j}^{k} f_j \right) e_k.$$

Applying the isomorphism ν_V to the both sides of this equality we obtain

$$\nu_V(\varepsilon_\alpha \circ f) = C_\alpha \cdot \nu_V(f), \tag{1.6}$$

where $C_\alpha = [c_{\alpha j}^{k}]_{j,k=1,\dots,n}$.

1.5 Note that each element $\varepsilon_\alpha \in S$ determines uniquely the matrix C_α and vice versa, and that the following diagram

$$\begin{array}{ccc} V & \xrightarrow{m(\varepsilon_\alpha)} & V \\ \Big\downarrow{\scriptstyle \nu_V} & & \Big\downarrow{\scriptstyle \nu_V} \\ R^n & \xrightarrow{m(C_\alpha)} & R^n \end{array}$$

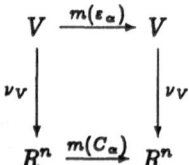

is commutative. Here $m(\varepsilon_\alpha) : f \mapsto \varepsilon_\alpha \circ f$; $m(C_\alpha) : \widetilde{f} \mapsto C_\alpha \cdot \widetilde{f}$.

1.6 Consider now the following system of axioms.

Let η be a $(p+1) \times (p+1)$ matrix of the form (1.1), κ be a symmetric or antisymmetric non singular $n \times n$ matrix and $\{C_\alpha\}$ be $(p+1)$ linearly independent $n \times n$ matrices such that

H*.1. if $\widehat{C}_\alpha := \kappa^{-1} C_\alpha^t \kappa$, then, for all $\{\alpha, \beta\} \subset N_p \cup \{0\}$ $\widehat{C}_\alpha C_\beta + \widehat{C}_\beta C_\alpha = 2\eta_{\alpha\beta} I_n$;

H*.2. the unit matrix I_n can be represented as a linear combination of the C_α;

H*.3. there is no proper subspace in R^n which is invariant under multiplication by all matrices C_α.

1.7 Theorem *The systems of axioms H1–H3 and H*.1–H*.3 are equivalent in the following sense: a Hurwitz pair $(S, V, \circ)$ generates canonically a unique set $\{C_\alpha\}$ with the properties H*.1–H*.3; a set $\{\kappa, \eta, C_\alpha\}$ with the properties H1–H3 generates canonically a unique Hurwitz pair.*

1.8 Remark. *Without loss of generality we may assume that the unit element ε coincides with ε_0, although the above is true without this assumption.*

Consider $p+1$ linearly independent $n \times n$ matrices $I_n, C_1, \dots, C_p$, and a symmetric or antisymmetric nonsingular $n \times n$ matrix κ such that:

H*e.1. for all $\{\alpha, \beta\} \in N_p$

$$C_\alpha C_\beta + C_\beta C_\alpha = -2\eta_{\alpha\beta} I_n,$$

where

$$\eta_{\alpha\beta} = \begin{cases} 0, & \alpha \neq \beta \\ 1, & \alpha = \beta = 1, \ldots, r \\ -1 & \alpha = \beta = r+1, \ldots, p \end{cases}$$

H*e.2. for all $\alpha \in N_p$

$$C_\alpha^t \kappa = -\kappa C_\alpha;$$

H*e.3. there is no proper subspace in R^n which is invariant under multiplication by all matrices C_α.

1.9 Theorem. *The system of axioms H.1–H.3 with the additional assumption:*

$$\varepsilon = \varepsilon_0$$

*is equivalent to the system of axioms H*e.1–H*e.3.*

2. Hurwitz pairs and Clifford algebras

2.1 For given integers r, s with $p := r + s$, consider the real Clifford algebra $Cl_{r,s}$, with the identity i_0, and generated by the elements $\{i_\alpha |\ \alpha \in N_p\}$ which satisfy only the following property:

$$i_\alpha \cdot i_\beta + i_\beta \cdot i_\alpha = -2\eta_{\alpha\beta} i_0, \tag{2.1}$$

where

$$\eta_{\alpha\beta} = \begin{cases} 0, & \alpha \neq \beta, \\ 1, & \alpha = \beta \in N_r, \\ -1, & \alpha = \beta \in N_p \backslash N_r, \end{cases}$$

Recall that the algebra $Cl_{r,s}$, as a real vector space, has dimension 2^p and basis consisting of all elements

$$i_{\alpha_1} \cdot i_{\alpha_2} \cdot \ldots \cdot i_{\alpha_q},$$

where $1 \leq \alpha_1 < \alpha_2 < \ldots < \alpha_q \leq p$ for $1 < q \leq p$ and $0 \leq \alpha_1 \leq r$ for $q = 1$; the dot "$\cdot$" means the Clifford multiplication.

2.2 Now let us return to a Hurwitz pair $(S, V, \circ)$ with $\varepsilon = \varepsilon_0$. By Theorem 2.6 there exist $(p+1)$ linearly independent $(n \times n)$-matrices $I_n, C_1, \ldots, C_p$ which satisfy axioms H*e.1–H*e.3.

Define the mapping $\widetilde{\pi}_n$ of generators of the Clifford algebra $Cl_{r,s}$ by the rule: for $\alpha \in N_p$

$$\widetilde{\pi}_n(i_\alpha) := C_\alpha. \tag{2.2}$$

Let π_n be its linear extension to the real algebra homomorphism

$$\pi_n : Cl_{r,s} \longrightarrow R(n).$$

2.3 Theorem. *(Hurwitz pairs and representations of Clifford algebras).* Let $(S, V, \circ)$ be a Hurwitz pair with $\varepsilon = \varepsilon_0$. Then there exists an n–dimensional irreducible representation π_n of the Clifford algebra $Cl_{r,s}(r+s=p)$, and this representation is generated by (2.2).

2.4 Fix a real Clifford algebra $Cl_{r,s}$ with the identity i_0 and unit elements $i_\alpha, \alpha \in N_p$, satisfying (3.1).

2.5 Theorem. *Let π_n be an n–dimensional real irreducible representation of the Clifford algebra $Cl_{r,s}$. Further let there exist a symmetric or antisymmetric nonsingular $n \times n$ matrix κ, such that the matrices $C_\alpha := \pi_n(i_\alpha), \alpha \in N_p$, satisfy the condition*

$$C_\alpha^t \cdot \kappa = -\kappa \cdot C_\alpha, \ \alpha \in N_p. \tag{2.3}$$

Then there exists a Hurwitz pair $(S, V, \circ)$ with dim $S = r+s+1$, dim $V = n$.

2.6 Remark. The existence of the matrix κ was considered in [LR3] and [LR4].

3. The algebra $Al(S, V, \circ)$ generated by a Hurwitz pair

3.1 Consider a fixed Hurwitz pair $(S, V, \circ)$ with $\varepsilon = \varepsilon_0$. Each element $a \in S$ generates in a natural way the operator of the "Hurwitz multiplication by a" acting on V by the rule

$$f \in V \longmapsto a \circ f \in V. \tag{3.1}$$

Denote this operator by $m(a)$. It is clear that for $a = \sum_{\alpha=0}^{p} a_\alpha \varepsilon_\alpha$,

$$m(a) = \sum_{\alpha=0}^{p} a_\alpha m(\varepsilon_\alpha), \tag{3.2}$$

and $m(\varepsilon_0) = I$ is the identity operator on V.

3.2 The mapping

$$\mu : a \longmapsto m(a) \tag{3.3}$$

gives a linear isomorphism between the space S and some linear subspace of Hom (V, V)

Denote by $Al(S, V, \circ)$ the algebra generated by all operators $m(a)$ acting on V. We have obviously

$$\mu(S) \subset Al(S, V, \circ) \subset \mathrm{Hom}\,(V, V). \tag{3.4}$$

3.3 The elements $\varepsilon_\alpha, \alpha \in N_p \cup \{0\}$, determine uniquely (see 1.6) the system of $n \times n$ matrices $C_\alpha, \alpha \in N_p \cup \{0\}$. Further, the correspondence

$$\pi_n : i_\alpha \longrightarrow C_\alpha \tag{3.5}$$

determines uniquely the n–dimensional irreducible representation π_n of the Clifford algebra $Cl_{r,s}$.

3.4 Theorem. *The algebra $Al(S, V, \circ)$ is isomorphic to the algebra $\pi_n(Cl_{r,s})$.*

In [RSV] explicit examples are computed for $Cl_{0,1}, Cl_{1,2}, Cl_{3,1}$ and $Cl_{0,4}$ applying the above described procedures.

References

[DSS] R. Delanghe, F. Sommen, V. Souček, Clifford Analysis and Spinor Valued Functions, Dordrecht, Kluwer Acad. Publ., 1992.

[H1] A. Hurwitz, Über die Komposition der quadratischen Formen von beliebig vielen Variablen, Nachrichten von der Königlichen Gesellschaft der Wissenschaften zu Göttingen, Math. phys. Kl. (1898), 308–316, reprinted in: A. Hurwitz, Mathematishe Werke II, Birkhäuser Verlag, Basel, 1933, 565-571.

[H2] ——— , Über die Komposition der quadratischen Formen, Math. Ann. 88(1923), 1–25, reprinted in: A. Hurwitz, Mathematische Werke II, Birkhäuser Verlag, Basel, 1933, 641–666.

[KR1] W. Królikovski and E. Ramírez de Arellano, Fueter–Hurwitz regular mappings and an integral representation, Clifford Algebras and their Applications in Mathematical Physics, A. Micali et al., eds., (1992) Kluwer Acad. Publ., Dordrecht, 221–237.

[KR2] ——— , Polynomial solutions of the Fueter–Hurwitz equation. Contemporary Mathematics 137, 1992, 297–305.

[LM] H.B. Lawson, M–L. Michelsohn, Spin Geometry. Princeton Mathematical Series: 38, Princeton University Press, 1989.

[LR1] J. Lawrynowicz and J. Rembieliński, Pseudo–euclidean Hurwitz pairs and generalized Fueter equations, in Clifford algebras and their applications in mathematical physics, J.S.R. Chrisholm and A.K. Common, eds., D. Reidel Publ. Co. Dordrecht, etc. (1986) 39–48.

[LR2] ——— Complete classification for pseudo euclidean Hurwitz pairs including symmetry applications, a) Inst. of Phys. Univ of Lódź Preprint 86–5(1986) 15pp. b) Bull Soc. Sc. Lettres Lódź 36, No. 29, 1986, 15pp.

[LR3] ——— , Pseudo–euclidean Hurwitz pairs and the Kaluza-Klein theories, J. Phys. A Math. Ser. 20(1987) 5831–5848.

[LR4] ——— , On the composition of nondegenerate quadratic forms with an arbitrary index, Ann. Fac Sci. Toulouse Math, (5) 10(1989), 141–168 (reprinted in the same journal because a printing error in 11(1990) 141–168).

[RSV] E. Ramírez de Arellano, M.V. Shapiro and N.L. Vasilevski, A canonical relation between Hurwitz Pairs and Clifford algebras. Reporte Interno 112, Departamento de Matemáticas, CINVESTAV–IPN, (34 pp.), 1993.

[Sh] D.B. Shapiro, Products of sums of squares. Expo. Math 2 (1984), 235–261.

[SY] T.L. Smith and P. Yiu, Construction of sums of squares formulae with integer coefficients. Preprint 1992.

ON THE BERGMANN KERNEL FUNCTION IN THE CLIFFORD ANALYSIS*

MICHAEL V. SHAPIRO and NIKOLAI L. VASILEVSKI
Department of Mathematics, Odessa State University,
Odessa, UKRAINE
and
Departamento de Matemáticas, CINVESTAV del IPN,
México, D.F., MEXICO

Abstract. We define and study the Bergmann kernel function and corresponding integral operator for Clifford algebra–valued hyperholomorphic functions. For any bounded domain Ω with a smooth enough boundary $\Gamma = \partial\Omega$ we have established the following facts:

The formula which gives an independent definition of ${}_{\psi}B$ by means of the Green function g of a domain Ω; ${}_{\psi}B$ is left-ψ-h.h. with respect to the first variable, is right–anti–ψ–h.h. with respect to the second one, is hermitian conjugate; Bergmannh.h. operator ${}^{\psi}B$ (i.e. the integral operator with the kernel ${}_{\psi}B$) reproduces left–ψ–h.h. functions; the integral representation of ${}_{\psi}B$ containing the integral over the boundary with the Clifford Cauchy kernel and the integral over Ω giving a continuous function.

Key words: Bergmann kernel function, reproducing property, integral representations, hyperholomorphic analysis

In the paper we define and study the Bergmann kernel function and corresponding integral operator for Clifford algebra–valued hyperholomorphic functions. The Bergmann function K_Ω of a domain $\Omega \subset \mathbb{C}$ in one-dimensional complex analysis has a lot of equivalent definitions given in different terms (see [B]). From our point of view the most productive definition is the following one:

$$K_\Omega(z,\overline{\zeta}) := -\frac{2}{\pi}\frac{\partial^2 G(z,\zeta)}{\partial z \partial\overline{\zeta}}, \tag{0.1}$$

where G is the Green function of a domain Ω.

This definition allows to obtain a lot of delicate and deep properties of K_Ω.

In the multidimensional complex analysis the simple connection between the Bergmann and Green functions is lost. To study the properties of the Bergmann kernel function one should use more powerful and more complex tools (see, for example, [R], [F1] [F2]).

The deep structural analogy between the one-dimensional complex analysis and the Clifford analysis allows to make use of the equality (0.1) to generalize the notion of the Bergmann kernel function and allows to follow in our investigation the general line of one-dimensional study of it.

* This work was partially supported by CONACYT project 1821-E9211

F. Brackx et al. (eds.), Clifford Algebras and their Applications in Mathematical Physics, 183–192.

This work is a natural development of our paper [SV] which treats with the quaternionic case.

We know only two papers more, related to the problem. In [BDS,§29] the Bergmann function in Clifford analysis had been constructed for the unit ball using the general approach of reproducing kernels in a Hilbert space. In [Dz] the author uses the language of $\mathbb{C}^2$-valued vectors and 2×2-matrices with operator entries which is, in fact, a matrix representation of a special case of our quaternionic considerations (see [SV], Section 8).

In our work we study the notion of the Bergmann ψ-hyperholomorphic kernel function ${}_\psi\mathcal{B}$.For any bounded domain Ω with a smooth enough boundary $\Gamma = \partial\Omega$ we have established the following facts:

1. The formula which gives an independent definition of ${}_\psi\mathcal{B}$ by means of the Green function g of a domain Ω.
2. ${}_\psi\mathcal{B}$ is left-ψ-h.h. with respect to the first variable, is right–anti–ψ–h.h. with respect to the second one; is hermitian conjugate.
3. Bergmannh.h. operator ${}^\psi B$ (i.e. the integral operator with the kernel ${}_\psi\mathcal{B}$) reproduces left–ψ–h.h. functions.
4. The integral representation of ${}_\psi\mathcal{B}$ containing the integral over the boundary with the Clifford Cauchy kernel and the integral over Ω giving a continuous function.

In the quaternionic case we have found quite sudden and unpredictable novelty: to investigate ${}_\psi\mathcal{B}$ for some fixed class of ψ–h.h. functions we necessarily have to use some other class of φ–h.h. functions. A new effect arises in general situation: it appears that the method of producing of integral representation does not work for m even.

1. Preliminaries

1.1 Let $\mathcal{A}_n^-$, denotes the universal real Clifford algebra with the quadratic form $-\sum_{k=1}^{n} t_k^2$. In particular, this means that if $i_1, \ldots, i_n$ is the basis of n–dimensional space on which the algebra $\mathcal{A}_n^-$ is built and i_0 is the unit of $\mathcal{A}_n^-$, then

$$i_k^2 = -i_0, \quad i_p i_q + i_q i_p = 0$$

for $k \in \{1, \ldots, n\}$, $p \neq q$. All necessary properties of this algebra one can find in [1] for instance. Also from [1] the terms and notations has been taken which are used without explanation. Let $m \in \{1, \ldots, n\}$. We shall consider functions with values in algebra $\mathcal{A}_n^-$ and defined in a domain $\Omega \subset \mathbb{R}^{m+1}$. If $n = 2$, i.e. $\mathcal{A}_2^-$ is the scew-field of the real quaternions, then all reasonings will remain true for $\Omega \subset \mathbb{R}^4$ also.

1.2 We denote by $\mathcal{A}_n^{-,k}$ the set of k–vectors from $\mathcal{A}_n^-$ and let $\psi := \{\psi^0, \psi^1, \ldots, \psi^m\} \subset \mathcal{A}_n^{-,0} \oplus \mathcal{A}_n^{-,1}$. On $C^1(\Omega, \mathcal{A}_n^-)$ we introduce the operator ${}^\psi D$ putting

$${}^\psi D[f] := \sum_{k=0}^{m} \psi^k \frac{\partial f}{\partial x_k}. \tag{1.1}$$

Let the conditions

$$\psi^p \cdot \overline{\psi^q} + \psi^q \cdot \overline{\psi^p} = \langle \psi^p, \psi^q \rangle_{\mathbb{R}^{m+1}} = 2\delta_{p,q}$$

be fulfilled for all $p, q \subset \{0, 1, \ldots, m\}$, where $\langle \cdot, \cdot \rangle$ is the scalar product in $\mathbb{R}^{m+1}$, $\delta_{p,q}$ is the Kronecker's delta. Of course this means that ψ is an orthonormalized system in $\mathbb{R}^{n+1} \approx \mathcal{A}_n^{-,0} \oplus \mathcal{A}_n^{-,1}$.

In this case the Laplace operator Δ has the following representation:

$$\Delta = {}^{\psi}D \cdot {}^{\bar{\psi}}D = {}^{\bar{\psi}}D \cdot {}^{\psi}D. \tag{1.2}$$

The operator ${}^{\psi}D$ will be called (left) Cauchy–Riemann operator for $\mathcal{A}_n^-$-valued functions, and the elements of the set ${}^{\psi}\mathfrak{M}(\Omega, \mathcal{A}_n^-) := ker^{\psi}D$ will be called left–ψ–hyperholomorphic in a domain Ω $\mathcal{A}_n^-$–valued functions of $(m+1)$–dimensional variable. Of course this name will be shortened when misunderstanding cannot arise. It follows from (1.2) that if $f \in {}^{\psi}\mathfrak{M}(\Omega)$ then its coordinate functions are harmonic in Ω.

2. Definition of the Hyperholomorphic Bergmann Kernel and Some its Principal Properties

2.1 Let Ω be a bounded multiconnected domain in $\mathbb{R}^{m+1}$ with the smooth boundary $\Gamma = \partial\Omega$. Denote by g the harmonic (or classical) Green function of a domain Ω. This means that g is a function of two ((m+1)-dimensional) variables x and ξ, $(x, \xi) \in \Omega \times \Omega \setminus$ diag, diag $:= \{(x, x) \mid x \in \Omega\}$, and it has the following properties:

1. $g(x, \xi) := \theta_{m+1}(\xi - x) + h(x, \xi)$, $(x, \xi) \in \Omega \times \Omega \setminus$ diag, *where*

$$\theta_{m+1}(x) = -\frac{1}{(m-1) \cdot \mid \mathbb{S}^m \mid} \mid x \mid^{-m+1} \quad (m \geq 3),$$

is the fundamental solution of the Laplace equation, $\mid \mathbb{S}^m \mid$ *is the area of the unit sphere in* $\mathbb{R}^{m+1}$, *h is harmonic in $\Omega \times \Omega$ and is such that*

$$\begin{aligned} h(x, \xi) &:= -\theta_{m+1}(\xi - x), \qquad (x, \xi) \in \Omega \times \partial\Omega, \\ h(x, \xi) &:= h(\xi, x), \qquad \text{on} \quad \Omega \times \Omega; \end{aligned}$$

2. *g is harmonic in $\Omega \times \Omega \setminus$ diag and* $g(x, \xi) = g(\xi, x)$ in $\Omega \times \Omega \setminus$ diag;
3. $g(x, \xi) = 0 \quad \forall (x, \xi) \in \Omega \times \Gamma$;
4. $g \in C(\bar{\Omega} \times \bar{\Omega} \setminus diag)$.

It is known (see, for instance, [HK], p.26) that such function exists and is unique.

For example, in the case of the unit ball, i.e. $\Omega = \mathbb{B}(0, 1)$, we have (see, [HK] p.26, [Do], p.14)

$$g(x, \xi) = -\frac{1}{(m-1) \mid \mathbb{S}^m \mid} \left(\frac{1}{|\xi - x|^{m-1}} - \frac{1}{(1 - 2\langle \xi, x \rangle + |\xi|^2 \cdot |x|^2)^{(m-1)/2}} \right),$$

where $\langle \xi, x \rangle = \sum_{k=0}^{m} \xi_k \cdot x_k$.

2.2 For an arbitrary structural set ψ introduce the function ${}_{\psi}\mathcal{B}$:

$${}_{\psi}\mathcal{B}(x, \xi) := D_{\xi}^{\psi} \cdot {}^{\bar{\psi}}D_x[g](x, \xi) = {}^{\bar{\psi}}D_x \cdot {}^{\psi}D_{\xi}[g](x, \xi), \quad (x, \xi) \in \bar{\Omega} \times \bar{\Omega} \setminus \text{diag}. \tag{2.1}$$

We shall call it the Bergmann left-ψ-hyperholomorphic kernel function and shall write the Bergmann h.h. kernel (or simply kern) function. Sometimes we shall use also the term the *Bergmann h.h. kernel.*

In particular, for the unit ball (compare with [BDS,§29]):

$$
{}_{\psi}\mathcal{B}(x,\xi)=\frac{m+1}{|\mathbb{S}^m|}\cdot\frac{1}{(1-2\langle\xi,x\rangle+|\xi|^2\cdot|x|^2)^{(m+1)/2}}-
$$

$$
-\frac{2}{|\mathbb{S}^m|}\cdot\frac{x_{\overline{\psi}}\cdot\xi_{\psi}}{(1-2\langle\xi,x\rangle+|\xi|^2\cdot|x|^2)^{(m+1)/2}}+
$$

$$
+\frac{m+1}{|\mathbb{S}^m|}\cdot\frac{(\xi-x\cdot|\xi|^2)_{\overline{\psi}}\cdot(x-\xi\cdot|x|^2)_{\psi}}{(1-2\langle\xi,x\rangle+|\xi|^2\cdot|x|^2)^{(m+3)/2}},
$$

here for $x\in\mathbb{R}^{m+1}$ $x_{\psi}:=\sum_{k=0}^{m}x_k\cdot\psi^k$.

2.3 Remark. *The function ${}_{\psi}\mathcal{B}$ can be continuously extended onto a diagonal of $\Omega\times\Omega$, and, moreover, in the definition of ${}_{\psi}\mathcal{B}$ the function g may be substituted by h.*

Really

$$
\begin{aligned}
{}_{\psi}\mathcal{B} &= D_{\xi}^{\psi}\cdot{}^{\bar\psi}D_x[g]=D_{\xi}^{\psi}\cdot{}^{\bar\psi}D_x[\theta_{m+1}(\xi-x)+h]\\
&= D_{\xi}^{\psi}\cdot{}^{\bar\psi}D_x[h]\quad on\quad \bar\Omega\times\Omega\setminus diag.
\end{aligned}
$$

Hence we assume further that ${}_{\psi}\mathcal{B}$ is defined in a whole $\Omega\times\Omega$.

2.4 Theorem. *Bergmann h.h. kern-function has the following properties:*

1. *for any fixed* $\xi\in\Omega$
$$ {}_{\psi}\mathcal{B}(\cdot,\xi)\in{}^{\psi}\mathfrak{M}(\Omega), $$
2. *for any fixed* $x\in\Omega$
$$ {}_{\psi}\mathcal{B}(x,\cdot)\in\mathfrak{M}^{\bar\psi}(\Omega), $$
3. ${}_{\psi}\mathcal{B}(x,\xi)=\overline{{}_{\psi}\mathcal{B}(\xi,x)}$.

Proof is the same as in the case of quaternionic functions [SV]. First two assertions are the simple corollaries of the factorization (1.2), of the fact that Δ is a "real" operator (i.e. it maps real-valued functions into real-valued ones) and of Remark 2.3 which allows substitute h instead of g.

And besides we have

$$
\overline{{}_{\psi}\mathcal{B}(\xi,x)}=\overline{D_x^{\psi}\cdot{}^{\bar\psi}D_{\xi}[g](\xi,x)}={}^{\bar\psi}D_x{}^{\psi}D_{\xi}[g]={}_{\psi}\mathcal{B}(x,\xi).
$$

2.5 Theorem. (Reproducing property of the Bergmann h.h. kernel). *Let $\Gamma=\partial\Omega$ be a smooth surface and $f\in{}^{\psi}\mathfrak{M}(\Omega,\mathcal{A}_n^-)\cap C(\bar\Omega,\mathcal{A}_n^-)$. Then*

$$
{}^{\psi}\mathcal{B}[f](x):=\int_{\Omega}{}_{\psi}\mathcal{B}(x,\xi)f(\xi)\,d\xi=f(x),\quad x\in\Omega. \tag{2.2}
$$

The natural name for ${}^{\psi}\mathcal{B}$ is the Bergmann h.h. operator.

Proof. For $\varepsilon > 0$ consider

$$A(x,\varepsilon) := \int_{\Omega_{x,\varepsilon}} {}_\psi\mathcal{B}(x,\xi) f(\xi)\, d\xi = \int_{\Omega_{x,\varepsilon}} D_\xi^\psi \cdot {}^{\bar\psi}D_x[g] \cdot f(\xi)\, d\xi.$$

Use the Stokes formula [BDS]:

$$\int_{\partial\Omega_{x,\varepsilon}} ({}^\psi D_x[g]) \cdot \sigma_\psi \cdot f = \int_{\Omega_{x,\varepsilon}} \left(D_\xi^\psi \cdot {}^{\bar\psi}D_x[g] \cdot f + {}^{\bar\psi}D_x[g] \cdot {}^\psi D_\xi[f] \right) d\xi$$

$$= \int_{\Omega_{x,\varepsilon}} D_\xi^\psi \cdot {}^{\bar\psi}D_x[g] \cdot f\, d\xi,$$

because of $f \in {}^\psi\mathfrak{M}(\Omega)$. Hence

$$A_{x,\varepsilon} = \int_{\partial\Omega_{x,\varepsilon}} {}^{\bar\psi}D[g] \cdot \sigma_\psi \cdot f.$$

By the property 3. from 2.1 we have

$${}^{\bar\psi}D_x[g] = 0 \qquad \forall \xi \in \Gamma.$$

As $\partial\Omega_{x,\varepsilon} = \Gamma \cup \mathbb{S}^-(x,\varepsilon)$ with negatively oriented sphere then

$$A_{x,\varepsilon} = -\int_{\mathbb{S}(x,\varepsilon)} {}^{\bar\psi}D[g] \cdot \sigma_\psi \cdot f.$$

Calculation

$${}^{\bar\psi}D_x[g](x,\xi) = {}^{\bar\psi}D_x[\theta_{m+1}(\xi - x)] + {}^{\bar\psi}D_x[h](x,\xi)$$

$$= -{}^{\bar\psi}D_\xi[\theta_{m+1}(\xi - x)] + h_1(x,\xi) = -\mathcal{K}_\psi(\xi - x) + h_1(x,\xi)$$

gives

$$A_{x,\varepsilon} = \int_{\mathbb{S}(x,\varepsilon)} \mathcal{K}_\psi(\xi - x) \cdot \sigma_{\psi,\xi} \cdot f(\xi) + o(\varepsilon) = f(x) + o(\varepsilon),$$

and tending $\varepsilon \to 0$ we obtain the necessary result.

2.6 Remark. *Theorem 2.4 and 2.5 justify the name* Bergmann kernel function *for ${}_\psi\mathcal{B}$, by our opinion.*

3. Integral Representation of the Bergmann Hyperholomorphic Kernel

3.1 Given arbitrary structural sets ψ and φ, introduce the function ${}_{\varphi,\psi}\mathcal{L}$ on $\bar\Omega \times \bar\Omega \setminus$ diag:

$${}_{\varphi,\psi}\mathcal{L} := {}^\varphi D_\xi \cdot {}^\psi D_x[g](x,\xi) = D_x^\psi \cdot {}^\varphi D_x[g](x,\xi).$$

It is clear that when $\varphi = \bar\psi$ we have

$${}_{\bar\psi,\psi}\mathcal{L} := {}_\psi\bar{\mathcal{B}} \quad \text{or} \quad {}_\psi\mathcal{B}(x,\xi) = {}_{\bar\psi,\psi}\mathcal{L}(\xi,x).$$

From subsection 2.1 it follows that

$$ {}_{\varphi,\psi}\mathcal{L}(x,\xi) = {}_{\varphi,\psi}\Lambda(x,\xi) + {}_{\varphi,\psi}\ell(x,\xi), \tag{3.1} $$

where

$$ \begin{aligned} {}_{\varphi,\psi}\Lambda(x,\xi) &= {}^{\varphi}D_\xi \cdot {}^{\psi}D_x[\theta_{m+1}(\xi - x)] = D_x^{\psi} \cdot {}^{\varphi}D_\xi[\theta_{m+1}(\xi - x)] \\ &= \frac{(m+1)\cdot(\xi - x)_\varphi \cdot (\xi - x)_\psi - |\,\xi - x\,|^2 \cdot \sum_{k=0}^{m} \varphi^k \cdot \psi^k}{|\,\mathbb{S}^m\,| \cdot |\,\xi - x\,|^{m+3}}, \\ {}_{\varphi,\psi}\ell(x,\xi) &= {}^{\varphi}D_\xi \cdot {}^{\psi}D_x[h](x,\xi) = D_x^{\psi} \cdot {}^{\varphi}D_\xi[h](x,\xi). \end{aligned} $$

3.2 Lemma. *Let $g_o\colon \Omega \to \mathbb{R}$ be a function, smooth enough, for which Γ is a level surface. Then for any structural set ψ and for any $\xi \in \Gamma$*

$$ \nu_\psi \cdot {}^{\bar\psi}D[g_0](\xi) = \varepsilon_0 \cdot |\, grad\, g_0(\xi)\,|, $$

and hence left hand side does not depend on a choice of ψ. Here $\varepsilon_0 = \pm 1$, $\vec\nu = (\nu_0, \nu_1, \nu_2, \nu_3)$ is a unit outer normal vector to a surface Γ in a point $\xi \in \Gamma$,

$$ \nu_\psi := \sum_{k=0}^{m} \nu_k \cdot \psi^k. $$

Proof. It is known from the standard calculus that under our assumptions $\vec\nu$ is collinear to $\operatorname{grad} g_0(\xi)$, hence

$$ \operatorname{grad} g_0(\xi) = \varepsilon_0 |\operatorname{grad} g_0(\xi)| \cdot \nu(\xi). $$

Thus

$$ \begin{aligned} \nu_\psi \cdot {}^{\bar\psi}D[g_0](\xi) &= \nu_\psi(\xi) \cdot \varepsilon_0 \cdot |\operatorname{grad} g_0(\xi)| \cdot \nu_{\bar\psi}(\xi) \\ &= \varepsilon_0 |\operatorname{grad} g_0(\xi)|. \end{aligned} $$

3.3 Corollary. *Functions ${}_\psi\mathcal{B}$ and ${}_{\varphi,\psi}\mathcal{L}$ are connected by the formula*

$$ \nu_\psi(\xi) \cdot {}_\psi\mathcal{B}(\xi, x) = \nu_{\bar\varphi} \cdot {}_{\varphi,\psi}\mathcal{L}(x,\xi), \qquad (x,\xi) \in \Omega \times \Gamma. \tag{3.2} $$

Proof. For $(x,\xi) \in \Omega \times \Gamma$ we have $g(x,\xi) = 0$, hence Γ is a level surface for

$$ \tilde g_k := \frac{\partial g}{\partial x_k}(x,\cdot), \qquad x \in \Omega,\ k \in \{0,1,\dots,m\}. $$

And so

$$ \begin{aligned} \nu_\psi(\xi) \cdot {}_\psi\mathcal{B}(\xi, x) &= \nu_\psi(\xi) \cdot {}^{\bar\psi}D_\xi \cdot {}^{\psi}D_x[g] \\ &= \sum_{k=0}^{m} \nu_\psi \cdot {}^{\bar\psi}D_\xi[\tilde g_k] \cdot \psi^k \\ &= \sum_{k=0}^{m} \nu_{\bar\varphi} \cdot {}^{\varphi}D_\xi[\tilde g_k] \cdot \psi^k \\ &= \nu_{\bar\varphi} \cdot {}^{\varphi}D_\xi \cdot {}^{\psi}D[g] = \nu_{\bar\varphi} \cdot {}_{\varphi\psi}\mathcal{L}(x,\xi). \end{aligned} $$

3.4 In the quaternionic case [SV] the following fact was of principal value: for any structural set ψ there exists a structural set φ such that

$$\sum_{k=0}^{3} \varphi^k \cdot \psi^k = 0.$$

The question arises whether the corresponding equality for an arbitrary m

$$\sum_{k=0}^{m} \varphi^k \cdot \psi^k = 0 \tag{3.3}$$

is true for some φ.

Let first m be odd, i.e. $(m+1)$ is even. Then, given ψ, define φ by the rule

$$\varphi^k := \begin{cases} \overline{\psi^k}, & k \in K_1, \\ -\overline{\psi^k}, & k \in K_2, \end{cases}$$

where $K_1 \cup K_2 = \{0, 1, \ldots, m\}$, $K_1 \cap K_2 = \emptyset$, and each of K_1, K_2 has $\frac{m+1}{2}$ elements. Clear that for such φ we have (3.3).

Let us show now that for m even the condition (3.3) is contradictive, and hence φ cannot be found. We can write $\psi^k = \psi_0^k + \vec{\psi^k}$, $\varphi^k = \varphi_0^k + \vec{\varphi^k}$, where $\psi_0^k, \varphi_0^k \in \mathcal{A}_n^{-,0}, \vec{\psi^k}, \vec{\varphi^k} \in \mathcal{A}_n^{-,1}$.

Equating k–vectors ($k \in \{0, 1, 2\}$) in (3.3) we arrive at the system

$$\sum_{k=0}^{m} \varphi_0^k \cdot \psi_0^k - \sum_{k=0}^{m} \langle \vec{\varphi^k}, \vec{\psi^k} \rangle_{\mathbb{R}^n} = 0, \tag{3.4}$$

$$\sum_{k=0}^{m} (\varphi_0^k \cdot \vec{\psi^k} + \psi_0^k \cdot \vec{\varphi^k}) = 0,$$

$$\sum_{k=0}^{m} \vec{\varphi^k} \wedge \vec{\psi^k} = 0, \tag{3.5}$$

which is equivalent to (3.3).

Denote by $\widetilde{\psi}$ a fixed basis in $\mathbb{R}^{n+1} \approx \mathcal{A}_n^{-,0} \oplus \mathcal{A}_n^{-,1}$ containing the set $\overline{\psi}$ (if $m = n$ then $\widetilde{\psi} := \overline{\psi}$).

To simplify calculations we assume that $\psi^0 := i_0$, then $\psi^k \in \mathcal{A}_n^{-,1}$ for $k \neq 0$, and $\widetilde{\psi}$ has a form

$$\widetilde{\psi} = \{i_0, -\psi^1, \ldots, -\psi^m, \widetilde{\psi}^{m+1}, \ldots, \widetilde{\psi}^n\}.$$

We may assume also that φ is given just in this basis: $\varphi^k = \sum_{j=0}^{n} \varphi_j^k \cdot \widetilde{\psi}^j$.

Condition (3.4) gives now

$$\sum_{k=0}^{m} \varphi_k^k = 0, \tag{3.6}$$

and (3.5) gives

$$\sum_{1\le k<j\le m} (\varphi_j^k - \varphi_k^j)\psi^k \wedge \psi^j = 0,$$

$$\varphi_j^k = 0 \quad \forall j > m.$$

So in fact $\varphi^k = \sum_{j=0}^{m} \varphi_j^k \cdot \overline{\psi}^j$. Introduce $(m+1)\times(m+1)$ matrix

$$\phi := \begin{pmatrix} \varphi^0 \\ \dots \\ \varphi^m \end{pmatrix}.$$

It is orthogonal by the definition of a structural set. And (3.6) means that ϕ has zero trace which is impossible.

3.5 Theorem. (Integral representation of the Bergmann hyperholomorphic kernel function). *Let m be odd and let ψ and φ be two structural sets with the only condition*

$$\sum_{k=0}^{m} \varphi^k \cdot \psi^k = 0,$$

then the following equality holds

$$_\psi\mathcal{B}(x,\xi) = \int_\Gamma \mathcal{K}_\psi(\tau - x)\cdot \sigma_{\bar\varphi,\tau}\cdot {}_{\varphi,\psi}\Lambda(\tau-\xi) + \int_\Omega \overline{{}_{\varphi,\psi}\ell(x,\tau)}\cdot {}_{\varphi,\psi}\ell(\xi,\tau)\, d\tau,$$

for $\forall (x,\xi)\in\Omega\times\Omega$.

Proof. For any fixed $\zeta\in\Omega$ the function $_\psi\mathcal{B}(\cdot,\zeta)$ is left-ψ-hyperholomorphic. Hence from the Cauchy formula we have

$$_\psi\mathcal{B}(x,\zeta) = \int_\Gamma \mathcal{K}_\psi(\xi - x)\cdot\sigma_{\psi,\xi}\cdot {}_\psi\mathcal{B}(\xi,\zeta), \quad (x,\zeta)\in\Omega\times\Gamma. \tag{3.7}$$

Multiplying (3.2) by ds and substituting into (3.7) we obtain

$$_\psi\mathcal{B}(x,\zeta) = \int_\Gamma \mathcal{K}_\psi(\xi - x)\cdot\sigma_{\bar\varphi,\xi}\cdot \mathcal{L}_{\varphi,\psi}(\zeta,\xi).$$

Using the representation (3.1) we arrive at

$$_\psi\mathcal{B}(x,\zeta) = \int_\Gamma \mathcal{K}_\psi(\xi - x)\cdot\sigma_{\bar\varphi,\xi}\cdot {}_{\varphi,\psi}\Lambda(\xi-\zeta)$$

$$+\int_\Gamma \mathcal{K}_\psi(\xi - x)\cdot\sigma_{\bar\varphi,\xi}\cdot {}_{\varphi,\psi}\ell(\zeta,\xi). \tag{3.8}$$

The general Stokes formula gives now

$$\int_\Gamma \mathcal{K}_\psi(\xi - x)\cdot\sigma_{\bar\varphi,\xi}\cdot {}_{\varphi,\psi}\ell(\zeta,\xi) = \int_{S(x,\varepsilon)} \mathcal{K}_\psi(\xi - x)\cdot\sigma_{\bar\varphi,\xi}\cdot {}_{\varphi,\psi}\ell(\zeta,\xi)$$

$$+\int_{\Omega_{x,\varepsilon}} D_\xi^{\bar\varphi}\left[\mathcal{K}_\psi(\xi-x)\right]\cdot {}_{\varphi,\psi}\ell(\zeta,\xi)\,d\xi+\int_{\Omega_{x,\varepsilon}} \mathcal{K}_\psi(\xi-x)\cdot {}^{\bar\varphi}D_\xi[{}_{\varphi,\psi}\ell](\zeta,\xi)\,d\xi. \quad (3.9)$$

The definition of ${}_{\varphi,\psi}\ell$ implies ${}^{\bar\varphi}D_\xi[{}_{\varphi,\psi}\ell]=0$. Calculate:

$$N_{x,\varepsilon}:=\int_{\mathbb{S}(x,\varepsilon)} \mathcal{K}_\psi(\xi-x)\cdot\sigma_{\bar\varphi,\xi}\cdot {}_{\varphi,\psi}\ell(\zeta,\xi)$$

$$=\frac{1}{(m-1)\mid\mathbb{S}^m\mid\varepsilon^{m+1}}\int_{\mathbb{S}(x,\varepsilon)}\sum_{k=0}^{m}\bar\psi_k(\xi_k-x_k)\cdot\sigma_{\bar\varphi,\xi}\cdot {}_{\varphi,\psi}\ell(\zeta,\xi).$$

Using

$$\sum_{k=0}^{m}\varphi^k\cdot\psi^k=0$$

we have

$$D_\xi^{\bar\varphi}\left(\sum_{k=0}^{m}\overline{\psi^k}(\xi_k-x_k)\right)=\sum_{k=0}^{m}\overline{\psi^k}\cdot\overline{\varphi^k}=0,$$

and consequently Cauchy integral theorem means that $N_{x,\varepsilon}=0\ \forall x,\varepsilon$. Thus the equality (3.9) turns into

$$\int_\Gamma \mathcal{K}_\psi(\xi-x)\cdot\sigma_{\bar\varphi,\xi}\cdot {}_{\varphi,\psi}\ell(\zeta,\xi)=\int_{\Omega_{x,\varepsilon}} D_\xi^{\bar\varphi}\left[\mathcal{K}_\psi(\xi-x)\right]\cdot {}_{\varphi,\psi}\ell(\zeta,\xi)\,d\xi. \quad (3.10)$$

Using the Stokes formula and the equality ${}^{\bar\varphi}D[{}_{\varphi\psi}\ell]=0$ we have for $(x,\zeta)\in\Omega\times\Omega$

$$\int_{\Omega_{x,\varepsilon}} {}_{\varphi,\psi}\overline{\mathcal{L}(x,\xi)}\cdot {}_{\varphi,\psi}\ell(\zeta,\xi)\,d\xi=\int_{\Omega_{x,\varepsilon}} D_\xi^{\bar\varphi}\cdot D_\xi^{\bar\psi}[g](x,\xi)\cdot {}_{\varphi,\psi}\ell(\zeta,\xi)\,d\xi$$

$$=\int_{\Gamma\cup\mathbb{S}(x,\varepsilon)} D_x^{\bar\psi}[g](x,\xi)\cdot\sigma_{\bar\varphi,\xi}\cdot {}_{\varphi,\psi}\ell(\zeta,\xi). \quad (3.11)$$

Substitute $g=\theta_{m+1}(\xi-x)+h$ into (3.11) and use $N_{x,\varepsilon}=0$:

$$\int_{\Omega_{x,\varepsilon}} D_\xi^{\bar\varphi}\,{}^{\bar\psi}D_x\left[\theta_{m+1}(\xi-x)\right]\cdot {}_{\varphi,\psi}\ell(\eta,\xi)\,d\xi+\int_{\Omega_{x,\varepsilon}} D_\xi^{\bar\varphi}\cdot{}^{\psi}D_x[h]\,{}_{\varphi,\psi}\ell(\eta,\xi)\,d\xi$$

$$=-\int_{\partial\mathbb{B}(x,\varepsilon)} \mathcal{K}_\psi(\xi-x)\cdot\sigma_{\bar\varphi,\xi}\cdot {}_{\varphi,\psi}\ell(\eta,\xi)+\alpha(\varepsilon)=\alpha(\varepsilon),$$

where

$$\alpha(\varepsilon)=-\int_{\partial\mathbb{B}_{(x,\varepsilon)}} D_x^{\psi}[h](x,\xi)\cdot\sigma_{\bar\varphi,\xi}\cdot {}_{\varphi,\psi}\ell(\eta,\xi),$$

or

$$\int_{\Omega_{x,\varepsilon}} D_\xi^{\bar\varphi}\left[\mathcal{K}_\psi(\xi-x)\right]\cdot {}_{\varphi,\psi}\ell(\zeta,\xi)\,d\xi=\int_{\Omega_{x,\varepsilon}} {}_{\varphi,\psi}\overline{\ell(x,\xi)}\cdot {}_{\varphi,\psi}\ell(\zeta,\xi)\,d\xi+o(\varepsilon). \quad (3.12)$$

From (3.10), (3.12) under $\varepsilon\to 0$ we obtain

$$\int_\Gamma \mathcal{K}_\psi(\xi-x)\cdot\sigma_{\bar\varphi,\zeta}\cdot {}_{\varphi,\psi}\ell(\zeta,\xi)=\int_\Omega {}_{\varphi,\psi}\overline{\ell(x,\xi)}{}_{\varphi,\psi}\ell(\zeta,\xi)\,d\xi. \quad (3.13)$$

Finally, comparing (3.8) and (3.13), we finish the proof.

3.6 Corollary. *Under the assumptions of Theorem 3.5 we have also*

$$_{\psi}\mathcal{B}(x,\xi) = \int_{\Gamma} {}_{\varphi,\psi}\overline{\Lambda(\tau - x)} \cdot \sigma_{\varphi,\tau} \cdot \mathcal{K}_{\bar{\psi}}(\tau - \xi) + \int_{\Omega} {}_{\varphi,\psi}\overline{\ell(x,\tau)} \cdot {}_{\varphi,\psi}\ell(\xi,\tau)\, d\tau.$$

It follows from the Hermitian simmetry of function $_{\psi}\mathcal{B}$.

References

[B] S. Bergmann. *The Kernel Function and Conformal Mappings.* AMS, Providence, R. I. 1970.

[BDS] F. Brackx, R. Delanghe, F. Sommen. *Clifford Analysis.* Boston, Pitman, 1982.

[Do] J.L. Doob. *Classical Potential Theory and Its Probabilistic Counterpart.* Springer: New York Berlin Heidelberg, 1984.

[Dz] A. Dzhuraev. *On kernel matrices and holomorphic vectors.* Complex Variables, v. 16, 1991, p. 43–57.

[F1] C. Fefferman. *The Bergman kernel and biholomorphic mappings of pseudoconvex domains.* Inventiones Math., v. 26, 1974, p. 1–65.

[F2] C. Fefferman. *Monge-Ampère equations, the Bergman kernel, and geometry of pseudoconvex domains.* Annals of Math., v. 103, 1976, p. 395–416.

[HK] W.K. Hayman, P.B. Kennedy. *Subharmonic Functions.* Academic Press; London New York San Francisco, 1976.

[R] R.M. Range. *Holomorphic Functions and Integral Representations in Several Complex Variables.* Springer: New York Berlin Heidelberg, 1986.

[SV] M.V. Shapiro, N.L. Vasilevski. *On the Bergman kernel function in hypercomplex analysis.* Reporte Interno No. 115, Departamento de Matemáticas, CINVESTAV del I.P.N., Mexico City, 1993, 35p.

$SO(M)$-INVARIANT OPERATORS ON CLIFFORD TENSORS

F. SOMMEN*
University of Gent
Department of Mathematical Analysis
Galglaan 2
B-9000 Gent/Belgium

Abstract. In this paper we characterize $SO(m)$-invariant operators on Clifford tensors. We show that a set of eight $Gl(m)$-invariant operators together with the pseudoscalar generate the algebra of all $SO(m)$-invariant operators.

Key words: Clifford analysis, Clifford tensors, invariant operators

Introduction

Tensors over a vector space V are multilinear functionals on that vector space. Tensor algebra can be used as a starting point for the construction of polynomials over V as well as forms (alternating tensors) on V. More in general, irreducible representations of $Gl(m)$ can be realized on suitable spaces of tensors by using symmetrization and alternation operators (see also [10]).

For the representation theory of $SO(m)$ we may consider tensors on $V = R^m$, provided with its inner product $\langle x, y\rangle = \sum x_j y_j$ and it seems more interesting to consider Clifford tensors, i.e. multilinear functionals on R^m taking values in the Clifford algebra R_m. In this way one can study Clifford polynomials as well as Clifford forms starting from Clifford tensors. Conversely, like in [5] one can represent Clifford tensors by R_m-valued multilinear polynomials $F(\underline{u}_1, \ldots, \underline{u}_k)$, where $\underline{u}_1, \ldots, \underline{u}_k$ are Clifford vectors and many basic operations on tensors involving vector variables $\underline{u}$ and Dirac type operators or vector derivatives $\partial_{\underline{u}}$ are immediately available. We'll use this way of representing tensors and call it R_m-valued notation. But as a vector space, R_m can be identified with a minimal left ideal in $End(R_m)$ which is isomorphic to the Clifford algebra $R_{m,m}$ over ultra-hyperbolic space $R^{m,m}$ and we can consider tensors with values in this ideal. Moreover we can consider them to depend on variables $u_1, \ldots, u_k, u_1', \ldots, u_l'$ belonging to suitable maximally isotropic subspaces V and V' of $R^{m,m}$ which are dual to one another.

This alternative way of representing tensors may seem a bit more complicated, but it enables us to represent $Gl(m)$ as a subgroup of $Spin(m, m)$ (see also [2], [4], [6], [8]) and to extend the action of $Spin(m)$ on Clifford tensors to the group $Gl(m)$. Moreover, the distinction between covariant and contravariant tensors is more clear.

* Senior Research Associate, NFWO, Belgium

F. Brackx et al. (eds.), Clifford Algebras and their Applications in Mathematical Physics, 193–202.

In this paper we characterize the $Gl(m)$- and $SO(m)$-invariant operators on Clifford tensors. Hereby we make use of results established in our papers [7], [8] about $SO(m)$-invariant operators on polynomials, which only need to be extended to the case of several vector variables in order to be available for Clifford tensors.

In the first section we describe the algebra $R_{m,m}$ of endomorphisms on R_m. Then we give a survey about our papers [7] and [8] and indicate how results may be generalized to the case of several vector variables. In the third section we construct eight basic operators on tensors which together with the pseudoscalar suffice to construct all other $Gl(m)$- and $SO(m)$-invariant operators.

In a forthcoming paper [9] we study Clifford tensors more in detail and indicate how irreducible representations of $SO(m)$ may be realized on spaces of Clifford tensors. Clifford differential forms were studied to some extent in [3]. A clear treatment of representation theory of $SO(m)$ involving representations on spaces of polynomials of a matrix variable was presented in [4].

1. Endomorphisms on Clifford algebras

The space R^m together with its inner product $\langle x, y\rangle = \sum x_j y_j$ leads to the Clifford algebra R_m, determined by the relations

$$xy + yx = -2\langle x, y\rangle$$

so that also $e_i e_j + e_j e_i = -2\delta_{ij}$. We know that R_m has dimension 2^m and $a \in R_m$ decomposes like

$$a = [a]_0 + \ldots + [a]_k + \ldots + [a]_m,$$

$[a]_k \in R^k_m$ being the k−vector part of a. The elements of R^m are identified with Clifford vectors $\underline{v} = [\underline{v}]_1$. The main involution and anti-involution on R_m are respectively given by $a \to \tilde{a}$ and $a \to \bar{a}$, where

$$\widetilde{ab} = \tilde{a}\tilde{b}, \quad \overline{ab} = \bar{b}\bar{a}, \quad \tilde{\underline{v}} = \bar{\underline{v}} = -\underline{v}, \quad \underline{v} \in R^m.$$

The eigenspaces of "$\tilde{\cdot}$" are the spaces R^+_m and R^-_m of even and odd Clifford numbers. The action of $SO(m)$ on R^m can be obtained directly from Clifford algebra by using the spingroup given by

$$Spin(m) = \left\{ s = \underline{\omega}_1 \ldots \underline{\omega}_{2l} : \quad \underline{\omega}_j \in S^{m-1} \right\}.$$

For $s \in Spin(m)$ and $a \in R_m$ we can consider the representations

$$l(s)a = sa, \quad h(s)a = sa\bar{s}.$$

$l(Spin(m))$ becomes irreducible on so called spinor spaces that can be realized as minimal left ideals in Clifford algebras. Under the representation $h(Spin(m))$, the spaces R^k_m of k-vectors are invariant subspaces and for $k < \frac{m}{2}$ they are irreducible. Next consider the map $a \to e_{1\ldots m}a$, $e_{1\ldots m}$ being the pseudoscalar; then this map commutes with $h(s)$, $s \in Spin(m)$ and maps R^k_m bijectively on R^{m-k}_m. Hence the action of $h(Spin(m))$ on R^{m-k}_m leads to irreducible representations of $Spin(m)$, equivalent to its action on R^k_m. For $m = 2p$, the space R^p_m splits into two inequivalent irreducible subspaces which are the eigenspaces of the map $a \rightarrow e_{1\ldots m}a$, which is equivalent to the Hodge star map.

The algebra $End(R_m)$ has also been considered by many authors (see [2], [4], [6], [8]). The main idea is Chevalley's construction of spinors (see [1]), i.e. one identifies R_m with the Grassmann algebra (spinor space) ΛR^m and then uses the fact that $End(\Lambda R^m) = R_{m,m}$. But it is better to use the following more direct approach, showing the beautiful structure.

Consider the basic endomorphisms

$$e_j[a] = e_j a, \quad e_j|[a] = \tilde{a}e_j, \quad a \in R_m.$$

Then we have within $End(R_m)$ the relations

$$e_j e_k + e_k e_j = -2\delta_{jk}, \quad e_j|\, e_k| + e_k|\, e_j| = 2\delta_{jk}$$

$$e_j e_k| = -e_k|e_j,$$

meaning that the algebra generated by the symbols e_j and $e_j|$ is $R_{m,m}$. As $R_{m,m}$ and $End(R_m)$ have the same dimension, the two algebras coincide. Next consider the basis elements

$$f_j = \frac{1}{2}(e_j - e_j|), \quad f_j' = \frac{1}{2}(e_j + e_j|),$$

then we have the relations

$$f_j f_k + f_k f_j = f_j' f_k' + f_k' f_j' = 0,$$

meaning that $V^m = span\{f_1, \ldots, f_m\}$ and $V^{m'} = span\{f_1', \ldots, f_m'\}$ are maximal isotropic spaces. Moreover, the relations

$$f_j f_k' + f_k' f_j = -\delta_{ij} = -\langle f_k', f_j\rangle,$$

mean that V^m and $V^{m'}$ are dual spaces. The subalgebras generated by V^m and $V^{m'}$ are Grassmann algebras V_m and V_m'.

Consider $I_j = -f_j f_j'$; then $I_j I_k = I_k I_j$ and $I = I_1 \ldots I_m$ is primitive idempotent. We also have the relations

$$e_j I = e_j|I = f_j' I.$$

Next consider the Euclidean spaces

$$E^m = span\{e_1, \ldots, e_m\}$$

and

$$E^m| = span\{e_1|, \ldots, e_m|\}$$

and the Clifford algebras E_m and $E_m|$ generated by these spaces. Then the above relations allow us to define the map $| : E_m \rightarrow E_m|$ by putting $aI = a|I$ and we have that

$$E_m I = V_m' I = E_m|I = R_{m,m} I,$$

which allows us to represent the vector space R_m either by E_m or V_m' i.e. the isomorphism between R_m and the Grassmann algebra is part of the picture. Of course also $IE_m = IV_m$, so that one could also identify R_m with V_m. Next consider the spingroup $Spin(m,m)$; then its subgroup

$$\widetilde{SO}(m) = \{s\bar{s}| : \quad s \in Spin(m)\},$$

is isomorphic to $SO(m)$. Moreover, for $S \in Spin(m,m)$ we can consider

$$l(S)aI = SaI, \quad h(S)a = Sa\bar{S}$$

and in case $a \in E_m$ and $S = s\bar{s}|$ we have that

$$l(S)aI = sa\bar{s}I, \quad h(S)a = sa\bar{s}, \quad h(S)a| = (h(S)a)|.$$

Next, let $\widetilde{Gl}(m)$ be the subgroup of $S \in Spin(m,m)$ such that $h(S)V^m = V^m$ and $h(S)V^{m'} = V^{m'}$, then clearly $h(S)$ determines an element $g \in Gl_+(V^m)$ and $g'' \in Gl_+(V^{m'})$ by restricting $h(S)$, where $Gl_+(V^m)$ is the group of general linear maps on V^m with positive determinant.

Lemma 1.1 *The map $h : \widetilde{Gl}(m) \to Gl_+(V^m)$ is onto. Moreover, for $u \in V^m$ and $u' \in V^{m'}$, $\langle h(S)u', h(S)u\rangle = \langle u', u\rangle$ so that $g'' = g'^{-1}$.*

Proof.
The first part follows by calculating Lie-algebras (see e.g. [2] and [8]). The second part follows from the fact that

$$\langle u', u\rangle = h(S)\langle u', u\rangle = -h(S)(u'u + uu')$$
$$= -\{h(S)u', h(S)u\} = \langle h(S)u', h(S)u\rangle = \langle g''[u'], g[u]\rangle$$

while on the other hand, $\langle u', u\rangle = \langle g'^{-1}[u'], g[u]\rangle$. □

In [8] we have shown that $\widetilde{Gl}(m)$ is the subgroup of $Spin(m,m)$ determined by the relations $h(S)I = SI\bar{S} = I$. The "fundamental bivector" is given by $B = \sum e_j e_j|$ and is linked to the primitive idempotent by means of $I = \sum_{k=0}^{\infty} \frac{1}{k!} B\wedge \ldots \wedge B = e^{\wedge B}$. Moreover, k-vectors $a \in R_m^k$ can be identified with k-vectors $a \in V_m^{k'}$ in V_m' and we have that $BaI = -(m-2k)aI$, so that the maps $a \to [a]_k$ are the eigenprojections of B. Note that a dilation $g : u \to \lambda u$ is represented by $S = exp(aB)$. From $Bf_j = f_j B = -2f_j$ it follows that $Sf_j\bar{S} = exp(-2a)f_j$ so that $\lambda = exp(-2a)$. On the other hand, $BI = -mI$, so that $SI = exp(-ma)I = \sqrt{det\, g}\, I$. Hence we obtain the formula

$$l(S)aI = SaI = Sa\bar{S}SI = \sqrt{det\, g}\, h(S)aI.$$

Note that the pseudoscalar $fI \to (-1)^{\frac{m(m+1)}{2}} e_{1\ldots m} e_1| \ldots e_m|fI$, corresponds to the main involution $a \to \tilde{a}$ on R_m.

2. Clifford polynomials

Let $\underline{x} = \sum e_j x_j$ be a vector variable; then a Clifford polynomial is an R_m-valued polynomial denoted by $P(\underline{x})$. Let $\mathcal{P}$ be the space of real polynomials, then on $\mathcal{P} \otimes R_m$ we can consider the representations of $Spin(m)$ given by

$$L(s)P(\underline{x}) = sP(\bar{s}\underline{x}s), \quad H(s)P(\underline{x}) = sP(\bar{s}\underline{x}s)\bar{s}.$$

The representation L makes sense for spinor valued functions while H makes sense for R_m^k-valued functions. A real polynomial operator is an element of $End(\mathcal{P})$. A Clifford polynomial operator is hence an element of $End(\mathcal{P}) \otimes R_m$. Such an operator A is called L-invariant if A commutes with $L(s)$, i.e. $AL(s) = L(s)A$, $s \in Spin(m)$. Differential operators with polynomial coefficients are special polynomial operators and on finite dimensional subspaces of $\mathcal{P}$, all polynomial operators can be represented by differential operators. For a differential operator $P(\underline{x}, \partial_{\underline{x}})$ the symbol is given by

$$P(\underline{x}, \partial_{\underline{x}})e^{\langle \underline{x}, \underline{u} \rangle} = P(\underline{x}, \underline{u})e^{\langle \underline{x}, \underline{u} \rangle}$$

and an operator is L-invariant iff $P(\underline{x}, \underline{u})$ is H-invariant, i.e.

$$P(\underline{x}, \underline{u}) = sP(\bar{s}\underline{x}s, \bar{s}\underline{u}s)\bar{s}.$$

In [7] we proved that

Theorem 2.1

(i) The algebra of H-invariant polynomials $P(\underline{x}, \underline{u})$ is generated by $\underline{x}$, $\underline{u}$ and the pseudoscalar $e_{1\ldots m}$.

(ii) The algebra of L-invariant operators $P(\underline{x}, \partial_{\underline{x}})$ is generated by $f \to \underline{x}f$, $f \to \partial_{\underline{x}}f$ and $f \to e_{1\ldots m}f$.

In [8] we investigated H-invariant polynomials. Hereby it is more natural to consider operators in $End(\mathcal{P} \otimes R_m) = End(\mathcal{P}) \otimes R_{m,m}$. In other words, instead of R_m-valued polynomials $P(\underline{x})$ we consider $V_m' I$-valued polynomials $P(x)I$, where $x = \frac{1}{2}(\underline{x} - \underline{x}|)$ belongs to V^m. Moreover, let $S = s\bar{s}| \in \widetilde{SO}(m)$, $s \in Spin(m)$, then we have that $L(S)$ corresponds to $H(S)$ in the following sense:

$$L(S)P(x)I = sP(1/2(s\underline{x}\bar{s} - (s\underline{x}\bar{s})|)\bar{s}I \to H(s)P(\underline{x}).$$

Note that this representation $L(S)$ extends to $\widetilde{Gl}(m)$. Hence we can investigate both $\widetilde{SO}(m)$- and $\widetilde{Gl}(m)$-invariant operators.

The algebra of differential operators acting on polynomials of the form $P(x)I$ is generated by $R_{m,m}$ together with the operators

$$fI \to x\,fI, \quad x = \frac{1}{2}(\underline{x} - \underline{x}|) = \sum x_j f_j;$$

$$fI \to \partial_x fI, \quad \partial_x = \frac{1}{2}(\partial_{\underline{x}} + \partial_{\underline{x}}|) = \sum \partial_{x_j} f_j',$$

which are $\widetilde{Gl}(m)$-invariant (i.e. commute with $L(S)$, $S \in \widetilde{Gl}(m)$). For general differential operators $P(x, \partial_x)$, the symbol is the $R_{m,m}$-valued polynomial $P(x, u')$, $(x, u') \in V^m \times V^{m'}$ given by

$$P(x, \partial_x)e^{\langle x, u' \rangle} = P(x, u')e^{\langle x, u' \rangle}.$$

It is clear that an operator $P(x, \partial_x)$ will be $L(S)$-invariant if its symbol $P(x, u')$ is $H(S)$-invariant, i.e.

$$P(x, u') = SP(\bar{S}xS, \bar{S}u'S)\bar{S}.$$

This enables us to characterize $\widetilde{SO}(m)$- and $\widetilde{Gl}(m)$-invariant operators. In [8] we proved the following

Theorem 2.2

(i) The algebra of $\widetilde{Gl}(m)$-invariant symbols $P(x,u')$ is generated by x, u' and $p_j = \sum_{|A|=j} f'_A I f_A$ where for $A = \{a_1,\dots,a_j\}$ with $a_1 < \dots < a_j$, $f_A = f_{a_1}\dots f_{a_j}, f'_A = f'_{a_1}\dots f'_{a_j}$.

(ii) The algebra of $\widetilde{SO}(m)$-invariant symbols $P(x,u')$ is generated by $\underline{x}$, $\underline{x}|$, $\underline{u}$, $\underline{u}|$ where $x = \frac{1}{2}(\underline{x} - \underline{x}|)$, $u' = \frac{1}{2}(\underline{u} + \underline{u}|)$, the numbers $E_j = \sum e_A e_A|$, $|A| = j$ and the pseudoscalar.

It takes a little exercise to see that the algebra of $\widetilde{SO}(m)$-invariant polynomials $P(x,u')$ is in fact generated by the $\widetilde{Gl}(m)$-invariant ones together with the pseudoscalar. We also proved

Theorem 2.3

(i) The algebra of $\widetilde{Gl}(m)$-invariant operators $P(x,\partial_x)$ is generated by

$$fI \to x\, fI,\ fI \to \partial_x\, fI,\ fI \to p_j\, fI.$$

(ii) The algebra of $\widetilde{SO}(m)$-invariant operators $P(x,\partial_x)$ is generated by the above ones together with the pseudoscalar $fI \to e_{1\dots m} fI$.

On going back to R_m-valued notation, the multiplication operators $fI \to p_j\, fI$ correspond (up to a sign) to the projections $f \to [f]_j$ of f on j-vectors. So the $H(Spin(m))$-invariant differential operators are in fact generated by

$$f \to \underline{x}f,\quad f \to f\underline{x},\quad f \to \partial_{\underline{x}}f,\quad f \to f\partial_{\underline{x}},\quad f \to [f]_k \quad \text{and} \quad f \to e_{1\dots m}f.$$

Among these operators the elements in the algebra generated by

$$f \to \underline{x}f - \tilde{f}\underline{x},\quad f \to \partial_{\underline{x}}f + \tilde{f}\partial_{\underline{x}},\quad f \to [f]_k$$

extend to $Gl(m)$-invariant operators and generate the algebra of such operators. Moreover, the fundamental bivector B is, as an operator, given by $B = \partial_{\underline{x}}\underline{x}| + \underline{x}|\partial_{\underline{x}}$ and $f \to [f]_k$ are the eigenprojections. Hence the algebra of $\widetilde{SO}(m)$-invariant operators is generated by $\underline{x}$, $\underline{x}|$, $\partial_{\underline{x}}$, $\partial_{\underline{x}}|$ and $e_{1\dots m}$. In this paper we need extensions of these results in two directions. First we need operators acting on polynomials $P(x,x')I$, $x \in V^m$ and $x' \in V^{m'}$, i.e. $R_{m,m'}$-valued operators of the form $P(x,x';\partial_x,\partial_{x'})$. The symbol of such an operator is given by

$$P(x,x';\partial_x,\partial_{x'})\, exp(\langle x,u'\rangle + \langle x',u\rangle) = P(x,x';u',u)\, exp(\langle x,u'\rangle + \langle x',u\rangle)$$

and the results obtained in [8] can be easily adapted to this more general situation. Instead of invariant polynomials of the form $P(x;u')$ we now have to characterize invariant polynomials of the form $P(x,x';u',u)$, which is only slightly more general. As basic $\widetilde{Gl}(m)$-invariant operators we now have

$$fI \to x\, fI,\quad fI \to x'\, fI,\quad fI \to \partial_x fI,\quad fI \to \partial_{x'} fI,\quad fI \to p_j\, fI$$

while as basic $\widetilde{SO}(m)$-invariant operators we have these ones together with the pseudoscalar $fI \to e_{1\dots m}fI$. It is interesting to note that the operators $fI \to e_{1\dots m}e_1|\dots e_m|fI$ together with

$$fI \to (x + x')fI, \quad fI \to (\partial_x + \partial_{x'})fI$$

are even invariant under the whole "phase group" $Spin(m,m)$. Back in R_m-valued notation these operators correspond to the map $f \to \tilde{f}$ together with

$$f \to \frac{1}{2}\left((\underline{x} + \underline{x}')f + \tilde{f}(-\underline{x} + \underline{x}')\right)$$

$$f \to \frac{1}{2}\left((\partial_{\underline{x}} + \partial_{\underline{x}'})f + \tilde{f}(\partial_{\underline{x}} - \partial_{\underline{x}'})\right).$$

Secondly we also need operators acting on polynomials $P(x_1,\dots,x_l)I$ of several vector variables $x_1,\dots,x_l$, i.e. $R_{m,m}$-valued polynomials of the form

$$P(x_j, \partial_{x_j}) = P(x_1,\dots,x_l; \partial_{x_1},\dots,\partial_{x_l}).$$

The symbol of such an operator is defined like in the one variable case as a polynomial of the form $P(x_j, u_j')$ and again we have to characterize invariant polynomials of several variables x_j, u_j'. Still more general, we need operators acting on polynomials of the form $P(x_j, x_j')I$, i.e. operators of the form $P(x_j, x_j'; \partial_{x_j}, \partial_{x_j'})$. Therefore we have to characterize invariant poynomials of the form $P(x_j, x_j'; u_j', u_j)$. It is clear that all these extensions are really similar and to translate our results from the one variable case to the several variable case one simply has to replace

$$x \to x_j, \quad x' \to x_j', \quad \partial_x \to \partial_{x_j}, \quad \partial_{x'} \to \partial_{x_j'}.$$

3. Clifford tensors

Like in [5], Clifford tensors can be simply defined as R_m-valued multilinear polynomials of the form $F(\underline{u}_1,\dots,\underline{u}_k)$. These are also called covariant k-tensors. To introduce contravariant k-tensors one can consider multilinear polynomials $F(\underline{u}_1',\dots,\underline{u}_k')$ depending on "dual" vector variables $\underline{u}_1',\dots,\underline{u}_k'$. But having our $\widetilde{Gl}(m)$-invariant picture available, it now seems better to redefine covariant and contravariant tensors as follows.

Definition 3.1 *A covariant Clifford k-tensor is a multilinear polynomial of the form*

$$F(u_1,\dots,u_k)\,I$$

with values in $V_m'I$, where $u_1,\dots,u_k$ are vector variables in V^m. A contravariant Clifford l-tensor is a multilinear polynomial of the form

$$F(u_1',\dots,u_l')\,I$$

with $u_1',\dots,u_l'$ vector variables in $V^{m'}$. A Clifford (k,l)-tensor is a multilinear polynomial of the form $F(u_1,\dots,u_k; u_1',\dots,u_l')\,I$.

What we have given here is the "standard representation" of tensors as multilinear polynomials. But for tensor calculus one also needs other polynomial representations. Let $a_1 < \ldots < a_k$, $b_1 < \ldots < b_l$; then we consider the multilinear polynomial

$$F(u_{a_1}, \ldots, u_{a_k}; u'_{b_1}, \ldots, u'_{b_l})\, I$$

to be the "same tensor as" the polynomial

$$F(u_1, \ldots, u_k; u'_1, \ldots, u'_l)\, I.$$

Hence tensors are in fact equivalence classes of multilinear polynomials.

Any polynomial operator transforming multilinear polynomials into multilinear polynomials hence projects down to tensors and all operators on Clifford tensors can be represented in this way by polynomial operators. The basic operators on Clifford tensors are those coming from differential operators. Their algebra is generated by $R_{m,m}$ together with the basic operators

$$\sigma_j : F(u_1, \ldots, u_k; u'_j)I \to u_j F(u_1, \ldots, u_{j-1}, u_{j+1}, \ldots, u_{k+1}; u'_j)I$$

$$\sigma'_j : F(u_j; u'_1, \ldots, u'_l)I \to u'_j F(u_j; u'_1, \ldots, u'_{j-1}, u'_{j+1}, \ldots, u'_{l+1})I$$

$$\partial_{\sigma_j} : F(u_1, \ldots, u_k; u'_j)I \to \partial_u F(u_1, \ldots, u_{j-1}, u, u_j, \ldots, u_{k-1}; u'_j)I$$

$$\partial_{\sigma'_j} : F(u_j; u'_1, \ldots, u'_l)I \to \partial_{u'} F(u_j; u'_1, \ldots, u'_{j-1}, u', u'j, \ldots, u'_{l-1})I.$$

The operators $\sigma = \sigma_1$, $\sigma' = \sigma'_1$ are some kind of "tensor variables". For example $\sigma = \sum \sigma(j) f_j$ where $\sigma(j)$ are linear functions given by $\sigma(j) : u \to \langle f'_j, u\rangle$ and multiplying like tensors, i.e.

$$\sigma(j)\sigma(k) : \quad (u_1, u_2) \to \langle f'_j, u_1\rangle\langle f'_k, u_2\rangle.$$

Using this multiplication rule, every (k, l)-tensor can be written in a unique way as

$$F(\sigma, \ldots, \sigma; \sigma', \ldots, \sigma')\, I.$$

The operators ∂_{σ_j} and $\partial_{\sigma'_j}$ can be used to define contraction of tensors with vectors. For example, if we write $\partial_\sigma = \partial_{\sigma_1}$, $\partial_{\sigma'} = \partial_{\sigma'_1}$, then for $v \in V^m$, the contraction $v \cdot FI$ is given by $\langle v, \partial_\sigma\rangle FI$. As to the invariant operators on tensors we have

Theorem 3.1 *The algebra of $\widetilde{Gl}(m)$-invariant operators on Clifford tensors is generated by*

$$\{\sigma_j, \sigma'_j, \partial_{\sigma_j}, \partial_{\sigma'_j} : j \in N\}.$$

The algebra of $\widetilde{SO}(m)$-invariant operators on Clifford tensors is generated by this together with the pseudoscalar $e_{1\ldots m}$.

Hereby the word "generators" refers to operators coming from differential operators on polynomials. The theorem follows from results established in the previous section together with

Lemma 3.1 *The Clifford numbers p_j are expressable as polynomials in $L = \partial_\sigma \sigma = \sum f'_j f_j$.*

Proof.
It is clear that $L = \partial_\sigma \sigma$. Moreover, as an operator on R_m,

$$LI = 0I,\ Lf_j'I = -f_j'I, \quad Lf_j'f_k'I = -2\, f_j'f_k'I$$

etc. so that the operators $[\cdot]_k$ are the eigenprojections of L. □

One can wonder whether all $\widetilde{Gl}(m)$-invariant operators on tensors are generated by $\{\sigma, \sigma', \partial_\sigma, \partial_{\sigma'}\}$, but this is not the case. To obtain a finite set of generators for the $\widetilde{Gl}(m)$- and $\widetilde{SO}(m)$-invariant algebras we introduce the notion of "bitensor" by using multilinear functions of the form

$$F(x_1, \ldots, x_l; x_1', \ldots, x_l'; u_1, \ldots, u_k; u_1', \ldots, u_k')I.$$

Hence we simply double the variables, an operation which may seem trivial, but it is enough to generate not only all $\widetilde{Gl}(m)$-invariant operators on tensors but also on tensor fields.

Similar to the basic operators σ_j, σ_j', ∂_{σ_j}, $\partial_{\sigma_j'}$ we can now introduce operators τ_j, τ_j', ∂_{τ_j}, $\partial_{\tau_j'}$, replacing the u_j, u_j' variables by the x_j, x_j' variables. In particular we put

$$\tau = \tau_1,\ \tau' = \tau_1',\ \partial_\tau = \partial_{\tau_1},\ \partial_{\tau'} = \partial_{\tau_1'}$$

and we can define the "gradient" and "anti-gradient" operators by

$$\begin{aligned}\langle \sigma, \partial_\tau \rangle : \quad F(x_1, \ldots, x_l; u_1, \ldots, u_k) &\to \langle u_1, \partial_x \rangle\, F(x, x_1, \ldots, x_{l-1}; u_1, \ldots, u_k) \\ &= F(u_1, x_1, \ldots, x_{l-1}; u_2, \ldots, u_{k+1})\end{aligned}$$

and

$$\langle \tau, \partial_\sigma \rangle : \quad F(x_1, \ldots, x_l; u_1, \ldots, u_k) \to F(x_2, \ldots, x_{l+1}; x_1, u_1, \ldots, u_{k-1}).$$

These operators can be used to store u-variables into x-variables and vice versa. Moreover we have

Theorem 3.2 *The algebra of $\widetilde{Gl}(m)$-invariant operators on bitensors is generated by*

$$\{\sigma, \sigma', \tau, \tau', \partial_\sigma, \partial_{\sigma'}, \partial_\tau, \partial_{\tau'}\}.$$

The algebra of $\widetilde{SO}m)$-invariant operators on bitensors is generated by this together with the pseudoscalar.

Proof.
It is easy to see that

$$\begin{aligned}\sigma_j &= \langle \sigma, \partial_\tau \rangle^{j-1} \sigma \langle \tau, \partial_\sigma \rangle^{j-1}, \\ \partial_{\sigma_j} &= \langle \sigma, \partial_\tau \rangle^{j-1} \partial_\sigma \langle \tau, \partial_\sigma \rangle^{j-1},\end{aligned}$$

while similar expressions hold for

$$\sigma_j', \partial_{\sigma_j'}, \tau_j, \tau_j', \partial_{\tau_j}, \partial_{\tau_j'}.$$

□

As to the operators on covariant tensors, it suffices to consider the operators σ, ∂_σ, τ, ∂_τ and the pseudoscalar $e_{1\ldots m}$ only. But to calculate zonal tensors (see [9]), one also needs the dual operators.

REFERENCES

[1] C. Chevalley. "The algebraic theory of spinors", Columbia Univ. Press, New York (1954).

[2] C. Doran, D. Hestenes, F. Sommen, N. Van Acker. "Lie groups as Spin groups", to appear in Foundations of Physics.

[3] R. Delanghe, F. Sommen and V. Souček. "Clifford Algebra and Spinor-valued functions: a function theory for the Dirac-operator", Mathematics and Its Applications 53, Kluwer Academic Publishers, Dordrecht (1992).

[4] J. Gilbert, M. Murray. "Clifford Algebras and Dirac operators in harmonic analysis", Cambridge University Press, Cambridge(1990).

[5] D. Hestenes, G. Sobczyk. "Clifford algebra to geometric calculus", D. Reidel, Dordrecht (1985).

[6] G. Sobczyk. "Unipotents, idempotents, and a spinor basis for matrices", Advances in Applied Clifford Algebras, 2(nr. 1) (1992) pp. 53-62.

[7] F. Sommen, N. Van Acker. "Monogenic differential operators", Results in Mathematics, 22 (1992) pp. 781-798.

[8] F. Sommen, N. Van Acker. "$SO(m)$-invariant operators acting on Clifford algebra-valued functions", submitted for the special issue dedicated to Hest. birthday .

[9] F. Sommen. "Clifford tensor calculus", in preparation.

[10] H. Weyl. "The theory of groups and quantum mechanics", Dover publications, Inc. (1950).

INVARIANT DIFFERENTIAL OPERATORS ON POLYNOMIAL- VALUED FUNCTIONS

F. SOMMEN* and N. VAN ACKER
University of Gent
Department of Mathematical Analysis
Galglaan 2
B-9000 Gent/Belgium

Abstract. Instead of spinor- or Clifford algebra-valued functions we are dealing with functions taking values in some space of spinor- or Clifford algebra-valued polynomials. We characterise the $Gl(m)$–invariant differential operators acting on these functions. The so-called monogenic decomposition of these operators leads to a generalised Fischer decomposition into so-called monogenic pieces, hyper-monogenic pieces or hypo-monogenic pieces.

Key words: Clifford analysis, differential operators, polynomial-valued functions

1. Introduction

Let $\Omega \subseteq R^m$ be open and let E be a vector space, then $\mathcal{E}(\Omega, E)$ denotes the space of E-valued smooth functions in Ω. In their paper [6], the authors consider first order differential operators with constant coefficients: $P : \mathcal{E}(\Omega, E) \to \mathcal{E}(\Omega, F)$ with $P = \sum_{j=1} \partial_{x_j} L_j, L_j \in L(E, F)$. For example they take the gradient-operator $d = \sum_{j=1}^m \partial_{x_j} dx_j$ where $\{dx_j, j = 1, \ldots, m\}$ is a basis for the space Λ^1 of 1-forms on R^m. It is clear that this operator is $Gl(m)$–invariant. Let $\mathcal{P}_k$ denote the space of scalar-valued homogeneous polynomials of degree k. When we consider operators of the form $P = \sum_{j=1}^m \partial_{x_j} L_j : \mathcal{E}(\Omega, \mathcal{P}_k) \to \mathcal{E}(\Omega, \mathcal{P}_{k+1})$, then it is clear that the only $Gl(m)$–invariant operator is the operator $\langle \underline{u}, \partial_{\underline{x}} \rangle = \sum_{j=1}^m u_j \partial_{x_j}$.

It is well-known that $\mathcal{P}_k$ has the following Fischer decomposition. Let $R_k(\underline{u})$ be a homogeneous polynomial of degree k; then $R_k(\underline{u})$ may be decomposed into spherical harmonics as

$$R_k(\underline{u}) = \sum_{2s \leq k} |\underline{u}|^{2s} H_{k-2s}(\underline{u}), \quad H_l \in \mathcal{H}_l$$

where $\mathcal{H}_l$ is the space of spherical harmonics of degree l. In his paper [7] Strasburger considers $\langle \underline{u}, \partial_{\underline{x}} \rangle : \mathcal{E}(\Omega, \mathcal{H}_k) \to \mathcal{E}(\Omega, \mathcal{H}_{k+1} + |\underline{u}|^2 \mathcal{H}_{k-1})$, leading to the two "harmonic" gradient operators d_+ and d_- .When considering spinor- or Clifford-algebra-valued functions (see [1]), this decomposition can be refined. Every $H_k \in \mathcal{H}_k$ admits a canonical splitting of the form $H_k(\underline{u}) = P_k(\underline{u}) + \underline{u} P_{k-1}(\underline{u})$, where P_k, P_{k-1} are (inner) spherical monogenics, i.e. homogeneous polynomial nullsolutions of the

* Senior Research Associate, NFWO, Belgium

F. Brackx et al. (eds.), *Clifford Algebras and their Applications in Mathematical Physics*, 203–212.

Dirac operator. We denote the module of these spherical monogenics of degree k by $\mathcal{M}_k$. Hence the space $\mathcal{P}_k$ splits up into irreducible polynomials as follows:

$$R_k(\underline{u}) = \sum_{s=0}^{k} \underline{u}^s P_{k-s}(\underline{u}), \quad P_{k-s} \in \mathcal{M}_{k-s}.$$

It is generally known that this decomposition is an orthogonal one w.r.t. the inner product for Clifford algebra-valued polynomials: $(P(\underline{u}), Q(\underline{u})) = \overline{P(\partial_{\underline{u}})}Q(\underline{u})|_{\underline{u}=0}$. Hence it is clear that, when dealing with polynomial-valued functions $f(\underline{x}, \underline{u})$, this inner product induces the following inner product on the space of Clifford algebra-valued polynomials of two vector variables:

$$(P(\underline{x}, \underline{u}), Q(\underline{x}, \underline{u})) = \overline{P(\partial_{\underline{x}}, \partial_{\underline{u}})}Q(\underline{x}, \underline{u})|_{(\underline{x},\underline{u})=(0,0)}.$$

In [3] functions with values in spaces of spherical monogenics of matrix variables are studied, and in the recent book [2] of Delanghe, Sommen, Souček, the Dirac-operator acting on spinor- or Clifford algebra-valued functions and Clifford differential forms are considered.

In this paper, instead of spinor- or Clifford algebra-valued functions, we consider functions with values in some space of spinor- or Clifford algebra-valued polynomials. Moreover we consider operators somewhat more general than the operators considered in the paper by Stein and Weiss, of the following form: $P : \mathcal{E}(\Omega, \mathcal{P}_k) \to \mathcal{E}(\Omega, \mathcal{P}_{k'})$. Hence, our theory is more refined than the theory using spherical harmonics as in [7] and the book [3] of Gilbert and Murray, and it is more general than the Stein-Weiss theory for polynomials. Indeed, by splitting up the above operator into homogeneous operators $P : \mathcal{P}_l(\mathcal{P}_k) \to \mathcal{P}_{l'}(\mathcal{P}_{k'})$ it suffices to consider

$$P(D) : \mathcal{P}_l(\underline{x}) \otimes \mathcal{P}_k(\underline{u}) \to \mathcal{P}_{l'}(\underline{x}) \otimes \mathcal{P}_{k'}(\underline{u}).$$

Such an operator can always be written as a linear combination of the basic operators $x_\alpha \partial_{x_\beta} u_\gamma \partial_{u_\delta}$ with $l + |\alpha| = |\beta| + l'$, $k + |\gamma| = |\delta| + k'$. We want to characterize the $Gl(m)$–invariant operators. The symbol of the operator $P(D) = x_\alpha \partial_{x_\beta} u_\gamma \partial_{u_\delta}$ is given by $P(\underline{x}, \underline{y}, \underline{u}, \underline{v}) = x_\alpha y_\beta u_\gamma v_\delta$. It is generally known that the algebra of $Gl(m)$–invariant polynomials in four vector variables is at most generated by (powers of) the several possible inner products between the vectors. So the only $Gl(m)$-invariant polynomials $P(\underline{x}, \underline{y}, \underline{u}, \underline{v})$ are generated by $\langle \underline{x}, \underline{y} \rangle$, $\langle \underline{x}, \underline{u} \rangle$, $\langle \underline{x}, \underline{v} \rangle$, $\langle \underline{y}, \underline{u} \rangle$, $\langle \underline{y}, \underline{v} \rangle$, $\langle \underline{u}, \underline{v} \rangle$. After re-identifying symbols and operators it is clear that the only $Gl(m)$-invariant differential operators $P(D) : \mathcal{P}_l \otimes \mathcal{P}_k \to \mathcal{P}_{l'} \otimes \mathcal{P}_{k'}$ are generated by

$$\langle \underline{x}, \partial_{\underline{x}} \rangle, \langle \underline{x}, \underline{u} \rangle, \langle \underline{x}, \partial_{\underline{u}} \rangle, \langle \underline{u}, \partial_{\underline{x}} \rangle, \langle \partial_{\underline{x}}, \partial_{\underline{u}} \rangle, \langle \underline{u}, \partial_{\underline{u}} \rangle.$$

But the Euler operators $\langle \underline{x}, \partial_{\underline{x}} \rangle$ and $\langle \underline{u}, \partial_{\underline{u}} \rangle$ are constant on $\mathcal{P}_l$ and $\mathcal{P}_k$, which leaves us with the remaining $Gl(m)$-invariant operators:

$$\langle \underline{x}, \underline{u} \rangle, \langle \underline{x}, \partial_{\underline{u}} \rangle, \langle \underline{u}, \partial_{\underline{x}} \rangle, \langle \partial_{\underline{x}}, \partial_{\underline{u}} \rangle.$$

Most of the work in the first section consists of establishing the so-called monogenic decomposition for these operators (see [5]) when acting on $f \in \mathcal{E}(\Omega, \mathcal{P}_k)$. In view of the classical commutation relations, these decompositions follow as soon as we know the decompositions of the operators $\underline{x}$ and $\partial_{\underline{x}}$. It turns out that we end up with six basic operators $D_{+,k}(\partial_{\underline{x}}), D_k(\partial_{\underline{x}}), D_{-,k}(\partial_{\underline{x}})$ and $D_{+,k}(\underline{x}), D_k(\underline{x}), D_{-,k}(\underline{x})$.

In the second section we introduce the concepts of monogenic, hyper-monogenic and hypo-monogenic functions $f \in \mathcal{E}(\Omega, \mathcal{M}_k)$. The monogenicity concept is of course an extension of the standard definition of monogenicity for functions $f(\underline{x})$ (see [1]). Next take a polynomial $R_l(\underline{x}, \underline{u}) \in \mathcal{P}_l(\mathcal{M}_k)$; the monogenic decompositions constructed in the first section then lead to orthogonal Fischer decompositions into "monogenic pieces", or "hyper-monogenic pieces" or "hypo-monogenic pieces". So far we considered the $\underline{u}$-variable merely as a tool to represent polynomial-valued functions. The case where we consider functions of two vector variables, leading to a Fischer decomposition into so-called simultaneously spherical monogenics, is presented in the forthcoming paper [4].

2. Monogenic decompositions

In our paper [5] we introduced the concept of monogenic operator using the monogenic decomposition of a homogeneous polynomial $R_k(\underline{x}) \in \mathcal{P}_k$ (as a refinement of its harmonic decomposition). This decomposition led to the corresponding monogenic decomposition of a homogeneous polynomial operator; each monogenic operator transforms $\underline{x}^s P_k(\underline{x})$ into $\underline{x}^{s'} P_{k'}(\underline{x})$, $P_{k'} \in \mathcal{M}_{k'}$ with fixed s' and k'. Now let $f \in \mathcal{E}(\Omega, \mathcal{P}_k)$, then we have that

$$f(\underline{x}, \underline{u}) = \sum_{j=0}^{k} \underline{u}^j f_{k-j}(\underline{x}, \underline{u})$$

where $f_{k-j}(\underline{x}, \cdot) \in \mathcal{M}_{k-j}$. We want to determine the monogenic decomposition of the operators under consideration, i.e. a decomposition into operators transforming $\underline{u}^j f_k(\underline{x}, \underline{u})$, $f_k(\underline{x}, \cdot) \in \mathcal{M}_k$ into $\underline{u}^{j'} f_{k'}(\underline{x}, \underline{u})$, $f_{k'}(\underline{x}, \cdot) \in \mathcal{M}_{k'}$. It is clear that for the gradient operator and for the operator $\langle \underline{u}, \underline{x} \rangle$ we have that

$$\langle \underline{u}, \partial_{\underline{x}} \rangle f(\underline{x}, \underline{u}) = \sum_{j=0}^{k} \underline{u}^j \langle \underline{u}, \partial_{\underline{x}} \rangle f_{k-j}(\underline{x}, \underline{u}), \quad \langle \underline{u}, \underline{x} \rangle f(\underline{x}, \underline{u}) = \sum_{j=0}^{k} \underline{u}^j \langle \underline{u}, \underline{x} \rangle f_{k-j}(\underline{x}, \underline{u}).$$

Hence it is sufficient to evaluate these operators for $\mathcal{M}_k$-valued functions. Of course, the other two operators $\langle \underline{x}, \partial_{\underline{u}} \rangle$ and $\langle \partial_{\underline{x}}, \partial_{\underline{u}} \rangle$ do not commute with $\underline{u}^j$. Let's start with the gradient operator $\langle \underline{u}, \partial_{\underline{x}} \rangle : \mathcal{E}(\Omega, \mathcal{P}_k) \to \mathcal{E}(\Omega, \mathcal{P}_{k+1})$. For $k = 0$ we have that $\mathcal{P}_0 = \mathcal{H}_0, \mathcal{P}_1 = \mathcal{H}_1$ so that $\langle \underline{u}, \partial_{\underline{x}} \rangle : \mathcal{E}(\Omega, \mathcal{H}_0) \to \mathcal{E}(\Omega, \mathcal{H}_1)$. However $\mathcal{H}_1$ has an orthogonal decomposition $\mathcal{H}_1 = \mathcal{M}_1 + \underline{u}\mathcal{M}_0$ w.r.t. the inner product on the unit sphere $(f, g) = \int_{S^{m-1}} \bar{f} g dS$. Let $f(\underline{x}) \in \mathcal{E}(\Omega) = \mathcal{E}(\Omega, \mathcal{H}_0)$, then $\langle \underline{u}, \partial_{\underline{x}} \rangle f(\underline{x})$ decomposes as

$$\left(P_+(\underline{u}, \partial_{\underline{x}}) + P_-(\underline{u}, \partial_{\underline{x}})\right) f(\underline{x})$$

where $P_+(\underline{u}, \partial_{\underline{x}}) = \langle \underline{u}, \partial_{\underline{x}} \rangle + \frac{1}{m} \underline{u} \partial_{\underline{x}}$ is $\mathcal{M}_1$-valued and $P_-(\underline{u}, \partial_{\underline{x}}) = -\frac{1}{m} \underline{u} \partial_{\underline{x}}$ is $(\underline{u}\mathcal{M}_0)$-valued. Thus the Dirac operator $\partial_{\underline{x}}$ arizes in this decomposition and we can say that the spinor- or Clifford algebra-valued function theory is a special case. The operator $P_+(\underline{u}, \partial_{\underline{x}})$ is called a monogenic gradient.

In general, let $f \in \mathcal{E}(\Omega, \mathcal{P}_k)$. To arrive at the monogenic decompositions of the considered operators we first prove the following

Lemma 2.1 *Let $p(\underline{x}, \underline{u}) \in \mathcal{P}_l(\mathcal{M}_k)$, i.e. $p(\underline{x}, \underline{u})$ is an inner spherical monogenic of degree k in $\underline{u}$ and a homogeneous polynomial of degree l in $\underline{x}$. Then*

(i) $\underline{x}\, p(\underline{x}, \underline{u})$ admits the Fischer decomposition

$$\underline{x}\, p(\underline{x}, \underline{u}) = a_k(\underline{x}, \underline{u}) + \underline{u} b_{k-1}(\underline{x}, \underline{u})$$

where

$$a_k(\underline{x}, \underline{u}) = \left(\frac{\underline{u}\partial_{\underline{u}}}{2k+m-2} + 1\right) \underline{x}\, p(\underline{x}, \underline{u}) \qquad \in \mathcal{P}_{l+1}(\mathcal{M}_k),$$

$$b_{k-1}(\underline{x}, \underline{u}) = -\frac{1}{2k+m-2} \partial_{\underline{u}} \underline{x}\, p(\underline{x}, \underline{u}) \qquad \in \mathcal{P}_{l+1}(\mathcal{M}_{k-1})$$

(ii) $\partial_{\underline{x}} p(\underline{x}, \underline{u})$ admits the Fischer decomposition

$$\partial_{\underline{x}} p(\underline{x}, \underline{u}) = c_k(\underline{x}, \underline{u}) + \underline{u} d_{k-1}(\underline{x}, \underline{u})$$

where

$$c_k(\underline{x}, \underline{u}) = \left(\frac{\underline{u}\partial_{\underline{u}}}{2k+m-2} + 1\right) \partial_{\underline{x}} p(\underline{x}, \underline{u}) \qquad \in \mathcal{P}_{l-1}(\mathcal{M}_k),$$

$$d_{k-1}(\underline{x}, \underline{u}) = -\frac{1}{2k+m-2} \partial_{\underline{u}} \partial_{\underline{x}} p(\underline{x}, \underline{u}) \qquad \in \mathcal{P}_{l-1}(\mathcal{M}_{k-1}).$$

Proof.
(i) As clearly $\underline{x}\, p(\underline{x}, \cdot) \in \mathcal{H}_k$, it immediately follows that its Fischer decomposition consists of two pieces only, and they are obtained as follows:

$$\partial_{\underline{u}}(\underline{x}\, p(\underline{x}, \underline{u})) = \partial_{\underline{u}}(\underline{u} b_{k-1}(\underline{x}, \underline{u})) = -(2k+m-2) b_{k-1}(\underline{x}, \underline{u}).$$

Hence we have that $a_k(\underline{x}, \underline{u}) = \underline{x}\, p(\underline{x}, \underline{u}) + \underline{u} \frac{1}{2k+m-2} \partial_{\underline{u}} \underline{x}\, p(\underline{x}, \underline{u})$ which leads to the desired result.
(ii) The decomposition of the Dirac operator is proved in a similar way, since it is clear that $\partial_{\underline{x}} p(\underline{x}, \cdot) \in \mathcal{H}_k$. □

Notice that this lemma still holds for $p(\underline{x}, \cdot) \in \mathcal{H}_k$.

Lemma 2.2 *Let $p(\underline{x}, \underline{u}) \in \mathcal{P}_l(\mathcal{M}_k)$. Then*

(i) $\underline{x}\, \underline{u}^j p(\underline{x}, \underline{u})$ admits the Fischer decomposition

$$\underline{x}\, \underline{u}^j p(\underline{x}, \underline{u}) = \begin{cases} \underline{u}^j p_k(\underline{x}, \underline{u}) + \underline{u}^{j+1} p_{k-1}(\underline{x}, \underline{u}), & j \text{ even} \\ \underline{u}^{j-1} p_{k+1}(\underline{x}, \underline{u}) + \underline{u}^j p_k^*(\underline{x}, \underline{u}), & j \text{ odd} \end{cases}$$

(ii) $\partial_{\underline{x}} \underline{u}^j p(\underline{x}, \underline{u})$ admits the Fischer decomposition

$$\partial_{\underline{x}} \underline{u}^j p(\underline{x}, \underline{u}) = \begin{cases} \underline{u}^j \tilde{p}_k(\underline{x}, \underline{u}) + \underline{u}^{j+1} \tilde{p}_{k-1}(\underline{x}, \underline{u}), & j \text{ even} \\ \underline{u}^{j-1} \tilde{p}_{k+1}(\underline{x}, \underline{u}) + \underline{u}^j \tilde{p}_k^*(\underline{x}, \underline{u}), & j \text{ odd.} \end{cases}$$

Proof.
(i) Clearly, if j is even, we have that $\underline{x}\,\underline{u}^j p(\underline{x},\underline{u}) = \underline{u}^j \underline{x}\, p(\underline{x},\underline{u})$, so from the previous lemma:

$$p_k(\underline{x},\underline{u}) = a_k(\underline{x},\underline{u}) \quad \in \mathcal{P}_{l+1}(\mathcal{M}_k), \quad p_{k-1}(\underline{x},\underline{u}) = b_{k-1}(\underline{x},\underline{u}) \quad \in \mathcal{P}_{l+1}(\mathcal{M}_{k-1}).$$

If j is odd, say $j = 2l+1$, we have that $\underline{x}\,\underline{u}^j p(\underline{x},\underline{u}) = \underline{u}^{2l}\underline{x}\,\underline{u}\, p(\underline{x},\underline{u}) = \underline{u}^{j-1}\underline{x}\,\underline{u}\, p(\underline{x},\underline{u})$ where $\underline{u}\, p(\underline{x},\underline{u}) \in \mathcal{P}_l(\mathcal{H}_{k+1})$. So the previous lemma guarantees that

$$\underline{x}\,\underline{u}^j p(\underline{x},\underline{u}) = \underline{u}^{j-1}\left[p_{k+1}(\underline{x},\underline{u}) + \underline{u}\, p_k^*(\underline{x},\underline{u})\right]$$

with

$$p_{k+1}(\underline{x},\underline{u}) = \left(\frac{\underline{u}\partial_{\underline{u}}}{2k+m} + 1\right)\underline{x}\,\underline{u}\, p(\underline{x},\underline{u}) \quad \in \mathcal{P}_{l+1}(\mathcal{M}_{k+1})$$

and

$$p_k^*(\underline{x},\underline{u}) = -\frac{1}{2k+m}\partial_{\underline{u}}\underline{x}\,\underline{u}\, p(\underline{x},\underline{u}) \quad \in \mathcal{P}_{l+1}(\mathcal{M}_k).$$

(ii) This is proved similar to (i). We end up with:

$$\tilde{p}_k(\underline{x},\underline{u}) = c_k(\underline{x},\underline{u}) = \left(\frac{\underline{u}\partial_{\underline{u}}}{2k+m-2} + 1\right)\partial_{\underline{x}} p(\underline{x},\underline{u}) \quad \in \mathcal{P}_{l-1}(\mathcal{M}_k),$$

$$\tilde{p}_{k-1}(\underline{x},\underline{u}) = d_{k-1}(\underline{x},\underline{u}) = -\frac{1}{2k+m-2}\partial_{\underline{u}}\partial_{\underline{x}} p(\underline{x},\underline{u}) \quad \in \mathcal{P}_{l-1}(\mathcal{M}_{k-1}),$$

$$\tilde{p}_{k+1}(\underline{x},\underline{u}) = \left(\frac{\underline{u}\partial_{\underline{u}}}{2k+m} + 1\right)\partial_{\underline{x}}\underline{u} p(\underline{x},\underline{u}) \quad \in \mathcal{P}_{l-1}(\mathcal{M}_{k+1}),$$

$$\tilde{p}_k^*(\underline{x},\underline{u}) = -\frac{1}{2k+m}\partial_{\underline{u}}\partial_{\underline{x}}\underline{u} p(\underline{x},\underline{u}) \quad \in \mathcal{P}_{l-1}(\mathcal{M}_k).$$

□

We then arrive at the monogenic decomposition of the gradient operator.

Theorem 2.1 *Let $f \in \mathcal{E}(\Omega, \mathcal{M}_k)$; then the gradient $\langle \underline{u}, \partial_{\underline{x}}\rangle f(\underline{x},\underline{u})$ decomposes as*

$$\left(D_{+,k}(\partial_{\underline{x}}) + \underline{u} D_k(\partial_{\underline{x}}) + \underline{u}^2 D_{-,k}(\partial_{\underline{x}})\right) f(\underline{x},\underline{u})$$

where $D_{+,k}(\partial_{\underline{x}}) : \mathcal{E}(\Omega, \mathcal{M}_k) \to \mathcal{E}(\Omega, \mathcal{M}_{k+1})$ is the so-called (spherical) monogenic gradient given by

$$D_{+,k}(\partial_{\underline{x}}) = \langle \underline{u}, \partial_{\underline{x}}\rangle + \frac{\underline{u}\partial_{\underline{x}}}{2k+m} - \frac{\underline{u}^2\partial_{\underline{u}}\partial_{\underline{x}}}{2(2k+m)},$$

$D_k(\partial_{\underline{x}}) : \mathcal{E}(\Omega, \mathcal{M}_k) \to \mathcal{E}(\Omega, \mathcal{M}_k)$ is the (spherical) monogenic Dirac-operator given by

$$D_k(\partial_{\underline{x}}) = \frac{-1}{2k+m}\left(\frac{\underline{u}\partial_{\underline{u}}}{2k+m-2} + 1\right)\partial_{\underline{x}}$$

and $D_{-,k}(\partial_{\underline{x}}) : \mathcal{E}(\Omega, \mathcal{M}_k) \to \mathcal{E}(\Omega, \mathcal{M}_{k-1})$ is the so-called monogenic divergence given by

$$D_{-,k}(\partial_{\underline{x}}) = \frac{1}{2(2k+m-2)}\partial_{\underline{u}}\partial_{\underline{x}}.$$

Proof.
As $\langle \underline{u}, \partial_{\underline{x}} \rangle = -\frac{1}{2}(\underline{u}\partial_{\underline{x}} + \partial_{\underline{x}}\underline{u})$ we immediately arrive at the decomposition

$$\begin{aligned} \langle \underline{u}, \partial_{\underline{x}} \rangle f(\underline{x}, \underline{u}) &= -\frac{1}{2}\underline{u}\partial_{\underline{x}} f(\underline{x}, \underline{u}) - \frac{1}{2}\partial_{\underline{x}}\underline{u} f(\underline{x}, \underline{u}) \\ &= -\frac{1}{2}\left(\frac{\underline{u}\partial_{\underline{u}}}{2k+m} + 1\right)\partial_{\underline{x}}\underline{u} f(\underline{x}, \underline{u}) - \frac{\underline{u}^2}{2}\left(-\frac{1}{2k+m-2}\partial_{\underline{u}}\partial_{\underline{x}} f(\underline{x}, \underline{u})\right) \\ &\quad -\frac{\underline{u}}{2}\left[\left(\frac{\underline{u}\partial_{\underline{u}}}{2k+m-2} + 1\right)\partial_{\underline{x}} - \frac{1}{2k+m}\partial_{\underline{u}}\partial_{\underline{x}}\underline{u}\right] f(\underline{x}, \underline{u}); \end{aligned}$$

it is only a technical exercise to verify that this indeed proves the desired result. □

Note that $(D_k(\partial_{\underline{x}}) + \underline{u}D_{-,k}(\partial_{\underline{x}}))$ is the Dirac operator as in the paper of Stein-Weiss. In a similar way we can write down the monogenic decomposition for the operator $\langle \underline{u}, \underline{x} \rangle$.

Theorem 2.2 *Let $f \in \mathcal{E}(\Omega, \mathcal{M}_k)$; then $\langle \underline{u}, \underline{x} \rangle f(\underline{x}, \underline{u})$ decomposes as follows*

$$\left(D_{+,k}(\underline{x}) + \underline{u}D_k(\underline{x}) + \underline{u}^2 D_{-,k}(\underline{x})\right) f(\underline{x}, \underline{u})$$

where $D_{+,k}(\underline{x}) : \mathcal{E}(\Omega, \mathcal{M}_k) \to \mathcal{E}(\Omega, \mathcal{M}_{k+1})$ is given by

$$D_{+,k}(\underline{x}) = \langle \underline{u}, \underline{x} \rangle + \frac{\underline{u}\,\underline{x}}{2k+m} - \frac{\underline{u}^2 \partial_{\underline{u}}\underline{x}}{2(2k+m)},$$

$D_k(\underline{x}) : \mathcal{E}(\Omega, \mathcal{M}_k) \to \mathcal{E}(\Omega, \mathcal{M}_k)$ is given by

$$D_k(\underline{x}) = \frac{-1}{2k+m}\left(\frac{\underline{u}\partial_{\underline{u}}}{2k+m-2} + 1\right)\underline{x}$$

and $D_{-,k}(\underline{x}) : \mathcal{E}(\Omega, \mathcal{M}_k) \to \mathcal{E}(\Omega, \mathcal{M}_{k-1})$ is given by

$$D_{-,k}(\underline{x}) = \frac{1}{2(2k+m-2)}\partial_{\underline{u}}\underline{x}.$$

Remark: Note that it is also possible to determine the monogenic decomposition of $\langle \underline{u}, \partial_{\underline{x}} \rangle = \sum_{j=1}^m u_j \partial_{x_j}$ and $\langle \underline{u}, \underline{x} \rangle = \sum_{j=1}^m u_j x_j$ using the Fischer decomposition of $u_j p(\underline{u})$, $p(\underline{u}) \in \mathcal{M}_k$:

$$u_j p(\underline{u}) = a(\underline{u}) + \underline{u} b(\underline{u}) + \underline{u}^2 c(\underline{u})$$

where $a(\underline{u}) \in \mathcal{M}_{k+1}$, $b(\underline{u}) \in \mathcal{M}_k$, $c(\underline{u}) \in \mathcal{M}_{k-1}$ can be constructed explicitly.

We still have to decompose the operators $\langle \partial_{\underline{x}}, \partial_{\underline{u}} \rangle$ and $\langle \underline{x}, \partial_{\underline{u}} \rangle$. A combination of previous results then yields the following

Theorem 2.3 *Let $f \in \mathcal{E}(\Omega, \mathcal{M}_k)$; then $\langle \underline{x}, \partial_{\underline{u}} \rangle \underline{u}^j f(\underline{x}, \underline{u})$ decomposes as follows*

$$\langle \underline{x}, \partial_{\underline{u}} \rangle \underline{u}^j f(\underline{x}, \underline{u}) = \begin{cases} j\underline{u}^{j-2} P_{k+1}(\underline{x}, \underline{u}) + j\underline{u}^{j-1} P_k(\underline{x}, \underline{u}) + \\ \qquad (j+m+2k-2)\underline{u}^j P_{k-1}(\underline{x}, \underline{u}), & j \text{ even} \\ (j-1)\underline{u}^{j-2} P_{k+1}(\underline{x}, \underline{u}) + (j+m+2k-1)\underline{u}^{j-1} P_k(\underline{x}, \underline{u}) + \\ \qquad (j+m+2k-1)\underline{u}^j P_{k-1}(\underline{x}, \underline{u}), & j \text{ odd} \end{cases}$$

where

$$P_{k+1}(\underline{x},\underline{u}) = \frac{1}{2}\left(\frac{\underline{u}\partial_{\underline{u}}}{2k+m}+1\right)\underline{x}\,\underline{u}f(\underline{x},\underline{u}) = -D_{+,k}(\underline{x})f(\underline{x},\underline{u}),$$

$$P_k(\underline{x},\underline{u}) = \frac{1}{2}\left(\frac{\underline{u}\partial_{\underline{u}}}{2k+m-2}+1\right)\underline{x}f(\underline{x},\underline{u}) - \frac{1}{2(2k+m)}\partial_{\underline{u}}\underline{x}\,\underline{u}f(\underline{x},\underline{u}) = -D_k(\underline{x})f(\underline{x},\underline{u})$$

$$P_{k-1}(\underline{x},\underline{u}) = -\frac{1}{2(2k+m-2)}\partial_{\underline{u}}\underline{x}f(\underline{x},\underline{u}) = -D_{-,k}(\underline{x})f(\underline{x},\underline{u}).$$

Proof.
It is clear that $\langle \underline{x},\partial_{\underline{u}}\rangle \underline{u}^j f(\underline{x},\underline{u}) = -\frac{1}{2}\partial_{\underline{u}}\underline{x}\,\underline{u}^j f(\underline{x},\underline{u}) - \frac{1}{2}\underline{x}\partial_{\underline{u}}\underline{u}^j f(\underline{x},\underline{u})$. If j is even , the right hand side of the previous equality decomposes as follows:

$$-\frac{1}{2}\partial_{\underline{u}}[\underline{u}^j p_k + \underline{u}^{j+1}p_{k-1}] + \frac{1}{2}j\underline{x}\,\underline{u}^{j-1}f$$

$$= \frac{1}{2}j\underline{u}^{j-1}p_k + \frac{1}{2}(j+m+2k-2)\underline{u}^j p_{k-1} + \frac{1}{2}j[\underline{u}^{j-2}p_{k+1} + \underline{u}^{j-1}p_k^*]$$

with p_k, p_{k-1}, p_{k+1} and p_k^* as in lemma 2.2(i), since $\partial_{\underline{u}}\underline{u}^l f = -l\underline{u}^{l-1}f$ for l even and $-(l-1+2k+m)\underline{u}^{l-1}f$ for l odd. This leads to

$$\langle \underline{x},\partial_{\underline{u}}\rangle \underline{u}^j f = \frac{1}{2}j\underline{u}^{j-2}p_{k+1} + \frac{1}{2}j\underline{u}^{j-1}[p_k + p_k^*] + \frac{1}{2}(j+m+2k-2)\underline{u}^j p_{k-1}$$

whence the desired result. If j is odd, the decomposition follows from

$$\langle \underline{x},\partial_{\underline{u}}\rangle \underline{u}^j f = -\frac{1}{2}\partial_{\underline{u}}(\underline{u}^{j-1}p_{k+1} + \underline{u}^j p_k^*) + \frac{1}{2}(j-1+m+2k)\underline{x}\,\underline{u}^{j-1}f$$

where $\underline{x}\,\underline{u}^{j-1}f$ again decomposes as in lemma 2.2(i). □

It is now easy to write down similarly the decomposition for the operator $\langle \partial_{\underline{x}},\partial_{\underline{u}}\rangle$ (using lemma 2.2(ii)).

Theorem 2.4 *Let $f \in \mathcal{E}(\Omega,\mathcal{M}_k)$; then $\langle \partial_{\underline{x}},\partial_{\underline{u}}\rangle \underline{u}^j f(\underline{x},\underline{u})$ decomposes as follows:*

$$\langle \partial_{\underline{x}},\partial_{\underline{u}}\rangle \underline{u}^j f(\underline{x},\underline{u}) = \begin{cases} j\underline{u}^{j-2}P_{k+1}(\underline{x},\underline{u}) + j\underline{u}^{j-1}P_k(\underline{x},\underline{u}) + \\ \qquad (j+m+2k-2)\underline{u}^j P_{k-1}(\underline{x},\underline{u}), & j \text{ even} \\ (j-1)\underline{u}^{j-2}P_{k+1}(\underline{x},\underline{u}) + (j+m+2k-1)\underline{u}^{j-1}P_k(\underline{x},\underline{u}) + \\ \qquad (j+m+2k-1)\underline{u}^j P_{k-1}(\underline{x},\underline{u}), & j \text{ odd} \end{cases}$$

where $P_{k+1}(\underline{x},\underline{u}) = -D_{+,k}(\partial_{\underline{x}})f(\underline{x},\underline{u})$ and
$P_k(\underline{x},\underline{u}) = -D_k(\partial_{\underline{x}})f$, $P_{k-1}(\underline{x},\underline{u}) = -D_{-,k}(\partial_{\underline{x}})f(\underline{x},\underline{u})$.

Note that the decompositions of $\langle \underline{x},\partial_{\underline{u}}\rangle$ and $\langle \partial_{\underline{x}},\partial_{\underline{u}}\rangle$ do not give rise to new monogenic operators! We recover the six basic operators $D_{-,k}(\partial_{\underline{x}})$, $D_k(\partial_{\underline{x}})$, $D_{+,k}(\partial_{\underline{x}})$ and $D_{-,k}(\underline{x})$, $D_k(\underline{x})$, $D_{+,k}(\underline{x})$ from the decompositions of $\langle \underline{u},\underline{x}\rangle$ and $\langle \underline{u},\partial_{\underline{x}}\rangle$, which act on $\mathcal{M}_k$-valued functions. This proves that in fact it suffices to consider $\mathcal{M}_k$-valued functions.

Note also that in the special case of $j=0$, the decomposition of $\langle \underline{x},\partial_{\underline{u}}\rangle f(\underline{x},\underline{u})$ and $\langle \partial_{\underline{x}},\partial_{\underline{u}}\rangle f(\underline{x},\underline{u})$ reduces to one term in $\mathcal{E}(\Omega,\mathcal{M}_{k-1})$ only:

$$\langle \underline{x},\partial_{\underline{u}}\rangle f(\underline{x},\underline{u}) = -\frac{1}{2}\partial_{\underline{u}}\underline{x}f(\underline{x},\underline{u}), \quad \langle \partial_{\underline{x}},\partial_{\underline{u}}\rangle f(\underline{x},\underline{u}) = -\frac{1}{2}\partial_{\underline{u}}\partial_{\underline{x}}f(\underline{x},\underline{u}).$$

3. Fischer decompositions

We now come to the following

Definition 3.1 *A function $f \in \mathcal{E}(\Omega, \mathcal{M}_k)$ is called left monogenic, resp. hyper-monogenic, hypo-monogenic if $D_k(\partial_{\underline{x}})f(\underline{x},\underline{u}) = 0$, resp. $D_{+,k}(\partial_{\underline{x}})f(\underline{x},\underline{u}) = 0$, $D_{-,k}(\partial_{\underline{x}})f(\underline{x},\underline{u}) = 0$.*

For a study of these functions, we refer to our paper [4]. As we are dealing with functions taking values in a vector space of polynomials, we already mentioned the inner product which is ready at hand:

$$(P(\underline{x},\underline{u}), Q(\underline{x},\underline{u})) = \left[\overline{P(\partial_{\underline{x}},\partial_{\underline{u}})}Q(\underline{x},\underline{u})\right]_{(\underline{x},\underline{u})=(0,0)}.$$

In the sequel we will use the orthogonality of the modules $\mathcal{P}_l(\mathcal{M}_k)$ and $\underline{u}\mathcal{P}_l(\mathcal{M}_{k'})$ (for arbitrary k and k') w.r.t. this inner product.

Take now $R_l(\underline{x},\underline{u}) \in \mathcal{P}_l(\mathcal{M}_k)$; then there are several possibilities to construct an orthogonal Fischer decomposition w.r.t. the above inner product for $R_l(\underline{x},\underline{u})$. The monogenic decompositions for the operators $\underline{x}$, $\langle \underline{u}, \underline{x}\rangle$, and $\langle \underline{x}, \partial_{\underline{u}}\rangle$, which we established in the first section, lead to Fischer decompositions into "monogenic pieces", or "hyper-monogenic pieces" or "hypo-monogenic pieces."

3.1. Monogenic Fischer decomposition

Consider the following submodule of $\mathcal{P}_l(\mathcal{M}_k)$:

$$\mathcal{S}_1 = \left\{q : q(\underline{x},\underline{u}) = (\frac{\underline{u}\partial_{\underline{u}}}{2k+m-2} + 1)\underline{x}\,p(\underline{x},\underline{u}), \quad p(\underline{x},\underline{u}) \in \mathcal{P}_{l-1}(\mathcal{M}_k)\right\}.$$

We characterize the orthogonal complement of $\mathcal{S}_1$ in $\mathcal{P}_l(\mathcal{M}_k)$, so for $r(\underline{x},\underline{u}) \in \mathcal{P}_l(\mathcal{M}_k)$ we claim

$$(q(\underline{x},\underline{u}), r(\underline{x},\underline{u})) = 0.$$

If we denote $\Pi_k = \frac{\underline{u}\partial_{\underline{u}}}{2k+m-2} + 1$, we find that

$$0 = (\Pi_k \underline{x}\,p(\underline{x},\underline{u}), r(\underline{x},\underline{u})) = (\underline{x}\,p(\underline{x},\underline{u}), r(\underline{x},\underline{u}))$$

in view of the monogenic decomposition of $\underline{x}\,p(\underline{x},\underline{u})$ (see lemma 2.1(i)). The above inner product hence simplifies to

$$0 = (p(\underline{x},\underline{u}), \partial_{\underline{x}} r(\underline{x},\underline{u})) = (p(\underline{x},\underline{u}), \Pi_k \partial_{\underline{x}} r(\underline{x},\underline{u}))$$

where we used lemma 2.1(ii) in the last step. But as $p(\underline{x},\underline{u})$ is an arbitrary polynomial of $\mathcal{P}_{l-1}(\mathcal{M}_k)$, we will only have orthogonality if $\Pi_k \partial_{\underline{x}} r(\underline{x},\underline{u}) = 0$, i.e. $r(\underline{x},\underline{u})$ is monogenic. Thus every $R_l(\underline{x},\underline{u}) \in \mathcal{P}_l(\mathcal{M}_k)$ can be written as

$$R_l(\underline{x},\underline{u}) = P_l(\underline{x},\underline{u}) + \Pi_k \underline{x} R_{l-1}(\underline{x},\underline{u})$$

where P_l is monogenic and $R_{l-1} \in \mathcal{P}_{l-1}(\mathcal{M}_k)$. It is clear that the above decomposition can be iterated, leading to

$$R_l(\underline{x},\underline{u}) = \sum_{s=0}^{l} \left[\left(\frac{\underline{u}\partial_{\underline{u}}}{2k+m-2} + 1\right)\underline{x}\right]^s P_{l-s}(\underline{x},\underline{u})$$

with

$$\left(\frac{\underline{u}\partial_{\underline{u}}}{2k+m-2}+1\right)\partial_{\underline{x}}P_j(\underline{x},\underline{u})=0, \quad j=0,\ldots,l.$$

3.2. Hypo-monogenic Fischer decomposition

Similarly we start from the following submodule of $\mathcal{P}_l(\mathcal{M}_k)$:

$$\mathcal{S}_2=\left\{q:\ q(\underline{x},\underline{u})=-\frac{1}{2}(\frac{\underline{u}\partial_{\underline{u}}}{2k+m-2}+1)\underline{x}\,\underline{u}\,p(\underline{x},\underline{u}), \quad p(\underline{x},\underline{u})\in\mathcal{P}_{l-1}(\mathcal{M}_{k-1})\right\}.$$

Let $r(\underline{x},\underline{u})$ be an element of the orthogonal complement of $\mathcal{S}_2$ in $\mathcal{P}_l(\mathcal{M}_k)$, then it follows that

$$\begin{aligned} 0 &= (q(\underline{x},\underline{u}),r(\underline{x},\underline{u}))=(D_{+,k-1}(\underline{x})p(\underline{x},\underline{u}),r(\underline{x},\underline{u}))\\ &= (\langle\underline{u},\underline{x}\rangle p-\underline{u}D_{k-1}(\underline{x})p-\underline{u}^2D_{-,k-1}(\underline{x})p,r(\underline{x},\underline{u}))\end{aligned}$$

where $D_{k-1}p$ is $\mathcal{M}_{k-1}$−valued and $D_{-,k-1}p$ is $\mathcal{M}_{k-2}$−valued. So in view of the orthogonality we end up with

$$(\langle\underline{u},\underline{x}\rangle p(\underline{x},\underline{u}),r(\underline{x},\underline{u}))=(p(\underline{x},\underline{u}),\langle\partial_{\underline{x}},\partial_{\underline{u}}\rangle r(\underline{x},\underline{u}))=(p(\underline{x},\underline{u}),-\frac{1}{2}\partial_{\underline{u}}\partial_{\underline{x}}r(\underline{x},\underline{u}))=0.$$

Hence the above inner product can only vanish for arbitrary $p\in\mathcal{P}_{l-1}(\mathcal{M}_{k-1})$ iff $\partial_{\underline{u}}\partial_{\underline{x}}r(\underline{x},\underline{u})=0$, i.e. $r(\underline{x},\underline{u})$ is hypo-monogenic. Every $R_l(\underline{x},\underline{u})\in\mathcal{P}_l(\mathcal{M}_k)$ can be written as follows:

$$R_l(\underline{x},\underline{u})=P_l(\underline{x},\underline{u})+\left(\frac{\underline{u}\partial_{\underline{u}}}{2k+m-2}+1\right)\underline{x}\,\underline{u}R_{l-1}(\underline{x},\underline{u})$$

where P_l is hypo-monogenic and $R_{l-1}\in\mathcal{P}_{l-1}(\mathcal{M}_{k-1})$. Iteration leads to the hypo-monogenic Fischer decomposition for $R_l(\underline{x},\underline{u})\in\mathcal{P}_l(\mathcal{M}_k)$:

$$R_l(\underline{x},\underline{u})=P_l(\underline{x},\underline{u})+\sum_{s=1}^{l}D_{+,k-1}D_{+,k-2}\ldots D_{+,k-s}P_{l-s}(\underline{x},\underline{u})$$

with

$$D_{+,j}=\langle\underline{u},\underline{x}\rangle+\frac{\underline{u}\,\underline{x}}{2j+m}-\frac{\underline{u}^2\partial_{\underline{u}}}{2(2j+m)}\underline{x}$$

and

$$\partial_{\underline{u}}\partial_{\underline{x}}P_j(\underline{x},\underline{u})=0, \quad j=0,\ldots,l.$$

3.3. Hyper-monogenic Fischer decomposition

Consider the submodule of $\mathcal{P}_l(\mathcal{M}_k)$ given by:

$$\mathcal{S}_3=\left\{q:\ q(\underline{x},\underline{u})=-\frac{1}{2}\partial_{\underline{u}}\underline{x}\,p(\underline{x},\underline{u}), \quad p(\underline{x},\underline{u})\in\mathcal{P}_{l-1}(\mathcal{M}_{k+1})\right\}.$$

The orthogonal complement is determined by

$$\begin{aligned} 0 &= (q(\underline{x},\underline{u}), r(\underline{x},\underline{u})) = (-\frac{1}{2}\partial_{\underline{u}}\underline{x}\, p(\underline{x},\underline{u}), r(\underline{x},\underline{u})) \\ &= (\langle \underline{x}, \partial_{\underline{u}}\rangle p(\underline{x},\underline{u}), r(\underline{x},\underline{u})) = (p(\underline{x},\underline{u}), \langle \underline{u}, \partial_{\underline{x}}\rangle r(\underline{x},\underline{u})) \\ &= (p(\underline{x},\underline{u}), D_{+,k}(\partial_{\underline{x}})r + \underline{u}D_k(\partial_{\underline{x}})r + \underline{u}^2 D_{-,k}(\partial_{\underline{x}})r) \end{aligned}$$

where $D_{+,k}(\partial_{\underline{x}})r(\underline{x},\underline{u})$ is $\mathcal{M}_{k+1}$–valued, while $D_k(\partial_{\underline{x}})r$ resp. $D_{-,k}(\partial_{\underline{x}})r$ are $\mathcal{M}_k$- resp. $\mathcal{M}_{k-1}$–valued. In view of the orthogonality, we have that

$$(p(\underline{x},\underline{u}), D_{+,k}r(\underline{x},\underline{u})) = 0$$

for arbitrary $p \in \mathcal{P}_{l-1}(\mathcal{M}_{k+1})$; hence $D_{+,k}r(\underline{x},\underline{u}) = 0$, so $r(\underline{x},\underline{u})$ is hyper-monogenic. Take now $R_l(\underline{x},\underline{u}) \in \mathcal{P}_l(\mathcal{M}_k)$; then R_l can be decomposed as follows:

$$R_l(\underline{x},\underline{u}) = P_l(\underline{x},\underline{u}) + \partial_{\underline{u}}\underline{x}R_{l-1}(\underline{x},\underline{u})$$

where $P_l(\underline{x},\underline{u})$ is hyper-monogenic and $R_{l-1}(\underline{x},\underline{u}) \in \mathcal{P}_{l-1}(\mathcal{M}_{k+1})$. The so-called hyper-monogenic Fischer decomposition is now given by

$$R_l(\underline{x},\underline{u}) = \sum_{s=0}^{l} (\partial_{\underline{u}}\underline{x})^s P_{l-s}(\underline{x},\underline{u})$$

with

$$\left(\langle \underline{u}, \partial_{\underline{x}}\rangle + \frac{\underline{u}\partial_{\underline{x}}}{2k+m} - \frac{\underline{u}^2\partial_{\underline{u}}\partial_{\underline{x}}}{2(2k+m)}\right) P_j(\underline{x},\underline{u}) = 0, \quad j = 0,\ldots,l.$$

References

1. F. Brackx, R. Delanghe and F. Sommen. "Clifford Analysis", Research Notes in Math. 76, Pitman, London (1982).
2. R. Delanghe, F. Sommen and V. Souček. "Clifford Algebra and Spinor-valued functions: a function theory for the Dirac-operator", Mathematics and Its Applications 53, Kluwer Academic Publishers, Dordrecht (1992).
3. J. Gilbert, M. Murray. "Clifford Algebras and Dirac operators in harmonic analysis", Cambridge University Press, 1990.
4. F. Sommen, N. Van Acker. "Functions of two vector variables", in preparation.
5. F. Sommen, N. Van Acker. "Monogenic differential operators", Results in Mathematics vol. 22 (1992), pp. 781-798.
6. E.M. Stein, G. Weiss. "Generalization of the Cauchy-Riemann equations and representations of the rotation group", Amer. J. Math. vol. 90 (1968), pp. 163-196.
7. A. Strasburger. "Invariant differential operators associated with spherical harmonics", preprint 1/89, University Warschau.

A DISTRIBUTIONAL APPROACH TO VECTOR MANIFOLDS

F. SOMMEN
Seminar for Algebra and Functional Analysis
Rijksuniversiteit Gent
Belgium

and

M. WATKINS
Institute of Mathematics and Statistics
University of Kent at Canterbury
United Kingdom

Abstract.
Every smooth manifold is shown to embed in its own space of compactly-supported distributions. A bilinear form is introduced, giving rise to a quotient space which under certain conditions will be finite dimensional and will also embed the manifold. The possibilities of introducing Clifford structures on both of these embedding spaces are then considered.

Key words: vector manifolds

1. Introduction

The basic idea here is a new approach to the treatment of vector manifolds. There will be no explicit Clifford algebraic or analytic content, but instead an interesting new setting will be created wherein Clifford analysis on manifolds can take place.

A vector manifold is any manifold realised as a surface in a vector space - its points are vectors. This is certainly not a new idea: For example, surfaces in Euclidean spaces have been thoroughly investigated. Also, the celebrated Nash embedding theorem demonstrates that any Riemannian manifold can be embedded in a finite dimensional vector space (see [1]). More recently Hestenes and Sobczyk have developed a whole theory of vector manifolds from a geometric (Clifford) algebraic standpoint (see [2]). However, this was developed in very abstract terms and lacks a general embedding algorithm.

Our approach, on the other hand, does include such an algorithm. It takes some simple functional analysis as its starting point, and provides a means to realise any smooth manifold as a surface in an infinite dimensional vector space. By introducing an appropriately chosen bilinear form on this, we are often able to reduce the embedding space to a finite dimensional quotient with a bilinear form which is non-degenerate both over the whole space, as well as over each embedded tangent space. A Clifford algebraic structure can then be established over this finite dimensional space as well as the original embedding space, generating some interesting new pos-

F. Brackx et al. (eds.), Clifford Algebras and their Applications in Mathematical Physics, 213–221.

sibilities.

Results shall be stated without proof, due to limited space and the context in which this work is being presented. For a fuller exposition of this theory, the reader is referred to [3].

2. Dual Space Embedding

2.1. Point sets

Suppose we have an arbitrary point set X. We define $F(X)$ to be the vector space of all real-valued functions on X, and $F'(X)$ to be its formal linear dual. If we think of $F(X)$ as a topological vector space with discrete topology, then all of the functionals in $F'(X)$ are continuous, so we can think of it as the *topological* dual of $F(X)$. To this vector space we apply the weak-dual topology, and then define the mapping

$$\delta : X \rightarrow F'(X)$$

where $\delta(p) : F(X) \rightarrow \mathbf{R}$ is the linear functional given by $\delta(p)[f] = f(p) \ \ \forall f \in F(X)$. This is the **point-evaluation** functional at p.

δ is some sense 'embeds' the point set X in the topological vector space $F'(X)$. The term **embedding** tends to suggest that some structure is preserved. The only structure on X is the separateness of its points. As δ is necessarily injective, this is preserved. So we can think of δ as a degenerate sort of embedding.

Result 1 *Every point set X is injectively 'embedded' in the topological vector space $F'(X)$ by the point-evaluation map $\delta : X \rightarrow F'(X)$. Here $F(X)$ is the space of all real-valued functions, with discrete topology, and $F'(X)$ its topological dual, with weak-dual topology.*

This is obviously a very simple, almost trivial construction, but we now proceed to adapt it, to embed more structured classes of point sets.

2.2. Topological Spaces

Suppose now that our point set X happens to be a topological space. We can then consider - does the mapping $\delta : X \rightarrow F'(X)$ preserve the topology of X?

The image set $\delta(X)$ will inherit a topology from its ambient space $F'(X)$. Recall that $F'(X)$ has been equipped with weak-dual topology (the reason for this choice will become clear as we proceed). Then for each point $q \in X$, $\{\delta(q)\} \subseteq \delta(X)$ constitutes an open set. This follows from the existence of such functions as $f : X \rightarrow \mathbf{R}$ where

$$f(x) = \begin{cases} 0 \ if \ x = q \\ 1 \ if \ x \neq q \end{cases}$$

and a simple seminorm argument. Hence the only case where $\delta : X \rightarrow \delta(X) \subseteq F'(X)$ preserves topological structure is that of X being a discrete topological space.

But as this really has no more structure than a point set, we dismiss it and consider how the situation could be improved. Looking closely at the embedding, we see that the difficulty arises from the non-continuity of most $F(X)$ functions. So suppose we were to restrict our interest to $C(X)$, the subspace of $F(X)$ consisting of all *continuous* real-valued functions on the topological space X.

We can equip $C(X)$ with compact-open topology, and then consider the topological dual $C'(X)$. This space is again given weak-dual topology, and we define the analagous point-evaluation map

$$\delta : X \to C'(X)$$

The image $\delta(X)$ now inherits its topology from $C'(X)$. Before we look at this, we must consider the possible non-injectivity of $\delta : X \to C'(X)$. $\delta(p) = \delta(q)$ only implies that all continuous functions agree at p and q. This doesn't necessarily mean that they are the same point. So we need to be sure the function space $C(X)$ *separates points* of X. To this end, we assume X is a *completely regular* space (see for example [4]). We then have that $\delta : X \to \delta(X) \subseteq C'(X)$ is a bijective mapping. Further, seminorm constructions can be used to show that both δ and δ^{-1} are continuous.

Result 2 *Every completely regular topological space X is homeomorphically embedded in the topological vector space $C'(X)$ by the point-evaluation map $\delta : X \to C'(X)$. Here $C(X)$ is the space of all continuous real-valued functions, with compact-open topology, and $C'(X)$ its topological dual, with weak-dual topology.*

2.3. Topological Manifolds

Suppose now that our topological space X has a manifold structure (that is, it's a locally Euclidean, 2nd countable Hausdorff space). Then X is necessarily completely regular, and we have a homeomorphic embedding $\delta : X \to \delta(X) \subseteq C'(X)$. It follows that the manifold structure of X is preserved by δ.

Result 3 *Every topological manifold X is homeomorphically embedded in the topological vector space $C'(X)$ by the point-evaluation map $\delta : X \to C'(X)$.*

2.4. Smooth Manifolds

We now consider the next layer of structure. Suppose we impose a differentiable structure (that is, a C^∞-compatible atlas) on our topological manifold X.

$\delta : X \to \delta(X) \subseteq C'(X)$ is still a homeomorphic embedding; however, it may not be a *diffeomorphic* embedding. We are interested in whether or not the smooth structure of X is preserved under the mapping δ. But we must stop and consider what is meant by this. Whereas the topological structure of $\delta(X)$ is inherited from the ambient space $C'(X)$ in a straightforward way, there is no obvious means to extract a differentiable structure. Of course, if we were free to impose any atlas we liked on $\delta(X)$, then the simple act of copying the X atlas, chart-by-chart, via δ, would give a smooth manifold which would be a diffeomorphic image of X. But recall that we want $\delta(X)$ *embedded* in $C'(X)$, so we really need to consider whether any given atlas on $\delta(X)$ is *compatible* with the ambient space. Recall that an atlas consists of homeomorphic chart maps from coordinate patches on X into regions of Euclidean space E^m. As $C'(X)$ has a topological structure, we can take limits, and consider whether the inverse chart maps, mapping from regions of E^m into the topological vector space $C'(X)$ are C^∞ as vector-valued mappings.

Although the proof is somewhat technical, it's possible to show that *any* choice of C^∞ atlas on $\delta(X) \subseteq C'(X)$ will contain at least one chart map whose inverse, mapping from some region of E^m into $C'(X)$, is not a smooth vector-valued map. In other words, no choice of smooth structure on $\delta(X)$ will be compatible with the ambient space $C'(X)$. By establishing the original atlas on X, we immediately generated a space of smooth functions on X, which we'll denote $E(X)$. The non-compatibility of any differentiable structure with $C'(X)$ is essentially due to the fact that not every element of $C(X)$ is in $E(X)$. In other words, not every continuous function is smooth.

This suggests that, as before, when we reduced our function space $F(X)$ to $C(X)$ in order to accomodate the newly imposed topological structure on the point set X, we should reduce $C(X)$ to $E(X)$ in order to accomodate the newly imposed differentiable structure on the manifold X.

$E(X)$ is a topological vector space equipped with the standard topology, and $E'(X)$ is the topological dual with weak-dual topology. Again, we define the analagous point-evaluation mapping $\delta : X \to E'(X)$.

It is possible to show that $\delta : X \to \delta(X) \subseteq E'(X)$ remains a homeomorphism. Further, it can be proven that there is a *unique* choice of atlas on the manifold $\delta(X)$ which is compatible with the ambient space $E'(X)$. It is the δ-image of the atlas on X, so if we treat $\delta(X)$ as a smooth manifold with this structure, then δ is of course a diffeomorphism.

Result 4 *Every smooth manifold X is diffeomorphically embedded in the topological vector space $E'(X)$ by the point-evaluation map $\delta : X \to E'(X)$. Here $E(X)$ is the space of all smooth real-valued functions on X, $E'(X)$ the topological dual equipped with weak-dual topology.*

$E'(X)$ is well-known to the functional analyst as the space of **compactly- supported distributions** on the smooth manifold X.

3. Reducing the Embedding Space

3.1. Structure Functions and Hyperspace

As we're interested in defining a Dirac-type operator on these vector manifolds, it would be helpful to have them embedded in spaces of finite dimension. A space of compactly-supported distributions are of *uncountably infinite* dimension, but as suggested earlier, by introducing an appropriately chosen bilinear form on our embedding space, we can sometimes reduce it to a finite-dimensional quotient.

Suppose then that M is a smooth manifold, and $E'(M)$ is the space of compactly-supported distributions thereupon. The most obvious means to introduce a continuous bilinear form on $E'(M)$ works as follows: We choose some symmetric function $F : M \times M \to \mathbf{R}$, smooth in both arguments, and then define $B(,)$ on $E'(M)$ by

$$B(S,T) = S_x[T_y[F(x,y)]]$$

Here the distribution T acts on F, treated as a function of y (x being treated as a dummy variable). This gives a smooth function in x, on which the distribution S acts, to return a real value. We call F a **structure function** on M.

$B(,)$ is symmetric, bilinear, and continuous, as required. Conversely, it can be shown with density arguments that *any* bilinear form over $E'(M)$ must be associated with some smooth symmetric function F in this way. However, for many choices of F it may well be degenerate. That is, there may exist some nonzero S in $E'(M)$ such that $B(S,T) = 0 \ \ \forall\, T \in E'(M)$. This possibility of degeneracy allows us to establish an equivalence relation $\sim$ on $E'(M)$ as follows:

$$\begin{aligned} S \sim T \ &\Leftrightarrow \ B(S,U) = B(T,U) \ \ \forall\, U \in E'(M) \\ &\Leftrightarrow \ B(S-T,U) = 0 \ \ \forall\, U \in E'(M) \end{aligned}$$

The quotient space $E'(M)/\!\sim$ is called the **hyperspace** of the vector manifold M, with respect to the choice of structure F, and is denoted $Hyp(M,F)$.

$X : E'(M) \to Hyp(M,F)$ is simply the projection which maps each distribution to its equivalence class. We are now able to define a nondegenerate bilinear form $<,>$ over the hyperspace:

$$< X(S), X(T) >= B(S,T).$$

This is clearly well-defined, symmetric, bilinear, continuous, and nondegenerate.

If the structure function F is chosen carefully, we are sometimes able to arrive at a finite dimensional hyperspace. The difficulty is that in the process of collapsing $E'(M)$ down to $Hyp(M,F)$ we may get a drastic reduction in dimensionality, but the diffeomorphic embedding of M is sometimes sacrificed. But before we look into this matter, we'll need to consider how the tangent bundle of M embeds in $E'(M)$.

3.2. The Embedded Tangent Bundle

We use the construction of tangent vectors to surfaces in Euclidean spaces as our guide. If $\alpha(t)$ is a curve on M such that $\alpha(0) = p$, then taking the limit

$$\lim_{\epsilon \to 0} \frac{\delta(\alpha(\epsilon)) - \delta(\alpha(0))}{\epsilon}$$

gives a compactly-supported distribution on M, namely a directional derivative on M at p, in the direction of α. Such distributions will be identified as tangent vectors

(this is in accordance with the Bourbaki notion of tangent vectors as first order differential operators).
By considering the action of $B(,)$ on pairs of tangent vectors to M, we are able to generate a Riemannian-type metric on the manifold. In fact we have the following result:

Result 5 *Suppose (M,g) is a Riemannian manifold, and $d(x,y)$ denotes the Riemannian distance between points x and y. Suppose further that F is some choice of structure function on M such that $F(x,y) = -\frac{1}{2}d^2(x,y)$ for points x,y suitably close together. Then F will induce a bilinear form $B(,)$ over $E'(M)$, which, when restricted to the embedded tangent bundle of M, gives a metric identical to g. In other words, every Riemannian manifold can be embedded metrically in its own space of compactly-supported distributions.*

3.3. Hyperspace Embeddings

We are now in a position to state the following results:

Result 6 *Suppose M is a smooth manifold, and F a choice of structure function on M such that*
(1) $X \circ \delta : M \to Hyp(M,F)$ is injective
(2) $B(,)$ is nondegenerate when restricted to each embedded tangent space
Then $X \circ \delta(M)$ is a homeomorphic embedding of the manifold in $Hyp(M,F)$. Further, there is a unique choice of smooth structure on the manifold $X \circ \delta(M)$ which is compatible with the ambient hyperspace, and given this structure, $X \circ \delta(M)$ is a diffeomorphic *embedding of M.*

Result 7 *Suppose M is a smooth manifold, and F a choice of structure on M such that*
(1) $X \circ \delta : M \to X \circ \delta(M) \subseteq Hyp(M,F)$ is topologically a covering map
(2) $B(,)$ is nondegenerate when restricted to each embedded tangent space.
Then $X \circ \delta(M)$ is a homeomorphic embedding of a quotient manifold of the universal covering space of M. Again, the embedded manifold $X \circ \delta(M)$ has a unique smooth structure which is compatible with $Hyp(M,F)$, and given this structure, $X \circ \delta(M)$ is a diffeomorphic embedding of the quotient manifold, formally taken as a smooth manifold with canonical atlas (see [5]).

These results open up the possibility of diffeomorphically embedding smooth manifolds in finite dimensional spaces with nondegenerate bilinear forms.

4. Establishing a Clifford Structure

4.1. Dirac-like Operators on Finite Dimensionally Embedded Vector Manifolds

In either of the cases treated by Results 6 and 7, the nondegeneracy of $B(,)$ on the tangent spaces embedded in $E'(M)$ implies the nondegeneracy of $<,>$ on the tangent spaces subsequently embedded in $Hyp(M,F)$. These are denoted $TX(M)(p)$, and we are then able to establish a basis $\{f_1(p),\ldots,f_m(P)\}$ for each, such that

$$< f_i(p), f_j(p) >= \pm\delta_{ij}.$$

If we have $dim\ Hyp(M,F) = k < \infty$, then starting with some basis $\{e_1,\ldots,e_k\}$, we can define both a Clifford algebraic structure, by putting

$$v^2 = - < v, v >,$$

and a Dirac operator acting on Clifford algebra-valued functions

$$\partial = \sum_{j=1}^{k} \frac{e_j}{< e_j, e_j >} \frac{\partial}{\partial x_j}.$$

Due to the nondegeneracy of the bilinear form on each tangent space, it's possible to project this operator onto each $TX(M)(p)$ as

$$\partial(p) = \sum_{j=1}^{m} \frac{f_j(p)}{< f_j(p), f_j(p) >} < f_j(p), \partial > .$$

This is the standard means to establish a Dirac-like operator on the embedded M.

4.2. An Example

We'll now look at an example of a finite dimensional embedding of the real projective space $\mathbb{R}P^{m-1}$. We shall make use of Result 7 by embedding it as a quotient of S^{m-1}. Starting with the usual distributional embedding

$$\delta : S^{m-1} \to E'(S^{m-1}),$$

we then choose the structure function

$$F(x,y) = -\frac{1}{2}(< x, y >^2 - 1)$$

where $<,>$ denotes the usual inner product on $\mathbb{E}^m$. It is then possible to calculate the hyperspace explicitly, and doing so we find

$$Hyp(S^{m-1}, F) = P_2{}^{m\prime}.$$

That is, it's the dual of the vector space P_2^m of homogeneous binomials on $\mathbb{E}^m$. $P_2^m = span\{x_ix_j : 1 \leq i,j \leq m\}$. It follow, then that

$$dim\ Hyp(S^{m-1}, F) = dim\ P_2^m = {}^mC_2 = \frac{m(m+1)}{2}.$$

$X \circ d : S^{m-1} \to Hyp(S^{m-1}, F)$ is not injective, since

$$\begin{aligned} F(x,y) = F(-x,y)\ \ \forall\ y \in S^{m-1} &\Rightarrow B(\delta(x), \delta(y)) = B(\delta(-x, \delta(y))\ \ \forall\ y \in S^{m-1} \\ &\Rightarrow B(\delta(x), Z) = B(\delta(-x), Z)\ \ \forall\ Z \in E'(M) \end{aligned}$$

The last step follows from the density of finite linear combinations of point-evaluation vectors in $E'(M)$, and we can now deduce

$$X \circ \delta(x) = X \circ \delta(-x) \ \ \forall\, x \in S^{m-1}$$

In other words, antipodal points of the sphere get identified in the hyperspace. In fact $X \circ \delta : S^{m-1} \to X \circ \delta(S^{m-1}) \subseteq Hyp(S^{m-1}, F)$ is easily seen to be a topological covering map, and so using Result 7 it is possible to demonstrate formally that $X \circ \delta(S^{m-1})$ is a diffeomorphic embedding of $\mathbf{R}P^{m-1}$ (that is, $\mathbf{R}P^{m-1}$ realised as a smooth quotient on S^{m-1}, via group action (See [5], for example).
It is left as an exercise for the reader to explicitly calculate the Dirac operator on the embedded $\mathbf{R}P^{m-1}$.

4.3. The Infinite Dimensional Case

Although finite-dimensional embeddings are desirable insofar as they provide a link with established, classical results concerning surfaces in vector spaces, the real strength of this theory lies in the distributional embedding. With this approach, whole realms of extrinsic geometry are opened up, bringing to light special manifolds which would otherwise be inaccessible.

An excellent example to illustrate this is given by the vector manifold embedding of E^m in $E'(\mathsf{E}^m)$ with the structure function

$$F(x, y) = e^{\frac{-|x-y|^2}{2}}.$$

The resulting hyperspace $Hyp(\mathsf{E}^m, F)$ is just $E'(\mathsf{E}^m)$; no reduction in dimension occurs here. This vector manifold has some remarkable properties: Intrinsically it is identical with E^m, but it's extrinsic geometry can be described in terms of Laguerre polynomials and spherical harmonics. Further, the bilinear form $B(,)$ can be shown to be positive-definite using properties of the Gaussian distribution, and with the resulting notion of distance $d(,)$ in $E'(\mathsf{E}^m)$, we find

$$\lim_{|x-y|\to\infty} d(\delta(x), \delta(y)) = \sqrt{2}$$

So as two points on E^m move further apart, the distance between them measured through the hyperspace approaces $\sqrt{2}$ asymptotically.

As we have a bilinear form over $E'(M)$ for a given vector manifold (M, F), we are always able to define an anticommutative, Clifford-like structure over the space. The complete linear independence of the point-evaluation vectors $\delta(p) \in E'(M)$ allows certain operations hitherto impossible. For example, given an oriented curve on M, through partitioning and limiting processes, we should be able to take a total wedge product, something like a path integral, along the curve. As yet we have not had time to fully explore such ideas, but it would appear that Clifford algebra and analysis in $E'(M)$ may bring forth valuable new insights.

References

1. Nash, J.: 1956, 'The Imbedding Problem for Riemannian Manifolds', *Annals of Mathematics* Vol. no. **63**, pp. 20–63
2. Hestenes, D. and Sobczyk, G.: 1984, 'Clifford Algebra to Geometric Calculus: A Unified Language for Mathematics and Physics', Reidel
3. Sommen, F. and Watkins, M.: 1993,'An Introduction to Vector Manifolds', *University of Kent Internal Report* UKC/IMS/A93/1
4. Gillman, L. and Jerison, M.: 1976, 'Rings of Continuous Functions' *Graduate Texts in Mathematics* **43**, Springer-Verlag, New York
5. Berger, M. and Gostiaux, B.: 1988, 'Differential Geometry: Manifolds, Curves, and Surfaces', *Graduate Texts in Mathematics* **115**, Springer, New York and London

CLIFFORD ANALYSIS FOR HIGHER SPINS

VLADIMÍR SOUČEK
Mathematical Institute
Charles University
Sokolovská 83
186 00 Praha
Czech Republic

Abstract. Higher spin analogues of the (massless) wave and Dirac equation on Minkowski space are well understood in dimension 4. They appear usually under the name massless field equations. In the paper, higher-dimensional analogues of the massless field equations are studied in Riemannian setting and basic results of the function theory are described. Another set of conformally invariant equations - the so called twistor equations for any spin - were studied in the physical case by R. Penrose and others. We are discussing the higher-dimensional analogues of these equations (in Riemannian situation) and their possible role in Clifford analysis. Multiplicative properties of solutions of the twistor equation are discussed.

Key words: Clifford analysis, massless field equations, twistor equations, higher spins

1. Introduction

The Dirac equation and its solutions studied in Clifford analysis are considered either for Clifford-valued or for spinor-valued functions (see (Delanghe *et al.* , 1992)). Higher spin analogues of the Dirac equation are well known in the physical situation (on four-dimensional Minkowski space), they are usually called massless field equations ((Penrose, Rindler , 1984)). They include, besides the vawe and the Dirac equation, Maxwell equations (spin 1 case) and linearized gravity (spin 2 case). Methods of Clifford analysis were applied already to Maxwell equations (see (Jancewicz , 1988)), but the function theory for higher spin and higher-dimensional cases were not studied yet. Higher-dimensional analogues of massless field equations were defined in an abstract representation theory language in (Fegan , 1975) or (Baston, Eastwood , 1991). The purpose of the paper is to pave a few first roads into this wast teritory and to show that all this subject should form an integral part of Clifford analysis. It indicates as well that a study of further conformally invariant equations with values in other spin representations from the point of view of function theory could be of interest.

Another important issue coming back again and again since Fueter's discovery of his equation is the fact that the space of solutions of the Dirac equation is not an algebra - product of two solutions need not be a solution. The basic reason behind it is clearly the fact that Dirac equation is (in principle) an equation for a spinor field. From that point of view it is understandable that only natural multiplication available for spinor fields is (a variant of) tensor multiplication which leads directly to higher spin fields. No version of a product for two spin $\frac{1}{2}$ fields is known which

F. Brackx et al. (eds.), Clifford Algebras and their Applications in Mathematical Physics , 223–232.

would garantee that the product of two such solutions is always a solution of spin 1 massless field equation.

On the other hand, the twistor equation (which is not very interesting in spin $\frac{1}{2}$ case, because solutions are linear) is becoming more interesting for higher spins; the space of its solutions is bigger and bigger (higher order polynomials). Moreover, it is possible to define a product in such a way that the product is commutative and product of two solutions is again a solution (of higher spin). Hence solutions of twistor equations (considered for all spins at the same time) offer an interesting generalization of Cauchy-Riemann equations to higher dimensions with the property that the space of solutions form a commutative algebra. The solutions with specific spins are analogues of homogeneous solutions of Cauchy-Riemann equations. All that together with a lot of relationships among solutions of massless field equations and twistor equations for any spin is well understood in dimension 4 due to the work of R. Penrose ((Penrose, Rindler , 1984), see also (Hitchin, 1980)), but higher dimensional analogues were not studied yet. So at the end of the paper we discuss basic multiplicative properties for solutions of the twistor equation in higher dimensions. THe paper gives a review of cartain results in this area but complete proofs of theorems given will be published in another paper.

2. The Cartan Powers of the Spinor Representation

Let us consider negative definite Euclidean space R^n, the corresponding Clifford algebra $R_{0,n} = C$ and its complexification C^c. As is standard in Clifford analysis, we consider functions defined on domains in R^n but their values will generally be in higher spin representations of the group Spin(n). In this section, we are going to describe these irreducible representations together with their elementary properties and we are setting the notation used in other sections.

It is necessary to distinguish even- and odd-dimensional cases.

The case $n = 2k$.

There is exactly one (up to isomorphism) irreducible representation S of the Clifford algebra C^c which splits, as a representation of the even part $(C^c)^+$, into two irreducible parts denoted by S_A and $S_{A'}$. The corresponding dual (contragredient) representations are denoted by S^A, resp. $S^{A'}$. All finite-dimensional irreducible representations of the group Spin(n) are classified by their highest weights $\mu \in \Lambda_W$, where

$$\Lambda_W = \{(\mu_1, \ldots, \mu_k) \in Z^k \cup (Z + \frac{1}{2})^k | \mu_1 \geq \mu_2 \geq \ldots \geq \mu_{k-1} \geq |\mu_k|\}$$

(a lot of useful details concerning the discussion in this section can be found in (Fulton, Harris , 1991), Chap.19). The basic spinor representations S_A, resp. $S_{A'}$ have the highest weights $\alpha = (\frac{1}{2}, \ldots, \frac{1}{2})$, resp. $\alpha' = (\frac{1}{2}, \ldots, \frac{1}{2}, -\frac{1}{2})$. The spinor representations $S_A, S_{A'}$ are selfdual for k even and they are dual to one another for k odd.

The higher spin representations we are interested in can be realized as invariant subspaces of suitable tensor products of S_A and $S_{A'}$. It is convenient to use the Penrose abstract index notation (for details see (Penrose, Rindler , 1984)) where indices are used as abstract (not coordinates) labels. Whenever same symbols appear up and down inside a tensor product, it will indicate the contraction in the corresponding pieces of the product, a round bracket around a set of indices means the symmetrisation. The tensor product $S_{A\ldots E}$ of p copies of the representation S_A decomposes into a sum of irreducible representations. The symmetric power $S_{(A\ldots E)} := Sym^p(S_A)$ is an invariant subspace but it is an irreducible representation only in dimension 4; it decomposes further in higher dimensions.

The Clifford algebra C^c can be identified with the space $End(S) \cong S \otimes S^*$. For the even and odd parts, we get $(C^c)^+ \cong S_A^A \oplus S_{A'}^{A'}$ and $(C^c)^- \cong S_A^{A'} \oplus S_{A'}^A$. Hence the complexification $C^n = R^n \otimes_C C$ is a subset of $S_A^{A'} \oplus S_{A'}^A$ and for a given vector $v \in C^n$, we denote by $v_A^{A'}$, resp. $v_{A'}^A$ the corresponding components of the sum. The projections of C^n into $S_A^{A'}$, resp. $S_{A'}^A$ will be denoted by $(C^n)_A^{A'}$ resp. $(C^n)_{A'}^A$. These both spaces are isomorphic (as representations) to the fundamental representation C^n.

The case $n = 2k+1$.

The situation is different in odd dimensions. There are two inequivalent irreducible representation S_A, resp. $S_{A'}$ of C^c but they are isomorphic as representations of $(C^c)^+$, hence as Spin(n)-representations. They are selfdual representations.

The set Λ_W of dominant weights is given by

$$\Lambda_W = \{(\mu_1, \ldots, \mu_k) \in Z^k \cup (Z + \frac{1}{2})^k | \mu_1 \geq \mu_2 \geq \ldots \geq \mu_{k-1} \geq \mu_k \geq 0\}.$$

The basic spinor representation S_A has the highest weight $\alpha = (\frac{1}{2}, \ldots, \frac{1}{2})$.

The Clifford algebra is described as $C^c \cong S_A^A \oplus S_{A'}^{A'}$. The projection of vectors in C^n into S_A^A are hence denoted as v_A^A. It is convenient to note that in the discussion below, a transition from even- to odd-dimensional case can be usually done simply by converting all primed indices to unprimed ones.

We need below a notion of the Cartan product (let us consider now any dimension, even or odd). If V_1 and V_2 are two irreducible representations of Spin(n) and if v_1, resp. v_2 are their highest weight vectors, then $v_1 \otimes v_2$ is a highest weight vector of an irreducible component of the product $V_1 \otimes V_2$ called the Cartan product of V_1 and V_2; let us denote it by $V_1 \boxtimes V_2$. In terms of highest weights, it is characterized by the sum of highest weights of V_1 and V_2. The corresponding orthogonal projection $\pi^{Car} : V_1 \otimes V_2 \mapsto V_1 \boxtimes V_2$ makes it possible to define the Cartan product of any two elements $v_1 \in V_1, v_2 \in V_2$; we define $v_1 \boxtimes v_2 = \pi^{Car}(v_1 \otimes v_2)$. The Cartan product is associative. Indeed, if V_i, $i =, 1, 2, 3$ are three irreducible representations and v_i, $i = 1, 2, 3$ are the corresponding highest weight vectors, then both $(V_1 \boxtimes V_2) \boxtimes V_3$ and $V_1 \boxtimes (V_2 \boxtimes V_3)$ have $v_1 \otimes v_2 \otimes v_3$ as their highest weight vector. It implies associativity of the product of vectors.

If $v \in S_A$ is a highest weight vector, then $v \otimes \ldots \otimes v \in S_A \otimes \ldots \otimes S_E$ is a highest weight vector of an irreducible piece of the product. This piece is characterised by its highest weight $p \cdot \alpha = (\frac{p}{2}, \ldots, \frac{p}{2})$ and will be denoted by $S_{\{A\ldots E\}}$ or shortly $(S_A)^p$; it is usually called the Cartan p-th power of S_A. The orthogonal projection (with respect to the Killing form) from $S_{A\ldots E}$ to $S_{\{A\ldots E\}}$ will be denoted again by π^{Car} and indicated by braces $\{\}$ around the corresponding set of indices. Using the projection, we can define the Cartan product of any two elements $s_{A\ldots E} \in$

$S_{\{A...E\}}, t_{F...H} \in S_{\{F...H\}}$ by

$$s_{A...E} \boxtimes t_{F...H} = s_{\{A...E}t_{F...H\}} = \pi^{Car}(s_{A...E} \otimes s_{F...H}).$$

Due to the fact that the Cartan power is a subspace of the symmetric power, it is clear that the multiplication

$$\boxtimes : (S_A)^p \times (S_A)^q \mapsto (S_A)^{p+q}$$

is commutative. This property is playing an important role in the discussion below. The same notation will can used in the primed version.

3. Conformally Invariant First Order Equations.

An important feature of the (massless) wave and Dirac equations is that they are conformally invariant. We are looking for their higher spin analogues and it is natural to require conformal invariance for these generalizations as well. There is a classification of all conformally invariant first order differential equations (see (Fegan , 1975)). We are going to consider equations for fields with values in the p-th Cartan power $(S_A)^p$ of the basic spinor representation (there is the corresponding primed case in even dimensions with the same theory). In this cases, there are two first order conformally invariant equations available. They are constructed as follows (we are going to treat only even-dimensional case, the odd-dimensional case means just to forget all primes).

Definition 3.1 *Let us consider the product $(C^n)_A^{A'} \otimes (S_A)^p$ of irreducible representations. Its decomposition into two irreducible parts is*

$$(C^n)_A^{A'} \otimes (S_A)^p \cong E_1 \oplus E_2, \tag{1}$$

where E_1, resp. E_2 are characterized by their highest weights $\beta_1 = (\frac{p}{2}+1, \frac{p}{2}, \dots, \frac{p}{2})$, resp. $\beta_2 = (\frac{p}{2}, \dots, \frac{p}{2}, \frac{p}{2}-1)$. Let us consider fields f on R^n with values in $(S_A)^p$. The de Rham differential d is applied to such a field componentwise. If π_1, resp. π_2 denote the corresponding projections onto parts with heighest weights β_1 and β_2, then we have the following two equations for f :

$$\pi_1(df) = 0 \tag{2}$$

and

$$\pi_2(df) = 0. \tag{3}$$

The first one is called the twistor equation and the second one the massless field equation for spin $\frac{p}{2}$.

The massless field equations and the twistor equations are defined here on a flat space but they are well defined, using Levi-Civita covariant derivative, on spin manifolds, see e.g. (Souček , 1986).

The coordinate description and basic properties of their solutions will be given in the paper. The equations are conformally invariant under the assumption that a suitable conformal weight for spinor fields, uniquely determined and different in each case, is considered but conformal properties of solutions will not be discussed here. This is a nice subject and it needs a more space than available here; it is one of places where the behaviour of individual cases for different spins is different so that information available from the standard case is not immediately applicable.

4. A Coordinate Description

To write down a coordinate form of these two types of equations is a delicate and a subtle task. An explicite decomposition proved in spin $\frac{1}{2}$ case in (Bureš, Souček , 1986) has to be used in the considered higher spin case. First it is necessary to understand the decomposition of the spinor spaces $S_{A\{B...E\}}$ and $S_{A'\{B...E\}}$ into irreducible components.

Lemma 4.1 *Let $n = 2k$ or $n = 2k+1$ and let us fix $p \in Z, p > 1$. The weights $\mu_j = (\frac{p}{2}, \ldots, \frac{p}{2}, \frac{p-2}{2}, \ldots, \frac{p-2}{2})$ with $k-j$ components equal to $\frac{p}{2}$ are dominant, the irreducible representations with highest weights μ_j are denoted by $F_j = F(\mu_j)$. Then the irreducible components of $S_{A\{B...E\}}$ and $S_{A'\{B...E\}}$ are characterized by:*

i) the case $n = 2k$:

$$S_{A\{B...E\}} \cong F_0 \oplus F_2 \oplus \ldots,$$

(the sum ends with F_k or F_{k-1}) and

$$S_{A'\{B...E\}} \cong F_1 \oplus F_3 \oplus \ldots,$$

(the sum ends with F_k or F_{k-1}).

ii) the case $n = 2k+1$:

$$S_{A\{B...E\}} \cong F_0 \oplus F_1 \oplus \ldots \oplus F_k.$$

Let $e \in R^n$ be a unit vector. In even dimension, the map e from $S_{A\{B...E\}}$ to $S_{A'\{B...E\}}$ given by $s_{A...E} \mapsto e^A_{A'} s_{AB...E}$ is an isomorphism and the inverse map $\tilde{e}$ is given by $s_{A'B...E} \mapsto -e^{A'}_A s_{A'B...E}$. In odd dimensions, the same is true when primed indices are substituted by unprimed ones.

The restriction of e (resp. of $\tilde{e}$) to F_j has values in F_{j+1} and F_{j-1} (where $F_{-1} = F_{k+1} = \{0\}$).

The following theorem is the key theorem showing an explicit form of the decomposition of the tensor product in the Def.3.1. It makes possible to write down the corresponding projections explicitely.

Theorem 4.1 *Let $e_1, \ldots, e_n$ be an orthonormal basis in R^n and let $\epsilon_1, \ldots, \epsilon_n$ be the dual basis. Then the components E_i in the decomposition (1) are given by (in odd dimensions all primes should be forgotten)*

$$E_1 = \left\{ \sum_j \epsilon_j \otimes (e_j)_{\{A}^{A'} s_{A'B...E\}} | s_{A'B...E} \in F_1 \right\}$$

and

$$E_2 = \Big\{\textstyle\sum_j \epsilon_j \otimes (s_j)_{AB\ldots E} \,\Big|\, (s_j)_{AB\ldots E} \in F_0 \text{ such that there exist } (s'_j)_{AB\ldots E} \in F_2 \text{ with } \sum_j (e_j)^A_{A'}[(s_j)_{AB\ldots E} - (s'_j)_{AB\ldots E}] = 0.\Big\}$$

The corresponding projections are given by

$$P_1\left(\sum_j \epsilon_j \otimes (s_j)_{A\ldots E}\right) = \frac{1}{n}\sum_j \epsilon_j \otimes (e_j)_{\{A}^{A'}\left(\sum_k (e_k)^F_{A'}(s_k)_{FB\ldots E\}}\right),$$

$$P_2\left(\sum_j \epsilon_j \otimes (s_j)_{A\ldots E}\right) = \sum_j \epsilon_j \otimes \left[(s_j)_{A\ldots E} + \frac{1}{n}(e_j)_{\{A}^{A'}\left(\sum_k (e_k)^F_{A'}(s_k)_{FB\ldots E\}}\right)\right].$$

Using Theorem 4.1., an explicit form of the twistor and massless field equations can be deduced.

Theorem 4.2 *The massless field equation (3) for a spinor field $f_{A\ldots E}$ with values in $(S_A)^n$ is given by*

$$(Df)_{A'B\ldots E} = \sum_j (e_j)^A_{A'} \frac{\partial f_{AB\ldots E}}{\partial x_j} = 0 \tag{4}$$

and the twistor equation (2) is equivalent to the set of equations ($j = 1, \ldots, n$)

$$\frac{\partial f_{AB\ldots E}}{\partial x_j} = \frac{1}{n}(e_j)_{\{A}^{A'}\left(D(f)_{A'B\ldots E\}}\right). \tag{5}$$

5. Massless field equations

It is easy to see that the spin $\frac{1}{2}$ case reduces back to the Dirac equation

$$D^A_{A'} f_A = \sum_{i=1}^{n} (e^i)^A_{A'} \frac{\partial f_A}{\partial x_i} = 0 \tag{6}$$

studied in Clifford analysis (see e.g. (Bureš, Souček , 1986)). Note that the contraction in the index A substitutes the usual Clifford multiplication.

To relate and to apply the wealth of results available in Clifford analysis to solutions of massless field equation with higher spin, it is important to understand the relation between the massless field equation and the so called twisted Dirac equation. By that we mean the equation

$$D(f)_{A'B\ldots E} = \sum_j (e_j)^A_{A'} \frac{\partial f_{AB\ldots E}}{\partial x_j} = 0 \tag{7}$$

for fields with values in $S_A \otimes (S_A)^{p-1}$. This system of equations decomposes into many copies of the standard Dirac equation, the term $(S_A)^{p-1}$ in the tensor product plays only auxiliary role and can be treated componentwise. The twisted Dirac equation is an elliptic system of equations, while the massless field equation is (a slightly) overdetermined system of equations. The form of the massless field equation shown in Theorem 4.2 confirms the important fact that solutionsof the massless field equations are special solutions of the twisted Dirac equation. constrained by the additional requirement that values of the field belong to $S^p \subset S_A \otimes S^{p-1}$. Let us state it explicitely.

Theorem 5.1 *The twisted massless field equation (7) restricted to the space of spinor fields with values in S^p coincides with the massless field equation (3).*

It is clear that Theorem 5.1 is a key information making possible to apply many results already known in Clifford analysis to higher spin case. We are going to show how the basic facts such as Cauchy theorem, Cauchy integral formula, Laurent series and Residue Theorem can be formulated for higher spins. We are going to formulate theorems only for even-dimensional case, small changes needed for the odd-dimensional case are easy to make.

In our notation, the basic $(n-1)$-form $d\sigma$ of Clifford analysis looks like

$$(d\sigma)^A_{A'} = \sum (-1)^{j+1}(e_j)^A_{A'} dx_1 \wedge \ldots \wedge d\widehat{x}_j \wedge \ldots \wedge dx_n. \tag{8}$$

The Cauchy theorem is an example where the classical result cannot be applied directly, but the computation goes in the same way as in the classical case.

Theorem 5.2 (Cauchy) *Let $f_{A\ldots E}$ be a function on a domain $\Omega \subset R^n$ with values in $(S_A)^p$. Then the form*

$$\omega = (d\sigma)^A_{A'} f_{A\ldots E} \tag{9}$$

is closed on Ω iff f is a solution of the massless field equation (3).

To formulate the Cauchy integral formula, let A_n denote the area of the unit sphere in R^n.

Theorem 5.3 (Cauchy integral formula) *Let $f_{A\ldots E}$ be a solution of the massless field equation (3) on a domain Ω and let $\Omega' \subset\subset \Omega$ be a relatively compact subdomain with a smooth boundary. Then for each point $x \in \Omega'$,*

$$f_{A\ldots E}(x) = \frac{1}{A_n} \int_{\partial\Omega'} \frac{(y-x)_{\{A}^{\;A'}}{|y-x|^n} (d\sigma_y)^F_{A'} f_{FB\ldots E\}}(y). \tag{10}$$

Cauchy integral formula is an example of a theorem which can be deduced by applying the theorem known in spin $\frac{1}{2}$ case (we can write the standard formula for fields twisted by S^{p-1} and to apply the projection π^{Car} to the both sides of the equation.

Weierstrass theorem or Mean Value Theorem, for example, hold without a change and they are immediate consequences of the standard version applied componentwise.

As for Taylor and Laurent series, let us recall the standard sets of monogenic functions $V_\alpha(x)$, $|\alpha| = k$ having values in the even part of the Clifford algebra (see

(Delanghe *et al.* , 1992)). In spinor notation, they belong either to S^B_A or to $S^{B'}_{A'}$. They form a basis for the space of inner spherical monogenics of degree k.

Similarly, the functions $W_\alpha, |\alpha| = k$ form a basis for outer spherical monogenics of degree k. They are vector valued, so they will be interpreted as functions with values in $S^{A'}_A \oplus S^A_{A'}$.

Theorem 5.4 (Laurent expansion) *Let a field $f_{A\ldots E}$ with values in $(S_A)^p$ be a solution of the spin $\frac{p}{2}$ massless field equation (3) in an annular domain $\Omega \subset R^n$. Then*

$$f_{A\ldots E}(x) = \sum_{k=0}^{\infty}\left(\sum_{|\alpha|=k}(V_\alpha)^F_{\{A}(x)(\mu_\alpha)_{FB\ldots E\}}\right) + \sum_{k=0}^{\infty}\left(\sum_{|\alpha|=k}(W_\alpha)^{A'}_{\{A}(x)(\mu_\alpha)_{A'B\ldots E\}}\right)$$

the convergence being normal on each closed annular subdomain. The coefficients in the expansion are given by

$$(\mu_\alpha)_{FB\ldots E} = \int_{\partial B}(W_\alpha(x))^{A'}_F(d\sigma_x)^G_{A'}f_{GB\ldots E}(x),$$

$$(\mu_\alpha)_{A'B\ldots E} = \int_{\partial B}(V_\alpha(x))^{B'}_{A'}(d\sigma_x)^F_{B'}f_{FB\ldots E}(x),$$

where the ball B is any concentric ball with its boundary inside Ω.

With Laurent series at our disposal, we can define the residue for a solution with a pointwise singularity and to prove the Residue Theorem.

Definition 5.1 *Let a field $f_{A\ldots E}$ with values in $(S_A)^p$ be a solution of the spin $\frac{p}{2}$ massless field equation (3) in $B(x)\setminus\{x\}$, where $B(x)$ is a ball with the center in x.*

Then the first coefficient $(\mu_\alpha)_{A'B\ldots E}$ of the negative part of the Laurent series is called the residue of f at the point x and denoted by $res_x(f)$.

Theorem 5.5 *Let $\Omega' \subset\subset \Omega$ be a relatively compact subdomain with a smooth boundary (oriented by its outer normal) and let $\{x_i\}_{i\in I}$ be a finite set of points in Ω.*

Then for every solution $f_{A\ldots E}$ of (3) in $\Omega \setminus (\cup_{i\in I})$ we have

$$\int_{\partial\Omega'}(d\sigma)^A_{A'}f_{A\ldots E} = \sum_{i\in I} res_{x_i}(f).$$

So we have shown how the basic amount of standard function theory can be extended to higher spin cases. It needs clearly a cartain amount of work to go through Clifford analysis and to try to extend its results to higher spins, it is not always straightforward (e.g. it is not easy to implement the standard proof of the Cauchy-Kowalewski theorem). Note also that ideas used for the proof of generalized integral formulae in (Bureš, Souček , 1985) are applicable in the case of massless field equations.

6. Twistor equations

Let us recall (see Sect.2) that the Cartan product of irreducible representations is associative and that the Cartan multiplication between Cartan powers of the spinor representations $\boxtimes : (S_A)^k \times (S_A)^l \mapsto (S_A)^{k+l}$ is commutative. Let $(S_A)^0$ be defined as the space of constants C. Then the sum $(S_A)^\infty = \oplus_{k=0}^\infty (S_A)^k$ (only finite number of terms nontrivial) is an infinite-dimensional commutative algebra.

We propose now to consider the twistor equation for functions with values in the space $(S_A)^\infty$.

Definition 6.1 *A function $f = \sum_{j=0}^\infty f_j, f_j \in (S_A)^j$ on a domain in R^n with values in $(S_A)^\infty$ is called a solution of the twistor equation (2), if $T(f) = 0$, i.e. if $T(f_j) = 0$ for all $j = 1, 2, \ldots$.*

The space of solutions of the twistor equations introduced above is an analogue of the space of all polynomials in complex function theory. Suitable completions of the space can be, of course, considered. A main and remarkable feature of this generalization of holomorphicity to higher dimensions is the fact that if the product used is the Cartan product, then the space of solutions is a commutative algebra.

Without going into a more detail study of the properties of solutions of the twistor equation, we need not use spinor algebra notation, so that formulas are kept simple and understandable.

Theorem 6.1 *If $f_j \in (S_A)^j$ and $g_k \in (S_A)^k$; $j, k = 0, 1, \ldots$ are two solutions of the twistor equation, then $f_j \boxtimes g_k = g_k \boxtimes f_j$ is again a solution of the twistor equation. Hence the space of all solutions of the twistor equation with values in V^∞ is a commutative algebra.*

To see why it is so, let $e_1, \ldots, e_n$ be an orthonormal basis, $(x_1, \ldots, x_n)$ the corresponding coordinates and $\varepsilon_1, \ldots, \varepsilon_n$ the dual basis, then

$$\begin{aligned} T(f_j \boxtimes g_k) &= \pi^{Car}(d(f_j \boxtimes g_k) = \pi^{Car}\left(\sum_i \varepsilon_i \otimes \frac{\partial(f_j \boxtimes g_k)}{\partial x_i}\right) \\ &= \left[\sum_i (\varepsilon_i \boxtimes \frac{\partial f_j}{\partial x_i}) \boxtimes g_k + \sum_i (\varepsilon_i \boxtimes (\frac{\partial g}{\partial x_i}) \boxtimes f\right] \\ &= T(f_j) \boxtimes g + T(g_k) \boxtimes f = 0 \end{aligned}$$

Few remarks are in order at this place. History of Clifford analysis started in the 30's with works of Fueter and his coworkers studying analogues of Cauchy -Riemann equations for quaternion functions. At that time and many times independently after, it was found that a natural generalization of holomorphicity in quaternionic case – namely the requirement of differentiability in quaternionic sense – is too strong condition; all differentiable functions in this sense are linear. It is worth to point out that the differentiability condition can be expressed using a set of partial differential equations similar to Cauchy-Riemann equations. This time, however, the set of equations is highly overdetermined and, as a consequence, the space of solutions is too small. In the quaternionic case, it is possible to identify functions with quaternionic values with spin $\frac{1}{2}$ fields (Fueter equation than coincides with the Dirac equation). It was pointed out some time ago ((Souček , 1986)) that (using this identification), the mentioned analogue of Cauchy-Riemann equations, describing differentiability in quaternionic sense, concides with the twistor equation for spin $\frac{1}{2}$ fields. Hence in dimension 4, we see that quaternionic differentiability, if considered for all possible spins, leads to a rich family of solutions and that the space of solutions has nice multiplicative properties. It brings back a question whether the twistor equation, considered for all spins, could not be a suitable generalization of Cauchy-Riemann equations to higher dimensions.

Another result proved in dimension 4 (see (Hitchin, 1980)) brings a new light to the discussion. It is shown in the paper that the space of solutions of the twistor equation of a given spin is the image of the 0-th order cohomology group on the twistor space of a certain homogeneity. The sum of these cohomology groups with different homogeneity forms clearly a commutative ring, so the algebra structure of the space of solutions of the twistor equation for all spins is translated by the Penrose transform to the commutative structure of polynomials in several complex variables.

There are many others interesting interlations among solutions of massless field equations, twistor equations and other conformally invariant equations which are well understood in dimension 4 (see (Penrose, Rindler , 1984)), it would be valuable to understand their generalization to higher dimensions.

References

R.Baston, M.Eastwood: 1991, *The Penrose transform: its interaction with representation theory*, Oxford University Press, Oxford

J.Bureš, V.Souček: 1985, Generalized hypercomplex analysis and its integral formulas, *Complex Variables: Theory and Applications*, 5, 53-70

J.Bureš, V.Souček: 1986, Regular spinor valued mappings, *Seminarii di Geometria*, Bologna 1984, ed. S.Coen, Bologna 1986, 7-22

R.Delanghe, F.Sommen, V.Souček: 1992, *Clifford Algebra and Spinor-Valued Functions*, Mathematics and its Applications 53, Kluwer, Dordrecht

H.D.Fegan: 1975, Conformally invariant first order differential operators, *Quat.J.Math.Oxford*, **27**, 2, 371-378

W.Fulton, J.Harris: 1991, *Representation Theory*, A First Course, GTM 129, Springer-Verlag, Heidelberg

N.J.Hitchin: 1980, Linear fields on self-dual spaces, *Proc.Royal Soc.London A*, **370**, 173-191

B.Janczewicz: 1988, *Multivectors and Clifford Algebra in Electrodynamics*, World Scientific, Singapore

R.Penrose, W.Rindler: 1984, *Spinors and Space-time I,II*, Cambridge University Press, Cambridge,

V.Souček: 1982, Holomorphicity in quaternionic analysis, Complex quaternionic analysis, connection to mathematical physics, Cauchy integral formula, 3 lectures in *Seminario di variabili complesse*, 1982, Bologna Istituto di Geometria, Univ. di Bologna (25 pages)

V.Souček: 1986, Generalized Cauchy-Riemann equations on manifolds, *Proc. of the Workshop Clifford algebra and their applications in mathematical physics*, Eds.J.Chisholm, A.Common, D.Reidel Publ. Comp., 219-227

QUATERNIONIC OPERATOR CALCULUS AND DOMAIN PERTURBATION PROBLEMS

W. SPRÖßIG
Freiberg University of Mining and Technology
Department of Mathematics
Institute of Applied Mathematics I
Bernhard-von-Cotta-Str. 2
D-09599 Freiberg

Abstract. By the aid of a special quaternionic operator calculus we consider the influence of a subdomain $G' \subset G$ on the solution of elliptic boundary value problems in the reduced domain $G \setminus \overline{G}'$ if we have on the boundary of the domain $G \setminus \overline{G}'$ corresponding conditions.

Key words: quaternionic analysis, perturbation problems, boundary value problems

1. Introduction

The numerical treatment of boundary value problems of partial differential equations makes it nessecary the given domain to approximate by a corresponding sequence of "more simple" domains. The choice of these approximating domains depends on the numerical procedure. After finite steps each procedure breaks off. In this way, we have to compare the solutions over different domains. Is the difference between the original domain and the approximating domain very small, than there are fundamental papers for instance by F. Stummel [3]. Another but also important question arises in the field of engineering sciences as follows: In what manner the solution of elliptic boundary value problems is changing after separation of a certain part of the material in the remaining part, if we determine on the new arisen pieces of the boundary corresponding boundary conditions? Similar situations we have in case of obstacles in a domain of streaming fluid. By using of quaternionic analysis we will compare the solutions of special chosen elliptic boundary value problems over different domains and give estimates.

2. Preliminaries

Let be $e_0 = 1, e_1, e_2, e_3$ the quaternionic units, which fulfil the conditions $e_i e_j + e_j e_i = -2\delta_{ij}, i, j = 1, 2, 3$. Furthermore let $G \subset \mathbb{R}^3$ be a bounded domain with the Liapunov boundary $\Gamma, u = \sum_{i=0}^{3} e_i u_i$ a quaternion-valued function and $D = \sum_{i=1}^{3} e_i D_i$ the Dirac operator with the partial derivatives D_i. Setting $\underline{u} =$ Vec $u = \sum_{i=1}^{3} e_i u_i$ then we have $Du =$ grad u_0+ rot $\underline{u}-$ div $\underline{u}$, where grad $u_0 := \sum_{i=1}^{3} D_i e_i$, rot $\underline{u} :=$

F. Brackx et al. (eds.), Clifford Algebras and their Applications in Mathematical Physics, 233–240.

$\sum_{i=1}^{3}(D_j u_k - D_k u_j)e_i, j = (i+1) \bmod 3, k = (j+1) \bmod 3$ and $\operatorname{div} \underline{u} := \sum_{i=1}^{3} D_i u_i$. The quaternion-valued functions from $\ker D$ are called H-regular. The quaternionic specimen of the functional-analytic spaces for instance $C(G), L_p(G), W_p^k(G)$ are denoted by $C(G,H), L_p(G,H) W_p^k(G,H)$ ... . Using the function $e(x) = -\frac{1}{4\pi}\sum_{i=1}^{3} x_i|x|^{-3}e_i$ we introduce the following integral operators

$$(T_G u)(x) : = \int_G e(x-y)u(y)dy, (F_G\, u)(x) := -\int_G e(x-y)n(y)u(y)d\Gamma_y \quad (x \notin \Gamma)$$

$$(S_\Gamma u)(x) : = -2\int_\Gamma e(x-y)n(y)u(y)d\Gamma_y \ (x \in \Gamma), \text{ where } n(y) = \sum_{i=1}^{3} e_i n_i(y)$$

is the unit vector of the outer normal at the point y. The integral which defines the operator S_Γ is to take in the sense of Cauchy's principle value. The operator $P_\Gamma := 1/2(I + S_\Gamma)$ denotes the projection onto the space of all H-valued functions which may be H-regular extended into the domain G. $Q_\Gamma := 1/2(I - S_\Gamma)$ denotes the projection onto the space of all H-valued functions which may be H-regular extended into the domain $\mathbb{R}^3 \setminus \overline{G}$ and vanish at infinity.

3. Some Elliptic Boundary Value Problems over Different Domains

In this part we will formulate these problems, which we have to consider afterwards. Let be $G_i, i = 1,2$ domains in $\mathbb{R}^3$ which are bounded by the Liapunov surfaces $\Gamma_i, i = 1,2$. Besides it shall be $G_2 \subset G_1$ and set $\overline{G}' := \overline{G}_1 \setminus G_2$ and $\partial G' := \Gamma'$.

3.1. Boundary Value Problem of the Dirac Operator

It is well-known (see [3]) that the first order boundary value problem $Du = f$ in G and $u = g$ on Γ is solvable if and only if the condition $tr_\Gamma T_G f = Q_\Gamma g$ is fulfilled. Now we consider this problem over the domains G_1 and G_2. Let be denoted by $g^{(i)}$ and $f^{(i)}$ the data and $u^{(i)}$ the solution of that problem over the domain G_i for $i = 1,2$. Additionally we assume that on the common boundary $\tilde{\Gamma} = \Gamma_1 \cap \Gamma_2$ the identity $g^{(1)} = g^{(2)} = \tilde{g}$ holds.

We have to calculate or estimate the quantity of the change of the solution $u^{(1)}$ after separation of the part $\overline{G}'$ at points of the remainder domain G_2.

3.2. Boundary Value Problem of First Kind of Class of Elliptic Differential Equations of Second Order

We consider the following problems

$$\begin{aligned} DM^{-1}D\,u^{(i)} &= f^{(i)} \text{ in } G_i \\ u^{(i)} &= g^{(i)} \text{ on } \Gamma_i \end{aligned}$$

for $i = 1,2$. Furthermore we suppose on the common boundary $\tilde{\Gamma} = \Gamma_1 \cap \Gamma_2$ $g^{(1)} = g^{(2)} = \tilde{g}$. The operator M acts as follows: $Mu := \sum_{k=0}^{3} m_k u_k e_k$, where m_k are real-valued functions over G_1 with $m_k(x) \neq 0$ for $x \in G_1$. Note that for $M = I$ the operator is the Laplacian and for $Mu = \frac{1}{2}\frac{m-2}{m-1}u_0 + \frac{1}{2}\sum_{k=1}^{3} u_k e_k$ DMD describes the equations of linear elasticity, where m is the Poisson number.

We wish to find the "magnitude of the changement" of the solution $u^{(1)}$ after separation of $\overline{G}'$ at points in G_2.

3.3. Stokes' Boundary Value Problem

We have it to do with the problems

$$\begin{aligned} \Delta \underline{u}^{(i)} + \beta \operatorname{grad} p^{(i)} &= f^{(i)} && \text{in } G_i \\ \operatorname{div} \underline{u}^{(i)} &= 0 && \text{in } G_i \\ \underline{u}^{(i)} &= 0 && \text{on } \Gamma_i \end{aligned}$$

for $i = 1, 2$.

We wish to express the influence of the data over the subdomain G' on the velocity $u^{(2)}$ in the subdomain G_2.

4. Treatment of Problem (3.1)

Let be (in G_2) $e := u^{(1)} - u^{(2)}$ and let suppose that our problem has a solution. Then we get

Theorem 1. Let $f^{(i)} \in L_1(G, H), g^{(i)} \in W_p^1(G, H), i = 1, 2, p > 1, f^{(2)} := f^{(1)}|_{G_2}$. Then we have

(i) $e \in \ker D(G_2)$

(ii) $e(x) = (F_{\Gamma'} g')(x) + (T_{G'} f')(x)$ with $g' := \begin{cases} g^{(1)} & \text{on } \Gamma_1 \\ g^{(2)} & \text{on } \Gamma_2 \end{cases}$ and $f' := f^{(1)}|_{G'}$

(iii) $|e(x)| \leq \frac{1}{4\pi} d^{-2}(x, \Gamma')\left[\|g'\|_{L_1(\Gamma', H)} + \|f'\|_{L_1(G', H)}\right]$ with $\Gamma' = \tilde{\Gamma} \cup (\Gamma_2 \setminus \Gamma_1)$.

Proof.

We know from [2] that in case of the solvability we have for $u^{(i)}$ the representation

$$u^{(i)} = (F_{\Gamma_i} g^{(i)})(x) + (T_{G_i} f^{(i)})(x).$$

Because of $T_{G'} f' := T_{G_1} f^{(1)} - T_{G_2} f^{(2)}$ and $F_{\Gamma'} g' := F_{\Gamma_1} g^{(1)} - F_{\Gamma_2} g^{(2)}$ we get immediately (ii). (i) follows from Theorem 2.3.1 and Theorem 2.5.5 in [1]. An easy calculation yields (iii).

Remark: There are big classes of boundary value problems of such type which can be solved. For instance we can take functions f which permit the representation $f = Dh$ with $tr_\Gamma h = 0$ and $g \in \ker P_\Gamma$. If we only assume the values of $g_0^{(i)}, i = 1, 2,$ on the boundaries and let Vect g arbitrarily then our condition can be always fulfilled.

Theorem 2: (with Gürlebeck) Let Γ' be a Liapunov surface, $f^{(i)} \in C_H(G_i), g^{(i)} \in C_H(\Gamma_i)$ for $i = 1, 2$. Then we get the estimate

$$\|e\|_{L_2(G_1, h)} \leq C_1 \|g'\|_{C(\Gamma', H)} + C_2 \|f'\|_{C(G', H)}.$$

The constants $C_i, i = 1, 2$, can be calculated explicitely by the formulas. For C_1 we get

$$C_1 = [(4\pi)^{-2}|\Gamma'|(\frac{2}{d})^{2-2\varepsilon} + \frac{|\sigma_1|}{\varepsilon}(d)^{2\varepsilon}]^{1/2}(\int_{G_1}\int_{\Gamma'} \frac{1}{|x-y|^{2+2\varepsilon}} d\Gamma_y dx)^{1/2},$$

where $0 < \varepsilon < 1/2, d$ the radius of the Liapunov ball $B_d(x')$ and $|\sigma_1|$ is the area of the projection from $\Gamma' \cap B(x')$ onto the tangential plane touching Γ' at the point $x' \in \Gamma'$ which has the shortest distance to x. For the second constant we find

$$C_2 = \frac{3}{-}\sqrt{\frac{3|G'|}{4\pi}}|G'|$$

Proof.
For the proof it is sufficient to estimate the norms of the operators $T_{G'}$ and $F_{\Gamma'}$ in the pair $[C(G', H), L_2(G', H)]$. At first we have $||T_{G'}u||_{L_2(G',H)} \leq ||T_{G'}u||_{C(G',H)}|G'|$. Because of Corollary 2.3.9 from [1] we get the constant C_2. The computation of the constant C_1 is more complicated. Let $x \in G_1$, then we find

$$|(F_{\Gamma'}g')(x)|^2 \leq \frac{1}{(4\pi)^2} \int_{\Gamma'} \frac{1}{|x-y|^{2-2\varepsilon}} d\Gamma'_y \int_{\Gamma'} \frac{|u(y)|^2}{|x-y|^{2+2\varepsilon}} d\Gamma'_y$$

where $0 < \varepsilon < 1/2$. Next we estimate the first integral. We take a point $x' \in \Gamma'$ which fulfils the equation $d(x, \Gamma') = d(x, x')$. Around x' we build a Liapunov ball with the radius d. For $d(x, \Gamma') > d/2$ it follows

$$\int_{\Gamma'} \frac{1}{|x-y|^{2-2\varepsilon}} d\Gamma_y \leq |\Gamma'|(\frac{2}{d})^{2-2\varepsilon}.$$

For $d(x, \Gamma') \leq \frac{d}{2}$ we get for the time being

$$\int_{\Gamma'} \frac{1}{|x-y|^{2-2\varepsilon}} d\Gamma'_y \leq \int_{\Gamma' \cap B_d(x')} \frac{1}{|x-y|^{2-2\varepsilon}} d\Gamma'_y + |\Gamma'|(\frac{2}{d})^{2-2\varepsilon}.$$

Further estimations require the use of local coordinates. Setting $\xi_3 := n_x$, and ξ_1, ξ_2 span the tangent plane at the point $x' \in \Gamma'$. Then after a straighforward computation we get

$$\int_{\Gamma' \cap B_d(x')} \frac{1}{|x-y|^{2-2\varepsilon}} d\Gamma'_y \leq \frac{|\sigma_1|}{\varepsilon} d^{2\varepsilon}$$

and therefore the wanted estimation for C_1.

Remark: The proof can also be given if Γ' is only a piecewise Liapunov surface.

5. Treatment of Problem (3.2)

In order to deal with this system of elliptic differential equations of second order we have to use the following orthogonal decomposition formula

$$L_2(G,H) = M(\ker D(G) \cap L_2(G,H)) \oplus_M D \overset{\circ}{W}{}_2^1 (G,H)$$

which one can find in Chapter 3 in [1]. The Hilbert space product is defined by $[u,v]_M = \int_G \overline{M^{-1}u}M^{-1}v dG$. Together with this orthogonal decomposition we also have the L_2-orthoprojections $\mathbb{P}_M, \mathcal{Q}_M$ onto the subspaces $M(\ker D(G) \cap L_2(G,H))$, $D \overset{\circ}{W}{}_2^1, (G,H)$, respectively. From Chapter 4.3 in [1] we achieve that the unique solution of Problem (B) permits the representation $Du = \mathbb{P}_M Dh + \mathcal{Q}_M Tf$, where h denotes a sufficient smooth extension of the boundary data into the domain G. Consequently we obtain for our domains G_1 and G_2 $(G_1 \supset G_2)$ the equations

$$\begin{aligned} Du^{(1)} &= \mathbb{P}_M^{(1)} Dh^{(1)} + \mathcal{Q}_M^{(1)} T_{G_1} f^{(1)} \\ Du^{(2)} &= \mathbb{P}_M^{(2)} Dh^{(2)} + \mathcal{Q}_M^{(2)} T_{G_2} f^{(2)}. \end{aligned}$$

Setting now $e := u^{(1)} - u^{(2)}$ in G_2 then we have

$$De = (\mathbb{P}_M^{(1)} Dh^{(1)} - \mathbb{P}_M^{(2)} Dh^{(2)}) + (\mathcal{Q}_M^{(1)} T_{G_1} f^{(1)} - \mathcal{Q}_M^{(2)} T_{G_2} f^{(2)}) \equiv A_1 + A_2 \quad (1)$$

At first we shall consider the item A_2. By the condition $f^{(1)}|_{G_2} = f^{(2)}$ we can easy get

$$A_2 = \mathcal{Q}_M^{(1)} T_{G'} f^{(1)} + (\mathcal{Q}_M^{(1)} - \mathcal{Q}_M^{(2)}) T_{G_2} f^{(2)}.$$

Next we will prove some elementary properties.

Lemma 1: Let $w \in W_p^k(G,H), p > 1, k \in N$. Then we have the following relations:
(i) $\mathcal{Q}_M^{(2)} \mathbb{P}_M^{(1)} w = 0$ (ii) $\mathcal{Q}_M^{(2)} \mathcal{Q}_M^{(1)} w = \mathcal{Q}_M^{(2)} w$ (iii) $\mathbb{P}_M^{(2)} \mathbb{P}_M^{(1)} w = \mathbb{P}_M^{(1)} w$
(iv) $\operatorname{im} \mathbb{P}_M^{(1)} \subset \operatorname{im} \mathbb{P}_M^{(2)}$ (v) $\operatorname{im} \mathcal{Q}_M^{(2)} \subset \operatorname{im} \mathcal{Q}_M^{(1)}$

Proof.
To prepare the proof of the first identity we shall verify that the identity $\mathcal{Q}_M^{(2)} M F_{\Gamma_2} M^{-1} \mathbb{P}_M^{(1)} w = 0$ is valid. Indeed, we have with the aid of the representation of the orthoprojection $\mathbb{P}_M^{(2)}$ by the expression $M F_{\Gamma_2} (tr_{\Gamma_2} T_{G_2} M F_{\Gamma_2})^{-1} tr_{\Gamma_2} T_{G_2}$ the following calculation

$$\begin{aligned} &\mathcal{Q}_M^{(2)} M F_{\Gamma_2} M^{-1} \mathbb{P}_M^{(1)} w = M F_{\Gamma_2} M^{-1} \mathbb{P}_M^{(1)} w - \mathbb{P}_M^{(2)} M F_{\Gamma_2} M^{-1} \mathbb{P}_M^{(1)} w = \\ &= M F_{\Gamma_2} M^{-1} \mathbb{P}_M^{(1)} w - M F_{\Gamma_2} (tr_{\Gamma_2} T_{G_2} M F_{\Gamma_2})^{-1} tr_{\Gamma_2} T_{G_2} M F_{\Gamma_2} M^{-1} \mathbb{P}_M^{(1)} w = 0. \end{aligned}$$

Applying Borel-Pompeiu's formula and using the fact $M^{-1} \mathbb{P}_M^{(1)} w \in (\ker D)(G_1)$ we obtain $\mathcal{Q}_M^{(2)} \mathbb{P}_M^{(1)} w = \mathcal{Q}_M^{(2)} M F_{\Gamma_1} M^{-1} \mathbb{P}_M^{(1)} w - \mathcal{Q}_M^{(2)} M F_{\Gamma_2} M^{-1} \mathbb{P}_M^{(1)} w =$
$\mathcal{Q}_M^{(2)} M (F_{\Gamma_1} - F_{\Gamma_2}) M^{-1} \mathbb{P}_M^{(1)} w = \mathcal{Q}_M^{(2)} M T_{G'} D M^{-1} \mathbb{P}_M^{(1)} w = 0$. The second relation

follows from (i) $\mathcal{Q}_M^{(2)}\mathcal{Q}_M^{(1)}w = \mathcal{Q}_M^{(2)}(I - \mathbb{P}_M^{(1)})w = \mathcal{Q}_M^{(2)}w$. (iii) follows immediately from (ii). The relations (iv) and (v) are obviously.

Corollary 1: Let $w \in W_p^k(G, H), p > 1, k \in N$. Then we get for the difference $(\mathcal{Q}_M^{(1)} - \mathcal{Q}_M^{(2)})w = \mathbb{P}_M^{(2)}\mathcal{Q}_M^{(1)}w$.

Proof.
The application of Lemma 1 leads for $w \in W_p^k(G, H)$ to

$$\mathcal{Q}_M^{(1)}w - \mathcal{Q}_M^{(2)}w = (\mathcal{Q}_M^{(1)} - \mathcal{Q}_M^{(2)})\mathcal{Q}_M^{(1)}w = (\mathbb{P}_M^{(2)} - \mathbb{P}_M^{(1)})\mathcal{Q}_M^{(1)}w = \mathbb{P}_M^{(2)}\mathcal{Q}_M^{(1)}w.$$

Remark:
The operators $\mathbb{P}_M^{(1)}\mathbb{P}_M^{(2)}$ and $\mathcal{Q}_M^{(1)}\mathcal{Q}_M^{(2)}$ are projections. Indeed we have $\mathbb{P}_M^{(1)}\mathbb{P}_M^{(2)}\mathbb{P}_M^{(1)}\mathbb{P}_M^{(2)} = \mathbb{P}_M^{(1)}\mathbb{P}_M^{(1)}\mathbb{P}_M^{(2)}\mathbb{P}_M^{(2)} = \mathbb{P}_M^{(1)}\mathbb{P}_M^{(2)}$ and $\mathcal{Q}_M^{(1)}\mathcal{Q}_M^{(2)}\mathcal{Q}_M^{(1)}\mathcal{Q}_M^{(2)} = \mathcal{Q}_M^{(1)}\mathcal{Q}_M^{(2)}\mathcal{Q}_M^{(2)} = \mathcal{Q}_M^{(1)}\mathcal{Q}_M^{(2)}$. Besides we have also $\mathbb{P}_M^{(2)}\mathbb{P}_M^{(1)}\mathcal{Q}_M^{(2)}\mathcal{Q}_M^{(1)} = 0$.

Lemma 2: Let u belongs to $\operatorname{im}(\mathcal{Q}_M^{(1)} - \mathcal{Q}_M^{(2)})$. Then u permits the representation $u = Dz$ with $z \in (\ker\Delta)(G_2)$ and vanishes on Γ_1.

Proof.
Let $u \in \operatorname{im}(\mathcal{Q}_M^{(1)} - \mathcal{Q}_M^{(2)})$. For $\operatorname{im}\mathcal{Q}_M^{(2)} \subset \operatorname{im}\mathcal{Q}_M^{(1)}$ we have $u \in \operatorname{im}\mathcal{Q}_M^{(1)}$. Hence there exists a function $z \in D(\overset{\circ}{W}{}_2^1(G_1, H)$ with $u = Dz$. On the other hand $u \in$ im $\mathbb{P}_M^{(2)}$ what means that u has the representation $u = Mz$ with a H-regular function z. Consequently we have $DM^{-1}u = 0$ on G_2 and therefore $DDz = -\Delta z = 0$ on the subdomain G_2.

Corollary: The image of the operator $\mathcal{Q}_M^{(1)} - \mathcal{Q}_M^{(2)}$ can be described by the relation

$$\operatorname{im}(\mathcal{Q}_M^{(1)} - \mathcal{Q}_M^{(2)}) = D(\overset{\circ}{W}{}_2^1(G_1, H)) \cap \ker\Delta(G_2).$$

Theorem 3: Let $f^{(1)} \in W_p^k(G_1, H), g^{(1)} \in W_p^{k+2-1/p}(\Gamma_1, H), g^{(2)} \in W_p^{k+2-1/p}(\Gamma_2, H)$, $g^{(1)}|_{\Gamma_2} = g^{(2)}$ and $f^{(1)}|_{G_2} = f^{(2)}$. On the subdomain G_2 we have for De the representation

$$De = \mathcal{Q}_M^{(1)}T_{G'}f^{(1)} + \mathbb{P}_M^{(2)}\mathcal{Q}_M^{(1)}(T_{G_2}f^{(1)} - Dh^{(1)}) + \mathbb{P}_M^{(2)}D(h^{(1)} - h^{(2)}), \tag{2}$$

where $h^{(i)}$ are smooth extensions of the boundary data $g^{(i)}$ into the domain G_i for $i = 1, 2$.

Proof.
Using decomposition (1) and property (iii) of Lemma 1 we obtain for A_1

$$A_1 = \mathbb{P}_M^{(2)}(\mathbb{P}_M^{(1)}Dh^{(1)} - Dh^{(2)}) = \mathbb{P}_M^{(2)}(D(h^{(1)} - h^{(2)}) - \mathbb{P}_M^{(2)}\mathcal{Q}_M^{(1)}Dh^{(1)}.$$

Together with Corollary 1 we finally get the assertion.

Remark: Assuming that in a small neighborhood of the line $\gamma = \partial\Gamma_2$ on Γ_1 as well as $g^{(1)}$ and $g^{(2)}$ are equal the same constant, then formula (2) in Theorem 3 simplifies. Indeed the extension $h^{(1)}$ can be chosen in such a manner that $h^{(1)}|_{\Gamma_2} = g_2$. Then we have instead of (2)

$$De = \mathcal{Q}_M^{(1)} T_{G'} f^{(1)} + \mathbb{P}_M^{(2)} \mathcal{Q}_M^{(1)} (T_{G_2} f^{(1)} - Dh^{(1)}) \tag{3}$$

Corollary: (i) For $g^{(1)} \equiv 0, g^{(2)} \equiv 0$ we get

$$De = \mathbb{P}_M^{(2)} \mathcal{Q}_M^{(1)} T_{G_2} f^{(2)} + \mathcal{Q}_M^{(1)} T_{G'} f^{(1)}$$

(ii) In case of $f^{(1)} \equiv 0$ in G then

$$De = -\mathbb{P}_M^{(2)} \mathcal{Q}_M^{(1)} Dh^{(1)} + \mathbb{P}_M^{(2)} D(h^{(1)} - h^{(2)})$$

Theorem 4: Let $f^{(1)} \in L_p(G, H), p > 3, g^{(2)} = C$ as well as $g^{(1)} = C$ in a neighborhood of $\partial\Gamma_2$ on Γ_1. Then we find the estimate

$$\|u^{(1)} - u^{(2)}\|_F = \|e\|_F \leq C(G_1, G_2)|G_1|^{\frac{2-3}{p-1}1/3} \|f'\|_{L_p(G',H)},$$

where $\|\cdot\|_F$ denotes the norm in the factor space $W_2^1(G_2, H)/(\ker \Delta \cap L_2)(G_2, H)$.

Proof.
As already shown one can find a in G_2 harmonic function z in that way, that the function

$$\varphi = \mathbb{P}_M^{(2)} (\mathcal{Q}_M^{(1)} (T_{G_2} f^{(1)} - Dh^{(1)}))$$

permits the representation $\varphi = Dz$ with $z \in (\ker \Delta)(G_2)$. Theorem 3 yields now

$$D(e - z) = \mathcal{Q}_M^{(1)} T_{G'} f^{(1)}$$

Let be $\lambda_1^{(2)}$ the first eigenvalue of Dirchlet's problem on G_2. Then it is easy to get the a-priori estimate

$$\|Du\|_{L_2(G_2,H)} \geq \left(\frac{\lambda_1^{(2)}}{1 + \lambda_1^{(2)}}\right)^{1/2} \|u\|_{W_2^1(G_2,H)}$$

and therefore

$$\|e - z\|_{W_2^1(G_2,H)} \leq \left(\frac{\lambda_1^{(2)} + 1}{\lambda_1^{(2)}}\right)^{1/2} \|\mathcal{Q}\|_{[C(G_2,H),L_2(G_2,H)]} \|T_{G'}\|_{[L_p(G',H),C(G_2,H)]} \|f'\|_{L_p(G'}$$

Using Proposition 2.3.10 in [1] and the factor norm we gain the statement.

Remark:
For $f^{(1)} \in L_2(G_1, H)$ we get the estimate

$$\|e\|_F \leq \|e - z\|_{W_2^1(G_2,H)} \leq \left(\frac{\lambda_1^{(2)} + 1}{\lambda_1^{(2)}}\right)^{1/2} 32\pi \text{ diam } G' \|f'\|_{L_2(G',H)}.$$

6. Treatment of Problem (3.3)

Now we have to deal with Stokes problem. Changes of the domain G_1 one can imagine as bumps of a container. From Formula 4.58 in [1] we get by setting

$$P_0^{(i)} := \text{Vec } F_{\Gamma_i}(tr_{\Gamma_i} T_{G_i} \text{ Vec } F_{\Gamma_i})^{-1} tr_{\Gamma_i} T_{G_i}$$

and $Q_0^{(i)} := I - P_0^{(i)}$ the following representation of the solution over the domain G_i

$$u^{(i)} = \rho\beta T_{G_i} \mathcal{Q}_0^{(i)} \text{ Vec } T_{G_i} f,$$

where $f = f^{(1)}$ and $f^{(2)} = f|_{G_2}$. It is easy to see that the operators $\mathbb{P}_0^{(i)}$ and $\mathcal{Q}_0^{(i)}$ are projections, where $\mathbb{P}_0^{(i)}$ denotes the $L_2(G_i, H)$-orthoprojection onto Vec (ker D) and $\mathcal{Q}_0^{(i)}$ is the orthoprojection onto Vec $D \overset{\circ}{W}{}_2^1 (G_i, H)$. After changing of the original domain G_1 to the new domain G_2 where $G_1 \supset G_2$ and $\underline{u} = 0$ on $\Gamma_2 \setminus \Gamma_1$ we obtain

$$D(u^{(1)} - u^{(2)}) = \rho\beta[\mathcal{Q}_0^{(1)} - \mathcal{Q}_0^{(2)}] \text{Vec } T_{G_1} f + \rho\beta[\mathcal{Q}_0^{(2)} \text{ Vec } T_{G'} f]$$

using Corollary 1 to Lemma 1 we have

$$D(u^{(1)} - u^{(2)}) = \rho\beta[\mathbb{P}_0^{(2)} \mathcal{Q}_0^{(1)} \text{ Vec } T_{G_1} f + \mathcal{Q}_0^{(2)} \text{ Vec } T_{G'} f]$$

Borel-Pompeiu's formula immediately yields by using the apriori-estimate in Part 5.

Theorem 5: After omission of the subdomain G' we get for the solution of Problem (3.3) over G_2 the representation

$$u^{(2)} = u^{(1)} + F_{\Gamma_2 \setminus \Gamma_1} u^{(1)} + \rho\beta T_{G_2} [\mathcal{Q}_0^{(2)} \text{ Vec } T_{G'} f + \mathbb{P}_0^{(2)} \mathcal{Q}_0^{(1)} \text{ Vec } T_{G_1} f]$$

where $F_{\Gamma_2 \setminus \Gamma_1} u^{(1)}$ denotes the Cauchy-type integral over the open surface $\Gamma_2 \setminus \Gamma_1$. Furthermore we have the estimate

$$\|u^{(1)} - u^{(2)}\|_{W_2^1(G_2,H)} \leq \sqrt{\frac{\lambda_1^{(2)} + 1}{\lambda_1^{(2)}}} [\|\mathcal{Q}_0^{(1)} \text{ Vec } T_{G_1} f\|_{L_2(G_2,H)} + \| \text{ Vec } T_{G'} f\|_{L_2(G_2,H)}]$$

where $\lambda_1^{(2)}$ denotes the first eigenvalue of Dirichlet's problem over G_2.

Remark. Similar problems arise, if it shall be investigated in numerical analysis the convergence of numerical procedures in case of boundary value problems.

References

1. Gürlebeck, K. and Sprössig, W.: 1990, "Quaternionic Analysis and Elliptic Boundary Value Problems", Birkhäuser-Verlag, Basel, Vol. 89.
2. Stummel, F.: 1976, "Perturbation of Domains in Elliptic Boundary Value Problems", In: Lecture Notes in Mathematics, Springer-Verlag, Berlin-Heidelberg-New York, Vol. 503, pp. 110-135.
3. Sprössig, W.: 1978, "Räumliches Analogon zum komplexen T-Operator, Beiträge zur Analysis 12, pp. 127-137.

CLASSICAL MECHANICS

A HAMILTONIAN MODEL OF DISSIPATION WITH CLIFFORD ALGEBRAIC GENERALIZATIONS

E.T.Y. ABOU EL DAHAB AND J. MCEWAN*
Institute of Mathematics and Statistics
University of Kent, Canterbury, Kent CT2 7NF, England.

Abstract.
This is a re-formulation, based on the real Clifford algebra $R_{1,1}$ over the real phase space $R^{1,1}$, of a recent complex unitary phase space formulation of a generalised Hamiltonian (or Birkhoffian) theory applied to the one-dimensional extended damped harmonic oscillator.

Several simplifications and clarifications are gained. There are new suggestions for generalized first quantization schemes. Generalizations to higher dimensional systems and to non-linear systems are also suggested.

1. Introduction

Dissipation is present to some extent in all practical dynamical systems. Only one of the simplest models of dissipation is studied here.

The extended one-dimensional damped harmonic oscillator (EDHO) is described by a system of second-order linear autonomous dynamical equations [1] of the form

$$\frac{d^2q_2}{dt^2} - k\frac{dq_2}{dt} + \omega_0^2 q_2 = 0$$

and

$$\frac{d^2q_1}{dt^2} + k\frac{dq_1}{dt} + \omega_0^2 q_1 = 0\,.$$

This system reduces to a first-order linear autonomous Hamiltonian system which is derivable from a conventional variational principle, is conservative and has other obvious invariance properties [2].

The damped harmonic oscillator is described by the above equation for q_1 alone, and this equation reduces to a first-order Hamiltonian system with none of the other 'nice' properties mentioned above for the EDHO. The above equation for q_2 describes runaway absorption of energy dissipated by the damped harmonic oscillator and consequently a consistent first-quantization scheme does not seem to exist for the EDHO [3].

Birkhoffian generalizations of Hamiltonian systems [4] have been developed to be applicable to non-conservative systems, and to non-local systems, and to retain many of the 'nice' properties mentioned above, and also to suggest generalized first-quantization schemes. A recent generalized Hamiltonian theory based on complex variables [5] seems to cover most of the Birkhoffian generalizations in a simpler way

* UKC/IMS/A93/5a

F. Brackx et al. (eds.), Clifford Algebras and their Applications in Mathematical Physics, 243–249.

and with clearer physical meaning. Variables with Clifford algebraic values are likely to include all the advantages of complex variables and some further advantages such as generalizability to more than two dimensions.

Section 2 introduces a formulation of the EDHO by $R_{1,1}$, and shows that the Hamiltonian consists entirely of terms associated with two constraints which also generate $GL(1)$ and $SO(2)$ symmetry groups.

Section 3 introduces a generalized Poisson bracket in order to consistently express Hamilton's equations for the EDHO in Clifford algebraic form, and hence generalized fundamental Poisson bracket relations follow.

Section 4 introduces an action functional from which Hamilton's equations and generalized Hamilton-Jacobi equations can be derived.

Section 5 suggests generalizations to higher dimensions and to non-linear systems.

2. EDHO by $R_{1,1}$

Clifford algebra $R_{1,1}$ has generators e_1, e_2 where $(e_1)^2 = -1$, $(e_2)^2 = 1$, $e_1e_2 = -e_2e_1$. Vector space $R^{1,1} \equiv span\{e_1, e_2\}$ and has elements of the forms

$$x = e_2x_2 + e_1x_1 \quad , \qquad \bar{x} = e_1xe_1 = e_2x_2 - e_1x_1 = e_2xe_2 \; .$$

The corresponding differential operators are respectively defined by

$$\partial_x = e_2\frac{\partial}{\partial x_2} + e_1\frac{\partial}{\partial x_1} \quad , \qquad \bar{\partial}_x = e_2\frac{\partial}{\partial x_2} - e_1\frac{\partial}{\partial x_1}$$

hence $\quad \partial_x x = (e_2)^2 + (e_1)^2 = 0$, $\partial_x\bar{x} = (e_2)^2 - (e_1)^2 = 2$ and $\bar{\partial}_x\bar{x} = 0$, $\bar{\partial}_x x = 2$.

The independent scalar-valued bilinear forms

$$x.y, \quad e_1e_2x\wedge y, \quad \bar{x}.y, \quad e_1e_2\bar{x}\wedge y$$

are invariant under the $GL(1)$ group of relative scale transformations defined by

$$x \to x(\alpha) = xexp\alpha, \quad y \to y(\alpha) = yexp(-\alpha), \quad 0 \le \alpha < \infty \, . \tag{1}$$

The transformations (1) are generated by $e_1e_2x\wedge y$ in the sense that

$$dx(\alpha) = x(\alpha)d\alpha = e_1e_2\bar{\partial}_y(e_1e_2x\wedge y)d\alpha \tag{2}$$

and

$$dy(\alpha) = -y(\alpha)d\alpha = e_1e_2\bar{\partial}_x(e_1e_2x\wedge y)d\alpha \, . \tag{3}$$

The bilinear forms $\bar{x}.y$ and $e_1e_2x\wedge y$ are also invariant under the $SO(2)$ group of rotations defined by

$$x \to x(\theta) = x\cos\theta - e_2xe_1\sin\theta \, ,$$

$$y \to y(\theta) = y\cos\theta - e_2ye_1\sin\theta \, , \quad 0 \le \theta < \infty \, . \tag{4}$$

The rotations (4) are generated by $\bar{x}.y$ in the sense that

$$dx(\theta) = e_2x(\theta)e_1d\theta = e_1e_2\bar{\partial}_y(\bar{x}.y)d\theta \tag{5}$$

and

$$dy(\theta) = e_2 y(\theta) e_1 d\theta = e_1 e_2 \bar{\partial}_y (\bar{x}.y) d\theta \,. \tag{6}$$

Let $\alpha = \frac{1}{2}kt$ and $\theta = \omega t$, where $0 \leq t < \infty$, and where k and ω are positive constants, then the composite of transformations (1) and rotations (4) defines the time evolution transformations

$$x \to x(t) = (x \cos \omega t - e_2 x e_1 \sin \omega t) exp(\tfrac{1}{2}kt) \,, \tag{7}$$

$$y \to y(t) = (y \cos \omega t - e_2 y e_1 \sin \omega t) exp(-\tfrac{1}{2}kt) \,, \quad 0 \leq t < \infty \tag{8}$$

which satisfy

$$\frac{dx(t)}{dt} = \frac{1}{2}kx(t) - \omega e_2 x(t) e_1 = e_1 e_2 \bar{\partial}_y (\frac{1}{2}k e_1 e_2 x \wedge y - \omega \bar{x}.y) \tag{9}$$

and

$$\frac{dy(t)}{dt} = -\frac{1}{2}ky(t) - \omega e_2 y(t) e_1 = e_1 e_2 \bar{\partial}_x (\frac{1}{2}k e_1 e_2 x \wedge y - \omega \bar{x}.y) \,. \tag{10}$$

The equations (9) and (10) for x_1, x_2, y_1 and y_2 are linearly equivalent to the conventional form of Hamilton's equations for the EDHO as follows

$$\frac{d}{dt}\begin{pmatrix} p_2 \\ q_2 \end{pmatrix} = \begin{pmatrix} -\frac{1}{2}kp_2 - \omega^2 q_1 \\ p_1 + \frac{1}{2}kq_2 \end{pmatrix} = \begin{pmatrix} 0 & -1 \\ 1 & 0 \end{pmatrix} \begin{pmatrix} \frac{\partial H}{\partial p_2} \\ \frac{\partial H}{\partial q_2} \end{pmatrix} \tag{11}$$

and

$$\frac{d}{dt}\begin{pmatrix} p_1 \\ q_1 \end{pmatrix} = \begin{pmatrix} \frac{1}{2}kp_1 - \omega^2 q_2 \\ p_2 - \frac{1}{2}kq_1 \end{pmatrix} = \begin{pmatrix} 0 & -1 \\ 1 & 0 \end{pmatrix} \begin{pmatrix} \frac{\partial H}{\partial p_1} \\ \frac{\partial H}{\partial q_1} \end{pmatrix} , \tag{12}$$

where $x = e_1 p_1 + e_2 \omega q_2$, $y = e_1 p_2 + e_2 \omega q_1$, $\omega^2 = \omega_0^2 - \frac{1}{4}k^2$ and where the Hamiltonian is

$$\begin{aligned} H &= \tfrac{1}{2}k(p_2 q_2 - p_1 q_1) + (p_1 p_2 + \omega^2 q_1 q_2) \\ &= \omega^{-1}(-\tfrac{1}{2}k e_1 e_2 x \wedge y + \omega \bar{x}.y) \,. \end{aligned} \tag{13}$$

From (9) and (10) it also follows that $\bar{x}.y$ and $e_1 e_2 x \wedge y$ are conserved in the sense that

$$\frac{d}{dt}\bar{x}.y = 0 \,, \quad \frac{d}{dt} e_1 e_2 x \wedge y = 0 \,, \tag{14}$$

and hence H in its Clifford algebraic formulation decomposes into a linear dependence on the two conserved forms $\bar{x}.y$ and $e_1 e_2 x \wedge y$.

H consists entirely of terms which could be regarded as accounting for two real constraints in the form of constant values for $\bar{x}.y$ and $e_1 e_2 x \wedge y$.

3. Generalised Symplectic Structure

If f and g are smooth functions on $R_{1,1}$ then a symplectic structure on $R_{1,1}$ is conventionally defined by a closed nondegenerate 2-form called the Poisson bracket defined by

$$[f,g] = \left(\frac{\partial f}{\partial q_1}\frac{\partial g}{\partial p_1} - \frac{\partial f}{\partial p_1}\frac{\partial g}{\partial q_1}\right) + \left(\frac{\partial f}{\partial q_2}\frac{\partial g}{\partial p_2} - \frac{\partial f}{\partial p_2}\frac{\partial g}{\partial q_2}\right)$$

$$= -\omega\left[(\bar{\partial}_x f).(e_1e_2\bar{\partial}_y g) + (\bar{\partial}_y f).(e_1e_2\bar{\partial}_x g)\right] . \tag{15}$$

For scalar functions f and g, the fundamental Poisson bracket relations follow in the conventional form

$$[q_i, p_j] = \delta_{ij} , \quad [p_1, p_2] = 0 = [q_1, q_2] \tag{16}$$

where $i, j = 1, 2$.

For vector function f and scalar function g, Hamilton's equations (9) and (10) take the forms

$$\frac{dx(t)}{dt} = \frac{1}{2}[x, H]_c , \quad \frac{dy(t)}{dt} = \frac{1}{2}[y, H]_c \tag{17}$$

where a generalised Poisson bracket is defined by

$$[f,g]_c = -\omega\left[(\bar{\partial}_x f)e_1e_2(\bar{\partial}_y g) + (\bar{\partial}_y f)e_1e_2(\bar{\partial}_x g)\right] . \tag{18}$$

The generalised fundamental Poisson bracket relations are

$$[q_i, p_j]_c = \delta_{ij} , \quad [p_1, p_2]_c = -\omega e_1e_2 = -\omega^2[q_1, q_2]_c \tag{19}$$

also $[x, y]_c = -4\omega e_1e_2$, $[x, x]_c = 0 = [y, y]_c$.

A generalised Poisson bracket with Lie-admissible properties is not new [4] but the above Clifford algebraic generalisations seem to be new and in particular the generalised fundamental Poisson bracket relations (19) seem to be new. These may be useful in the search for consistent canonical first quantization rules for dissipative systems.

4. Generalized Hamilton-Jacobi Equations

The action functional

$$S = (-\omega)^{-1}\int_{t_0}^{t}\left[\frac{1}{2}y.(e_1e_2\dot{x}) + \frac{1}{2}x.(e_1e_2\dot{y}) + \omega H\right] dt \tag{20}$$

is stationary for

$$e_1e_2\frac{dx}{dt} = \bar{\partial}_y(-\omega H) \tag{21}$$

and

$$e_1e_2\frac{dy}{dt} = \bar{\partial}_x(-\omega H) \tag{22}$$

by the conventional variational principle. Hence Hamilton's equations (9) and (10) follow from (21), (22) and (13) since (e_1e_2) is invertible. The Lagrangian in (20) includes a $GL(1)$ gauge transformation generated by $e_1e_2\bar{x} \wedge y$ and defined by

$$y.(e_1e_2\dot{x}) + x.(e_1e_2\dot{y}) = (y - \bar{y}).(e_1e_2\dot{x}) + (x - \bar{x}).(e_1e_2\dot{y}) + \frac{d}{dt}(e_1e_2\bar{x} \wedge y) \tag{23}$$

then H is the conventional Legendre transform of the conventional Lagrangian

$$\begin{aligned} L &= \frac{1}{2\omega}[(\bar{y} - y).(e_1e_2\dot{x}) + (\bar{x} - x).(e_1e\dot{y})] - H \\ &= p_1\dot{q}_1 + p_2\dot{q}_2 - H\,. \end{aligned} \tag{24}$$

The Hamilton-Jacobi problem requires the existence of an invertible transformation of variables $x(t) \to x_0(t)$, $y(t) \to y_0(t)$ and Hamiltonian $H \to 0$ defined by

$$\begin{aligned} -\omega dS &= y.(e_1e_2dx) + x.(e_1e_2dy) \\ &\quad - y_0.(e_1e_2dx_0) - x_0.(e_1e_2dy_0) + \omega H dt\,. \end{aligned} \tag{25}$$

Hence the generalized Hamilton-Jacobi equations are of the form

$$\frac{\partial S}{\partial t} + H(x, y) = 0\,, \quad \omega\bar{\partial}_x S = e_1e_2y\,, \quad \omega\bar{\partial}_y S = e_1e_2x \tag{26}$$

therefore (13) implies

$$\frac{\partial S}{\partial t} + \frac{1}{2}\omega k(\bar{\partial}_y S) \wedge (\bar{\partial}_x S) + \omega^2(\partial_y S).(\bar{\partial}_x S) = 0\,. \tag{27}$$

The bilinear $x.y$ generates an $SO(1,1)$ group of transformations hence

$$S = -(2\omega)^{-1}e_1e_2x.y - Et \tag{28}$$

solves the above Hamiltonian-Jacobi problem for any given constant value E of the Hamiltonian H.

Since $SO(1,1)$ has no finite dimensional unitary representations, conventional first quantization rules are not consistent but conventional second-quantization rules may be consistent [3].

5. Final Remarks

Only the underdamped dynamics for $\omega^2 > 0$ has been considered here although the critical case $\omega^2 = 0$ is not singular. The overdamped dynamics for $\omega^2 < 0$ introduces effectively the larger Clifford algebra $R_{1,2}$ which has a realization of the $\sqrt{-1}$ so essential for conventional quantization rules [6].

Any number n of EDHO's can be described by

$$\frac{dx(t)}{dt} = \frac{1}{2n}[x, G]_c\,, \quad \frac{dy(t)}{dt} = \frac{1}{2n}[y, G]_c\,, \quad G = \frac{1}{2}K(x \wedge y) + \bar{x}.y$$

where

$$x = \sum_{j=1}^{n}(e_{n+j}\omega_j q_j^{(2)} + e_j p_j^{(1)}) \in R^{n,n}, \quad y = \sum_{j=1}^{n}(e_{n+j}\omega_j q_j^{(1)} + e_j p_j^{(2)}) \in R^{n,n},$$

$$[f,g]_c = -[(\bar{\partial}_x f)(\bar{\bar{\partial}}_y g) + (\bar{\partial}_y f)(\bar{\bar{\partial}}_x g)],$$

$$\bar{\partial}_x = \sum_{j=1}^{n}\left(e_{n+j}\frac{\partial}{\partial x_{n+j}} - e_j\frac{\partial}{\partial x_j}\right), \quad \bar{\bar{\partial}}_x = \sum_{j=1}^{n}\omega_j\left(e_j\frac{\partial}{\partial x_{n+j}} - e_{n+j}\frac{\partial}{\partial x_j}\right),$$

and

$$K(x \wedge y) = \sum_{j=1}^{n}\omega_j^{-1}k_j(x_j y_{n+j} - x_{n+j}y_j), \quad (\bar{x}.y) = \sum_{j=1}^{n}(x_{n+j}y_{n+j} + x_j y_j).$$

If any EDHO in the above is overdamped then a further extension to $R^{n,n+1}$ is required.

A non-linear (Birkhoffian) generalization for $R_{1,1}$ introduces Birkhoff's most general scalar Pfaffian [7] of the form

$$\int_{t_0}^{t}\left(\frac{d\theta}{dt}.A(\theta) - B(\theta)\right)dt$$

which is stationary for

$$(\partial_\theta \wedge A(\theta)\frac{d\theta}{dt} = \partial_\theta B(\theta)$$

by a conventional variational principle, where B is the Birkhoffian generalization of a Hamiltonian, and A is a vector field which may be non-linear in $\theta = x \oplus y$ unlike the EDHO model for which $A(x \oplus y) = \frac{1}{2\omega}(e_1e_2y) \oplus (e_1e_2x)$. The generalized Poisson bracket for smooth scalar functions becomes

$$(\partial_\theta f)(\partial_\theta \wedge A)(\partial_\theta g)$$

for which a Jacobi identity is satisfied.

Note that $\partial_\theta \wedge A$ must be invertible to obtain a generalized form of Hamilton's equations.

Acknowledgements

Grateful thanks are due to Professor D. Hestenes for a helpful discussion on the subject of Hamiltonian mechanics by Clifford algebra. Thanks are due to The Royal Society and the University of Kent for financial support.

References

[1] Bateman, S.: 1931 *Physical Review*, Vol.38, p.815.
[2] Morse, P.M. and Feshbach, H.: 1953 *Methods of Theoretical Physics*, McGraw-Hill, New York.
[3] Celeghini, E., Rosetti, M. and Vitiello, G.: 1992 *Annals of Physics*, Vol.215, p.156.
[4] Santilli, M.: 1978, 1983 *Foundations of Theoretical Mechanics*, Vols.I,II, Springer, New York.
[5] McEwan, J.: 1993 *Foundations of Physics*, Vol.23 no.2, p.313.
Abou El Dahab, E.T.Y.: 1990 Unitary Space Formulation of Generalized Hamiltonian (Birkhoffian) Theory, Ph.D. Thesis, University of Kent, Canterbury, U.K.
[6] Adler, S.L.: 1988 *Niels Bohr, Physics and the World*: Proceedings of the Niels Bohr Centennial Symposium, Hermann Feshbach, eds., Harwood Publications, New York, p.213.
[7] Birkhoff, E.D.: 1927 *Dynamical Systems*, American Mathematical Society, Providence, Rhode Island.

A FORMULATION OF HAMILTONIAN MECHANICS USING GEOMETRIC CALCULUS

RICHARD C. PAPPAS
Department of Mathematics
Widener University
Chester, PA 19013 USA

Abstract. The Clifford-algebra-based geometric calculus developed by Hestenes and Sobczyk is used to formulate some of the key concepts and theorems of Hamiltonian mechanics. A brief comparison with other formulations is made.

1. Introduction

The Hamiltonian formalism provides the most sophisticated approach to many problems in classical mechanics, statistical mechanics, ergodic theory, optics, and quantum mechanics. There are currently three different mathematical versions of the Hamiltonian formalism, which we may designate as (1) the "analytical" version, (2) the "symplectic" version, and (3) the "differential forms" version. Each has some advantages, but also some drawbacks, which are briefly summarized below. We then present an outline of a new mathematical framework for Hamiltonian mechanics: (4) the "geometric calculus" version.

(1) The *analytical* version studies Hamilton's equations

$$\dot{q}_i = \frac{\partial H}{\partial p_i}, \quad \dot{p}_i = -\frac{\partial H}{\partial q_i} \tag{1}$$

as first order differential equations, obtaining integrals by using the invariance of the equations under canonical transformations. Geometric concepts, such as "phase space", may be introduced, but they often remain little more than convenient terminology. By the standards of modern theoretical physics, this version of Hamiltonian mechanics does not adequately exploit the geometric content of the subject. The analytical version is ably expounded by Whittaker [1].

(2) The *symplectic* version identifies the geometric structure of the phase space of a Hamiltonian system more explicitly. Hamilton's equations appear in matrix form as

$$\dot{z} = K\frac{\partial H}{\partial z}, \tag{2}$$

where z is a vector in a $2n$-dimensional space, K is the $2n \times 2n$ matrix

$$K = \begin{pmatrix} 0 & I \\ -I & 0 \end{pmatrix}$$

F. Brackx et al. (eds.), Clifford Algebras and their Applications in Mathematical Physics, 251–258.

and I is the $n \times n$ identity matrix. One studies transformations whose Jacobian matrix J satisfies

$$J^T K J = K;$$

i.e., J is a symplectic matrix. Corben and Stehle [2] discuss Hamiltonian mechanics in this way. Although geometric concepts receive more attention than in the analytical version, the technical tools (matrices, etc) do not provide the coordinate-independent formulation that modern differential geometers and theoretical physicists prefer.

(3) The *differential forms* version defines the symplectic structure of phase space by a closed nondegenerate differential 2-form $\omega = \sum dp_i \wedge dq_i$. If $\imath$ is an isomorphism associating one-forms with vectors, then Hamilton's equations can be written in the coordinate-independent form

$$\dot{x} = \imath(dH(x)). \tag{3}$$

Canonical transformations are mappings of the phase space that preserve ω. The differential forms version is now widely held to be the most appropriate one; indeed, V. Arnold has stated ([3], p. 163) that "Hamiltonian mechanics cannot be understood without differential forms"! Presumably, this is because the phase space of many physical systems is a manifold, rather than simply (a subset of) a euclidean space, and the modern approach to calculus on manifolds is via differential forms.

However, there are both conceptual problems and practical difficulties with exterior differential calculus as a tool in mechanics, as has been emphasized by Hestenes ([4], [5]). It can be argued, for example, that the standard treatment of calculus on manifolds, with its charts, atlases, tangent bundles, tensor fields, metrics, etc, is not a particularly economical formulation. Moreover, it is not truly coordinate-free, since the basic definition of a manifold involves mappings to open sets in R^n. Finally, the calculus of differential forms does not contain the vector calculus in terms of which Newtonian mechanics is formulated in a natural way.

Geometric calculus, developed by Hestenes and Sobczyk [5], does contain all of vector calculus and also permits a coordinate-free approach to manifolds. In the remainder of this paper, we briefly recall some elements of the Clifford-algebra-based concept of a *vector manifold*, and then show how a symplectic structure suitable for Hamiltonian mechanics can be defined on such a manifold. Limitations of space allow only the treatment of selected topics; see also the independent discussion by Hestenes in [4].

2. Symplectic Structure on Vector Manifolds

Recall that a *vector manifold* $\mathcal{M}$ is a set whose points are assumed to be vectors on which a Clifford algebra has been defined. The difference of two vectors is called a *chord*, and the limit of a sequence of chords defines a *tangent vector*. The set of all vectors tangent to $\mathcal{M}$ at a point x is the *tangent space* at x. This tangent space itself generates a Clifford algebra, the *tangent algebra* of $\mathcal{M}$ at x. The projection of any multivector A into the tangent algebra at x is $P(A) = [A \cdot I(x)] \cdot I^{-1}(x)$, where $I(x)$ is the unit pseudoscalar of the tangent algebra. If $P(A) = A$, then A is said to be tangent to $\mathcal{M}$ at x. A multivector function which is tangent to $\mathcal{M}$ at every point is called a *field* on $\mathcal{M}$. (Further details about the italicized terms may be found in [5].)

Let $\mathcal{M}^{2m}$ be the phase space of a physical system; each vector x represents a state of the system. Then a bivector field $G(x)$ determines a skew-symmetric bilinear form g on $\mathcal{M}^{2m}$, namely,

$$\begin{aligned} g(x,y) &= x\cdot(y\cdot G) \\ &= (x\wedge y)\cdot G \end{aligned} \tag{4}$$

The bivector field G plays the role of the 2-form ω in defining a symplectic structure.

3. Differential Structure and Hamilton's Equations

The algebraic structure of a vector manifold makes it possible to define differentiation with respect to a vector, denoted by ∂ or ∂_x . If A is a field, ∂A is not necessarily tangent to $\mathcal{M}^{2m}$; applying the projection operator P, we define the *coderivative* as

$$\begin{aligned} \nabla A &\equiv P(\partial A) \\ &= P(\partial{\cdot}A + \partial\wedge A) \\ &\equiv \nabla A + \nabla\wedge A \end{aligned} \tag{5}$$

For scalar fields, $\nabla f = \partial f$. The *cocurl* of A, $\nabla\wedge A$, corresponds to the exterior differential. (See [5], pp. 155f.)

Note that ∂ is assumed to have the algebraic properties of a vector field on $\mathcal{M}$. This allows us to deduce a large number of useful differential identities from corresponding algebraic identities, giving geometric calculus considerable computational power. As an example of this process, we prove the first of the following Lemmas, which state identities that we will need in section 4. G is a bivector field and T is a trivector field.

Lemma 1: $G\wedge(G\cdot\partial) = -\frac{1}{2}(G\wedge G)\cdot\partial$

Proof: From equation (1-1.42) on p. 12 of [5], we have

$$a\cdot(A_r\wedge B_s) = (a\cdot A_r)\wedge B_s + (-1)^r A_r\wedge(a\cdot B_s)$$

where a is a vector, A_r is an r-vector, and B_s is an s-vector. This can be rearranged, using (1.23a,b) of [5], as

$$(A_r\wedge B_s)\cdot a = (-1)^{rs} B_s\wedge(A_r\cdot a) + A_r\wedge(B_s\cdot a)$$

Now put $A_r = B_s = G$, $a = \partial$, $r = s = 2$. □

Similarly, the following can be deduced from equation (1-1.25b) of [5].

Lemma 2: $(G\wedge G)\cdot(\partial\wedge T) = ((G\wedge G)\cdot\partial)\cdot T$

The Hamiltonian $H(x)$ determines a *gradient system*, that is, a system of autonomous differential equations

$$\dot{x} = \partial H{\cdot}G, \tag{6}$$

where the overdot indicates the derivative with respect to a parameter, t. If $x = x(t)$ is an integral curve of (6), and $f(x)$ is any function, one easily shows that

$$\dot{f} = (\partial f \wedge \partial H) \cdot G. \tag{7}$$

Thus, $\dot{H} = 0$ ("conservation of energy").

In order to see that (6) represents Hamilton's equations, consider the case $\mathcal{M}^{2m} = R^{2m}$. Then the bivector G can be decomposed into a sum

$$G = \sum_i e_{p_i} \wedge e_{q_i} \tag{8}$$

where the sets of vectors $\{e_p\}$ and $\{e_q\}$ are orthonormal and mutually orthogonal. Writing

$$\partial H = \sum_i \left(\frac{\partial H}{\partial q_i} e_{q_i} + \frac{\partial H}{\partial p_i} e_{p_i} \right) \tag{9}$$

the gradient system (6) assumes the form (1).

4. Poissson Brackets

Let $f(x)$ and $g(x)$ be scalar-valued functions on $\mathcal{M}^{2m}$. The Poisson bracket of f and g is defined as

$$\{f, g\} = (\partial f \wedge \partial g) \cdot G \tag{10}$$

The following properties can easily be demonstrated:

$$\begin{aligned} \{f, g\} &= -\{g, f\} && (11) \\ \{\lambda f + \mu g, h\} &= \lambda\{f, h\} + \mu\{g, h\} && (12) \\ \{fg, h\} &= f\{g, h\} + g\{f, h\} && (13) \end{aligned}$$

where λ and μ are constants.

However, in order to have the Poisson bracket of scalar-valued functions form a Lie algebra, we also need the Jacobi identity

$$\{\{f, g\}, h\} + \{\{g, h\}, f\} + \{\{h, f\}, g\} = 0. \tag{14}$$

This imposes a condition on the bivector field G, the derivation of which using geometric calculus is instructive.

The definition (10) may be rearranged as follows:

$$\begin{aligned} \{f, g\} &= \partial f \cdot (\partial g \cdot G) \\ &= (\partial g \cdot G) \cdot \partial f \\ &= -(G \cdot \partial g) \cdot \partial f \\ &\equiv -b \cdot \partial f && (15) \end{aligned}$$

where we have temporarily put $b = (G \cdot \partial g)$. Another useful rearrangement is

$$\begin{aligned} \{f, g\} &= [\partial \wedge (f \partial g)] \cdot G \\ &= G \cdot [\partial \wedge (f \partial g)] \\ &= (G \cdot \partial) \cdot (f \partial g) && (16) \end{aligned}$$

A typical term on the left side of the Jacobi identity (14) is then

$$\begin{aligned}\{\{f,g\},h\} &= \{-b\cdot\partial f, h\}\\ &= -(G\cdot\partial)\cdot[(b\cdot\partial f)\partial h],\end{aligned}\tag{17}$$

using both (15) and (16).

Similarly, with $a \equiv G\cdot\partial f$ and $c \equiv G\cdot\partial h$, we obtain for the entire left-hand side of (14)

$$-(G\cdot\partial)\cdot[(b\cdot\partial f)\partial h + (c\cdot\partial g)\partial f + (a\cdot\partial h)\partial g],$$

which can be rewritten as

$$-(G\cdot\partial)\cdot[G\cdot\{(\partial g\wedge\partial f)\partial h + (\partial h\wedge\partial g)\partial f + (\partial f\wedge\partial h)\partial g\}].$$

The three terms in square brackets can be collected using the Clifford algebra identity

$$G\cdot(p\wedge q\wedge r) = G\cdot(p\wedge q)r + G\cdot(q\wedge r)p + G\cdot(r\wedge p)q,\tag{18}$$

for bivector G and vectors p,q,r. Thus, putting $T \equiv \partial f\wedge\partial g\wedge\partial h$, the left side of (14) becomes

$$-(G\cdot\partial)\cdot(G\cdot T),$$

which can be expanded using Lemmas 1 and 2 above as follows:

$$\begin{aligned}-(G\cdot\partial)\cdot(G\cdot T) &= T\cdot[(G\cdot\partial)\wedge G] + [G\wedge(G\cdot\partial)]\cdot T\\ &= T\cdot[(G\cdot\partial)\wedge G] - \frac{1}{2}(G\wedge G)\cdot(\partial\wedge T)\\ &= T\cdot[(G\cdot\partial)\wedge G].\end{aligned}\tag{19}$$

The last line uses the fact that $\partial\wedge T = 0$. Hence the Jacobi identity holds iff (18) vanishes; i.e. iff

$$P((G\cdot\partial)\wedge G) = (G\cdot\nabla)\wedge G = 0.\tag{20}$$

Note that this result is equivalent to the vanishing of the so-called "Schouten-Nijenhuis bracket" of G with itself. Lichnerowicz [6] uses this to define a "Poisson manifold", which is more general than a symplectic manifold in that G need not be of maximal rank. A cursory examination of his work suggests that geometric calculus should be a useful tool in discussing Poisson manifolds and their Lie algebras.

To see that (10) is equivalent to the Poisson bracket as defined in the analytical version of Hamiltonian mechanics, it is sufficient to use (8) and the equivalent of (9) for ∂f and ∂g:

$$\begin{aligned}\{f,g\} &= (\partial f\wedge\partial g)\cdot G\\ &= \partial f\cdot(\partial g\cdot G)\\ &= \sum_i\left(\frac{\partial f}{\partial q_i}\frac{\partial g}{\partial p_i} - \frac{\partial g}{\partial q_i}\frac{\partial f}{\partial p_i}\right)\end{aligned}$$

5. Integral Invariants

Suppose that M is a vector field on $\mathcal{M}^{2m}$ and γ is a parametrized curve. Then we will say that the integral

$$J = \int_\gamma dx \cdot M = \int_\gamma M \cdot dx \tag{21}$$

is an integral invariant of a vector field X if $dJ/dt = 0$ along the flow of X, defined by $\dot{x} = X$; that is, if the value of J along a curve γ is the same as along the image γ' under the flow of X.

The condition on M for J to be an integral invariant may be derived as follows. Let s be a parameter chosen so that, for fixed t, γ is defined as s goes from 0 to 1; and for fixed s, the same point is located on the trajectory for all t. Then $x = x(s,t)$ and $\partial x/\partial t = X$. Writing the integral as

$$J = \int_0^1 M \cdot \frac{\partial x}{\partial s} ds, \tag{22}$$

we compute

$$\begin{aligned} \frac{dJ}{dt} &= \int_0^1 \frac{\partial}{\partial t}\left(M \cdot \frac{\partial x}{\partial s}\right) ds \\ &= \int_0^1 \left(\frac{\partial M}{\partial t} \cdot \frac{\partial x}{\partial s} + M \cdot \frac{\partial X}{\partial s}\right) ds \\ &= \int_0^1 [(X \cdot \partial)M + \dot{\partial}(M \cdot \dot{X})] \cdot \frac{\partial x}{\partial s} ds \\ &= \int_\gamma [(X \cdot \partial)M + \dot{\partial}(M \cdot \dot{X})] \cdot dx, \end{aligned}$$

where the overdot indicates the action of ∂. But we have ([5], p. 53)

$$\partial(M \cdot X) = X \cdot \partial M + M \cdot \partial X - M \cdot (\partial \wedge X) - X \cdot (\partial \wedge M) \tag{23}$$

so that

$$\dot{\partial}(M \cdot \dot{X}) = M \cdot \partial X - M \cdot (\partial \wedge X),$$

which allows us to write the integrand above as

$$\begin{aligned} X \cdot \partial M + \dot{\partial}(M \cdot \dot{X}) &= X \cdot \partial M + M \cdot \partial X - M \cdot (\partial \wedge X) \\ &= \partial(M \cdot X) + X \cdot (\partial \wedge M). \end{aligned}$$

Therefore,

$$\frac{dJ}{dt} = M \cdot X \,|_\gamma + \int_\gamma [X \cdot (\partial \wedge M)] \cdot dx \tag{24}$$

and M is an integral invariant of X iff this expression is zero for every γ. Sufficient conditions are $M \cdot X = 0$ and $X \cdot (\partial \wedge M) = 0$.

If γ is closed, the first term in (24) vanishes and we have the result that M is a (relative) integral invariant of X iff

$$\oint [X \cdot (\partial \wedge M)] \cdot dx = 0 \qquad \text{for every closed } \gamma, \tag{25}$$

or iff

$$X \cdot (\partial \wedge M) = \partial \phi \qquad \text{for some function } \phi. \tag{26}$$

As an example, consider the Hamiltonian $H = \frac{1}{2}(u^2 + v^2) + \frac{1}{2}y^2$ on R^4 . Then $M = \langle u, v, 0, 0\rangle$ is a relative integral invariant of the field $X = \nabla H \cdot G$. In fact, $X = \langle u, v, 0, -y\rangle$ and

$$X \cdot (\nabla \wedge M) = -\nabla H.$$

This example of course is a special case of the general theorem:

Theorem 1: If p_i and q_i are canonical coordinates for a phase space, then $M = \langle p_1, p_2, \ldots, 0, \ldots, 0\rangle$ is a relative integral invariant of the Hamiltonian field ∇H.

Recalling that Stokes' theorem can be written as

$$\int dV \cdot (\nabla \wedge M) = \oint dS \cdot M$$

where dV and dS are the directed volume elements on $\mathcal{M}$ and $\partial\mathcal{M}$ (see [5], p. 258), we deduce the well-known result

Theorem 2: If M is a relative integral invariant of a vector field X, then $\nabla \wedge M$ is an absolute integral invariant of X.

Theorems 1 and 2 together imply

Theorem 3: The bivector field G defining the symplectic structure on $\mathcal{M}$ is an absolute integral invariant of any Hamiltonian vector field.

6. Canonical Transformations

A mapping of $\mathcal{M}^{2m}$ into itself which has G as an integral invariant is called a canonical transformation. One can easily show that if M and N are integral invariants of a vector field X, then $M \wedge N$ is also an integral invariant of X. But the volume element on $\mathcal{M}^{2m}$ is proportional to $G^m = G \wedge \ldots \wedge G$ (m factors). Hence we have

Liouville's Theorem: Volume in phase space is preserved under canonical transformations.

7. Comments

We have indicated how the geometric calculus used to describe Newtonian mechanics in [7] can be used to formulate Hamiltonian mechanics. The Clifford-algebra-based geometric calculus seems more economical that the differential forms approach, in that it eliminates the complication of considering both the tangent space and the cotangent space; the duality between forms and vectors does not really contribute anything essential to the understanding of mechanics. And although our presentation has emphasized the coordinate-free formulation, we have shown that standard coordinate-dependent results are easily recoverable.

Further work along these lines might involve (a) use of geometric calculus to construct a canonical formalism for the spinor mechanics in [7] and applying the results to perturbation theory in celestial mechanics; (b) investigation of the transition from classical Hamiltonian mechanics to quantum mechanics using geometric calculus; or (c) pursuing the geometric structure of Poisson manifolds using geometric calculus.

8. Acknowledgements

The author would like to thank the Provost and President of Widener University for a partial leave of absence during which this work was completed, and J. Lysko for indispensable help in the preparation of the manuscript.

References

1. E.T. Whittaker (1937), *A Treatise on the Analytical Dynamics of Particles and Rigid Bodies*, Cambridge Univ. Press, London/New York.
2. H.C. Corben and P. Stehle (1960), *Classical Mechanics*, John Wiley & Sons, Inc., New York.
3. V.I. Arnold (1978), *Mathematical Methods of Classical Mechanics*, Springer-Verlag, New York.
4. D. Hestenes (1993), "Hamiltonian Mechanics with Geometric Calculus", in *Spinors, Twistors and Clifford Algebras*, Z. Iziewicz, A. Borowicz, and B. Jancewicz (eds) [To be published by Kluwer Acad. Publishers.]
5. D. Hestenes and G. Sobczyk (1984), *Clifford Algebra to Geometric Calculus*, D. Reidel Publishing Co., Dordrecht/Boston.
6. A Lichnerowicz (1977), "Les variétés de Poisson et leurs algèbres de Lie associées", *J. Differential Geometry* 12, 253-300.
7. D. Hestenes (1986), *New Foundations for Classical Mechanics*, D. Reidel Publishing Co., Dordrecht/Boston.

MATHEMATICAL PHYSICS

LOCAL AUTOMORPHISM INVARIANCE: A GENERALIZATION OF GENERAL RELATIVITY

JAMES P. CRAWFORD
Department of Physics
Penn State - Fayette
Uniontown, PA 15401 USA

Abstract. I propose a generalization of the Principle of Equivalence, upon which General Relativity is founded. General Relativity may be viewed as a gauge theory of local Lorentz invariance in the tangent space. The condition that the covariant derivatives of the Clifford algebra generators vanish, $\nabla_\mu \gamma_\nu = 0$, then insures that the Riemann condition is satisfied, $\nabla_\mu g_{\rho\sigma} = 0$, and that the Riemann curvature tensor is identical to the field strength tensor for the gauge fields. I propose to extend local Lorentz invariance to the further demand of local automorphism invariance of the Clifford algebra. Local automorphism invariance is equivalent to allowing the basis spinors of the spinor space to be chosen locally, just as local Lorentz invariance is equivalent to allowing the basis vectors of the tangent space to be chosen locally. Consequently, the Clifford algebra generators are found not to have vanishing total covariant derivative, that is $\nabla_\mu \gamma_\nu \neq 0$, in contrast to the case of General Relativity. The additional dynamical degrees of freedom of the Clifford algebra generators may be conveniently prescribed to reside in *drehbeins* ("spin-legs"), these constituting the transformation matrix between the local spinor basis and an arbitrarily assigned but constant spinor basis. For the case of four-dimensional spacetime considered herein, the automorphism group is the conformal group, and therefore this theory may be viewed as a spinor inspired conformal extension of General Relativity. However, if we treat this theory as an ordinary gauge theory, then the action does not contain the Einstein-Hilbert term (that is, a term linear in the Ricci scalar). Nevertheless, I argue that the Einstein-Hilbert action, along with the mass of the spinor field and masses for some of the gauge fields, may all arise together as a symmetry breaking phenomenon in which the ground state is only invariant under Lorentz transformations.

Key words: Clifford algebra, Automorphism Invariance, General Relativity

1. Introduction: Clifford Algebra in General Relativity

The incorporation of spinors into general relativity [1] was originally considered in 1929 separately by Schrödinger [2], Weyl [3], and Fock and Ivanenko [4]. The idea of general relativity as a gauge theory based on local Poincaré invariance was initiated by Utiyama [5], Kibble [6], and Sciama [7], and more recently developed by several groups [8]. In addition, the past few decades have witnessed a wide variety of related activity including the work on local $Sl(2, C)$ invariance [9], twistors [10], Ashtekar variables [11], and spin-gauge theory [12].

Clifford algebras occur in any theory which involves fermionic (spinor) fields, including electrons, neutrinos, and quarks, and are therefore ubiquitous in fundamental

F. Brackx et al. (eds.), *Clifford Algebras and their Applications in Mathematical Physics*, 261–268.

physical theories. Their basic defining relation is given by the familiar expression:

$$\{\gamma_\alpha, \gamma_\beta\} = 2\eta_{\alpha\beta}1 \qquad (1)$$

where the γ_α are the generators of the algebra, and the $\eta_{\alpha\beta}$ are the components of the metric of the tangent space. For a suitable choice of basis vectors the metric components may be taken to be in Minkowskian form. We will refer to such a basis as orthonormal, and for the specific case of four-dimensional spacetime considered here we have $\eta_{\alpha\beta} = diag(1, -1, -1, -1)$.

If we consider a matrix representation of this algebra, we may envision the spinor fields to be lying in the carrier space of the representation, and in this case the spinor fields contain four complex components. Then the Clifford algebra generators may be considered to be the objects which connect a spinor basis of the spinor space to an orthonormal basis of the tangent space of spacetime, and hence are sometimes referred to as "soldering forms." Once a particular matrix representation for the Clifford algebra generators has been chosen it is canonically assumed to be the same at each point in spacetime, thereby fixing the relationship between the tangent space basis and the spinor space basis. The various matrix representations of the Clifford algebra are related by similarity transformations; in this case the similarity matrices are elements of $\mathcal{G}l(4, C)$.

In General Relativity we consider both the spacetime metric components and the Clifford algebra generators to be functions of spacetime:

$$\{\gamma_\mu(x), \gamma_\nu(x)\} = 2g_{\mu\nu}(x)1 \qquad (2)$$

However, the spacetime dependence of both may be conveniently assumed to reside entirely in the vierbeins:

$$\gamma_\mu(x) = h^\alpha_{\ \mu}(x)\gamma_\alpha, \quad g_{\mu\nu}(x) = h^\alpha_{\ \mu}(x)h^\beta_{\ \nu}(x)\eta_{\alpha\beta} \qquad (1.3a,b)$$

The vierbeins may be viewed as constituting the transformation matrix between the coordinate basis and an orthonormal basis of the tangent space, and may therefore be used to transform components of a vector from one basis to another:

$$V_\mu = h^\alpha_{\ \mu}V_\alpha, \quad V_\alpha = h_\alpha^{\ \mu}V_\mu \qquad (1.4a,b)$$

Consequently, just as the Clifford algebra generators γ_α "solder" the spinor space basis to the orthonormal tangent space basis, the vierbeins "solder" the orthonormal basis to the coordinate basis of the tangent space, and are accordingly also referred to as "soldering forms."

The Theory of General Relativity is based upon the complementary notions of general coordinate covariance and local Lorentz invariance. General coordinate covariance follows from the demand that any physical theory cannot depend on the particular choice of coordinates, and local Lorentz invariance follows from the demand that any physical theory cannot depend on the particular orientation of the orthonormal basis of the tangent space. These conditions require introducing the affine connection $\Gamma^\rho_{\ \nu\mu}$ to define the covariant derivatives of the vector components in the coordinate basis, and the spin connection $\omega^\beta_{\ \alpha\mu}$ to define the covariant derivatives of the vector components in the orthonormal basis:

$$\nabla_\mu V_\nu \equiv \partial_\mu V_\nu - \Gamma^\rho_{\ \nu\mu}V_\rho, \quad \nabla_\mu V_\alpha \equiv \partial_\mu V_\alpha + \omega^\beta_{\ \alpha\mu}V_\beta \qquad (1.5a,b)$$

Equivalently, the spin connection may be considered to be the gauge field of local Lorentz invariance.

Both the Riemann curvature tensor and the field strength tensor of the gauge field may be obtained by considering the effects of parallel transport of a vector around a closed loop:

$$R^{\sigma}{}_{\rho\mu\nu} = -\partial_\mu \Gamma^{\sigma}{}_{\rho\nu} + \partial_\nu \Gamma^{\sigma}{}_{\rho\mu} - \Gamma^{\sigma}{}_{\lambda\mu}\Gamma^{\lambda}{}_{\rho\nu} + \Gamma^{\sigma}{}_{\lambda\nu}\Gamma^{\lambda}{}_{\rho\mu} \tag{1.6a}$$

$$\Omega^{\alpha}{}_{\gamma\mu\nu} = \partial_\mu \omega^{\alpha}{}_{\gamma\nu} - \partial_\nu \omega^{\alpha}{}_{\gamma\mu} + \omega^{\alpha}{}_{\delta\mu}\omega^{\delta}{}_{\gamma\nu} - \omega^{\alpha}{}_{\delta\nu}\omega^{\delta}{}_{\gamma\mu} \tag{1.6b}$$

The Riemann condition, $\nabla_\mu g_{\rho\sigma} = 0$, or more precisely the vanishing of the covariant derivative of the vierbein, yields the identification of these two tensors:

$$\nabla_\mu h^{\alpha}_{\nu} = 0 \Rightarrow R^{\alpha}{}_{\rho\mu\nu} = \Omega^{\alpha}{}_{\rho\mu\nu} \tag{7}$$

This establishes General Relativity as a gauge theory based on local Lorentz invariance.

The covariant derivative of a spinor field is written as:

$$\nabla_\mu \Psi = \partial_\mu \Psi + i\omega_\mu \Psi, \quad \omega_\mu \equiv \frac{1}{2}\omega^{\alpha\beta}{}_{\mu}\gamma_{\alpha\beta}, \quad \gamma_{\alpha\beta} \equiv \frac{i}{2}[\gamma_\alpha, \gamma_\beta] \tag{1.8a,b,c}$$

where $\gamma_{\alpha\beta}$ are the generators of the Lorentz transformations in the spinor basis. For the covariant derivatives of the Clifford algebra generators we have:

$$\nabla_\mu \gamma_\alpha = \partial_\mu \gamma_\alpha + \omega^{\beta}{}_{\alpha\mu}\gamma_\beta + i[\omega_\mu, \gamma_\alpha] = \partial_\mu \gamma_\alpha \tag{9}$$

which says that the Clifford algebra generators are Lorentz scalars; stated differently, under a Lorentz transformation the spinor part of the transformation exactly cancels the vector part of the transformation. Therefore the Clifford algebra generators (written in orthonormal basis) may be taken to be constant (non-dynamic) quantities. Then with equation (1.7) we have an alternate statement of the Riemann condition: $\nabla_\mu \gamma_\nu = 0$.

The action for gravity coupled to a massive spinor field may be written as:

$$S = \int \sqrt{-g}\, d^4x \{ -\frac{1}{16\pi G} R + \overline{\Psi}(i\gamma^\mu \nabla_\mu - m)\Psi \} \tag{10}$$

where the Ricci curvature scalar is defined in the usual way: $R = R^{\mu\nu}{}_{\mu\nu}$, g is the determinant of the metric tensor: $g = det[g_{\mu\nu}]$, and G is the gravitational constant. Notice, however, that this action is rather peculiar from the point of view of gauge theory. In particular, the field strength tensor appears linearly, and not quadratically, in the action. In the local automorphism invariant generalization, we will begin with a quadratic action, and recover the linear action as a symmetry breaking effect.

2. Local Automorphism Invariance

The spinor space may be considered to be endowed with a metric γ [13]. In the case of four-dimensional spacetime being considered here, this metric is given by $\gamma = \gamma_0$

as may be seen by inspection of the scalar bilinear covariant: $\Psi^{\not{+}}\gamma_0\Psi \equiv \overline{\Psi}\Psi$. Note that the symbol $\not{+}$ represents hermitian conjugation, and we have defined the Dirac conjugate (bar) in the usual way.

Under a local Lorentz transformation, all representations of the Lorentz group transform. Specifically, the Lorentz transformation not only acts on the orthonormal basis of the tangent space, but it also acts on the spinor basis of the spinor space, and this is reflected in equations (1.5b), (1.8a) and (1.9). In addition, the Lorentz transformations not only leave the Minkowski metric $\eta_{\alpha\beta}$ invariant (by definition) but also leave the metric on the spinor space γ invariant. Now although the Lorentz group is the largest group of transformations leaving the Minkowski metric invariant, there is a larger group which leaves the spinor metric invariant: $SU(2,2)$ [13,14]. This group is the metric preserving automorphism group, since the action of this group is to "mix up" the basis elements of the full Clifford algebra. Henceforth we shall refer to this group simply as the automorphism group. Of course, this group is contained in the full group of similarity transformations $\mathcal{G}l(4,C)$, but an arbitrary element of $\mathcal{G}l(4,C)$ will not in general leave the spinor metric invariant. Finally note that for four-dimensional spacetime the automorphism group is the conformal group.

Now we may also consider the action of the automorphism group to be a transformation on the basis spinors of the spinor space, just as the action of the Lorentz group may be considered to be a transformation of the basis vectors of the tangent space. Surely just as any reasonable physical theory cannot depend on the particular orientation of the basis vectors of the tangent space, it also should not depend on the particular orientation of the basis spinors of the spinor space. This is equivalent to the statement that the particular matrix representation chosen for the Clifford algebra generators cannot effect the physical predictions of a theory. In fact, one can show that the usual Dirac lagrangian density (the second term in equation (1.10)) is indeed invariant under global automorphism transformations. However, the basis vectors of the tangent space may be *locally* assigned. I propose that the basis spinors should also be allowed to be *locally* assigned.

This demand of local automorphism invariance [15] requires the introduction of a gauge field. This gauge field may also be considered to be a connection, and I propose that it be referred to as the Clifford connection. Then the covariant derivative of a spinor may be written:

$$\nabla_\mu\Psi = (\partial_\mu + i\omega_\mu + iA_\mu)\Psi \equiv (\partial_\mu + i\Phi_\mu)\Psi \tag{1}$$

where the Clifford connection has the following structure [16]:

$$A_\mu = a^\alpha{}_\mu\gamma_\alpha + \frac{1}{2}a^{\alpha\beta}{}_\mu\gamma_{\alpha\beta} - b^\alpha{}_\mu\tilde{\gamma}_\alpha - b_\mu\tilde{\gamma} \equiv A^\alpha{}_\mu\Gamma_\alpha \tag{2}$$

Notice that the spin connection and the bivector component of the Clifford connection contribute in exactly the same way (see equation (1.8)):

$$\Phi_\mu = a^\alpha{}_\mu\gamma_\alpha + \frac{1}{2}(\omega^{\alpha\beta}{}_\mu + a^{\alpha\beta}{}_\mu)\gamma_{\alpha\beta} - b^\alpha{}_\mu\tilde{\gamma}_\alpha - b_\mu\tilde{\gamma} \equiv \Phi^\alpha{}_\mu\Gamma_\alpha \tag{3}$$

Furthermore, $a^{\alpha\beta}{}_\mu$ is the spin gauge field of Chisholm and Farwell [12].

For the covariant derivative of the Clifford algebra generator we have:

$$\nabla_\mu \gamma_\alpha = \partial_\mu \gamma_\alpha + \omega^\beta{}_{\alpha\mu}\gamma_\beta + i[\Phi_\mu, \gamma_\alpha] = \partial_\mu \gamma_\alpha + i[A_\mu, \gamma_\alpha] \tag{4}$$

Now if we were to demand $\nabla_\mu \gamma_\alpha = 0$ we would find that the only allowed automorphism transformations are those generated by the bivectors [2], and we would thereby retain only spin gauge theory. Hence we assume $\nabla_\mu \gamma_\alpha \neq 0$, and consequently we find $\nabla_\mu \gamma_\nu \neq 0$. Herein lies a major distinction between local automorphism invariance and the spin gauge theory of Chisholm and Farwell [12]. This result should not be surprising since allowing spacetime dependent automorphism transformations implies that the Clifford algebra generators must be treated as dynamical variables. This may be implemented by introducing the *drehbein* fields $\Delta(x)$ ("spin-legs") [15] which constitute the transformation matrix connecting an arbitrarily assigned constant representation of the Clifford algebra generators $\mathring{\gamma}_\alpha$ to the spacetime dependent representation $\gamma_\alpha(x)$, and thereby the drehbeins contain the dynamic content of the Clifford algebra generators induced by the automorphism transformations. Note that the drehbeins play a role in the spinor space similar to that of the vierbeins in the tangent space. In particular, we have:

$$\gamma_\alpha(x) \equiv \Delta(x)\mathring{\gamma}_\alpha\overline{\Delta}(x) \Rightarrow \gamma_\mu(x) = h_\mu^{\ \alpha}(x)\Delta(x)\mathring{\gamma}_\alpha\overline{\Delta}(x) \tag{5}$$

where the Clifford algebra generators in coordinate basis $\gamma_\mu(x)$ satisfy equation (1.2). In addition, the drehbiens satisfy the constraint:

$$\Delta\overline{\Delta} = 1 \Leftrightarrow \overline{\Delta}\Delta = 1 \tag{6}$$

This insures that equation (1.1) remains valid.

The field strength tensor may be obtained by considering parallel transport of a spinor around a closed loop [15,17]:

$$\Phi_{\mu\nu} = \partial_\mu \Phi_\nu - \partial_\nu \Phi_\mu + i[\Phi_\mu, \Phi_\nu] \equiv \Phi^\alpha_{\ \mu\nu}\Gamma_\alpha \tag{7}$$

Note that this tensor obtains contributions from derivatives of the drehbeins, in addition to the usual derivatives of the gauge fields, so that although this field strength tensor looks quite canonical, in fact it contains many interesting additional pieces. Space does not allow a display of this tensor in its full glory, but since we will need to consider parts of the bivector piece in detail, I record them here:

$$\Phi^{\alpha\beta}_{\ \ \mu\nu} = \Omega^{\alpha\beta}_{\ \ \mu\nu} + a^\alpha_{\ \mu}a^\beta_{\ \nu} - a^\alpha_{\ \nu}a^\beta_{\ \mu} - b^\alpha_{\ \mu}b^\beta_{\ \nu} + b^\alpha_{\ \nu}b^\beta_{\ \mu} +$$

$$\partial_\mu a^\alpha_{\ \gamma\nu} - \partial_\nu a^\alpha_{\ \gamma\mu} + a^\alpha_{\ \delta\mu}a^\delta_{\ \gamma\nu} - a^\alpha_{\ \delta\nu}a^\delta_{\ \gamma\mu} + \textit{drehbein derivative terms} \tag{8}$$

It is the presence of both the Riemann curvature term (in the guise of the field strength tensor for the spin connection, see equation (1.7)) and the vector part of the automorphism gauge potential $a^\alpha_{\ \mu}$ which admits the possibility of obtaining the Einstein-Hilbert action as a symmetry breaking effect.

The lagrangian density for this theory may be written as the sum of three parts: $L = L_\Phi + L_\lambda + L_\Psi$ [15,17]. For the lagrangian density of the gauge fields we will take the standard quadratic form:

$$L_\Phi = -\frac{1}{16g^2}tr[\Phi_{\mu\nu}\Phi^{\mu\nu}] = -\frac{1}{8g^2}\Phi^{\alpha\beta}_{\ \ \mu\nu}\Phi_{\alpha\beta}^{\ \ \mu\nu} + \textit{other terms} \tag{9}$$

where g is the coupling constant. Notice that this does not contain the canonical form for the gravitational part of the action (see equations (1.10) and (2.8)). However, this lagrangian density does include drehbein derivative terms, so we need not introduce a separate "kinetic" term for the drehbeins. However, we do need to introduce a lagrange multiplier field λ to guarantee the drehbein constraint (2.6):

$$L_\lambda = tr(\overline{\Delta}\lambda\Delta - \lambda) \tag{10}$$

The construction of the Dirac lagrangian density is a bit more involved than in General Relativity. In particular, the ordering of the covariant derivative and the Clifford algebra generators is unimportant in General Relativity since the covariant derivative of the Clifford algebra generators vanishes, but for the local automorphism invariant theory this is no longer the case. It turns out that the unique "minimal substitution" that yields a real lagrangian density is the following [18]:

$$\begin{aligned} L_\Psi &= \overline{\Psi}[\frac{i}{2}(\gamma^\mu \overrightarrow{\nabla}_\mu - \overleftarrow{\nabla}_\mu \gamma^\mu)]\Psi = \overline{\Psi}[\frac{i}{2}(\gamma^\mu \overrightarrow{\partial}_\mu - \overleftarrow{\partial}_\mu \gamma^\mu) - \frac{1}{2}\{\gamma^\mu, \Phi_\mu\}]\Psi \\ &= \overline{\Psi}[\frac{i}{2}(\gamma^\mu \overrightarrow{\partial}_\mu - \overleftarrow{\partial}_\mu \gamma^\mu) - \phi - 3\tilde{\omega}_\mu\tilde{\gamma}^\mu + \tilde{b}_{\mu\nu}\gamma^{\mu\nu}]\Psi \end{aligned} \tag{11}$$

where we have introduced the convenient definitions:

$$\phi \equiv a^\alpha_{\ \mu} h^\mu_{\ \alpha}, \quad \tilde{\omega}_\mu \equiv \frac{1}{3!}\varepsilon_{\mu\alpha\beta}^{\ \ \ \ \nu}\omega^{\alpha\beta}_{\ \ \ \nu}, \quad \tilde{b}_{\mu\nu} \equiv \frac{1}{2}\varepsilon_{\mu\nu\alpha}^{\ \ \ \ \rho} b^\alpha_{\ \rho} \tag{2.12a,b,c}$$

We recover the normal coupling between the fermion and the spin connection. In addition there is a bivector coupling, and the "trace" of the vector part of the Clifford connection ϕ generates a scalar (Yukawa) coupling to the fermion. Notice that a fermion mass term has not been included. This insures that the lagrangian density contains only one dimensionless parameter - the gauge coupling constant g.

3. Mass and the Einstein-Hilbert Action

The Einstein-Hilbert action may possibly be recovered through a Higgs [19] type mechanism. To see this, first observe that since only the trace of the vector part of the Clifford connection couples to the fermion, the following definition is indicated:

$$a^\alpha_{\ \mu} \equiv \frac{1}{4}\phi h^\alpha_{\ \mu} + c^\alpha_{\ \mu}, \quad c^\alpha_{\ \mu} h^\mu_{\ \alpha} = 0 \tag{3.1a,b}$$

Note that this equation is consistent with (2.12a). Substitution into the gauge field lagrangian density yields an avalanche of terms, including the following:

$$L_\Phi = -\frac{1}{32g^2}\phi^2\Omega^{\mu\nu}_{\ \ \mu\nu} - \frac{3}{64g^2}\phi^4 + everything\ else \tag{2}$$

as may be seen by inspection of equations (2.8) and (2.9). Consequently, if the scalar field ϕ develops a non-zero vacuum expectation value we recover the Einstein-Hilbert action:

$$\langle\phi\rangle = M, \quad G = \frac{2g^2}{\pi M^2} \equiv \frac{1}{M_P^2} \tag{3}$$

were the second expression follows from inspection of equation (1.10). Furthermore, the fermion acquires mass M as may be seen by inspection of equation (2.11). Notice, however, that if the gauge coupling is equal to the electromagnetic coupling, then M is on the order of the Planck mass: $M = (.24)M_P \approx 3 \times 10^{27}\,ev$. On the other hand, if we take the mass to be the electron mass we find that the coupling must be very small: $g^2/4\pi = 1.73 \times 10^{-45}$. In addition, all of the gauge fields except the spin connection acquire mass on the order of M. In this case, the automorphism gauge group $SU(2,2)$ is not a symmetry of the ground state, the remaining symmetry group of the ground state being the Lorentz group $SO(1,3)$. This result should not be surprising since conformal symmetry is only exhibited in massless theories, yet the presence of a mass term does not spoil Lorentz invariance.

Since the lagrangian contains a ϕ^4 term (see equation (2.14)), the scalar field ϕ may develop a non-zero vacuum expectation value via the dynamical symmetry breaking mechanism of Coleman and Weinberg ("dimensional transmutation") [20]. However, to explore this possibility in detail the quantized version of this theory must be fully developed.

Finally note that when the scalar field develops a non-zero vacuum expectation value, this mass term in the covariant derivative of a spinor (see equations (2.1) and (2.2)) is precisely in the form of the "frame field" introduced in the spin gauge approach of Chisholm and Farwell [12]. Therefore, if the scalar field indeed develops a vacuum expectation value, I expect to recover the results of spin gauge theory, with possible modifications arising from the additional gauge fields.

4. Conclusions and Further Development

Local automorphism invariance arises from the demand that the physical predictions of a theory involving spinor fields should not depend on the particular local representation of the Clifford algebra generators, this being equivalent to the demand that the theory not depend on the particular local orientation of the spinor basis. This is analogous to the situation in General Relativity where we demand that the physical predictions of the theory should not depend on the particular orientation of the basis vectors in the tangent space. Therefore, this theory constitutes a generalization of General Relativity. However, if we treat local automorphism invariant theory as a standard gauge theory, the Einstein-Hilbert action does not result. Nevertheless, if the scalar field (the trace of the vector part of the Clifford connection) develops a non-zero vacuum expectation value, we recover the Einstein-Hilbert action, and the fermion gains mass as well as some of the gauge particles. Under these circumstances, the ground state of the theory is not invariant under the full automorphism group, but only the Lorentz subgroup.

Certainly one of the major open questions concerns the quantization of the theory. Since the gauge group $SU(2,2)$ is non-compact, the theory may be plagued with problems of non-unitarity, negative norm states, and tachyonic states. There is, however, reason for optimism. In particular, the gauge group by definition preserves the spinor metric and also preserves the metric on the Clifford algebra. Therefore the gauge fields should not induce transitions between states on positive and negative norm, and absence of these transitions may ameliorate the problems of non-unitarity and ghost states [21].

The classical solutions of this theory should also be worth examining. For example, the second order curvature terms and the presence of the additional interactions should induce modifications of the standard black hole solutions. Since the field equations of this theory are so complicated, it may be worthwhile to examine the lower dimensional cases first. On the other hand, the higher dimensional cases should also be interesting. In particular, if compactification occurs on some symmetric space, we will expect this to induce mass terms for the gauge fields which are no longer associated with symmetries of the ground state. This yields the possibility of generating unified field theories which incorporate gravitation in a natural and compelling way.

Acknowledgments

This project has been under development for several years. During this time I have benefied from conversations with Professors A.O. Barut, J.S.R. Chisholm, John Collins, Ivko Dimitric, Ruth Farwell, Steve Gull, Anthony Lasenby, Alan MacFarlene, K.T. Mahanthappa, Hohn McBwan, Ted Newman, Bill Passaglia, and Carlo Rovelli.

This work was supported in part by a Pennsylvania State University Research Development Grant.

References

1. For a review of Clifford algebras in General Relativity see: J.P. Crawford, *Advances in Applied Clifford Algebras* 2 (1992) 75.
2. E. Schrödinger, *Sitzungsh. Akad. f. Physik* 57, 261 (1929).
3. H. Weyl, *Zeitsch. f. Physik* 56, 330 (1929).
4. V. Fock and D. Ivanenko, *C.R. Acad. Sci.* 188, 1470 (1929).
5. R. Utiyama, *Phys. Rev.* 101, 1597 (1956).
6. T.W.B. Kibble, *J. Math. Phys.* 2, 212 (1961).
7. D.W. Sciama, in *Recent Developments in General Relativity* (Permagon, Oxford, England, 1962).
8. A. Chamseddine and P. West, *Nucl. Phys.* B129, 39 (1977).
S.W. MacDowell and F. Mansouri, *Phys. Rev. Lett.* 38, 739 (1977).
G. Grignani and G. Nardelli, *Phys. Rev. D* 45, 2719 (1992).
9. E.T. Newman and R. Penrose, *J. Math. Phys.* 3, 566 (1962).
See also: M. Carmeli, *J. Math. Phys.* 11, 2728 (1970).
10. R. Penrose, *J. Math. Phys.* 8, 345 (1967).
See also: R. Penrose and W. Rindler, *Spinors and Space-time*, Volumes 1 and 2 (Cambridge University Press, Cambridge, England, 1984).
11. A. Ashtekar, *Phys. Rev. Lett.* 57, 2244 (1986).
12. J.S.R. Chisholm and R. Farwell, "Unified Spin Gauge Theories of the Four Fundamental Interactions," in: *The Interface of Mathematics and Particle Physics*, D.G. Quillen, G.B. Segal, and Tsou S.T. editors. (Oxford University Press, Oxford, England, 1990); and references therein.
13. J.P. Crawford, *J. Math. Phys.* 32, 576 (1991) .
14. J.P. Crawford, *J. Math. Phys.* 31, 1991 (1990).
15. J.P. Crawford, "Local Automorphism Invariance," in preparation.
16. For the definitions of my notational conventions see references 13 and 14.
17. J.P. Crawford, "A Generalization of the Principle of Equivalence," in preparation.
18. A.O. Barut and J. McEwan, *Phys. Lett. B* 135, 172 (1984).
19. P.W. Higgs, *Phys. Lett.* 12 , 132 (1964) ; *Phys. Rev. Lett.* 13, 508 (1964) ;
F. Englert and R. Brout, *Phys. Rev. Lett.* 13, 321 (1964) ;
G.S. Guralnik, C.R. Hagen, and T.W.B. Kibble, *Phys. Rev. Lett.* 13, 585 (1964) .
20. S. Coleman and E. Weinberg, *Phys. Rev. D* 7, 1888 (1973) .
21. J.P. Crawford and A.O. Barut, *Phys. Rev. D* 27, 2493 (1983) .

DIFFERENTIAL FORMS IN GEOMETRIC CALCULUS

DAVID HESTENES
Department of Physics and Astronomy
Arizona State University
Tempe, Arizona 85287-1504, USA

Abstract. Geometric calculus and the calculus of differential forms have common origins in Grassmann algebra but different lines of historical development, so mathematicians have been slow to recognize that they belong together in a single mathematical system. This paper reviews the rationale for embedding differential forms in the more comprehensive system of Geometric Calculus. The most significant application of the system is to relativistic physics where it is referred to as Spacetime Calculus. The fundamental integral theorems are discussed along with applications to physics, especially electrodynamics.

Key words: Differential forms, geometric calculus, Clifford algebra, Dirac operator, Stokes' Theorem, manifolds.

1. Introduction

Contrary to the myth of the solitary scientific genius, science (including mathematics) is a social activity. We all feed on one another's ideas. Without our scientific culture the greatest mathematical genius among us could not progress beyond systematic counting recorded on fingers and toes. The primary means for communicating new ideas is the scientific literature. However, it is extremely difficult to read that literature without learning how through direct contact with others who already can. Even so, important ideas in the literature are overlooked or misconstrued more often than not. The history of mathematical ideas (especially those of Hermann Grassmann) shows this conclusively.

A workshop like this one, bringing together scientists with common interests but divergent backgrounds, provides a uniquely valuable opportunity to set the written record straight—to clarify and debate crucial ideas—to progress toward a consensus. We owe an immense debt of gratitude to the Workshop organizers who made this possible: *Professors Fred Brackx, Richard Delanghe*, and *Herman Serras.* This is also a good opportunity to pay special tribute to *Professor Roy Chisholm*, who, with uncommon insight into the social dimension of science, conceived, organized and directed the *First International Workshop on Clifford Algebras and Their Applications* in 1986. He set the standard for Workshops to follow. Without his leadership we would not be here today.

As in previous *Clifford Algebra Workshops* [1–4] my purpose here is to foment debate and discussion about fundamental mathematical concepts. This necessarily overflows into debate about the terminology and notations adopted to designate those concepts. At the outset, I want it understood that I intend no offense toward my esteemed colleagues who hold contrary opinions. Nevertheless, I will not mince

F. Brackx et al. (eds.), Clifford Algebras and their Applications in Mathematical Physics, 269–285.

words, as I could not take the subject more seriously. At stake is the very *integrity of mathematics*. I will strive to formulate and defend my position as clearly and forcefully as possible. At the same time, I welcome rational opposition, as I know that common understanding and consensus is forged in the dialectic struggle among incompatible ideas. Let the debate proceed!

I reiterate my contention that the subject of this conference should be called *Geometric Algebra* rather than *Clifford Algebra*. This is not a mere quibble over names, but a brazen claim to vast intellectual property. What's in these names? To the few mathematicians familiar with the term, "Clifford Algebra" refers to a minor mathematical subspecialty concerned with quadratic forms, just one more algebra among many other algebras. We should not bow to such a myopic view of our discipline.

I invite you, instead, to join me in proclaiming that *Geometric Algebra* is no less than a universal mathematical language for precisely expressing and reasoning with geometric concepts. "Clifford Algebra" may be a suitable term for the grammar of this language, but there is far more to the language than the grammar, and this has been largely overlooked by the strictly formal approach to Clifford Algebra.

Let me remind you that Clifford himself suggested the term *Geometric Algebra*, and he described his own contribution as an application of Grassmann's *extensive algebra* [3]. In fact, all the crucial geometric and algebraic ideas were originally set forth by Grassmann. What is called "Grassmann Algebra" today is only a fragment of Grassmann's system. His entire system is closer to what we call "Clifford Algebra." Though we should remember and admire the contributions of both Grassmann and Clifford, I contend that the conceptual system in question is too universal to be attached to the name of any one individual. Though Grassmann himself called it the *Algebra of Extension*, I believe he would be satisfied with the name *Geometric Algebra*. He was quite explicit about his intention to give geometry a suitable mathematical formulation.

Like the real number system, Geometric Algebra is our common heritage, and many individuals besides Grassmann and Clifford have contributed to its development. The system continues to evolve and has expanded to embrace differentiation, integration, and mathematical analysis. No consensus has appeared on a name for this expanded mathematical system, so I hope you will join me in calling it *Geometric Calculus*.

Under the leadership of *Richard Delanghe*, mathematical analysis with Clifford Algebra has become a recognized and active branch of mathematics called *Clifford Analysis*. I submit, though, that this name fails to do justice to the subject. Clifford analysis should not be regarded as just one more branch of analysis, along side real and complex analysis. Clifford analysis, properly construed, generalizes, subsumes, and unifies all branches of analysis; it is the whole of analysis. To proclaim that fact, workers in the field should set modesty aside and unite in adopting a name that boldly announces claim to the territory. At one time I suggested the name *Geometric Function Theory* [5], but I am not particularly partial to it. However, I insist on the term *Geometric Calculus* for the broader conceptual system which integrates analysis with the theory of manifolds, differential geometry, Lie groups, and Lie algebras.

The proclamation of a universal *Geometric Calculus* [1,5] has met with some skepticism [3], but the main objection has now been decisively answered in [6], which shows that, not only does embedding a vector space with its dual into a common geometric algebra not suffer a loss of generality, but there are positive advantages to it as well. Indeed, physicists and mathematicians have been doing just that for some time without recognizing the fact. I believe that the remaining barriers to establishing a consensus on *Geometric Calculus* are more psychological or sociological than substantive. My intention in this article is to keep hammering away at those barriers with hope for a breakthrough.

The literature relating Clifford algebra to fiber bundles and differential forms is rapidly growing into a monstrous, muddled mountain. I hold that the muddle arises mainly from the convergence of mathematical traditions in domains where they are uncritically mixed by individuals who are not fully cognizant of their conceptual and historical roots. As I have noted before [1], the result is a highly redundant literature, with the same results appearing over and over again in different notational guises. The only way out of this muddle, I think, is to establish a consensus on the issues. Toward that end, I now present my own views on the issues. I include some personal history on the evolution of my views with the hope that it will highlight the most important ideas. I will presume that the reader has some familiarity with the notation and nomenclature I use from my other publications.

2. What is a manifold?

The formalism for modern differential geometry (as expounded, for example, by O'Neill [7]) was developed without the insights of Geometric Algebra, except for a fragment of Grassmann's system incorporated into the calculus of differential forms. Can the formalism of differential geometry be improved by a *new synthesis* which incorporates Geometric Algebra in a fundamental way? My answer is a resounding YES! Moreover, I recommend the Geometric Calculus found in [5] as the way to do it. I am afraid, however, that the essential reasons for this new synthesis have been widely overlooked, so my purpose is to emphasize them today. Readers who want more mathematical details can find them in [5].

Everyone agrees, I suppose, that the concept of a (*differentiable*) *manifold* is the foundation for differential geometry. However, the very definition of "manifold" raises a question. In the standard definition [7] coordinates play an essential role, but it is proved that the choice of well-defined coordinates is arbitrary. In other words, the concept of a manifold is really independent of its representation by coordinates. Why, then, is the clumsy apparatus of coordinate systems used to define the concept? The reason, I submit, is historical: no better means for describing the structure of a manifold was available to the developers of the concept. Futhermore, I claim that Geometric Algebra alone provides the complete system of algebraic tools needed for an intrinsic characterization of manifolds to replace the extrinsic characterization with coordinates. This is not to say that coordinates are without interest. It merely displaces coordinates from a central place in manifold theory to the periphery where they can be employed when convenient.

Now to get more specific, let x be a generic point in a m-dimensional manifold

$\mathcal{M}$, and suppose that a patch of the manifold is parameterized by a set of coordinates $\{x^\mu\}$, as expressed by

$$x = x(x^1, x^2, \ldots, x^m). \tag{1.1}$$

If the manifold is embedded in a vector space, so x is vector-valued, then the vector fields $e_\mu = e_\mu(x)$ of tangent vectors to the coordinate curves parameterized by x_μ are given by

$$e_\mu = \partial_\mu x = \frac{\partial x}{\partial x^\mu}. \tag{1.2}$$

I recall that when I was a graduate student reading Cartan's work on differential geometry, I was mystified by the fact that Cartan wrote down (1.2) for any manifold without saying anything about the values of x. This violated the prohibition against algebraic operations among different points on a general manifold which I found in all the textbooks; for the very meaning of (1.2) is supplied by its definition as the limit of a difference quotient:

$$\partial_\mu x = \lim \frac{\Delta x}{\Delta x^\mu}. \tag{1.3}$$

Certainly Δx^μ is well defined as a scalar quantity, but what is the meaning of Δx if it is not a "difference vector," and what meaning can be attributed to the limit process if no measure $|\Delta x|$ of the magnitude of Δx is specified? I concluded that (1.2) was merely a heuristic device for Cartan, for he never appealed to it in any arguments.

Evidently, others came to the same conclusion, for in modern books on differential geometry [7] the mysterious x has been expunged from (1.2) so e_μ is identified with ∂_μ; in other words, *tangent vectors are identified with differential operators.* I think this is a *bad idea* which has complicated the subject unnecessarily. It is all very well to treat differential operators abstractly and express some properties of manifolds by their commutation relations, but this does not adequately characterize the properties of tangent vectors. The usual way to remedy this is to impose additional mathematical structure, for example, by defining a metric tensor by

$$g_{\mu\nu} = g(\partial_\mu, \partial_\nu). \tag{1.4}$$

Geometric algebra gives us another option which I maintain is more fundamental. As has been explained many times elsewhere, the very meaning of being a vector entails defining the *geometric product*

$$e_\mu e_\nu = e_\mu \cdot e_\nu + e_\mu \wedge e_\nu. \tag{1.5}$$

The inner product defines a metric tensor by

$$g_{\mu\nu} = e_\mu \cdot e_\nu \tag{1.6}$$

This has the huge advantage over (1.4) of integrating the metric tensor into algebraic structures at the ground floor. Of course, the geometric product (1.5) is incompatible with the identification $e_\mu = \partial_\mu$ of vectors with differential operators. This lead me eventually to what I believe is a deeper approach to differentiation as explained below.

Adopting (1.5) requires that we regard e_μ as a vector, so (1.2) and (1.3) are meaningful only if the point x is a vector so Δx is a vector difference. I call such a manifold, whose points are vectors, a *vector manifold*. Now this seems to subvert our original intention of developing a general theory of manifolds by limiting us to a special case. It took me many years to realize that this is not the case, so I am sympathetic of colleagues who are skeptical of my claim that *the theory of vector manifolds is a general theory of manifolds*, especially since all details of the theory are not fully worked out. I would like to convince some of you, at least, that the claim is plausible and invite you to join me working out the details. I believe the payoff will be great, because the effort has been very productive already, and I believe the work is essential to establishing a truly Universal Geometric Calculus.

As explained in [3], I believe that skepticism about Geometric Calculus in general and vector manifolds in particular can be attributed to the prevalence of certain *mathematical viruses*, beliefs that limit or otherwise impair our understanding of mathematics. These include the beliefs that a vector manifold cannot be well defined without embedding it in a vector space, and it is necessarily a metric manifold, thus being too specialized for general manifold theory. As I have treated these viruses in [3] and [5], I will not address them here. I merely wish to describe my own struggle with these viral infections in the hope that it will motivate others to seek treatment. Let me mention, though, that [6] contains some potent new medicine for such treatment.

Though we want a coordinate-free theory, it is worth noting that the geometric product (1.5) facilitates calculations with coordinates. For example, it enables the construction of the pseudoscalar for the coordinate system:

$$e_{(m)} = e_1 \wedge e_2 \wedge \ldots \wedge e_m . \tag{1.7}$$

For a metric manifold we can write

$$e_{(m)} = |e_{(m)}| I_m , \tag{1.8}$$

where $I_m = I_m(x)$ is a unit pseudoscalar for the manifold, and its modulus

$$|e_{(m)}| = |\det g_{\mu\nu}|^{1/2} \tag{1.9}$$

can be calculated from (1.7) using (1.6)

Instead of beginning with coordinate systems, the coordinate-free approach to vector manifolds in [5] begins by assuming the existence of a pseudoscalar field $I_m = I_m(x)$ and characterizing the manifold by specifying its properties. At each point x, $I(x)$ is a pseudoscalar for the tangent space. If the manifold is smooth and orientable, the field $I_m(x)$ is single-valued. If the mainfold is not orientable, I is double-valued. Self-intersections and discontinuities in a manifold can be described by making I_m and its derivatives multivalued. This brings us back to the question of how to define differentiation without using coordinates. But let us address it first by reconsidering coordinates.

The inverse of the mapping (1.1) is a set of scalar-valued functions

$$x^\mu = x^\mu(x) \tag{1.10}$$

defined on the manifold $\mathcal{M}$. The gradients of these functions are vector fields

$$e^\mu = \partial x^\mu \tag{1.11}$$

on $\mathcal{M}$, and this entails the existence of a "vectorial" *gradient operator* $\partial = \partial_x$. But how to define it? If we take the e^μ as given, then it can be defined in terms of coordinates by

$$\partial = e^\mu \partial_\mu, \tag{1.12}$$

where

$$\partial_\mu = e_\mu \cdot \partial \tag{1.13}$$

provided

$$e^\mu \cdot e_\nu = \delta^\mu_\nu. \tag{1.14}$$

But how can we define ∂ without using coordinates?

Before continuing, I want to make it clear that I do not claim that vector manifolds are the only manifolds of interest. My claim is that *every manifold is isomorphic to a vector manifold*, so any manifold can be handled in a coordinate-free way by defining its relation to a suitable vector manifold instead of defining a coordinate covering for it. Of course, coordinate coverings have the practical value that they have been extensively developed and applied in the literature. We should take advantage of this, but my experience suggests that new insight can be gained from a coordinate-free approach in nearly every case.

It is often of interest to work directly with a given manifold instead of indirectly with a vector manifold isomorph. For example, the spin groups treated in [6] are multivector manifolds, so if (1.1) is applied directly, the point x is a spinor not a vector. In that case, it is easily shown that the tangents e_μ defined by (1.2) are not vectors but, when evaluated at the identity, they are bivectors comprising a basis for the Lie algebra of the group. This is good to know, but the drawback to working with e_μ which are bivectors or multivectors of other kind is that the pseudoscalar (1.7) is not defined, and that complicates analysis. The advantage of mapping even such well-behaved entities as spin groups into vector manifolds is that it facilitates differential and integral calculus on the manifold.

3. What is a derivative?

The differential operator defined by (1.12), where the e^μ are tangent vectors generating a Clifford algebra on the manifold, is often called the *Dirac operator*. With no offence intended to my respected colleagues, I think that name is a *bad choice!*—not in the least justified by the fact that it has been widely used in recent years. Worse, it betrays a failure to understand what makes that operator so significant, not to mention its insensitivity to the historical fact that the idea for such an operator originated with Hamilton nearly a century before Dirac.

Whether they recognize it or not, everyone using the Dirac operator is working directly with functions defined on a vector manifold or indirectly with some mapping into a vector manifold. I hold that the *Dirac operator is a vectorial operator precisely because it is the derivative with respect to a vector*. It is the derivative with respect to a vector variable, so I propose to call it simply the *derivative* when the variable is understood, or the *vector derivative* when emphasis on the vectorial nature of the variable is appropriate. This is to claim, then, that the operator has a universal significance transcending applications to relativistic quantum mechanics where Dirac introduced it.

The strong claim that the operator $\partial = \partial_x$ is *the derivative* needs justification. If it is so fundamental, why is this not widely recognized and accepted as such? My answer is: Because the universality of Geometric Algebra and the primacy of vector manifolds have not been recognized. When Geometric Calculus is suitably formulated, the conclusion is obvious. Let me describe how I arrived at a formulation. At the same time we will learn how to define the vector derivative without resorting to coordinates, something that took me some years to discover.

The fundamental significance of the vector derivative is revealed by Stokes' theorem. Incidentally, I think the only virtue of attaching Stokes' name to the theorem is brevity and custom. His only role in originating the theorem was setting it as a problem in a Cambridge exam after learning about it in a letter from Kelvin. He may, however, have been the first person to demonstrate that he did not fully understand the theorem in a published article: where he made the blunder of assuming that the double cross product $v \times (\partial \times v)$ vanishes for any vector-valued function $v = v(x)$. The one-dimensional version of Stokes' theorem is widely known as the *fundamental theorem of integral calculus*, so it may be surprising that this name is not often adopted for the general case. I am afraid, though, that many mathematicians have not recognized the connection. Using different names for theorems differing only in dimension certainly doesn't help. I suggest that the *Boundary Theorem of Calculus* would be a better name, because it refers explicitly to a key feature of the theorem. Let me use it here.

My first formulation of the Boundary Theorem [8] entirely in the language of Geometric Calculus had the form

$$\int d\omega \cdot \partial A = \oint d\sigma A, \tag{2.1}$$

where the integral on the left is over an m-dimensional oriented vector manifold $\mathcal{M}$ and the integral on the right is over its boundary $\partial\mathcal{M}$. The integrand $A = A(x)$ has values in the Geometric Algebra, and $\partial = \partial_x$ is the derivative with respect to the vector variable x.

The most stiking and innovative feature of (2.1) is that the differential $d\omega = d\omega(x)$ is m-vector-valued; in other words, it is a pseudoscalar for the tangent space of $\mathcal{M}$ at x. Likewise, $d\sigma = d\sigma(x)$ is an $(m-1)$-vector-valued pseudoscalar for $\partial\mathcal{M}$. Later I decided to refer to $d\omega$ as a *directed measure* and call the integrals with respect to such a measure *directed integrals*. In formulating (2.1) it became absolutely clear to me that *it is the use of directed integrals along with the vector derivative that makes the Boundary Theorem work*. This fact is thoroughly disguised in other formulations of Stokes' Theorem. As far as I know it was first made explicit in [8]. It seems to me that hardly anyone else recognizes this fact even today, and the consequence is unnecessary redundancy and complexity throughout the literature.

When I showed in [8] that the scalar part of (2.1) is fully equivalent to the standard formulation of the "Generalized Stokes' Theorem" in terms of differential forms, I wondered if (2.1) is a genuine generalization of that theorem. It took me several years to decide that, properly construed, this is so. I was impressed in [8] with the fact that (2.1) combined nine different integral theorems of conventional vector calculus into one, but I haven't seen anyone take note of that since. In any case, the deeper significance of directed measure appears in the definition of the derivative.

For a long time I was bothered by the appearance of the inner product on the left side of (2.1). I thought that in a fundamental formulation of the Boundary Theorem only the geometric product should apppear. I recognized in [8], though, that if $d\omega \wedge \partial = 0$ then $d\omega \cdot \partial = d\omega\partial$, and, with the appropriate limit process, the *vector derivative can be defined by*

$$\partial A = \lim_{d\omega \to 0} \frac{1}{d\omega} \oint d\sigma A. \tag{2.2}$$

This definition is indeed coordinate-free as desired, but considerable thinking and experience was required to see that it is *the best way to define the vector derivative.* The clincher was the fact that it simplifies the proof of the Boundary Theorem almost to a triviality. The Boundary Theorem is so fundamental that we should design the vector derivative to make it as simple and obvious as possible. The definition (2.2) does just that! The answer to the question of when the inner product $d\omega \cdot \partial$ in eqn. (2.1) can be dropped in favor of the geometric product $d\omega\partial$ is inherent in what has already been said. Those who want it spelled out should refer to [5] or [10].

I should say that the general idea of an integral definition is an old one—I do not know how old—I learned about it from [9], where it is used to define gradient, divergence, and curl. The standard definition of a derivative is so heavily emphasized that few mathematicians seem to realize the advantages of an integral definition. The fact that the right side of (2.2) reduces to a difference quotient in the one-dimensional case supports the view that the integral definition is the best one.

The next advance in my understanding of the vector derivative and the Boundary Theorem began in 1966 when I started teaching graduate electrodynamics entirely in Geometric Algebra. As I reformulated the subject in this language, I was delighted to discover fresh insights at every turn. There is no substitute for detailed calculation and problem solving to deepen and consolidate mathematical and physical understanding. During this period I developed the necessary techniques for performing *completely coordinate-free calculations* with the vector derivative. The basic ideas were published in two brief papers which I still consider as among my best work. The first paper [10] refined, expanded and generalized my formulations of the vector derivative, directed integration, and the Boundary Theorem. It was there that I was finally convinced that the integral definition for the vector derivative is fundamental.

The second paper [11] derived a generalization of Cauchy's integral formula for n dimensions. I believe that this is one of the most important results in mathematics —so important that it has been independently discovered by several others, most notably Richard Delanghe [12] because he, with the help of brilliant students like Fred Brackx and Frank Sommen, has been responsible for energetically developing the implications of this result into the rich new mathematical domain of *Clifford Analysis*. As my paper is seldom mentioned in this domain, perhaps you will forgive me for pointing out that it contains significant features which are not appreciated in most of the literature even today. Besides the fact that the formulation and derivations are completely coordinate-free, my integral formula is actually more general than usual one, because it applies to any differentiable function or distribution, not just monogenic functions. That has too many consequences to discuss here.

In these two brief papers [10,11] on the foundations of Geometric Calculus, I made the mistake of not working out enough examples. There were so many applications to choose from that I naively assumed that anyone could generate examples easily. Subsequent years teaching graduate students disabusd me of that assumption. I found that it was not an inherent difficulty of the subject so much as misconceptions from prior training that limited their learning [3].

My work on the foundations of Geometric Calculus continued into 1975, though the resulting manuscript was not published as a book [5] until 1984. That book includes and extends the previous work. It contains many other new developments in Geometric Calculus, but let me point out what is most relevant to the topics of present interest. In my previous work I restricted my formulation of the Boundary Theorem (2.1) and the vector derivative (2.2) to manifolds embedded in a vector space, though I had the strong belief that the restriction was unnecessary. It was primarily to remove that restriction that I developed the concept of vector manifolds in [5]. I was still not convinced that (2.2) applies without modification to such general vector manifolds until the relation between the vector *derivative* ∂ and the *coderivative* ∇ was thoroughly worked out in [5]. The operator ∂ can be regarded as a coordinate-free generalization of the "partial derivative," while ∇ is the same for the "covariant derivative." Though the Boundary Theorem is formulated for general vector manifolds in [5], and its scalar part is shown to be equivalent to Stokes' Theorem in terms of differential forms, most of its applications are restricted to manifolds in a vector space, because it's only for that case that explicit Green's functions are known. Nevertheless, I am convinced that there are beautiful applications waiting to be discovered in the general case. This is especially relevant to cohomology theory which has not yet been fully reformulated in terms of Geometric Calculus, though I am confident that it will be enlightening to do so.

For a final remark about foundations, let me call your attention to the article [13] by Garret Sobczyk. Triangulation by simplexes is an alternative to coordinates for a rigorous characterization of manifolds, and it is especially valuable as an approach to calculations on vector manifolds. Garret and I talked about this a lot while preparing [5], so I am glad he finally got around to writing out the details and illustrating the method with some applications. I believe this method is potentially of great value for treating finite difference equations with Geometric Algebra. Anyone who wants to apply Geometric Calculus should put it in his tool box.

4. What is a differential form?

The concept of differential needs some explication, because it comes with many guises in the literature. I believe that the concept is best captured by defining a *differential of grade k* to be a k-blade in the tangent algebra of a given vector manifold. Recall from [5] that a k-blade is a simple k-vector. Readers who are unfamiliar with other technical terms in this article will find full explanations in [5]. Of course, differentials have usually been employed without any reference to Geometric Algebra or vector manifolds, but I maintain that they can always be reformulated to do so. The point of the present formulation is that the property of a direction in a tangent space is inherent in the concept of a differential, and this property should be given an explicit formulation by representing the differential as a blade.

For the differential in a directed integral such as (2.1), I often prefer the notation

$$d\omega = d^m x, \tag{4.1}$$

because it has the advantage of designating explicitly both the differential's grade and the point to which it is attached. The differential of a coordinate curve through x is a tangent vector which, using (1.2), can be expressed in terms of the coordinates by

$$d_\mu x = e_\mu dx^\mu \tag{4.2}$$

(no sum on μ). Note the placement of the subscript on the left to avoid contusion between dx^μ, a scalar differential for the scalar variable x^μ, and the vector differential $d_\mu x$ for the vector variable x. We can use (3.2) to express (3.1) in terms of coordinates:

$$d^m x = d_1 x \wedge d_2 x \wedge \ldots \wedge d_m x = e_1 \wedge e_2 \wedge \ldots \wedge e_m \, dx^1 dx^2 \ldots dx^m. \tag{4.3}$$

This is appropriate when one wants to reduce a directed integral to an iterated integral on the coordinates. However, it is often simpler to evaluate integrals directly without using coordinates. (Examples are given in [5].)

On a metric manifold, a differential $d^m x$ can be resolved into its magnitude $|d^m x|$ and its direction represented by a unit m-blade I_m:

$$d^m x = I_m |d^m x|. \tag{4.4}$$

Then, according to (3.3) and (1.9),

$$|d^m x| = |\det g_{\mu\nu}|^{1/2} dx^1 dx^2 \ldots dx^m. \tag{4.5}$$

This is a familiar expression for the "volume element" in a "multiple integral," and it is really all one needs to establish my contention that any integral can be reformulated as a directed integral, for

$$|d^m x| = I_m^{-1} d^m x, \tag{4.6}$$

so we can switch from one integral with the "scalar measure" $|d^m x|$ to one with "directed measure" $d^m x$ simply by inserting $I_m^{-1}(x)$ in the integrand. Of course, this is not always desirable, but you may be surprised how often it is when you know about it!

A *differential k-form*

$$L = L(d^k x) = L(x, d^k x) \tag{4.7}$$

can be defined on a given vector manifold as a linear function of a differential of grade k with values in the Geometric Algebra. To indicate that its values may vary over the manifold, dependence on the point x is made explicit on the right side of (3.7). As explained in [5], the *exterior differential* of L can be defined in terms of the vector derivative $\partial = \partial_x$ by

$$dL = \grave{L}(d^k x \cdot \grave{\partial}) = L(\grave{x}, (d^k x) \cdot \grave{\partial}), \tag{4.8}$$

where the accent on $\grave{\partial}$ indicates that it differentiates the variable $\grave{x}$.

Now we can write down the *Boundary Theorem* in its most general form:

$$\int dL = \oint L. \tag{4.9}$$

This generalizes (2.1), to which it reduces when $L = d^{m-1}xA$. The formulation (4.9) has been deliberately chosen to look like the standard "Generalized Stokes' Theorem," but it is actually more general because L is not restricted to scalar values, and this, as has been mentioned leads to such powerful new results as the "generalized Cauchy integral formula."

Equally important, (4.7) makes the fundamental dependence of a k-form on the k-vector variable explicit, and (4.8) shows how *the exterior derivative derives from the vector derivative* (or Dirac operator, if you will). All this is hidden in the abbreviated formulation (4.9) and, in fact, throughout the standard calculus of differential forms. A detailed discussion and critique of this standard calculus is given in [5]. A huge literature has arisen in recent years combining differential forms with Clifford algebras and the Dirac operator. By failing to understand how all these things fit together in a unified Geometric Calculus, this literature is burdened by a *gross excess of formalism*, which, when stripped away, reveals much of it as trivial.

There is an alternative formulation of the Boundary Theorem which is often more convenient in physics and Clifford analysis. We use (4.4) and the fact that on the boundary the interior pseudoscalar I_m is related to the boundary pseudoscalar I_{m-1} by

$$I_m = I_{m-1}n, \tag{4.10}$$

where $n = n(x)$ is the *unit* outward normal (null vectors not allowed here). Indeed, (4.10) can be adopted as a definition of the outward normal. We define a tensor field $T(n) = T(x, n(x))$, by

$$T(n) = L(I_m n), \tag{4.11}$$

and its *divergence* by

$$\grave{T}(\grave{\partial}) = \grave{L}(I_m \grave{\partial}) + L(\grave{I}_m \cdot \grave{\partial}). \tag{4.12}$$

The last term vanishes if

$$\partial \cdot I_m = 0, \tag{4.13}$$

in which case, using (3.4), the Boundary Theorem can be rewritten in the form

$$\int \grave{T}(\grave{\partial})|d^m x| = \oint T(n^{-1})|d^{m-1}x|. \tag{4.14}$$

This version can fairly be called *Gauss' Theorem*, since it includes theorems with that name as a special case. It has the advantage of exhibiting the role of the vector derivative explicitly. This theorem applies to spaces of any signature, including the indefinite signature of spacetime. The effect of signature in the theorem is incorporated in the n^{-1}, which becomes $n^{-1} = n$ if $n^2 = 1$ or $n^{-1} = -n$ if $n^2 = -1$.

As an application of great importance, suppose we have a Green's function $G = G(y, x)$ defined on our manifold $\mathcal{M}$ and satisfying the differential equation

$$\partial_y G(y, x) = -G(y, \dot{x})\dot{\partial}_x = \delta^m(y - x), \tag{4.15}$$

where the right side is the m-dimensional delta function. Let $T(n)$ be given by

$$T(n) = GnF, \tag{4.16}$$

where $F = F(x)$ is any differentiable function. If y is an interior point of $\mathcal{M}$, substitution of (4.16) into (4.14) yields

$$F(y) = \int G(y, x)\partial F(x)|d^m x| - \oint G(y, x)n^{-1}F(x)|d^{m-1}x|. \tag{4.17}$$

This great formula allows us to calculated F inside $\mathcal{M}$ from its derivative ∂F and its values on the boundary if G is known.

The specific form of the function G, when it can be found, depends on the manifold. If $\mathcal{M}$ is embedded in an m-dimensional vector space, G is the so-called *Cauchy Kernal*:

$$G(y, x) = \frac{\Gamma(m/2)}{2\pi^{m/2}} \frac{x - y}{|x - y|^m}, \tag{4.18}$$

and (4.17) yields the generalization of Cauchy's Integral formula originally found in [11]. The Γ in (4.18) denotes the gamma function. The function $F = F(x)$ is said to be *monogenic* if $\partial F = 0$, in which case the first term on the right side of (4.17) vanishes. It is a good exercise for beginners to show that, in this case, (4.17) really does reduce to the famous Cauchy integral when $m = 2$.

5. Spacetime Calculus

When applied to a *spacetime manifold*, that is, a 4-dimensional vector manifold modeling physical spacetime, the Geometric Algebra is called *Spacetime Algebra* [8], and Geometric Calculus is called *Spacetime Calculus*. The preceding results have many applications to *spacetime physics*. Note that I did not say "relativistic physics," because the spacetime calculus provides us with an *invariant* (coordinate-free) formulation of physical equations generally, and it enables us to calculate without introducing inertial frames and Lorentz transformations among them. True, it is important to relate invariant physical quantities to some reference frame in order to interpret experimental results, but that is done better with a *spacetime split* [14] than with Lorentz transformations. An example is given below.

We limit our considerations here to *Minkowski spacetime*, modeled with the elements $\{x\}$ of a 4-dimensional vector space. Let u be a constant, unit, timelike vector (field) directed in the forward light cone. The assumption $u^2 = 1$ fixes the signature of the spacetime metric. The vector u determines a 1-parameter family of spacetime hyperplanes $\mathcal{S}(t)$ satisfying the equation

$$u \cdot x = t. \tag{5.1}$$

The vector u thus determines an inertial frame with time variable t, so $\mathcal{S}(t)$ is a surface of simultaneous t.

Let $\mathcal{V}(t)$ be a convex 3-dimensional region (submanifold) in $\mathcal{S}(t)$ which sweeps out a 4-dimensional region $\mathcal{M}$ in the time interval $t_1 \leq t \leq t_2$. In this interval the 2-dimensional boundary $\partial\mathcal{V}(t)$ sweeps out a 3-dimensional wall $\mathcal{W}$, so $\mathcal{M}$ is bounded by $\partial\mathcal{M} = \mathcal{V}(t_1) + \mathcal{V}(t_2) + \mathcal{W}$. We can use the integral formula (3.17) to solve *Maxwell's equation*

$$\partial F = J \tag{5.2}$$

in the region $\mathcal{M}$ for the *electromagnetic field* $F = F(x)$ "produced by" the *charge current* (density) $J = J(x)$. The field F is bivector-valued while the current J is vector-valued. For simplicity, let us enlarge $\mathcal{V}(t)$ to coincide with $\mathcal{S}(t)$ and assume that the integral of F over $\partial\mathcal{V}$ is vanishingly small at spatial infinity. Then $\mathcal{M}$ is the entire region between the hyperplanes $\mathcal{S}_1 = \mathcal{S}(t_1)$ and $\mathcal{S}_2 = \mathcal{S}(t_2)$, and (4.17) gives us

$$F(y) = \int_{\mathcal{M}} G(y, x)J(x)|d^4x| + F_1 - F_2, \tag{5.3}$$

where

$$F_k(y) = \int_{\mathcal{S}_k} G(y, x)uF(x)|d^3x|. \tag{5.4}$$

Because of the condition (3.15) on the Green's function, the F_k satisfy the homogeneous equation

$$\partial F_k = 0. \tag{5.5}$$

A *retarded Green's function* G_k can be found which vanishes on $\mathcal{S}_2$, in which case F_1 solves the *Cauchy problem* for the homogeneous Maxwell equation (5.5).

Green's functions for spacetime have been extensively studied by physicists and the results, contained in many books, are easily adapted to the present formulation. Thus, from [15] we find the following Green's function for (5.3) and (5.4):

$$G(r) = \frac{1}{4\pi}\partial_r\delta(r^2) = \frac{1}{2\pi}r\delta'(r^2), \tag{5.6}$$

where $r = x - y$ and δ denotes a 1-dimensional delta function with derivative δ'. The analysis of retarded and advanced parts of G and their implications is standard, so it need not be discussed here.

Taking $\mathcal{M}$ to be all of spacetime so F_1 and F_2 can be set to zero, equation (5.3) with (5.6) can be integrated to get the field produced by point charge. For a particle with charge q and world line $z = z(\tau)$ with proper time τ, the charge current can be expressed by

$$J(x) = q\int_{-\infty}^{\infty} d\tau\, v\,\delta^4(x - z(\tau)), \tag{5.7}$$

where $v = v(\tau) = dz/d\tau$. Inserting this into (5.3) and integrating, we find that the retarded field can be expressed in the following explicit form

$$F(x) = \frac{q}{4\pi}\frac{r\wedge[v + r\cdot(v\wedge\dot{v})]}{(r\cdot v)^3} = \frac{q}{4\pi(r\cdot v)^2}\left[\frac{r\wedge v}{|r\wedge v|} + \frac{1}{2}\frac{r\dot{v}vr}{r\cdot v}\right], \tag{5.8}$$

where $r = x - z$ satisfies $r^2 = 0$ and z, v, $\dot{v} = dv/d\tau$ are all evaluated at the intersection of the backward light cone with vertex at x. This elegant invariant form for the classical *Lienard-Wiechart field* has been found independently by Steve Gull.

As another important example, we show that (4.14) gives us an immediate integral formulation of any physics conservation law with a suitable choice of $T(n)$. Introducing the notations

$$\dot{T}(\dot{\partial}) = f \tag{5.9}$$

and

$$\mathcal{I} = \int_{\mathcal{M}} f|d^4x| = \int_{t_1}^{t_2}\int_{\mathcal{V}(t)} f|d^3x|, \tag{5.10}$$

for the region $\mathcal{M}$ defined above, we can write (3.14) in the form

$$\mathcal{I} = P(t_2) - P(t_1) - \int_{t_1}^{t_2} dt \oint_{\partial\mathcal{V}(t)} T(n)|d^2x|, \tag{5.11}$$

where

$$P(t) = \int_{\mathcal{V}(t)} T(u)|d^3x|. \tag{5.12}$$

Now for some applications.

Energy-Momentum Conservation:

We first suppose that $T(n)$ is the energy-momentum tensor for some *physical system*, which could be a material medium, an electromagnetic field, or some combination of the two, and it could be either classical or quantum mechanical. For example, the usual energy-momentum tensor for the electromagnetic field is given by

$$T(n) = -\tfrac{1}{2}FnF. \tag{5.13}$$

In general, the tensor $T(n)$ represents the flux of energy-momentum through a hypersurface with normal n.

For the vector field $f = f(x)$ specified independently of the tensor field $T(n) = T(x, n(x))$, equation (5.9) is the *local energy-momentum conservation law*, where the *work-force density* f characterizes the effect of external influences on the system in question. Equation (5.11) is then the *integral energy-momentum conservation law* for the system. The vector $P(t)$ given by (5.12) is the *total energy-momentum* of the system contained in $\mathcal{V}(t)$ at time t. The quantity $\mathcal{I}$ is the total *Impulse* delivered to the system in the region $\mathcal{M}$.

In the limit $t_2 \to t_1 = t$, the conservation law (5.11) can be written

$$\frac{dP}{dt} = \mathcal{F} + \oint_{\partial\mathcal{V}} T(n)|d^2x|, \tag{5.14}$$

where

$$\mathcal{F}(t) = \int_{\mathcal{V}(t)} f|d^3x| \tag{5.15}$$

is the total work-force on the system. We can decompose (5.14) into separate energy and momentum conservation laws by using a spacetime split: we write

$$Pu = E + p, \tag{5.16}$$

where $E = P \cdot u$ is the energy and $p = P \wedge u$ is the momentum of the system. Similarly we write

$$\mathcal{F}u = W + F, \tag{5.17}$$

where $W = \mathcal{F} \cdot u$ is the work done on the system and $F = \mathcal{F} \wedge u$ is the force exerted on it. We write

$$T(n)u = n \cdot s + T(n), \tag{5.18}$$

where $n \cdot s = u \cdot T(n)$ is the *energy flux*, $T(n) = T(n) \wedge u$ is the *stress tensor*, and $n = n \wedge u = nu$ represents the normal as a "relative vector." We also note that

$$xu = t + x \tag{5.19}$$

splits x into a time $t = x \cdot u$ and a position vector $x = x \wedge u$. Finally, we multiply (5.14) by u and separate scalar and relative vector parts to get the *energy conservation law*

$$\frac{dE}{dt} = W + \oint s \cdot n|d^2x| \tag{5.20}$$

and the *momentum conservation* law

$$\frac{dP}{dt} = F + \oint T(n)|d^2x| \tag{5.21}$$

These are universal laws applying to all physical systems.

Angular Momentum Conservation:

The "generalized orbital angular momentum tensor" for the system just considered is defined by

$$L(n) = T(n) \wedge x. \tag{5.22}$$

With (5.9), its divergence is

$$\dot{L}(\dot{\partial}) = f \wedge x + T(\dot{\partial}) \wedge \dot{x}. \tag{5.23}$$

For a symmetric tensor such as (5.13) the last term vanishes. But, in general, there exits a bivector-valued tensor $S(n)$, the *spin tensor* for the system, which satisfies

$$\dot{S}(\dot{\partial}) = \dot{x} \wedge T(\dot{\partial}). \tag{5.24}$$

Now define the *total angular momentum tensor*

$$M(n) = T(n) \wedge x + S(n). \tag{5.25}$$

Then the *local angular momentum conservation* law for the system is

$$\dot{M}(\dot{\partial}) = f \wedge x. \tag{5.26}$$

replacing (5.9) by (5.26), we can reinterpret (5.11) as an integral law for angular momentum and analyze it the way we did energy-momentum.

Charge Conservation:

From Maxwell's equation we derive the local charge conservation law

$$\partial \cdot J = \partial \cdot (\partial \cdot F) = (\partial \wedge \partial) \cdot F = 0. \tag{5.27}$$

Now write $T(n) = n \cdot J$ and change the notion of (5.12) to

$$Q(t) = \int_{\mathcal{V}(t)} u \cdot J |d^3x|, \tag{5.28}$$

an expression for the total charge contained in $\mathcal{V}(t)$. Then (5.11) becomes

$$Q(t_2) - Q(t_1) = \int_{t_1}^{t_2} dt \oint_{\partial\mathcal{V}(t)} n \cdot J |d^2x|. \tag{5.29}$$

This is the charge conservation equation, telling us that the total charge in $\mathcal{V}(t)$ changes only by flowing through the boundary $\partial\mathcal{V}(t)$.

To dispel any impression that only the Gaussian form (4.14) of the Boundary Theorem is of interest in spacetime physics, I present one more important example: an integral formulation of Maxwell's equation (5.2), which can be decomposed into trivector and vector parts:

$$\partial \wedge F = 0, \tag{5.30}$$

$$\partial \cdot F = J. \tag{5.31}$$

Using the algebraic identity $(d^3x) \cdot (\partial \wedge F) = (d^3x) \cdot \partial F$, we deduce immediately from (2.1) that

$$\oint d^2x \cdot F = 0 \tag{5.32}$$

for any closed 2-dimensional submanifold $\mathcal{B}$ in spacetime. A spacetime split shows that this integral formula is equivalent to Faraday's Law or "the absence of magnetic poles," or a mixture of the two, depending on the choice of $\mathcal{B}$.

To derived a similar integral formula for the vector part (5.31) of Maxwell's equation, in analogy to (4.10), define a unit normal n by writing

$$d^3x = in|d^3x|, \tag{5.33}$$

where i is the unit dextral pseudoscalar for spacetime, and the use of the identity $(\partial \cdot F)i = \partial \wedge (Fi)$ to establish

$$d^3x \cdot (\partial \wedge (Fi)) = d^3x \cdot (Ji) = J \cdot n|d^3x|, \tag{5.34}$$

Insertion of this into (2.1) yields the integral equation

$$\oint d^2x \cdot (Fi) = \int J \cdot n|d^3x|, \tag{5.35}$$

which, like (5.32), holds for *any* closed 2-manifold $\mathcal{B}$, where the integral on the right is over any 3-manifold with boundary $\mathcal{B}$. Again a spacetime split reveals that (5.35) is equivalent to Ampere's Law, Gauss' Law, or a combination of the two, depending on the choice of $\mathcal{B}$.

The two integral equations (5.32) and (5.35) are fully equivalent to the two parts of Maxwell's equations (5.30) and (5.31). They can be combined into a single equation. First multiply (5.35) by i and use (5.34) to put it in the less familiar form

$$\oint (d^2x) \wedge F = \int (d^3x) \wedge J. \tag{5.36}$$

Adding (5.32) to (5.36), we can write the integral version of the whole Maxwell's equation (5.2) in the form

$$\oint < d^2xF >_I = \int < d^3xJ >_I, \tag{5.37}$$

where $< \ldots >_I$ selects only the "invariant (= scalar+pseudoscalar) parts." I have not seen Maxwell's equation in the form (5.37) before. I wonder if this form has some slick physical applications.

References

1. Hestenes, D.: 1986, 'A Unified Language for Mathematics and Physics,' *Clifford Algebras and their Applications in Mathematical Physics*, J.S.R. Chisholm/A.K. Common (eds.), Reidel, Dordrecht/Boston, pp. 1–23.
2. Hestenes, D.: 1988, 'Universal Geometric Algebra,' *Simon Stevin* **62**, pp. 253–274.
3. Hestenes, D.: 1991, 'Mathematical Viruses,' *Clifford Algebras and their Applications in Mathematical Physics*, A. Micali, R. Boudet, J. Helmstetter (eds.), Kluwer, Dordrecht/Boston, pp. 3–16.
4. Hestenes, D.: 1993, 'Hamiltonian Mechanics with Geometric Calculus,' Z. Oziewicz, A. Borowiec, B. Jancewicz (eds.), *Spinors, Twistors and Clifford Algebras* Kluwer, Dordrecht/Boston.
5. Hestenes, D. and Sobczyk, G.: 1984, *CLIFFORD ALGEBRA TO GEOMETRIC CALCULUS, A Unified Language fo Mathematics and Physics*, D. Reidel Publ. Co., Dordrecht, paperback 1985, Third printing 1992.
6. Doran, C., Hestenes, D., Sommen, F. & Van Acker, N.: 'Lie Groups as Spin Groups,' *Journal of Mathematical Physics* (accepted).
7. O'Neill, B.: 1983, *Semi-Riemannian Geometry*, Academic Press, London.
8. Hestenes, D.: 1966, *Space-Time Algebra*, Gordon and Breach, New York.
9. Wills, A.P.: 1958, *Vector Analysis with an Introduction to Tensor Analysis*, Dover, New York.
10. Hestenes, D.: 1968, 'Multivector Calculus,' *J. Math. Anal. and Appl.* **24**, pp. 313–325.
11. Hestenes, D.: 1968, 'Multivector Functions,' *J. Math. Anal. and Appl.* **24**, pp. 467–473.
12. Delanghe, R.: 1970, 'On regular-analytic functions with values in a Clifford algebra,' *Math. Ann.* **185**, pp. 91–111.
13. Sobczyk, G.: 1992, 'Simplicial Calculus with Geometric Algebra,' *Clifford Algebras and Their Applications in Mathematical Physics*, A. Micali, R. Boudet and J. Helmstetter (eds.), Kluwer, Dordrecht/Boston, pp. 279–292.
14. Hestenes, D.: 1974, 'Proper Particle Mechanics,' *J. Math. Phys.* **15**, 1768-1777.
15. Barut, A.: 1980, *Electrodynamics and the classical theory of fields and particles*, Dover, New York.

CLIFFORD VALUED CONVOLUTION OPERATOR ALGEBRAS ON THE HEISENBERG GROUP

A Quantum Field Theory Model

VLADIMIR V. KISIL*
Department of Mathematics
Odessa State University
ul. Petra Velikogo, 2
Odessa, 270000, UKRAINE.

Abstract. We study the pseudodifferential convolution type operator algebras. These convolutions are defined on the Heisenberg group $\mathbb{H}^n$ (which is a step 2 nilpotent Lie group) and act on Clifford algebra $\mathrm{Cl}(p,q)$ valued functions. The quantum field theory leads to the studied algebras exactly. The physical interpretations of the Heisenberg group $\mathbb{H}^n$ and Clifford algebras $\mathrm{Cl}(p,g)$ are given in Section 1. This problem is related with the so-called superanalysis (analysis of commutative and anticommutative variables), quickly developing now. The irreducible unitary representations of our algebras are described in Section 2.

Key words: Clifford algebras, Heisenberg group, Representation theory, Quantum field theory

1. Physical Background

In this section the detailed description of our algebras physical origin is given. The consideration of quantum operators in subsection 1.1 and 1.2 leads to convolution operator algebras on the Heisenberg group. State of identical Fermi particles ensemble is written as a Clifford algebra element in subsection 1.3. There is new description of matter and anti-matter duality in subsection 1.4.

1.1. Quantum Mechanics: Coordinates + Impulses = the Heisenberg Group

Let us consider a particle on n-dimension space. Postulate in quantum mechanics says that observables corresponding to classic coordinates and impulses can be identified with some operators Q_i and P_i $(i = 1, \ldots, n)$. These operators have to satisfy to the famous Heisenberg-Weyl commutation relations [3]:

$$[Q_i, P_j] = i\hbar\delta_{ij}I, \ [Q_i, Q_j] = [P_i, P_j] = 0. \tag{1}$$

We do not consider any dependence of our theory upon Plank constant $\hbar$, so we can set $\hbar = 1$ here. If we also change P_k for $-iP_k$, more simple and suitable relations

* I am grateful to *University of Gent* (België) and the *Ukrainian Renaissance foundation* offered grants me for participation at the Conference.

F. Brackx et al. (eds.), Clifford Algebras and their Applications in Mathematical Physics, 287–294.

are obtained:

$$[Q_i, P_j] = \delta_{ij} I,\ [Q_i, Q_j] = [P_i, P_j] = 0. \tag{2}$$

Operators Q_i, P_i and I with the commutator defined by (2) are generators of a nilpotent Lie algebra $\mathfrak{h}(n)$, called the Weyl algebra. The nilpotent Lie group, associated with Lie algebra $\mathfrak{h}(n)$, is called the Heisenberg(–Weyl) group and usually denoted by $\mathbb{H}^n$. As a C^∞-manifold it coincides with $\mathbb{R}^{2n+1}$. If an element of it is given in the form $g = (u, v) \in \mathbb{H}^n$, where $u \in \mathbb{R}$ and $v = (v_1, \ldots, v_n) \in \mathbb{C}^n$, then the group law on $\mathbb{H}^n$ can be written as

$$(u, v) * (u', v') = \left(u + u' - 2\,\mathrm{Im} \sum_1^n v'_k \overline{v_k},\ v_1 + v'_1, \ldots, v_n + v'_n\right). \tag{3}$$

We single out on $\mathbb{H}^n$ the group of nonisotropic dilations $\{\delta_\tau\}$, $\tau \in \mathbb{R}_+$:

$$\delta_\tau(u, v) = (\tau^2 u, \tau v).$$

Functions with the property $(f \circ \delta_\tau)(g) = \tau^k f(g)$ will be called *δ_τ-homogeneous functions of degree k*. The class of such functions having continuous restriction to the nonisotropic unit sphere $\Omega^{2n} := \{(u, v) \in \mathbb{H}^n \mid u^4 + |v|^2 = 1\}$ is denoted by $H_\delta^k(C(\Omega^{2n}))$.

The left and the right Haar measure on the Heisenberg group coincides with the Lebesgue measure. The operators of right, left [13] and two-sided [9, 15] convolution on the Heisenberg group with kernel $k(g)$ are introduced as the integrals of the shift operators $\pi_l(g)$ and $\pi_r(g)$ giving rise to the regular representation of the Heisenberg group $\mathbb{H}^n$ on the space $L_2(\mathbb{H}^n)$:

$$K_r = (2\pi)^{-N/2} \int_{\mathbb{H}^n} k(g) \pi_r(g)\, dg, \tag{4}$$

$$K_l = (2\pi)^{-N/2} \int_{\mathbb{H}^n} k(g) \pi_l(g)\, dg, \tag{5}$$

$$K = (2\pi)^{-N} \int_{\mathbb{H}^n} \int_{\mathbb{H}^n} k(g_1, g_2) \pi_l(g_1) \pi_r(g_2)\, dg_1\, dg_2, \tag{6}$$

where $N = 2n + 1$. We do not discuss properties of the kernel $k(g)$ now, it can be taken from various classes. For example, convolution type operators (4)–(6) with singular kernels from $H_\delta^0(C(\Omega^{2n}))$ [15] or L_1 [13] kernels are bounded.

1.2. Quantum Mechanics: Other Observables

In this context the other observables can be written as functions of the operators Q_i and P_i. The algebra of convolution type operators (4) or (5) on the Heisenberg group $\mathbb{H}^n$ is a usual way to achieve this observables. Indeed, there is (unbounded) representation of $\mathfrak{h}(n)$ on $L_2(\mathbb{R}^n)$, defined by formulas:

$$Q_i \mapsto x_i I,\ P_i \mapsto \frac{\partial}{\partial x_i}, \tag{7}$$

This representation can be expanded to a convolution algebra on the Heisenberg group $\mathbb{H}^n$. If a convolution K_r on $\mathbb{H}^n$ with a kernel $k(x_1, \ldots, x_n, y_1, \ldots, y_n)$ corresponds to an observable $k(q_1, \ldots, q_n, p_1, \ldots, p_n)$ of a classic system, then our representation maps the convolution K_r to the pseudodifferential operator $k(x_1, \ldots, x_n,$

$\frac{\partial}{\partial x_1}, \ldots, \frac{\partial}{\partial x_n})$. Reader may easily recognize the method of quantization by Schrödinger in the last formula.

There is an unitary form of representation (7), moreover, this form is the unique one (in sense of unitarily equivalence) by Stone–von Neumann theorem:

$$\pi_{\pm\lambda}(t, x, y) = e^{i(\pm\lambda t I \pm \lambda^{1/2} yX + \lambda^{1/2} xD)}, \tag{8}$$
$$\pi_{(u,v)}(t, x, y) = e^{i(ux+vy)}. \tag{9}$$

where:

$$(yX)u(z) = \sum y_j z_j u(z),$$
$$(xD)u(z) = (\frac{1}{i}) \sum x_j \frac{\partial u}{\partial z_j},$$
$$(t, x, y) \in \mathbb{H}^n, \qquad (u, v) \in \mathbb{R}^{2n}.$$

Then relative to (8)–(9) representations of convolution algebra are expressed by formulas:

$$\pi_{\pm\lambda}[k(t, x, y)] = \hat{k}(\pm\lambda, \pm\lambda^{1/2} X, \lambda^{1/2} D), \tag{10}$$
$$\pi_{(u,v)}[k(t, x, y)] = \hat{k}(0, u, v). \tag{11}$$

Let us remind the quantization scheme by Berezin [1]. They say that there is a quantization, if

1. There is a family of operator algebras $\{\Omega_\hbar \mid \hbar \in \mathbb{R}^+\}$, where
   a) algebras $\Omega_\hbar$ for $\hbar \neq 0$ are non-commutative algebras of operators on some Hilbert spaces;
   b) the algebra $\Omega_\hbar$ for $\hbar = 0$ is a commutative algebra of functions on $\mathbb{R}^{2n}$.
2. There is a topology on $\mathbb{R}^+$ such that:
   a) there are limits[1] $\lim_{\hbar \to 0} A_\hbar = A_0$, for $A_\hbar \in \Omega_\hbar$;
   b) for any $A_\hbar$ and $B_\hbar$ the following equalities hold:

$$\lim_{\hbar \to 0} A_\hbar \circ_\hbar B_\hbar = A_0 \cdot B_0 \tag{12}$$
$$\lim_{\hbar \to 0} \frac{i}{\hbar}[A_\hbar, B_\hbar]_\hbar = \{A_0, B_0\}. \tag{13}$$

Here $\circ_\hbar$ and $\cdot$ in (12) denote the operator composition in $\Omega_\hbar$ and usual product in $L_2(\mathbb{R}^{2n})$ correspondingly. As usual we denote in (13) the commutator of two operators by $[\cdot,\cdot]_\hbar$, and $\{\cdot,\cdot\}$ is the Poisson brackets of two functions.

Let π_0 denote the map $k(t, x, y) \mapsto \hat{k}(0, u, v)$ from (11). Then we have the family of mapping $\pi_\hbar$ from (10)–(11) from a space of function on $\mathbb{H}^n$ to a space of operators on $L_2(\mathbb{R}^n)$ depended on the parameter $\hbar (:= \lambda) \in \mathbb{R}^+$. For $\hbar = 0$ the map $\pi_\hbar$ acts to a (commutative) algebra of functions on $\mathbb{R}^{2n}$. It was showed in papers [5, 7, 9, 14], that for kernels $k(g) \in L_1(\mathbb{H}^n)$ or $H^0_\delta(C(\Omega^{2n}))$ we have $\pi_\hbar \to \pi_0$ in Jacobson topology

[1] The more exact meaning of these limits is following: $\lim_{\epsilon \to 0} a_\epsilon = b$ iff b belong to the closure of the set $\cup_{t \leq \epsilon} a_t$ for all ϵ.

sense when $\hbar \to 0$. The validity of equation (12) easily follows from general properties of representations. Formula (13) may be checked by use of the symbol calculus developed in [8, 14, 15] and the usual technique of pseudodifferential operators (see, for example, [4, 12]). Thus we may conclude:

1. *Algebra of convolutions on the Heisenberg group contains both classical and quantum descriptions of a particle on n-dimension space;*
2. *When we studying the spectrum and representations of convolution operator algebras on the Heisenberg group, we have found a quantization scheme by the Berezin's meaning.*

1.3. Quantum Field Theory and Clifford Algebras

The passing from a single particle system to a quantum ensemble consisting of m ones depends on kind of particles we have. For example [3], photons satisfy to Bose-Einstein statistics, but electrons satisfy to Fermi-Dirac statistics. This means that observables of a photon ensemble are symmetrical operators relative to equivalent particles; but observables of an electron ensemble are antisymmetric one. This involves, in particular, the Pauli's exclusion principle: there is only one or none Fermi particle in each of different states. Such distributions of particles among different states can be easily described with the help of Clifford algebra $\mathbf{Cl}(m, 0)$. The notation $\mathbf{Cl}(p, q)$ means the complex algebra generated by the identity 1 and imaginary units e_i $(1 \leq i \leq p+q)$ with following relations [6, 13]:

$$e_i e_j = -e_i e_j,\ e_i^2 = 1\ (1 \leq i \leq p),\ e_i^2 = -1\ (p+1 \leq i \leq p+q). \tag{14}$$

Then an element of $\mathbf{Cl}(m, 0)$ can be written as the linear combination of the monomials:

$$\sum_{i_k = 0 \text{ or } 1} a_{i_1 i_2 \ldots i_m} e_1^{i_1} e_2^{i_2} \cdots e_m^{i_m}. \tag{15}$$

The monomial $e_1^{i_1} e_2^{i_2} \cdots e_m^{i_m}$ can be considered as the description of a physical system, in which the only states enumerated by k, for which $i_k \neq 0$, are filled by a Fermi particle. The general element (15) of $\mathbf{Cl}(m, 0)$ can be interpreted [3] as some probability distribution among "pure" states defined by such monomials[2].

1.4. The Matter and Anti-Matter Duality and Two-Sided Convolution Type Operators

The usual tools in quantum field theory are not operators of coordinates and impulses from (1) but creation and annihilation operators of particles. For Bose particles they may be defined as follows:

$$\eta_k^+ = (2\hbar)^{-1/2}(P_k + iQ_k),\ \eta_k^- = (2\hbar)^{-1/2}(P_k - iQ_k). \tag{16}$$

If operators P_k and Q_k satisfy to relations (1) then operators η_k^+ and η_k^- satisfy to (2) with substitution instead of P_k and Q_k correspondingly. It is well known, there is another set of operators μ_k^+ and μ_k^- are also satisfying to (16) and are commuting

[2] We study now the case of finite, but arbitrary number of different states only. The question on infiniteness of our world is more philosophical, than mathematical one.

with both η_k^+ and η_k^-. These operators may be interpreted as creation and annihilation operators of corresponding anti-particles. Thus operators $\eta_k^+, \eta_k^-, \mu_k^+$ and μ_k^- form the algebra $\mathfrak{h}(2n)$ and, obviously, generate group $\mathbb{H}^{2n}$. But there is another way to produce these operators: we may identify the algebra generated by η_k^+ and η_k^- (μ_k^+ and μ_k^-) with algebra of left (5) (right (4)) convolutions on $\mathbb{H}^n$ correspondingly. Thus an algebra of two-sided convolutions (6) realizes observables of a system containing particles and corresponding anti-particles. The relations between symbols and kernels of convolutions in these two descriptions on $\mathbb{H}^n$ and $\mathbb{H}^{2n}$ was studied in [7]. We have following interpretation of the main result from [7]:

– *a system with matter and anti-matter is equivalent to a system of a higher dimension containing only matter.*

Different symmetries (transformations, which do not change the algebra of observables) are very important in physics. It is well known (see, for example [8, 9]), that an algebra of convolutions on $\mathbb{H}^n$ does not invariant under actions of contact transformations, but algebras generated by convolutions and multiplications by continuous functions do. So we have to extend the algebra under our consideration by operators of multiplications by functions continuous on one-point compactification of the Heisenberg group $\mathbb{H}^n$.

One can summarize the consideration of this section by the following way: to describe of an ensemble of Fermi particles we have to study the algebra of operators generated by two-sided convolution type operators like (6) with kernels from $H_\delta^k(C(\Omega^{2n})) \otimes H_\delta^k(C(\Omega^{2n}))$ acting on the following functions:

$$f(h) : \mathbb{H}^n \to \mathbf{Cl}(m, 0).$$

Let us introduce the algebra $\tilde{\mathfrak{S}}$, which is the algebra of two-sided convolutions (6) on the Heisenberg group $\mathbb{H}^n$ with kernels from $H_\delta^{-2n-2}(C(\Omega^{2n})) \otimes H_\delta^{-2n-2}(C(\Omega^{2n}))$. Let algebra $\mathfrak{S}$ be generated by two-sided convolutions (6) and by operators $a(h)I$ of multiplication by continuous functions on one-point compactification $\mathbb{H}^n$,whish are acting on $\mathbb{C}$-valued functions. Let algebra $\mathfrak{C}$ be the same algebra but acting on $\mathbf{Cl}(m, 0)$-valued functions. The description of algebra $\mathfrak{C}$ representations is main goal of this paper.

2. The Spectrum and Irreducible Unitary Representations of Algebra $\mathfrak{C}$

The important property of our problem is the noncommutativity of both the group operation (3), which generates convolutions (6), and of the multiplication (14) on Clifford algebra $\mathbf{Cl}(m, 0)$. It is well known that each of these noncommutativity leads up to serious difficulties [10, 15].

The C^*-algebra $\mathfrak{C}$ generated by our operators is the crossed product of $\mathbf{Cl}(m, 0)$ by trivial action of $\mathbb{H}^n$, but this is then nothing but the tensorproduct of $\mathbf{Cl}(m,)$ by the C^*-algebra $\mathfrak{S}$ of two-sided convolutions [11]. So we write the well known representation of $\mathbf{Cl}(m, 0)$ in Subsection 2.1 and repeat results [9, 14] concerned to the algebra $\mathfrak{S}$ in Subsection 2.2.

2.1. The Representations of $\mathbf{Cl}(2l, 0)$

In this subsection we discuss the physical meaning of representations of $\mathbf{Cl}(m, 0)$. It was be shown above, that this algebra generated by the creation and annihilation operators. The total number of such operators is even, of course. So by physical reason, we could consider the case of even $m = 2l$ only. It is coincidence or not, but this case is a little bit technically simpler than the odd one.

Decomposition on irreducible representations of observables algebra corresponds to decomposition of a physical system on subsystems, which is independent one from anothers. It is well known, that Clifford algebra $\mathbf{Cl}(2l, 0)$ is a full matrix algebra $M(2^l)$ in 2^l-dimension space [2]. Thus there is not any invariant subspace under action of $\mathbf{Cl}(2l, 0)$ and this representation is irreducible.

However not only representations of an algebra itself are important, but representations of different groups in this algebra are significant too. The representation of a group on an operator algebra shows us, that the algebra is invariant under the group action. Such symmetry groups play an important role in the physics. For example, Dirac introduced the Clifford algebra in his famous electron equation to preserve the Lorentz form. We remind only one representation on $\mathbf{Cl}(p, q)$, so-called the spinors one [2, 13]. It is a representation of $\mathrm{O}(p, q)$ group, which preserved the quadratic form $x_1^2 + x_2^2 + \cdots + x_p^2 - x_{p+1}^2 - \cdots - x_{p+q}^2$ on $\mathbb{R}^{p+q}$.

2.2. The Representations of Algebra $\mathfrak{S}$

The description of algebra $\mathfrak{S}$ may be obtained by "local principle". Let us repeat here main result from [8, 9, 14].

Theorem 1 *[9] At the point $(t^0, x^0, y^0) \in \mathbb{H}^n$ $((x^0, y^0) \neq 0)$ the algebra of two-sided convolutions that is generated by the right (4) and left (5) convolutions on the Heisenberg group is locally equivalent to the algebra of triples $(\omega_l, \omega_r, \omega_e)$ of symbols corresponding to the left, right, and ordinary* (Euclidean) *convolutions. The localization mapping acts as follows on the generators of the algebra:*

$$\begin{aligned} K_r &\mapsto (\hat{k}(-\lambda, -\xi - 2\lambda y, -\nu + 2\lambda x),\ \hat{k}(0, -y^0, x^0), -\hat{k}(0, -y^0, x^0)),\\ K_l &\mapsto (\hat{k}(0, -y^0, x^0), \hat{k}(\lambda, \xi - 2\lambda y, \nu + 2\lambda x), -\hat{k}(0, -y^0, x^0)). \end{aligned}$$

The algebraic operations act as follows on triples:

$$\begin{aligned} (\omega_l, \omega_r, \omega_e) &+ (\omega_l', \omega_r', \omega_e')\\ &= (\omega_l + \omega_l', \omega_r + \omega_r', \omega_e + \omega_e'),\\ (\omega_l, \omega_r, \omega_e) &\# (\omega_l', \omega_r', \omega_e')\\ &= (\omega_l \# \omega_l', \omega_r \# \omega_r', (\mathring{\omega}_r - a)(\mathring{\omega}_l' - b) + (\mathring{\omega}_l - a)(\mathring{\omega}_r' - b) + \omega_e \cdot \omega_e'), \end{aligned}$$

where $\mathring{\omega}$ denotes the freezing of the direct variable $h = h_0$ in the function $\omega(h, \nu)$.

We remark that the law of composition of operator triples is similar to the group law (3) defined in $\mathbb{H}^n$.

Let $\mathfrak{S}_{(x_0, y_0)}$ be the algebra of triples described in Theorem 1 and $\Psi(CS(\mathbb{R}^{2n}))$ be the algebra of pseudodifferential operators on $L_2(\mathbb{R}^n)$ with continuous on $\mathbb{R}^{2n}$ Weyl symbols $a(h,\xi)$ such that there are all limits $\lim_{t\to\infty} a(th,t\xi)$ [4, 12, 14]. Then we can write the spectrum of two algebras $\tilde{\mathfrak{S}}$ and $\mathfrak{S}_{(x_0,y_0)}$ by following way:

Theorem 2 *[14] There are all irreducible representations of algebra $\tilde{\mathfrak{S}}$:*
1. Two infinite-dimensional representations (see (10)–(11))

$$\pi_{i\pm} = \pi_{\pm\lambda} \times \pi_{\pm\lambda};$$

2. Four sets of infinite-dimensional representations depending on point $(y,\nu) \in S^{2n-1}$

$$\pi_{1\pm}(y,\nu) = \pi_{\pm\lambda} \times \pi_{(y,\nu)},\quad \pi_{2\pm}(y,\nu) = \pi_{(y,\nu)} \times \pi_{\pm\lambda};$$

3. The set of one-dimensional representations depending on point $(x,\xi,y,\nu) \in S^{2n-1} \times S^{2n-1}$

$$\pi_{(x,\xi,y,\nu)} = \pi_{(x,\xi)} \times \pi_{(y,\nu)}.$$

So spectrum of algebra $\tilde{\mathfrak{S}}$ may be identified with set[3]

$$sp\tilde{\mathfrak{S}} = \{i_+, i_-\} \cup S_{1-}^{2n-1} \cup S_{1+}^{2n-1} \cup S_{2-}^{2n-1} \cup S_{2+}^{2n-1} \cup \{S^{2n-1} \times S^{2n-1}\}. \tag{17}$$

Theorem 3 *[8] There are all representations of algebra $\mathfrak{S}_{(x_0,y_0)}$*
1. Four infinite-dimensional representations

$$\pi_{1\pm}(y,\nu) = \pi_{\pm\lambda} \times \pi_{(-y_0,x_0)},\quad \pi_{2\pm}(y,\nu) = \pi_{(-y_0,x_0)} \times \pi_{\pm\lambda};$$

2. One-dimensional representations depending on point $(x,\xi,y,\nu) \in S^{2n-1} \times S^{2n-1}$ such that three points (x,ξ), (y,ν) and $(-y_0,x_0)$ belong to a 2-dimensional plane

$$\pi_{(x,\xi,y,\nu)} = \pi_{(x,\xi)} \times \pi_{(y,\nu)}.$$

Thus spectrum of algebra $\mathfrak{S}_{(x_0,y_0)}$ is

$$sp\mathfrak{S}_{(x_0,y_0)} = \{i_{1-}, i_{1+}, i_{2-}, i_{2+}\} \cup S, \tag{18}$$

where S is the subset of $S^{2n-1} \times S^{2n-1}$ defined in 2.

It was mentioned above, that algebra $\mathfrak{S}$ locally coincides
- with an algebra of two-sided convolutions $\tilde{\mathfrak{S}}$ at the points $(t_0,0,0) \in \mathbb{H}^n$,
- and with an algebra $\mathfrak{S}_{(x_0,y_0)}$ of triples from Theorem 1 at the points $(t_0,x_0,y_0) \in \mathbb{H}^n$, where $(x_0,y_0) \neq 0$.

So we can easily describe the spectrum of $\mathfrak{S}$ through spectrums of algebras $\tilde{\mathfrak{S}}$ and $\mathfrak{S}_{(x_0,y_0)}$:

Theorem 4 *[8] The spectrum of $\mathfrak{S}$ is the set*

$$sp\mathfrak{S} = (\mathbb{R} \times sp\tilde{\mathfrak{S}}) \cup \big((\mathbb{H}^n \backslash \mathbb{R}) \times sp\mathfrak{S}_{(x_0,y_0)}\big).$$

[3] We do not describe Jacobson topology on the spectrum here, it showed at [8, 14].

Acknowledgements

I am grateful to Professor N. L. Vasilevski for guidance of this research and Professor O. Bratteli for helpful discussion of this paper. I would like also to express my gratitude to Professor M. V. Shapiro attracted my attention to beauty of Clifford algebras theory at first time.

References

1. F. Berezin, *Method of second quantization*, "Nauka", Moscow, 1988. (russian)
2. R. Brauer and H. Weyl, *Spinors in n-dimensions space*, Amer. Journ. Math. **57** (1935), no. 425.
3. P. Dirac, *Lectures on quantum field theory*, Yeshiva University, New York, 1967.
4. L. Hörmander, *The analysis of linear partial differential operators III: Pseudodifferential operators*, Springer–Verlag, Berlin Heidelberg New York Tokyo, 1985.
5. D. S. Kalyuzhny, *The C^*-algebra, generated by the operators of two-sided convolution on Heisenberg group*, Dokl. Acad. Nauk SSSR. (to appear)
6. A. A. Kirillov, *Elements of the theory of representations*, Springer, New York, 1974.
7. V. V. Kisil, *Connection between two-sided and one-sided convolution type operators on non-commutative group*, Math. Nachr. (to appear)
8. ———, *Algebras of pseudodifferential operators associated with the Heisenberg group*, PhD dissertation, Odessa State University, Department of Mathematics, 1991. (russian)
9. ———, *Two-sided convolution operators algebra on Heisenberg group*, Dokl. Acad. Nauk SSSR **325** (1992), no. 1, 20–23. (russian)
10. M. V. Shapiro and N. L. Vasilevski, *Quaternionic ψ-hyperholomorphic functions, singular operator with quaternionic Cauchy kernel and analogies of the Riemann boundary value problem*, Reporto Interno #102, Departamento de Matematicas, CINVESTAV del I.P.N., Mexico City, 1992.
11. M. Takesaki, *Duality for crossed products and the structure of von Neumann algebras of type III*, Acta Math. **131** (1973), no. 3-4, 249–310.
12. M. E. Taylor, *Pseudodifferential operators*, Princeton University Press, Princeton, New Jersey, 1981.
13. ———, *Noncommutative harmonic analysis*, Math. Surv. and Monographs, vol. 22, American Mathematical Society, Providence, Rhode Island, 1986.
14. N. L. Vasilevski and R. Trujillo, *Group convolutions on standard Cauchy-Riemann manifolds*, Preprint 17, Ser. B, Math. Inst. Polish Acad. Sci, 1988. (russian)
15. ———, *Group convolution operators on standard CR-manifold. I. Structural properties*, Reporto Interno #103, Departamento de Matematicas, CINVESTAV del I.P.N., Mexico City, 1992.

CLASSICAL SOLUTIONS OF THE DIRAC EQUATION: BOUND COULOMB STATES IN AHARONOV–BOHM AND ZEEMAN FIELDS

HEINZ KRÜGER
Fachbereich Physik der Universität
Postfach 3049
D-6750 Kaiserslautern
West Germany

Abstract. Exact solutions ψ of the real Dirac equation are called classical, if the biquaternion ψ satisfies $\psi\tilde{\psi} \in R$, i.e., the angle of Yvon and Takabayasi is restricted to the values 0 and π. *On a point set, defined by* $\psi\tilde{\psi} = 0$, *the conventional polar decomposition* $\psi = (\psi\tilde{\psi})^{1/2}R$, $R\tilde{R} = 1$, $iR = Ri$, *does not hold.* This zero–divisor problem is solved in terms of a new decomposition of the radial spinor in the algebra $C^{+}(1,2) \simeq C(2)$. A Riccati equation is obtained, whose general solution determines the radial spinor completely. By means of (spin) bivector–valued Mbius transformations of the dependent Riccati–variable the corresponding simplest linear differential equation of the second order may be found.

With this new method, classical bound Coulomb states in the presence of particular magnetostatic fields are treated exactly. The magnetic field strength has to be directed everywhere parallel or antiparallel to the local angular momentum $\vec{r} \times \vec{v}$ of a streamline $\dot{\vec{r}} \times \vec{v} = \vec{0}$, and its magnitude has to be restricted to depend on the distance $r = |\vec{r}|$ from the center $\vec{r} = \vec{0}$ only. An Aharonov–Bohm field microlocally concentrated on a polar axis and other less–singular supported fields are contained in this family. The influence of the Aharonov–Bohm field on the fine–structure levels results in a shift of the angular momentum quantum number to a non–integer effective value. The less–singular supported fields induce Zeeman–like energy shifts and splittings.

1. Introduction

A complete separation of the Dirac equation in spherical polar coordinates was published by Erwin Schrödinger [1]. Without knowing Schrödinger's work, I repeated essentially the same steps [3] in the real form of the Dirac theory as developed by David Hestenes [2]. On page 353 of [1], Schrödinger concludes "Wir dürfen hoffen, keine Eigenlösungen unseres Problems übersehen zu haben (eine vollständige Diskussion wäre freilich erwünscht)". — In fact, Schrödinger first of all overlooked and finally discarded a whole set of square–integrable, two–valued solutions, namely those defined by eqs. (3.1) and (3.2) in [3].

In section 2 of the present article it is shown, that the approximate classical solutions obtained from quantized Sommerfeld orbits in section 6 of [4], precisely lead to those overlooked exact classical solutions. One only needs to replace the approximate radial spinor part by an exact solution of the radial eq. (4.1) in [3]. The task of constructing semiclassical central field solutions for the Dirac eq. therefore may be restricted to the radial algebra $C(2)$. The finding that the relativistic classical equations of motion for a pointcharge in a Coulomb field remain completely integrable when a particular magnetic as the one described in the abstract is added,

F. Brackx et al. (eds.), Clifford Algebras and their Applications in Mathematical Physics, 295–306.

leads to *the conjecture that exact classical solutions of the Dirac eq. also may exist in this case.* The proof of this conjecture supplies the main result of section 2.

A new decomposition of the radial spinor $\eta \in C(2)$ is presented in section 3. In contrast to the polar form $\eta = \sqrt{\eta\tilde{\eta}}R$, this decomposition also holds for $\eta\tilde{\eta} = 0$. Making use of the new decomposition, the radial spinor eq. can be reduced to a single Riccati eq.. Equivalent other Riccati eqs. are obtained from the first one if (spin) bivector–valued Mbius transformations are applied to the dependent Riccati–variable. This is exploited in order to find the corresponding simplest linear differential eq. of the second order, which decouples the radial spinor eq..

In section 4, this new decoupling method is applied to the calculation of bound Coulomb states in the presence of the above mentioned Aharonov–Bohm field and a simple Zeeman–like field. A modified fine–structure formula is derived, which displays the effect of these additional fields on the energy eigenvalues. The degeneracy of the solutions is investigated.

Finally, in section 5, the radii of the spheres on which $\psi\tilde{\psi} = 0$ are calculated for a few of the lowest excited states. The ground state turns out to be the only one with the property that $\psi\tilde{\psi} > 0$ everywhere. Sign changes of $\psi\tilde{\psi}$ are obtained for the excited states in the regions between the spheres on which $\psi\tilde{\psi} = 0$.

2. Classical Solutions of the Real Dirac Equation

A classical limit for stationary central field solutions of the real Dirac eq. has been derived in [4], sections 4–6. In terms of the spherical coordinates r, ϑ, φ defined by [4] eqs. (79 b,c) or [3], eqs. (2.6) – (2.8)

$$\begin{gathered}\vec{r} = r\hat{r} = rS\vec{\sigma}_3\tilde{S}, \quad S = e^{-i_3\varphi/2}e^{-i_2\vartheta/2}, \quad i_k = i\vec{\sigma}_k = i\gamma_k\gamma_0, \quad k = 1,2,3, \\ 0 < r < \infty, \quad 0 < \vartheta < \pi, \quad 0 < \varphi < 2\pi,\end{gathered} \tag{2.1}$$

the spin–quantized *approximate* classical solutions of the real Dirac eq. (1) in [4] may be written in the form

$$\psi_c = \frac{L(\vec{\sigma}_3 \pm \hat{l})\vec{\sigma}_3}{r\sqrt{2\sin\vartheta}}\, e^{\frac{-i_3}{\lambda}[\varepsilon r_0 + l\vartheta + \phi_r + \phi_0(r,\varphi)]}, \tag{2.2}$$

where

$$\hat{l} = -S\vec{\sigma}_2\tilde{S} = -\vec{\sigma}_2 e^{i_3\varphi} = \frac{\vec{r}\times\vec{\sigma}_3}{|\vec{r}\times\vec{\sigma}_3|} \tag{2.3}$$

is the classical orbital angular momentum unitvector [4], eq. (76b), and

$$\phi_r = \int |dr|\sqrt{(\varepsilon - V)^2 - 1 - \frac{l^2}{r^2}} = \int |dr|\,|\dot{r}| \tag{2.4}$$

is the radial action integral [4], eq. (81). The phase ϕ_0 will appropriately be chosen below. According to [4], eqs. (78a), (79e), the spinor L in (2.2) is given by

$$L = \frac{1 + \varepsilon - V + \hat{r}\left(\dot{r} + i\frac{l}{r}\hat{l}\right)}{\sqrt{2(1+\varepsilon - V)}} = S\eta_c(r)\tilde{S}. \tag{2.5}$$

With the help of $(\vec{\sigma}_3 \pm \hat{l})\vec{\sigma}_3 = \sqrt{2}e^{-i_3\varphi/2}e^{\mp i_1\pi/4}e^{i_3\varphi/2}$ and

$$e^{\mp i_1\pi/4}\vec{\sigma}_3 = \mp\vec{\sigma}_2 e^{\mp i_1\pi/4}, \tag{2.6}$$

eq. (2.2) may be rearranged to

$$\psi_c r\sqrt{\sin\vartheta} = S\eta_c\, e^{\pm i_2\phi_r/\lambda}\, e^{\pm i_2\vartheta\left(\frac{l}{\lambda}\pm\frac{1}{2}\right)}\, e^{\mp i_1\pi/4}\, e^{-\frac{i_3}{\lambda}(\varepsilon r_0 + \phi_0 - \lambda\varphi/2)}. \tag{2.7}$$

Comparing this form of ψ_c with [3], eqs. (2.14) – (2.16), (3.1) – (3.2) for $\lambda = 0$, one notes, that ψ_c becomes an exact classical solution ψ, $\psi\tilde{\psi} \in I\!R$, of the real Dirac eq.

$$\lambda\partial\psi i_3 - \mathcal{A}\psi = \psi\gamma_0, \quad \lambda = \frac{\hbar}{mc}, \quad mc^2\mathcal{A} = q(A_0 + \vec{A})\gamma_0, \tag{2.8}$$

provided one chooses $2\phi_0 = \lambda\varphi$ and replaces the approximate radial spinor $\eta_c(r)e^{\pm i_2\phi_r/\lambda}$ in (2.7) by an *exact solution* η of the radial eq. (2.19) in [3]. This state of affairs will now be investigated in detail with the ***inclusion of the special magnetic field as defined in the abstract.***

Guided by (2.7), I put

$$\psi r\sqrt{\sin\vartheta} = S\eta(r)\, e^{\pm i_2 k\vartheta}\, e^{\mp i_1\pi/4}\, e^{-\frac{i_3}{\lambda}\varepsilon r_0}, \quad k \in I\!N, \tag{2.9}$$

and infer from (2.8) a radial eq. for η by imposing the condition that (2.9) be an exact solution for

$$mc^2\mathcal{A} = \gamma_0\left[mc^2V(r) - \hbar c\frac{\beta(r)}{r}S\vec{\sigma}_1\tilde{S}\right]. \tag{2.10}$$

So, the four–potential $(A_0+\vec{A})\gamma_0$ is composed of the central potential $qA_0 = mc^2V(r)$ and of the vector potential

$$qr\vec{A} = -\hbar c\beta(r)\hat{l}\times\hat{r} = \hbar c\beta(r)S\vec{\sigma}_1\tilde{S}, \tag{2.11}$$

corresponding to the magnetic field $\vec{B} = \vec{\partial}\times\vec{A}$ according to

$$qr\vec{B} = -\hbar c\frac{d\beta}{dr}\hat{l}, \quad qr^2\vec{\partial}\cdot\vec{A} = \hbar c\beta\frac{\vec{r}\cdot\vec{\sigma}_3}{|\vec{r}\cdot\vec{\sigma}_3|}. \tag{2.12}$$

For completenes, let me repeat those formulae from [3], section 2, which are relevant for the deduction of the desired radial eq.. The unitary spinor S, which by eqs. (2.1) defines a Spin (3) representation of the spherical coordinates, induces a spinor factorization of the vector–valued derivative (gradient operator) both on $I\!E^3$

$$\vec{\partial} = \Omega\vec{\partial}'\Omega^{-1}, \quad \Omega r\sqrt{\sin\vartheta} = S, \tag{2.13}$$

$$\vec{\partial}' = \vec{\sigma}_3\partial_r + \frac{\vec{\sigma}_1}{r}\partial_\vartheta + \frac{\vec{\sigma}_2\partial_\varphi}{r\sin\vartheta}, \quad \partial_r \equiv \frac{\partial}{\partial r}, \quad \partial_\vartheta \equiv \frac{\partial}{\partial\vartheta}, \quad \partial_\varphi \equiv \frac{\partial}{\partial\varphi}, \tag{2.14}$$

and on the Minkowski space

$$\partial = \Omega \partial' \Omega^{-1}, \quad \partial' = \gamma_0(\partial_0 + \vec{\partial}'), \quad \partial_0 \equiv \frac{\partial}{\partial r_0}, \quad r_0 = ct. \tag{2.15}$$

It is straightforward to insert (2.9), (2.10) and (2.13) – (2.15) into (2.8). The result is, that (2.9) is an exact solution of (2.8), if and only if η is a solution of

$$\mp\lambda\vec{\sigma}_3\partial_r\eta i_2 + \left[\varepsilon - V(r) + \vec{\sigma}_1\frac{\lambda}{r}(k + \beta(r))\right]\eta = \eta^* \equiv \gamma_0\eta\gamma_0, \quad k \in \mathbb{N}, \tag{2.16}$$

which has to fulfill

$$i\eta = \eta i, \quad i = \gamma_0\gamma_1\gamma_2\gamma_3 = \vec{\sigma}_1\vec{\sigma}_2\vec{\sigma}_3. \tag{2.17}$$

As discussed already in [3], section 4, a solution of (2.16) must be composed of at least two real components η_0, η_1 according to $\eta = \eta_0 + \vec{\sigma}_1\eta_1$. This however would be too special as to cover the full range of the classical approximation (2.5) and (2.7). Throughout this article, therefore, the following "complexified" form of η will be preferred (cf. [3], eq. (4.2))

$$\eta = \eta_0 + i_2\eta_2 + \vec{\sigma}_1(\eta_1 + i_2\eta_3), \quad \eta_\mu \in \mathbb{R}, \quad \mu = 0, 1, 2, 3. \tag{2.18}$$

It is evident that if η is a solution of (2.16), then, its complex conjugate $\bar{\eta} = \vec{\sigma}_1\eta\vec{\sigma}_1$, its real part $\frac{1}{2}(\bar{\eta} + \eta)$ and its imaginary part $\frac{1}{2}(\bar{\eta} - \eta)i_2$ will solve (2.16) as well.

This section ends with a brief discussion of the form taken by $\psi\tilde{\psi}$, the Dirac current $\psi\gamma_0\tilde{\psi}$ and the spin $\psi\gamma_3\tilde{\psi}$ on account of (2.18), (2.9) and (2.3), (2.1). One finds

$$\psi\tilde{\psi}r^2 \sin\vartheta = S\eta\bar{\eta}\tilde{S} = \eta\bar{\eta} = \eta_0^2 + \eta_2^2 - (\eta_1^2 + \eta_3^2), \tag{2.19}$$

$$\begin{aligned} \psi\gamma_0\tilde{\psi}\gamma_0 r^2 \sin\vartheta &= S\eta\eta^\dagger\tilde{S} \\ &= \sum_{\mu=0}^{3}\eta_\mu^2 + 2S\vec{\sigma}_1(\eta_0 - i_2\eta_2)(\eta_1 + i_2\eta_3)\tilde{S} \\ &= \varrho(\nu_0 + \vec{\nu})r^2 \sin\vartheta, \end{aligned} \tag{2.20}$$

$$\hat{l} \times (\hat{r} \times \vec{\nu}) = \vec{0}, \tag{2.21}$$

$$\psi\gamma_3\tilde{\psi}\gamma_0 r^2 \sin\vartheta = \psi\vec{\sigma}_3\psi^\dagger r^2 \sin\vartheta = \mp\eta\bar{\eta}S\vec{\sigma}_2\tilde{S} = \pm\eta\bar{\eta}\hat{l} = \vec{s}r^2 \sin\vartheta. \tag{2.22}$$

It should be noted that, as indicated in the abstract, $\psi\tilde{\psi}$ *may vanish and change its sign.* Therefore the applicability of the conventional polar decomposition of ψ is restricted to only local portions of $\mathbb{E}^3$. As in [4], section 6, the spinvector $\vec{s}$, eq. (2.22), is directional quantized to point either parallel or antiparallel to the local angular momentum $\hat{l}$ of a streamline defined by $\dot{\vec{r}} \times \vec{\nu} = \vec{0}$.

3. Decoupling of the Radial Equation and Mbius Transformations

In terms of the reduced distance x,

$$|\vec{r}| \equiv r = \lambda x, \quad \lambda = \frac{\hbar}{mc}, \tag{3.1}$$

and the quantities

$$W_0 = \varepsilon - V(r), \quad W_1 = \frac{1}{x}\left[k + \beta(r)\right], \quad k \in \mathbb{N}, \tag{3.2}$$

the radial eq. (2.16) takes the form

$$\mp\vec{\sigma}_3\eta' i_2 + (W_0 + \vec{\sigma}_1 W_1)\eta = \eta^* \equiv \gamma_0\eta\gamma_0, \eta' \equiv \frac{d\eta}{dx}. \tag{3.3}$$

In order to display the announced decomposition of η concisely, it is worthwhile to introduce the projectors

$$P_\pm = \frac{1}{2}(1 \pm \vec{\sigma}_1) = P_\mp^* = P_\pm^\dagger, \quad P_\pm^2 = P_\pm, \tag{3.4}$$

with the properties

$$P_+P_- = 0, \quad P_+ + P_- = 1, \quad \vec{\sigma}_1 P_\pm = \pm P_\pm, \quad \vec{\sigma}_3 P_\pm i_2 = \mp P_\mp. \tag{3.5}$$

The decomposition then is defined by

$$\eta = (P_+ + P_- g)e^{-\phi}, \tag{3.6}$$

where the functions g and ϕ in general are bivector–valued complex, i.e.,

$$g = g_1 + i_2 g_2, \quad g_{\substack{1\\2}} \in \mathbb{R}, \quad \phi = \phi_1 + i_2\phi_2, \quad \phi_{\substack{1\\2}} \in \mathbb{R}. \tag{3.7}$$

Inserting (3.6) into (3.3) and making use of (3.5), (3.4) one finds

$$\pm\phi' = Ag - 1, \quad A = W_0 - W_1 = \varepsilon - V(r) - \frac{1}{x}\left[k + \beta(r)\right], \tag{3.8}$$

and for g the two Riccati eqs.

$$\pm g' = Ag^2 - 2g + B, \quad B = W_0 + W_1 = \varepsilon - V(r) + \frac{1}{x}\left[k + \beta(r)\right]. \tag{3.9}$$

Note, that the quantity $\eta\tilde{\eta}$ occurring in (2.19) and (2.22) according to (3.6) is given by

$$2\eta\tilde{\eta} = (g + g^*)e^{-(\phi + \phi^*)} = 2g_1 e^{-2\phi_1}, \tag{3.10}$$

which changes its sign whenever $2g_1 = g + g^* \equiv 2\mathrm{Re}(g)$ changes its sign. With (3.7) the quantity $\eta\eta^\dagger$ in (2.20) becomes

$$2\eta\eta^\dagger = e^{-2\phi_1}\left[1 + |g|^2 + \vec{\sigma}_1(1 - |g|^2) + 2\vec{\sigma}_3 g_2\right], \quad |g|^2 \equiv gg^* = g_1^2 + g_2^2. \tag{3.11}$$

The importance of the Mbius transformation

$$g=\frac{af+b}{cf+d},\quad D=ad-bc\neq 0,\quad f=\frac{dg-b}{-cg+a},\tag{3.12}$$

in the theory of Riccati eqs. is wellknown [5]. If (3.12) is inserted into (3.9), the following Riccati eqs. for f are obtained

$$\pm Df'=a_2f^2+a_1f+a_0,\tag{3.13}$$

$$a_0 = b^2A-2bd+d^2B\pm(bd'-b'd),\tag{3.14}$$
$$a_1 = 2abA+2cdB-2(ad+bc)\pm(ad'-a'd+bc'-b'c),\tag{3.15}$$
$$a_2 = a^2A-2ac+c^2B\pm(ac'-a'c).\tag{3.16}$$

Eq. (3.6) may be expressed in terms of f according to

$$\eta=[P_+(cf+d)+P_-(af+b)]\,\varrho F,\tag{3.17}$$

where

$$\varrho F(cf+d)=e^{-\phi}.\tag{3.18}$$

In order to derive for F a linear differential eq. of the second order, I calculate

$$\frac{F'}{F}=-\frac{\varrho'}{\varrho}-\Psi',\quad \Psi'=\phi'+\frac{(cf+d)'}{cf+d}\tag{3.19}$$

and

$$\frac{F''}{F}=\Psi'^2-\Psi''+2\frac{\varrho'}{\varrho}\Psi'+\left(\frac{\varrho'}{\varrho}\right)^2-\left(\frac{\varrho'}{\varrho}\right)'.\tag{3.20}$$

With (3.8) and

$$(cf+d)(a-cg)=D,\tag{3.21}$$

the quantity Ψ' becomes

$$\pm\Psi'=\pm\phi'\pm\frac{(cf+d)'}{cf+d}=Ag-1\pm\frac{D'}{D}-\frac{(a-cg)'}{a-cg},\tag{3.22}$$

from which g' may be eliminated with the help of (3.9). Then, reinserting f, one obtains

$$\pm\Psi'=p+\frac{a_2}{D}f,\quad Dp=D\pm D'+abA-2ad+cdB\pm(bc'-a'd).\tag{3.23}$$

Making use of (3.23), the r.h.s. of (3.20) is seen to contain f', f^2 and f. The terms f' and f^2 can be eliminated with the help of (3.13). So, (3.20) finally may be arranged to contain f only, viz.

$$\begin{aligned}\frac{F''}{F} &= \left(p\pm\frac{\varrho'}{\varrho}\right)^2\mp\left(p\pm\frac{\varrho'}{\varrho}\right)'-\frac{a_0a_2}{D^2}\\ &+\frac{a_2f}{D}\left[\pm2\frac{\varrho'}{\varrho}+2p\mp\frac{D}{a_2}\left(\frac{a_2}{D}\right)'-\frac{a_1}{D}\right].\end{aligned}\tag{3.24}$$

Now, there is the freedom to choose the "integrating factor" ϱ introduced in (3.18) in such a way that the square bracket contribution to (3.24) vanishes. Replacing a_1 by (3.15), one thus notes that ϱ has to be fixed according to

$$(\varrho D)^2 = a_2, \tag{3.25}$$

whence the differential eq. (3.24) becomes

$$\frac{F''}{F} = \left(p \pm \frac{\varrho'}{\varrho}\right)^2 \mp \left(p \pm \frac{\varrho'}{\varrho}\right)' - \frac{a_0 a_2}{D^2}, \quad \frac{\varrho'}{\varrho} = \frac{a_2'}{2a_2} - \frac{D'}{D}. \tag{3.26}$$

Eqs. (3.23) and (3.19) allow to express f in terms of the logarithmic derivative $\frac{F'}{F}$,

$$a_2 f = -D\left[p \pm \left(\frac{\varrho'}{\varrho} + \frac{F'}{F}\right)\right]. \tag{3.27}$$

Provided that $a_2 \neq 0$, the solution f of Riccati eqs. (3.13) thus may be eliminated from (3.17), which finally leads to

$$\begin{aligned} \eta = {} & P_+\varrho\left[d - \frac{c}{a_2}D\left(p \pm \frac{\varrho'}{\varrho} \pm \frac{d}{dx}\right)\right]F \\ & + P_-\varrho\left[b - \frac{a}{a_2}D\left(p \pm \frac{\varrho'}{\varrho} \pm \frac{d}{dx}\right)\right]F. \end{aligned} \tag{3.28}$$

This is a new decomposition of the radial spinor (3.3) into the solutions of the linear second-order differential eq. (3.26). By exploiting the freedom of choosing the Mbius transformation (3.12) appropriately, the most convenient form of (3.26) may be achieved.

4. Bound Coulomb States in Aharonov–Bohm and Zeeman Fields

In this section, bound state solutions are derived for eqs. (3.2), (3.3), where the quantity W_0 is given by the attractive Coulomb potential

$$W_0 = \varepsilon - V(r) = \varepsilon + \frac{\bar{\alpha}}{x}, \quad \bar{\alpha} = Z\alpha, \quad Z \geq 1, \quad \frac{1}{\alpha} = \frac{\hbar c}{q^2} = 137.036, \tag{4.1}$$

and

$$W_1 = \frac{1}{x}(k + \beta_0 + \beta_1 x), \quad k \in \mathbb{N}, \quad \beta_{\substack{0\\1}} \in \mathbb{R}. \tag{4.2}$$

The quantity W_1 is composed of the centrifugal potential $\frac{k}{x}$, the Aharonov–Bohm potential $\frac{\beta_0}{x}$ and a constant Zeeman-like contribution β_1. According to eqs. (3.8), (3.9), the quantities A and B are

$$A = W_0 - W_1 = \frac{\bar{\alpha} - l}{x} + \varepsilon - \beta_1, \quad B = \frac{\bar{\alpha} + l}{x} + \varepsilon + \beta_1, \tag{4.3}$$

where

$$l = k + \beta_0 \tag{4.4}$$

is the effective angular momentum quantum number resulting from k via an Aharonov–Bohm potential–dependent shift by β_0.

The form of A and B is ideally suited for the demonstration of a simple, non-standard application of the decoupling method developed in the last section. In fact, putting

$$a' = b' = c' = d' = 0, \tag{4.5}$$

and

$$b = -a, \quad d = c, \tag{4.6}$$

whence

$$D = ad - bc = 2ac, \quad D' = 0, \tag{4.7}$$

one finds from (3.14) and (3.16)

$$a_{\substack{0\\2}} = \frac{1}{x}\left[\bar{\alpha}(a^2 + c^2) + l(c^2 - a^2)\right] + \varepsilon(a^2 + c^2) + \beta_1(c^2 - a^2) \pm 2ac. \tag{4.8}$$

So, this $\frac{1}{x}$–contribution can be removed by choosing

$$a^2 + c^2 = 2l, \quad a^2 - c^2 = 2\bar{\alpha}, \tag{4.9}$$

or

$$a^2 = l + \bar{\alpha} = b^2, \quad c^2 = l - \bar{\alpha} = d^2, \tag{4.10}$$

which in (4.8) leads to

$$a_{\substack{0\\2}} = 2(\varepsilon l - \bar{\alpha}\beta_1 \pm ac), \quad a'_{\substack{0\\2}} = 0. \tag{4.11}$$

Since a_2 and D are constants, the integrating factor ϱ according to (3.25) also is a constant and eq. (3.26) has the form

$$\frac{F''}{F} = p^2 \mp p' - \frac{a_0 a_2}{D^2}, \tag{4.12}$$

where, as seen from (3.23),

$$acp = \frac{a^2c^2}{x} - \bar{\alpha}\varepsilon + \beta_1 l. \tag{4.13}$$

Making use of eqs. (4.7), (4.10) and (4.11), eq. (4.12) may be written

$$\frac{F''}{F} = 1 + \beta_1^2 - \varepsilon^2 - \frac{2}{x}(\bar{\alpha}\varepsilon - \beta_1 l) + \frac{ac(ac \pm 1)}{x^2}. \tag{4.14}$$

The radial spinor η is obtained in terms of F by means of eq. (3.28). Apart from a postmultiplication factor, which in general may be i_2–complex, the expression

$$\begin{aligned}\eta = & P_+ \left[a(\beta_1 - \varepsilon) + c\left(1 + \frac{ac}{x} \pm \frac{d}{dx}\right)\right] F \\ + & P_- \left[c(\beta_1 + \varepsilon) - a\left(1 - \frac{ac}{x} \mp \frac{d}{dx}\right)\right] F\end{aligned} \tag{4.15}$$

follows. Only the squares of the coefficients a and c are fixed by eqs. (4.10). So, one may choose

$$-a = b = \pm\sqrt{l + \bar{\alpha}}, \quad c = d = \sqrt{l - \bar{\alpha}}. \tag{4.16}$$

With the definition

$$\gamma = \sqrt{l + \bar{\alpha}}\sqrt{l - \bar{\alpha}} = \mp ac = \pm bc, \tag{4.17}$$

eq. (4.15) becomes

$$\begin{aligned}\eta x^{-\gamma} = & \left\{P_+ \left[b(\varepsilon - \beta_1) + c\left(1 \pm \frac{d}{dx}\right)\right]\right. \\ + & \left.P_- \left[c(\varepsilon + \beta_1) + b\left(1 \mp \frac{d}{dx}\right)\right]\right\}(x^{-\gamma} F),\end{aligned} \tag{4.18}$$

and the differential eq. (4.14) takes the form

$$\left[\frac{d^2}{dx^2} + \varepsilon^2 - 1 - \beta_1^2 + \frac{2}{x}(\bar{\alpha}\varepsilon - \beta_1 l) - \frac{\gamma(\gamma - 1)}{x^2}\right] F = 0. \tag{4.19}$$

Changing from the independent variable x to y by means of the linear (i_2–complex) substitution

$$y = 2\omega x, \quad \omega = \sqrt{1 + \beta_1^2 - \varepsilon^2}, \tag{4.20}$$

one obtains from (4.19) Whittakter's differential eq. [6]

$$\left[\frac{d^2}{dy^2} - \frac{1}{4} + \frac{K}{y} + \frac{\frac{1}{4} - \mu^2}{y^2}\right] F = 0, \tag{4.21}$$

where the parameters K and μ are given by

$$\omega K = \bar{\alpha}\varepsilon - \beta_1 l, \quad l = k + \beta_0, \quad k \in \mathbb{N}, \tag{4.22}$$

$$\mu = \gamma - \frac{1}{2}. \tag{4.23}$$

For bound states, a necessary condition is , that the solution of (4.21) be exponentially decaying for $x \to \infty$. Therefore, F must be proportional to Whittaker's function $W_{K,\mu}(y)$, [6], viz.

$$F = \mathcal{N} W_{K,\mu}(y). \tag{4.24}$$

Sufficient regularity at the origin $x \to 0_+$ then implies the quantization condition

$$K = \nu + \mu + \frac{1}{2} = \nu + \gamma, \quad \nu \in I\!N_0 \equiv \{0, 1, 2, \ldots, \infty\}, \tag{4.25}$$

and

$$F = \frac{(-1)^\nu}{\nu!} W_{\nu+\gamma,\gamma-1/2}(y) = e^{-\frac{y}{2}} y^\gamma L_\nu^{(2\gamma-1)}(y), \tag{4.26}$$

where $L_\nu^{(\alpha)}(y)$ denotes the generalized Laguerre polynomials [6] defined by the recurrence relation

$$\begin{aligned}(\nu+1)L_{\nu+1}^{(\alpha)}(y) &= (2\nu+\alpha+1-y)L_\nu^{(\alpha)}(y) \\ &\quad - (\nu+\alpha)L_{\nu-1}^{(\alpha)}(y), \quad \nu \geq 0, \quad \alpha > -1,\end{aligned} \tag{4.27}$$

and the initial values

$$L_{-1}^{(\alpha)}(y) = 0, \quad L_0^{(\alpha)}(y) = 1. \tag{4.28}$$

When (4.26) is inserted into (4.18), the derivative of the Laguerre polynomials can be eliminated with the help of

$$\frac{d}{dy} L_\nu^{(\alpha)}(y) = -L_{\nu-1}^{(\alpha+1)}(y), \quad \nu \geq 0. \tag{4.29}$$

Then, making use of

$$L_\nu^{(\alpha)}(y) = L_\nu^{(\alpha+1)}(y) - L_{\nu-1}^{(\alpha+1)}(y), \quad \nu \geq 0, \tag{4.30}$$

the radial spinor η may be written in the unnormalized form

$$\eta = (P_+\eta_+ + P_-\eta_-)e^{-\frac{y}{2}} y^\gamma, \tag{4.31}$$

where

$$\eta_+ = [b(\varepsilon - \beta_1) + c(1 \mp \omega)]\, L_\nu^{(2\gamma)}(y) - [b(\varepsilon - \beta_1) + c(1 \pm \omega)]\, L_{\nu-1}^{(2\gamma)}(y), \tag{4.32}$$

and

$$\eta_- = [c(\varepsilon + \beta_1) + b(1 \pm \omega)]\, L_\nu^{(2\gamma)}(y) - [c(\varepsilon + \beta_1) + b(1 \mp \omega)]\, L_{\nu-1}^{(2\gamma)}(y). \tag{4.33}$$

Recall that according to eqs. (2.7), (2.9) and (2.17), the upper (lower) signs in eqs. (4.32) – (4.33) lead to solutions of (2.8), where in the approximate classical limit [4], the spinvector $\vec{s}$ is directed parallel (antiparallel) to the orbital angular momentum unitvector $\hat{l}$ in the allowed classical region.

The quantization condition (4.25) implies the modified fine–structure formula

$$\varepsilon\left[(\nu+\gamma)^2 + \bar{\alpha}^2\right] = (\nu+\gamma)\sqrt{(\nu+\gamma)^2 + \bar{\alpha}^2 + \beta_1^2\nu(\nu+2\gamma)} + \bar{\alpha}\beta_1 l, \quad \nu \in I\!N_0, \tag{4.34}$$

provided that

$$l = k + \beta_0 > 0, \quad k \in I\!N, \quad \beta_1 \gamma \leq \bar{\alpha}, \quad \gamma = \sqrt{l^2 - \bar{\alpha}^2} > 0. \tag{4.35}$$

Note, that an Aharonov–Bohm effect [7] results on the energy levels (4.34) from the shift $l = k + \beta_0$ of the angular momentum quantum number $k \in I\!N$ to the non–integer value l.

In complete analogy to a pure Coulomb field, i.e., $\beta_0 = \beta_1 = 0$, eqs. (4.32) and (4.33) define a twofold degenerated set of solutions ($\vec{s}$ parallel or antiparallel to $\hat{l}$) for the energy eigenvalues (4.34) if $\nu \geq 1$. Only for $\nu = 0$, which includes the ground state, the lower sign in (4.32) and (4.33) ($\vec{s}$ antiparallel to $\hat{l}$) leads to the trivial solution $\eta = 0$, because

$$l\eta_+ = \sqrt{l - \bar{\alpha}}\,[l \mp \bar{\alpha} \pm (l + \bar{\alpha})], \quad l\eta_- = \sqrt{l + \bar{\alpha}}\,[l - \bar{\alpha} \pm (l \pm \bar{\alpha})] \tag{4.36}$$

if $\nu = 0$. So, for all states with $\nu = 0$, the spin $\vec{s}$ is parallel to $\hat{l}$, since according to (2.22) $\vec{s} r^2 \sin\vartheta = \eta\tilde{\eta}\hat{l}$ and $\eta\tilde{\eta} > 0$, as is shown in the following section.

5. The Spheres $\psi\tilde{\psi} = 0$

This article ends with a brief discussion of the set of points on which $\psi\tilde{\psi} = 0$. Also, the sign of $\psi\tilde{\psi}$ is investigated in the complementary regions to which the sets $\psi\tilde{\psi} = 0$ are boundaries. From eqs. (2.19) and (4.31) – (4.35) one notes

$$\psi\tilde{\psi} r^2 \sin\vartheta = \eta\tilde{\eta} = e^{-y} y^{2\gamma} (P_+ \eta_+ \tilde{\eta}_- P_+ + P_- \eta_- \tilde{\eta}_+ P_-) = e^{-y} y^{2\gamma} \eta_+ \eta_- . \tag{5.1}$$

Inserting (4.36) into (5.1), one finds

$$\psi\tilde{\psi} r^2 \sin\vartheta = 4\gamma y^{2\gamma} e^{-y} > 0 \tag{5.2}$$

for all $\nu = 0$ states. Therfore, $\nu \geq 1$ is necessary in order that $\psi\tilde{\psi}$ vanishes in the domain $0 < r < \infty$, $0 < \vartheta < \pi$. For simplicity, I only treat the special case $\nu = 1$ under the restriction that $\beta_1 = 0$ and $l = k + \beta_0 = 1$. Putting $\nu = 0$ in eq. (4.27) and making use of (4.28), one obtains

$$L_1^{(2\gamma)}(y) = 2\gamma + 1 - y, \tag{5.3}$$

whence a simple calulation of the zeros of η_+ and η_- according to eqs. (4.32), (4.33) leads to

$$\eta_+(y_1) = 0, \quad y_1 = 2\gamma + 1 - \frac{b\varepsilon + c(1 \pm \omega)}{b\varepsilon + c(1 \mp \omega)}, \tag{5.4}$$

and

$$\eta_-(y_2) = 0, \quad y_2 = 2\gamma + 1 - \frac{c\varepsilon + b(1 \mp \omega)}{c\varepsilon + b(1 \pm \omega)}, \tag{5.5}$$

where

$$b = \pm\sqrt{1 + \bar{\alpha}}, \quad c = \sqrt{1 - \bar{\alpha}}, \quad \gamma = \sqrt{1 - \bar{\alpha}^2}, \tag{5.6}$$

$$\varepsilon = \sqrt{\frac{1+\gamma}{2}}, \quad \omega = \frac{\bar{\alpha}\varepsilon}{1+\gamma}. \tag{5.7}$$

In order to single out positive zeros, a careful evaluation of y_1 and y_2 can not be avoided. Displaying an expansion into powers of $\bar{\alpha}$ for the upper (lower) sign forms of eqs. (5.4) and (5.5) separately, the result is

$$\text{upper sign} \qquad y_{\frac{1}{2}} = 2 \mp \frac{\bar{\alpha}}{2} - \frac{7}{8}\bar{\alpha}^2 + \mathcal{O}(\bar{\alpha}^3), \tag{5.8}$$

$$\text{lower sign} \qquad y_{\frac{1}{2}} = \pm\frac{3}{2}\bar{\alpha} - \frac{9}{8}\bar{\alpha}^2 + \mathcal{O}(\bar{\alpha}^3). \tag{5.9}$$

Hence, according to (5.1), there are *two positive zeros of* $\psi\tilde{\psi}$ *for the upper sign* and only *one positive zero of* $\psi\tilde{\psi}$ *for the lower sign.* For $\nu = 1$, the quantity $\eta_+\eta_-$ is a quadratic form in the variable y, which satisfies

$$\frac{d^2}{dy^2}(\eta_+\eta_-) = 4\varepsilon^2\left(\sqrt{\frac{2}{1+\gamma}} \pm 1\right) > 0. \tag{5.10}$$

Combined with eq. (5.8), this condition of definiteness implies that the solution ψ for the upper sign is such that $\psi\tilde{\psi} < 0$ for $y_1 < y < y_2$ and that $\psi\tilde{\psi} > 0$ for $0 < y < y_1$ and $y_2 < y < \infty$. Correspondingly one concludes from eq. (5.9) that the solution for the lower sign has the property, that $\psi\tilde{\psi} < 0$ for $0 < y < y_1$ and that $\psi\tilde{\psi} > 0$ for $y_1 < y < \infty$.

This discovery renders the conception of the angle of Yvon and Takabayasi to an even more mysterious one than it had been before.

Acknowledgement

In grateful appreciation for her encouraging support, I dedicate this article to Edith Guillaume.

References

1. Schrödinger, E. (1938) 'Eigenschwingungen des sphärischen Raumes', Pontificia Academia Scientiarvm, Commentationes Vol.II..N.9, 321–364, section 7, p. 346–357, see also: *Zusatz b. d. Korrektur (Juli 1938)*, on p. 364.
2. Hestenes, D. (1975) 'Observables, Operators, and Complex Numbers in the Dirac Theory', J. Math. Phys. 16, 556–572.
3. Krüger, H. (1991) 'New solutions of the Dirac Equation for Central Fields', in D. Hestenes and A. Weingartshofer (eds.), The Electron, Kluwer Academic Publishers, Dordrecht, pp. 49–81.
4. Krüger, H. (1993) 'Classical Limit of Real Dirac Theory: Quantization of Relativistic Central Field Orbits', Found. Phys., in print.
5. Ince, E.L. (1956) Ordinary Differential Equations, Dover, New York, pp. 293–295.
6. Magnus, W., Oberhettinger, F., and Soni, R.P. (1966) Formulas and Theorems for the Special Functions of Mathematical Physics, Springer, Berlin chapters 7.1, 7.6 and 5.5.
7. Aharonov, Y. and Bohm, D. (1959) 'Significance of Electromagnetic Potentials in the Quantum Theory', Phys. Rev. 115, 485–491.

GEOMETRIC ALGEBRA VERSUS NUMERICAL CARTESIANISM

The historical trend behind Clifford's algebra

JOSEP M. PARRA
Departament de Física Fonamental and Lab. Física Matemática (I.E.C)
Universitat de Barcelona Diagonal 647, E-08028 Barcelona (Spain)

Abstract. Clifford's fundamental paper of 1878 completed the task of creating a true geometrical analysis by a successful embedding of Hamilton's quaternions into the Grassmann system. Exactly 150 years after the discovery of quaternions and 100 years after its rejection by physicists, physics still needs to accept as fundamental the Rodrigues–Hamilton-Cayley–Clifford expression for the rotation group. It seems time to perform this step, which will bring classical and quantum theories much closer and may lead to their eventual fusion and resolution in a more complete and understandable physical theory.

Key words: Clifford, Dirac, Gibbs, Grassmann, Leibniz, Quaternion, Rodrigues, Rotation, Spinor

1. Introduction

The year 1993 is the 150th anniversary of Hamilton's discovery of quaternions. In 1893, exactly 50 years after the discovery, the struggle for life between quaternions and the vectorial system of Gibbs-Heaviside reached a climax in a set of letters published in Nature [2]. Since then quaternions have played a minor (one can say negligible) role in physics. In pure mathematics, being the largest associative field extension of the real number system, they have been the object of sustained development. As a counterpart the vector analysis of physicists has been largely ignored by mathematicians, who have considered it an ill–formulated, small part of the more complete tensor analysis. We are not able to change this past history of mutual ignorance but, as active workers in the field of Clifford's algebra, it is our responsibility to ensure that mathematical and physical science no longer neglects Clifford's fundamental paper of 1878 [4]. This is because Clifford's paper sets the gauge against which we can measure the historical importance of both the quaternion and the vector systems. In it Clifford establishes beyond any doubt that Hamilton's quaternionic units are but particular instances (binary products) of Grassmann geometric units, and establishes a unique system of geometric (multi-)vector algebra:

> "In order, therefore, to bring the quaternion algebra within that of the Ausdehnungslehre, we have to make the square of each of our units equal to -1, as pointed out by Grassmann. But I venture to differ from his authority in thinking that the quaternion symbols do not in first place answer to the "Elementargrösse" of the Ausdehnungslehre, but to binary products of them; from which supposition, as we have seen, the laws of their multiplication follow at

F. Brackx et al. (eds.), Clifford Algebras and their Applications in Mathematical Physics, 307–316.

once." [4]

As a geometric algebra competing for a fundamental place in physics, Hamilton's quaternions could have lost their place in 1878 if due attention to Clifford's paper had been paid. As a consequence, the debate in the 1890's was strictly nonsensical, forcing a choice between two incomplete alternatives. It is most remarkable that Gibbs paid only lip–service to Grassmann's system, and was unable to understand it. In the preserved draft of a letter to Schlegel, dated August 1, 1888, Gibbs says:

> "At all events I saw that the methods wh I was using, while nearly those of Hamilton, were almost exactly those of Grassmann. I procured two Ed. of the Ausd. but I cannot say that I found them easy reading. In fact I have never had the perseverance to get through either of them, and have perhaps got more ideas from his miscellaneous memoirs than from those works.
>
> I am not however conscious that Grassmann's writings exerted any particular influence on my V-A, although I was glad enough in the introductory paragraph to shelter myself behind one or two distinguished names in making changes of notation wh I felt would be distasteful to quaternionists.
>
> I have no doubt that you consider, as I do, the methods of Grassmann to be superior to those of Hamilton. ... I was led essentially to Grassmann's algebra of vectors, independently of any influence from him or any other else." [19] [6]

Words that sharply contrast with the complete understanding and sincere valuation of Grassmann's major work that we find in Clifford's opening paragraph:

> "Until recently I was unacquainted with the Ausdehnungslehre, and knew only so much of it as is contained in the author's geometrical papers in *Crelle's Journal* and in *Hankel's Lectures on Complex Numbers*. I may, perhaps, therefore be permitted to express my profound admiration of that extraordinary work, and my conviction that its principles will exercise a vast influence upon the future of mathematical science." [4]

Of special relevance is that the second distinguished name in Gibbs' introductory paragraph was Clifford. And that the work quoted was not Clifford's article in the Amer. J. of Math. "Applications of Grassmann extensive algebra", which directly addresses and wholly solves the problem of the relationship between Grassmann's and Hamilton's work in four pages, but the first textbook *Kinematics* of an intended series planned by Clifford.

2. Leibniz's dream and Grassmann-Clifford's achievement

Clifford's high esteem and praise of Grassmann's work were not an exaggeration. With his "Ausdehnungslehre" [10] Grassmann fully accomplished Leibniz's dream of building geometric algebra. This was repeatedly considered by Leibniz as the missing part of mathematics to express directly geometrical and physical entities and relationships directly. In several letters, from 1678 on, Leibniz wrote:

> "Je ne cherche presque plus rien en Geometrie, que l'art de trouver d'abord les belles constructions. Je voy de plus en plus que l'Algebre n'est pas la voye naturelle pour y arriver, et qu'il y a moyen de faire une autre characteristique

propre aux lignes, et naturelle pour les solutions lineaires: au lieu que l'Algebre est commune à toutes les grandeurs." (to Galloys, 1678)

"Mais apres tous les progres que j'ay faits en ces matieres, je ne suis pas encore content de l'Algebre, en ce qu'elle ne donne ny les plus courtes voyes, ni les plus belles constructions de Geometrie. C'est pour quoy ... je croy qu'il nous faut encor une autre analyse proprement géométrique linéaire, qui nous exprime directement *situm*, comme l'Algebre exprime *magnitudinem* ... je croy qu'on pourroit manier par ce moyen la mécanique presque comme la Géometrie ... Enfin je n'espère pas qu'on puisse aller assez loin en Physique avant que d'avoir trouvé un tel abrégé pour soulager l'imagination" (to Huygens, 1679)

"Car premierement je puis exprimer parfaitement par ce calcul toute la nature ou definition de la figure (ce que l'Algebre ne fait jamais, car disant que x^2+y^2 aeq. a^2 est l'equation du cercle, il faut expliquer par la figure ce que c'est x et y)." (to Huygens, 1679)

"J'ai même le projet d'une Analyse géométrique toute nouvelle, entièrement différente de l'Algèbre, qui sert *pro situ exprimendo* comme l'Algèbre est *pro magnitudine exprimenda*; et les calculs y sont des véritables représentations de la figure et donnent directement les constructions." (to L'Hospital,1693)

"Mon Analyse de la situation paroist bien plus curieuse encore ... Il faut que je m'attache un jour à en commencer des Elemens. Un très habile homme de mes amis, Geometre insigne d'ailleurs, y estoit entré, mais sa mort nous a privé de ce qu'il auroit pu faire." (to L'Hospital, 1701)

"J'ay parlé de ma Specieuse generale à Mr.le Marquis de l'Hospital, et à d'autres; mais ils n'y ont point donné plus d'attention que si je leur avois conté un songe." (to Remond 1714) [5, Chap. IX]

The great merit of Clifford consisted in a precise demonstration that the Grassmann and Hamilton systems were not competing systems, and that the inclusion of quaternions into the more extended system of Grassmann amounted to an enrichment of it, as they were the bivector units expressions for the generators of space rotations. In this way the i,j,k of Hamilton were explained as being the e_2e_3, e_3e_1, e_1e_2 of Grassmann. It is also in this way that Rodrigues' fundamental treatment of the rotation group, a silenced "pre–invention" of quaternions dated 1840 [1], and from which Cayley drew "his" qvq^{-1} formula [3], acquires a definitive geometrical status valid in any dimension.

Essential to the Leibniz-Grassmann-Clifford geometric calculus is the presence in it of symbols that cannot in any way be considered as "numbers" or "pure magnitudes". These literal symbols are related by a fundamental specific relationship: the geometric product. Once this is clearly understood, it becomes clear that the opposition between geometric calculus and Cartesian analytic geometry has, since Leibniz's time, polarized a great deal of the inner history of mathematics and physics. The object of our communication is to apply this historically grounded "polarization" analysis upon two selected critical episodes. These are the transition to Vector Analysis mentioned before and the creation of spinor analysis after the works of Pauli and Dirac. In so doing we are conscious of offering a new and perhaps conflicting view upon the fundamental relationship between mathematics and physics.

The resolution of this conflict may affect the future of physical mathematics, a field distinctly different from both pure and applied mathematics [17]. Hence we close this section with F.J. Dyson's views on the historical relevance of the very subject of this communication. Concisely expressed in his memorable J.W. Gibbs Lecture of 1972 entitled ***Missed Opportunities***, they provide a stimulating reference point for a lively discussion of what has been said and what follows.

> "When Grassmann's work finally became known, mathematicians were divided into quaternionists and antiquaternionists, and were spending more energy in polemical arguments for and against quaternions than in trying to understand how Grassmann and Hamilton might be fitted together into a larger scheme of things. So it was left to the physicist Gibbs to present for the first time in his 1886 lecture the essential ideas of Grassmann and Hamilton side by side. The last words of his lecture are, "We begin by studying multiple algebras; we end, I think, by studying MULTIPLE ALGEBRA."
>
> I do not know how many pure mathematicians heard or read Gibbs' lecture. If they had studied it carefully, they would soon have noticed that Gibbs had not really succeeded in unifying the notions of quaternion and vector. On the contrary, by putting the two notions side by side he had made explicit the lack of any real compatibility between them. His lecture ought to have suggested to any attentive mathematician the question, "How can it happen that the properties of three–dimensional space are represented equally well by two quite different and incompatible algebraic structures?" If this question had once been clearly asked, the answer would almost certainly have been forthcoming. And the answer would have led inevitably to a complete theory of the single–valued and double–valued representations of the three–dimensional rotation group. The vectors are the simplest nontrivial single–valued representation, and the quaternions are the simplest double–valued representation. Also, the quaternions are the prototype of what later were called spinor representations. The development of spinor representations, which was actually begun by Elie Cartan in 1913 and completed during the 1930's with substantial help from the physicists Pauli and Dirac, might have been accelerated by approximately 40 years. It is impossible to say what effects such an accelerated development would have had on other branches of pure mathematics, but the effects could hardly have failed to be substantial." [7]

3. A mathematical fact: Are spinors needed in physics?

Spinors are needed in physics because they are the mathematical field entities that appear in the wave equations of quantum mechanics. More precisely, the assumed and experimentally established rotational and Lorentz invariance of the equations implies specific transformation laws for the wave function. These laws are characterized by the fact that a 2π rotation does not necessarily correspond to the identity transformation, this being achieved only after a 4π rotation. This has been accommodated into the theory saying that all observables should be bilinear in the wave function (the sandwich recipe for the expected values); for them a 2π rotation acts as the identity. Now, this peculiar form of the calculus of observables, the true justification for quantum mechanics according to the empiricist principles under which it developed, can be shown to be intimately related to a real Clifford–Grassmann calculus performed upon space or space–time. This is a surprising fact, almost unbelievable after so many years of indoctrination upon "inner complex spinor spaces" and fully–fledged theories upon sigma and gamma matrices.

But the mathematical fact is that these matrices are but square numerical arrays of numbers whose row–column matrix product merely reproduces the symbolic calculus of Grassmann–Clifford upon the geometric units. Interpreting the sigma and gamma matrices upon which the spinor calculus of physics is based as the unit vectors of ordinary space and four dimensional space–time we gain a geometrical picture of a hitherto "cryptic" domain that "proudly" claimed to be inaccessible to space–time physico–geometrical description.

The emerging picture clearly explains in a classical language very deep-rooted facts about quantum mechanics. We take as the most important the clear explanation of the role of the observer in establishing the wave function of the physical system under scrutiny. We cannot speak of a wave function without reference to an observer, just because the wave function is the geometric operation that relates the physical structure of the system to the physical structure of the system of observation in which the wave equation is written [15] [16].

This bilinearity, the key feature of all quantum mechanical formalism, is precisely the universal law of the geometric action of the relevant orthogonal group. Gauss, Rodrigues, Hamilton and Cayley discovered it for the rotation group. "Everybody" knows and quotes that unit quaternions are isomorphic to the SU(2), and that $-i\sigma_1, -i\sigma_2, -i\sigma_3$ correspond to Hamilton's quaternionic units i,j,k. Why, then, between two isomorphic theories aiming to describe physical reality, the one that seems to need an additional complex two dimensional space has been selected? The reason we see in a twentieth century extension of Cartesian analytic geometry, trying to reduce geometric entities and relationships to pure numeric algebra. The complex "inner" space was not needed in Hamilton's time and, as a direct consequence of Clifford's work, is not needed today.

4. A historical question: Why physics needed spinors?

The microsociological root for this flagrant violation of Occam's Razor we find in the defeat of the quaternionic system in 1893 that excluded quaternions from the mainstream of physical science. In itself not a major loss, as Clifford had shown that quaternions were not really fundamental. But the specific way in which its rejection took place implied also the rejection of the geometric system of analysis developed by Grassmann. Let us explain why.

That Grassmann and Hamilton systems are compatible was demonstrated once and for all by Clifford in 1878; we need only to read the paper. That the Grassmann system is essentially Gibbs' vector analysis was Gibbs' belief transmitted to the whole of scientific community. But it is not. Conceived as a linear calculus, and completed by matrix algebra, it is unable to cover the multilinear structure to which the geometric product of Grassmann-Clifford gives direct access. In particular, naming his orthonormal unit vectors of positive (scalar) square with the same letters as those which Hamilton had given his units, that square to -1, Gibbs eliminated not only an essential arithmetical feature of the quaternionic system that he judged inconvenient; he eliminated also the geometric meaning of Hamilton's units as being rotation operators in Euclidean space. But that geometric interpretation was one of the keys that allowed Clifford to embed successfully the quaternions into the Grassmann system, amending the erroneous identification with the e_1, e_2, e_3 made by the late Grassmann [9].

It is remarkable that Clifford committed also a small error, easily amendable, around this point: following Grassmann he stated that the embedding required that the unit vectors of Grassmann system square to -1. This was obviously necessary for Grassmann, who tried to identify his first order units e_1, e_2, e_3 with the quaternionic symbols without altering their square (as Gibbs did!). But it was not necessary in Clifford's correspondence with bivectors, that square to -1 in both positive and negative definite signature. To cope with the positive definite three–dimensional space considered by Gibbs, Clifford would only have had to change from e_2e_3, e_3e_1, e_1e_2 to the reversed binary products e_3e_2, e_1e_3, e_2e_1. It is also curious that in his recent book Altmann [1, pp. 219–221] endeavours to reestablish this last connection, quoting Clifford's paper but without commenting that this was an important outcome of Clifford's paper. The result is a "non–Grassmannian" geometrical presentation of Clifford algebra (the product of two vectors is said to have "*nothing to do with* $a \times b$", along with the arithmetically true but geometrically wrong association $i \leftrightarrow e_2e_1, j \leftrightarrow e_3e_2, k \leftrightarrow e_1e_3$.

The three bivector units e_3e_2, e_1e_3, e_2e_1, in this specific order, are the transparent geometric representation of the quaternionic units in three–dimensional Euclidean space. But the defective Gibbs–Heaviside algebraic system (a non–associative structure with two products) gives exactly the same expression to vectors and bivectors which can only be distinguished by the non–algorithmic words "polar" and "axial". As all (polar or axial) vectors have positive square, Gibbs' system offers no room to accommodate the quaternionic geometrical operators. It is most relevant to notice that it was through the use of the "Ergänzung" or duality operation that relates bivectors and vectors that Grassmann came to relate the unit quaternions to his geometrical units of first order. This duality was represented by Grassmann using a vertical bar [9, eq.(5)]: $e_1 = |[e_2e_3]$, $[e_2e_3] = |e_1$, and cyclical permutations. This $|$ operation corresponds exactly to the geometric product with the element $-e_1e_2e_3$ iff $e_i^2 = -1$, being a twisted change of sign in the positive definite case. Then Grassmann was able to express the product of two pure quaternions (without the scalar part) in the form $ab = -[a|b] + |[ab]$: that is, as a sum of the "negative definite scalar product" and the "Ergänzung" or dual of the exterior product (today's most current admitted version of the "vector product" of two vectors a and b). But it was precisely Clifford's rejection of the fundamental character of the "Ergänzung", which led him to dissent from Grassmann's solution and, incidentally, from the full line of thinking that finally led to Gibbs' Vector Analysis. Just after having "corrected" Grassmann Clifford says [4]:

> "It is quite true that in process of time the conception of a product as derived from factors of the same kind, and so of the product of two vectors as a thing which might be thought of without regarding them as rectangular versos, grew

upon Hamilton's mind, and led to the gradual replacement of the units i, j, k by the more general symbols S and V. To explain the laws of multiplication of i, j, k on this view, we must have recourse to the theory of "Ergänzung," or, which comes to the same thing, *represent* an area ij by a vector k perpendicular to it. But the explanation in this case is by no means so easy; and it is instructive to observe that the distinction between a quantity and its "Ergänzung," *i.e.* between an area and its representative vector, which, for some purposes it is so convenient to ignore, has to be reïntroduced in physics. Thus Maxwell specially distinguishes the two kinds of vectors which he calls *force* and *flow*, and which in fact are respectively linear functions of the units and of their binary products."

Thus, physical thinking, easiness or naturality of explanation, and strict observance of the rules of the geometric algebra, without allowing foreign thoughts to come into play, guided Clifford to a true synthesis of the works of Hamilton and Grassmann, and not only a formal embedding or rephrasing of the different product laws. The use of the "Ergänzung" in Grassmann's expression above blocks the identification between the quaternionic and the geometric product understood as the sum of the inner and outer products of Grassmann. And the use of a positive-definite metric forced Gibbs to separate again the two terms of the right hand side. Complementing rather than dissenting from Couturat's appreciation [5, pp. 538], we express our conviction that only with Clifford's fundamental paper of 1878 was Leibniz's program of developing a geometric analysis led to completion. And for this, Clifford's mind followed, more strictly than anybody else, the guidelines and values so clearly expressed in Leibniz's unsuccessful correspondence.

The momentous achievement of Clifford still waits for its incorporation to the basic mathematical tools of physics. Gibbs' vectorial system, enlarged with the whole of linear algebra, can only accommodate a non-operational representation of the rotation group by means of orthogonal matrices. From our point of view this is only the Cartesian numerical form of the vectorial representation, and not the group itself. In particular the matrix does not discriminate between the rotation of angle θ and the one of angle $-(2\pi - \theta)$. Being the rotation group essential to the characterization of the physical space and to the expression of the mutual relationships between different objects contained in it, the "classical" mathematical representation that makes no use of the geometric product is severely limited. Now empirical facts emerged from the experimental study of the intimate structure of matter that forced physicists to represent "faithfully" the geometry of the physical relationships between different objects and between the objects and the experimental arrangement. Believing that the only possible geometrical image is the Gibbs–branded one they have declared that it is not possible to build a physico–geometrical model of microphysics, and that abstract, pure numerical mathematics deprived of "images" are the only possible tools.

It is not a silly idea that the Grassmann-Clifford geometric algebra may be the tool to overcome the great divide in fundamental physics: classical physics, including relativity, based on geometry but with an unrecognized defective geometrical analysis and quantum theory, with an improved analysis in a totally de–geometrized form. The positive–energy theorem in General Relativity and the multivector formulation of Dirac's equation [12] [13] [14] [15] [16] are two instances showing that either language cannot be considered complete. Classical physics must renounce its defective orthogonal matrix representation of rotations. Quantum theory must renounce its baroque construction of "inner" numerical and non–geometrical spaces as a substitute for geometrical understanding.

¿From Clifford's unifying geometric viewpoint, the two sides are no longer opposite clashing viewpoints, as they are sometimes presented. Distinguished by Altmann [1, pp. 22] as the geometrical and physical approaches, both appear as the offspring of the numerical Cartesianism which Leibniz opposed. SO(3) and SU(2) matrices, SO(3,1) and SL(2,C) matrices obviously led to incompatible theories when they are taken as giving the fundamental group-geometric structure of the space and space–time according to the Erlangen Program. Not only high-energy physics, but inexpensive macroscopic devices such as Dirac's cube and scissors, show that there are plenty of situations in which the real orthogonal matrix approach is defective. But there is an obvious resistance to replace, in macroscopic directly visualizable situations, the real three–dimensional space by an abstract two dimensional complex space. Explicit, paradigmatic renunciation of a space–time representation of quantum (ultramicroscopic and ultraquick) spinning processes, has provoked less resistance, and allowed Rodrigues–Grassmann–Clifford fundamental findings to find their way into the actual science in divested non–geometrical clothing. Grassmann anticipated that in the preface of its second Ausdehnungslehre:

> "But I know and feel obliged to state (though I run the risk of seeming arrogant) that even if this work should again remain unused for another seventeen years or even longer, without entering into the actual development of science, still that time will come when it will be brought forth from the dust of oblivion and when ideas now dormant will bring forth fruit ... there will come a time when these ideas, perhaps in a new form, will arise anew and will enter into living communication with contemporary developments ... truth remains, even though the garment in which poor mortals clothe it may fall to dust." [6, p. 89]

5. Conclusions

Van der Waerden, one of the greatest contributors to the development of spinor calculus, said

> "In a textbook, the exposition of physical theories is bound to be, at least to a certain extent, *dogmatic* ... On the other hand, the *historical method* follows, step by step, the development of ideas in the minds of theorists, as far as they can be traced in their publications. In my opinion, a thorough understanding of a physical theory can be reached only by the historical method. This method enables us to judge whether a certain hypothesis is really necessary to explain the phenomena, whether it can be modified and under what conditions a theory is valid." [18, p. 199]

We see in an irrational "religious faith" on being in the right track, in the ensuing dogmatic and ahistorical training tradition (others say pragmatic), and in the logicist degeometrization of mathematics three main obstacles to accept Clifford's legacy: a complete mathematical theory, fulfilling Leibniz's expectations, that "can be explained to the first man you meet in the street". Complete understanding, reached through geometry, was Clifford's faith, aim and gift. But, as he put, *"geometry, you know, is the gate of science, and the gate is so low and small that one can only enter it as a little child."* Pure abstract mathematics and higher applied or computational mathematics may not be the best paths for physical science. The absolute inability to understand what is going on in fundamental physics [8] is "homeomorphic" to the inability to grasp the meaning of Clifford's paper starting either from Bourbaki's or from Dirac's definition of what is known as "Clifford" algebra. Both, for different reasons, fail to meet Leibniz's requirement of being a direct geometrical language.

This is why the "poor men in the street" can gain a better perspective than the pure mathematician or applied mathematician (theoretical physicist) have from his ivory tower of abstract constructs. He can, as a little child, receive the historical legacy of these scarce true geniuses that were able to perform decisive steps in the human understanding of nature. A decisive "Galilean" step was the Leibniz–Hamilton–Grassmann–Clifford disclosure of the GEOMETRIC ALGEBRA.

With its use there is hope that the stagnation process that theoretical physics has suffered should come to an end, rebuilding the path through the geometrical gate to physical science, removing the short–lived materials of Gibbs' time and using instead the "eternal" discoveries of Grassmann. Only then will a way out be found to Feynman's blind alley for those who cannot refrain from asking **physical** questions referring to the fundamental theory of matter. Only then will the heavy scientific work done in the quantum domain will be relieved of "blind (or abstract) -though equally true- mathematical exercising upon numerical matrix components" and proceed "with alive formulations that call up an immediate geometrical picture of the physics symbolized". It is but a minor irony that to follow Heaviside's principles [11, pp. 132–134] we have to demolish Gibbs–Heaviside's classical vector calculus. Adherence to principles can never be mere lip-service, but its critical application to each new situation. The Galilean principle of geometrical intelligibility is a foundational principle of modern physics. Clifford's 1878 synthesis still appears today as an essential part of the language we need to read the Great Book of Nature.

Acknowledgements

The author acknowledges the financial support of the Spanish Ministry of Education received through the D.G.I.C. y T. contract No. PB90-0482-c02-01.

References

1. Altmann, S.L.: 1986, *Rotations, Quaternions, and Double Groups*, (Clarendon Press, Oxford)
2. Bork, A.M.: 1966, ' "Vectors Versus Quaternions"- The Letters in Nature', *Am. J. Phys* Vol. no. 34, pp. 202–211
3. Cayley A.: 1845, 'On certain results relating to quaternions', *Phil. Mag.* Vol. no. XXVI, pp. 141–145
4. Clifford, W.K.: 1878, 'Applications of Grassmann's extensive algebra', *Amer. J. Math.* Vol. no. 1, 350–358
5. Couturat, L.: 1901, *La logique de Leibniz,d'après des documents inédits* (Georg Olms Verlagsbuchhandlung, Hildesheim, 1961 (Paris, 1901))
6. Crowe, M.J.: 1967, *A History of Vector Analysis*, (Univ. Notre Dame Press, Notre Dame) and (Dover, N. York, 1985).
7. Dyson, F.J.: 1972, 'Missed Opportunities', *Bull. A.M.S.* Vol. no. 78, 635–652

8. Feynman, R.: 1965, *The Character of Physical Law*, (The M.I.T. Press), page 129
9. Grassmann, H.: 1877, 'Der Ort der Hamilton'schen Quaternionen in der Ausdehnungslehre', *Math. Annalen* Vol. no. **12**, pp. 375–386
10. Grassmann, H.: 1844, *Teoria de la Extension* (Espasa-Calpe Argentina, Buenos Aires, 1947) (Die Lineale Ausdehnungslehre, Berlin)
11. Heaviside, O.: 1893, *Electromagnetic Theory, vol. 1*, (Chelsea P.C., N.Y., 1971)
12. Hestenes D.: 1966, Space-Time Algebra (Gordon and Breach, New York, 1987).
13. Hestenes, D.:1990, 'The Zitterbewegung Interpretation of Quantum Mechanics', *Found. Phys.* Vol. no. **20**, 1213–1232
14. Parra, J.M.: 1992, 'On Dirac and Dirac-Darwin-Hestenes equations', pp. 443–467 in A. Micali, R. Boudet and J. Helmstetter (eds.) *Proceedings of the Second International Conference on Clifford Algebras and Their Applications to Physics* (Kluwer, Dordrecht/Boston)
15. Parra, J.M.: 1992, 'Relativistic invariance of Dirac's equation revisited', pp. 233–242 in *Recent Developments in Gravitation*, A. Feinstein and J. Ibáñez eds. (World Scientific)
16. Parra, J.M.: 1993, 'Intrinsic non Invariant Forms of Dirac Equation' in *Twistors and Clifford Algebras* Oziewicz, Z., Borowiec, A. and Jancewicz, B. eds., Kluwer
17. Parra, J.M.: 1992, 'Clifford Algebras. Towards a Common Language for Mathematicians and Physicists. Some Suggested Notations', *Advances in Applied Clifford Algebras* Vol. no. **2**, 145–168
18. van der Waerden, B.L.: 1958, 'Exclusion Principle and Spin', pp. 199 in *Theoretical Physics in the XXth. Century, Pauli Memorial Volume*, (Fierz and Weisskopf eds., Interscience).
19. Wheeler, L.P.: 1951, *Josiah W. Gibbs. The History of a Great Mind*, (Archon Books, 1970)

Appendix

A. English translation of Leibniz's quotations

"I do not seek in Geometry but the art of finding the beautiful constructions. I realize more and more that the Algebra is not the natural way to achieve them, and that there is possible to build another characteristic proper to the lines, and natural for the linear solutions: while te Algebra is common to all magnitudes." (to Galloys, 1678)

"But after all progress I have achieved in these matters, I am not yet satisfied with the algebra, inasmuch as it gives neither the shortest paths nor the most beautiful constructions in Geometry. This is why ... I believe we still need another analysis rightly geometric and linear that will allow us to express directly *situm* as the algebra expresses *magnitudinem*... I believe that in that way one will be able to deal with mechanics almost the same as with Geometry ... In conclusion, I do not hope one will be able to go ahead in physics, in a significant way, before having found such an abridgement to relieve the imagination." (to Huygens, 1679)

"Because, above all, with this calculus I can express completely all the nature or definition of a figure (and this Algebra never does because saying that $x^2 + y^2$ aeq. a^2 is the equation of a circle is still needed to explain with a figure what are x and y." (to Huygens, 1679)

"I even have the project of an altogether new geometrical Analysis, completely different from Algebra, that will be as useful *pro situ exprimendo* as is the algebra *pro magnitudine exprimenda*; and the calculus are true representations of the figure and give directly the constructions." (to L'Hospital, 1693)

"My Analysis of the situation still seems to me a very curious thing ... It is necessary that some day I apply myself to begin its Elements. A very able man among my friends, and outstanding Geometer, has entered this work, but his death has deprived us of all that he could have done." (to L'Hospital, 1701)

"I have told Mr.le Marquis de l'Hospital and others about my "Specieuse generale"; but they have paid no more attention as if I had told them a dream." (to Remond, 1714)

CLASSIFICATION OF MULTIVECTOR THEORIES AND THE MODIFICATION OF THE POSTULATES OF PHYSICS

WILLIAM M. PEZZAGLIA JR.
Department of Mathematical Sciences
St. Marys College of California
PO Box 3517
Moraga, CA 94575
U.S.A.

Abstract.
We propose a graded classification of the entire field of multivector physics, including all alternative points of view. The (often tacit) postulates of different types of formulations are contrasted, summarizing their consequences. Specifically, spin-gauge formulations of gravitation and GUT which assume standard column spinors will require unnecessarily large matrix algebras. An extreme generalization is introduced, where wavefunctions are assumed to be multivectors, in which multiple generations of particle families naturally appear without increasing the size of the algebra. Further, this allows for two-sided (bilateral) operators, which can accomodate in excess of 10 times more gauge fields without increasing the algebraic representation. As this generalization encompasses all the essential features of the other categories, it is proposed to be the best path to new physics.

Key words: spin-gauge, multivector, clifford

1. Introduction

Collectively, all work in this field can be divided into two broad categories: *Mathematics* and *Physics* (and the relationships between these disciplines). Before we concentrate on the latter, we note that the former can be further subdivided into two areas, *Algebra* and *Analysis*. The structure of Clifford algebra, its relation to other systems (e.g. Grassmann, Lie or Cayley algebras) is the concern of an algebraist with an ultimate goal of creating a single unified math language based on geometry. The analyst is concerned with the calculus of functions of multivectors, with the potential of providing coordinate-free formulations of fields in curved space-time.

In contrast, only a small minority of published papers actually address the role of using Clifford Algebras in physical models. The category of *multivector physics* can be subdivided into two areas, *Philosophy* and *Formulation*. Generically, the former is concerned with guiding principles, which has been ignored by all but a few authors. This is unfortunate because basic postulates can not be deduced from logical reasoning or pure math (e.g. from the Clifford algebra). There has been more attention to formulation, i.e. the association of physical phenomena with mathematical symbols, however no "new" physics will be obtained by applying Clifford algebra while tacitly assuming the old postulates.

The intent of this paper is to focus more upon the different underlying philoso-

F. Brackx et al. (eds.), Clifford Algebras and their Applications in Mathematical Physics, 317–323.

phies involved in how multivector physics are formulated. We introduce two broad classifications for comparison and contrast, organized along classic greek lines, the *Platonistic* and *Aristotelian.*

2. Platonistic

Generally, standard theory has the classical Platonistic view that abstract mathematical entities exist. We are forbidden (due to the imperfection of physical existence) to directly experience (i.e. "measure") these ideal forms (e.g. "quantum phase"). They are only knowable through mathematical formulation. The physical structure of space is hence really mathematical structure in disguise.

2.1. Standard Theory

Specifically, in quantum mechanics the fundamental structure of "real" particles is represented by abstract multicomponent spinors which exist in transcendental complex *spin space,* which has no concrete physical analog. The unobservable nature of spin space is built into the postulates of quantum mechanics, wherein the wavefunction's "phase" angle (in complex phasor space) is ordained to be unmeasurable. Hence the theory (e.g. the Lagrangian) must be formally invariant with respect to changes in phase, which are generated by the $U(1)$ Lie group. This symmetry *gauge transformation* induces (via Noether's theorem) bilinear covariant forms independent of the unobservable phase. Obeying a conservation law, these forms are associated via the correspondence principle with concrete measurable quantities (e.g. $\psi^\dagger\psi$ is the observable probability density).

Requiring the formulation to be invariant under local changes in phase (different at each point in space) is called the *principle of local gauge invariance.* Through an inductive generalization called *minimal coupling,* a gauge covariant derivative is induced which contains a connection function interpreted to be the electromagnetic 4-vector potential. The curvature of this connection is the electromagnetic field strength tensor, whose source is found to be the bilinear covariant current induced by the gauge transformation on the Lagrangian.

Requiring invariance with respect to larger non-abelian symmetry groups induces other gauge fields e.g. "gluons" from $SU(3)$. The emphasis of modern physics has shifted to choosing the right "magical" symmetry group of the universe which will correctly describe the known four fundamental forces in one "unified theory". Generally the group is expressed as a complex matrix which operates on the column spinor wavefunction. The size of the representation induces the number of degrees of freedom in the "spinor-space", which in order to incorporate all the known fermions must unfortunately be in excess of 90.

2.2. Operationalist

The most conservative use of Clifford Algebra in quantum physics is to view its elements only as abstract operators on a spinor space. Only those symmetry groups which are embedded within a global Clifford algebra are to be considered for gauge field generators. Unfortunately this restricts one to groups such as $SU(2^d)$, where d is an integer, eliminating the standard $SU(3)$ and $SU(5)$ [except as subgroups of a larger algebra]. Further, since the representation size is restricted to $2d$, this forces global multispinors of either probably two few (64) or too many (128) components for the estimated 90 or 96 known degrees of freedom needed to represent all the fermions of the standard model.

One argument for using Clifford algebra is that it naturally appears in relativistic quantum mechanics. Four mutually anticommuting algebraic entities were required to factor the Klein-Gordon equation to the first order Dirac equation. These four elements generate the 16 element Majorana algebra, which has the lowest order representation of 4 by 4 matrices, inducing 4 component *bispinor* wavefunctions. These bispinors successfully correspond to a Dirac fermion (e.g. electron) with "half-integral spin". In order for the formulation to be generally covariant, the 4 basis generators obey the defining condition of a Clifford algebra, $\frac{1}{2}\{\gamma_\mu, \gamma_\nu\} = g_{\mu\nu} = e_\mu \bullet e_\nu$ where e_μ is a coordinate basis vector (not part of a Clifford algebra). Hence the γ_μ can be determined from the Riemann space metric $g_{\mu\nu}$ up to a similarity transformation, $\gamma'_\mu = S\gamma_\mu S^{-1}$, i.e. a change in matrix representation, called a *spin transformation.*

Spin gauge theories [1] invoke the principle of *local representation invariance*, i.e. require the theory to be invariant under similarity transformations which are a function of position. This induces a spin covariant derivative ∇_μ where the *spin connection*: $\Omega_\mu = \Omega_\mu{}^i\Gamma_i$ (Fock-Ivanenko coefficient) is interpreted to be a Clifford aggregate (Γ^i is an element of the algebra) of new spin gauge fields,

$$\psi' = S\psi, \tag{1a}$$

$$\nabla_\mu\psi^\alpha = \partial_\mu\psi^\alpha + \Omega_\mu{}^{\alpha\beta}\psi_\beta, \tag{1b}$$

$$\Omega'_\mu = S\Omega_\mu S^{-1} + (\partial_\mu S)S^{-1}, \tag{1c}$$

$$K_{\mu\nu} = [\nabla_\mu, \nabla_\nu] = \Omega_{\mu;\nu} - \Omega_{\nu;\mu} + [\Omega_\mu, \Omega_\nu] = K_{\mu\nu}{}^i\Gamma_i, \tag{1d}$$

$$\gamma^\nu_{;\mu} - [\Omega_\mu, \gamma^\nu] = 0 \tag{1e}$$

The last equation, requiring the spin covariant derivative of any element of the Clifford algebra to vanish is arbitrarily imposed. It is a sufficient (but not necessary) condition to insure $g_{\mu\nu;\alpha} = 0$, i.e. the coordinate covariant derivative of the metric to vanish defining a Riemann space. If it is desired that $\gamma^\nu_{;\mu}$ is grade preserving, then the spin connection, and the resulting field strength tensor $K_{\mu\nu}$ must be a bivector (or the commuting center of the algebra).

2.3. Structuralist

The connection between abstract spin space and tangible coordinate space can be made more clear if we propose a geometric structure to spin a space. A *spin transformation* can now be associated with a change of spin basis. A spin vector $\Psi = \psi^\alpha\xi_\alpha$ will be manifestly invariant, where the basis spin vectors ξ_α and spinor components ψ^α will be modified by the transformation. The Fock-Ivanenko coefficients $\Omega_\mu{}^{\alpha\beta}$, are now interpretable as the spin space analogy of the Christoffel symbols, describing how the spin vectors change with position in coordinate space, $\partial_\mu\xi_\alpha = \xi_\beta\Omega_\mu{}^\beta{}_\alpha$. A non-vanishing field strength tensor $K_{\mu\nu}$ (eq. 1d) is now interpretable as intrinsic spin space curvature. If one demands that the spin metric $\eta_{\alpha\beta} = \eta^*_{\alpha\beta} = \xi_\alpha{}^\dagger\xi_\beta$ be coordinate independent then it follows that the spin connection must be restricted $\Omega_\mu^\dagger = -\Omega_\mu$. In the particular case of Minkowski spacetime with the associated Majorana Clifford algebra the spin connection would be restricted to Dirac bar-negative elements, the vectors and bivectors, which are generators of the Poincare group. Hence it is tempting to pursue this as a gauge derivation of gravity [2].

An element $\gamma^\mu_{\alpha\beta}$ of the matrix Clifford algebra is viewed as a *second rank spinor.* The manifestly spin representation independent bilinear form $e^\mu = \xi_\alpha \gamma^\mu_{\alpha\beta} \xi_\beta{}^\dagger$ can be concretely interpreted as the observable basis vector of Riemann space. This forces a relationship between spin and coordinate space, in particular eq. (1e) is built into this definition. More importantly, this will force a decomposition of the spin gauge field when the Clifford gradient replaces the standard in Lagrangians and equations of motion. For example, consider the non-relativistic Pauli-Schrodinger equation for a two-component spinor. The associated Clifford geometry of three dimensional space is the Pauli algebra. The generators of the unitary spin transformations form the $U(1) \otimes SU(2)$ group with elements $\Gamma^j = \{i, i\sigma_1, i\sigma_2, i\sigma_3\}$. The induced spin connection $\Omega_\mu = \Omega^k_\mu \Gamma_k = iA_\mu + i\sigma_k Z^k_\mu$ apparently contains the usual electromagnetic vector field A_μ plus three other vectors [3]. However, it is $\sigma^k \nabla_k$ that appears in the Lagrangian (and hence equation of motion) which forces a decomposition of the gauge fields,

$$-i\sigma^k \nabla_k = -i\nabla + A + \phi + iC, \tag{2a}$$

$$A = A^k \sigma_k, \tag{2b}$$

$$\phi = Z^k_k, \tag{2c}$$

$$C^n = \epsilon^{jkn} Z_{jk}. \tag{2d}$$

Apparently the symmetric part of the spin field Z^{jk} does not directly couple to the particle! The remaining components enter as scalar and pseudovector interactions (e.g. analogous to f_0 and f_1 mesons respectively). The sources of these fields will be spin currents; in particular it is interesting to note that the scalar field ϕ couples to the helicity density: $J_\phi = i\psi^\dagger \sigma^k \partial_k \psi$.

3. Aristotelian

In contrast, consider an Aristotelian view that abstract entities (e.g. spin space) do not exist, rather the structure of particle fields is (represented by) tangible physical geometry. Both operators and structure will be described by a single unifying, geometrically interpretable Clifford Algebra. We make a distinction as to the degree by which the theory is committed to the full exploitation of the geometric algebra.

3.1. Pragmatism

The *pragmatist* notes that restricted combinations of the algebra called minimal ideals can be used to replace the column spinor. For example, in the Pauli equation the basis spin vectors of a two component spin space could be replaced by $\xi_2 = \sigma_1 \xi_1$ and $\xi_1 = \frac{1}{2}(1 + \sigma_3)$. The quadratic form of the wavefunction $\psi^\dagger \psi$ may no longer be a real scalar, but have other geometric pieces, however the pramatist does not feel obligated to utilize these extra pieces. In fact, he may restrict the geometric content of the wavefuntion such that the non scalar portions disappear. For example, consider a Dirac-Hestenes equation of the form $\gamma^{\mu}\partial_{\mu}\psi = m\psi\Gamma$, where Γ, is an element of the algebra such that $\Gamma^2 = +1$ if in the $(-+++)$ metric, else $\Gamma^2 = -1$ for the $(+---)$ metric). For $\tilde{\Gamma} = -\Gamma$, (the tilda represents the reversion anti-involution) the multivector current equation is,

$$\partial_{\mu} = m[\tilde{\psi}\psi, \Gamma], \tag{3a}$$

$$j^{\mu} = \tilde{\psi}\gamma^{\mu}\psi = s^{\mu} + T^{\mu\nu}\gamma_{\nu} + \rho^{\mu}\gamma_0\gamma_1\gamma_2\gamma_3. \tag{3b}$$

For a unrestricted multivector wavefunction ψ only the scalar current s^{μ} is conserved, so the pragmatist could interpret it as an observable and ignore the parts of eq. (3b) by taking the scalar part of eq. (3a). Alternatively, he would restrict the form of the wavefunction such that the commutator of eq. (3a) vanishes, hence all three of the currents of eq. (3b) are conserved and hence interpretable as observables.

Such formulations tend to be calculationally isomorphic to standard theory, hence one can argue that it is mere reformulation. One can adopt a "minimalistic" principle, where the formulation uses the least number of geometric degrees of freedom to represent the phenomena. Hence the postulates of quantum mechanics are mildly challenged in that for example there is no imaginary "i" present in the 4 dimensional real Clifford algebra. Specifically the standard bias that the "i" is required for charged fields is clearly not true.

3.2. Radicalism

In contrast, the *Radicalist* holds to the extreme position that in formulation we are ontologically committed to use and interpret EVERY geometric degree of freedom within the given dimension of the space. For example, consider the multivector Dirac equation $\gamma^{\mu}\partial_{\mu}\psi = m\psi$ (only in the $-+++$ metric) which has Greider [4] multivector current,

$$\partial_{\mu}j^{\mu} = 0, \tag{4a}$$

$$j^{\mu} = \bar{\psi}\gamma^{\mu}\psi = T^{\mu\nu}\gamma_{\nu} + M^{\mu\alpha\beta}\gamma_{\alpha}\wedge\gamma_{\beta}, \tag{4b}$$

where the Dirac "bar" operator is the main algebra anti-involution (reversion followed by inversion of the basis vector elements).

In contrast to eq. (3a), each part of the Greider current is conserved (and hence interpretable as an observable) without any restriction on the wavefunction ψ. Hence the general multivector wavefunction is no longer constrained to be a minimal ideal describing one particle, but contains multiple generations of particles [5]. This general structure may violate the standard Fierz identities [6], and introduce new quantum (hidden?) parameters. A Lagrangian constructed of the general multivector wavefunction may have non-scalar parts, which could contribute in the path integral formulation of quantum mechanics.

Now allowable *dextral* (right-side applied)[6] operators on the wavefunction must be included, which brings a challenge to the postulate of compatible observables. Two operators may not commute but will represent compatible observables if one is a dextral operator while the other a *sinistral* one (left-side applied, e.g. a spin-transformation). Further, the same operator when applied to the right has an entirely different physical effect (observable) than when applied on the left. For example, in the Pauli algebra, the spin would be given by $S_j = SP(\psi^\dagger \sigma_j \psi) = SP(\psi\psi^\dagger \sigma_j)$ while the isospin by $T_j = SP(\psi^\dagger \psi \sigma_j) = SP(\psi \sigma_j \psi^\dagger)$. The SP notation means to take the scalar part (matrix trace) as the fundamental bilinear covariant form $\psi^\dagger \sigma_j \psi$ will in general no longer be a pure scalar, but a multivector. Each piece of such a multigeometric current must be given interpretation; we cannot just pick the ones we like and exclude the rest. In particular, we have to consider new *bilateral currents* [6] of the form $R_{jk} = SP(\psi^\dagger \sigma_j \psi \sigma_k) = SP(\psi \sigma_k \psi^\dagger \sigma_j)$.

Finally one needs to address a generalized *principle of geometric covariance*, which considers left, right and both-sided operations on the wavefunction. This will give interactions which couple to both the right and left side of the multivector wavefunction. The generalized *bilateral multivector covariant derivative* would have the form,

$$\nabla_\mu(\psi) = \partial_\mu \psi + \Omega_\mu^{\ \alpha\beta} \Gamma_\alpha \psi \Gamma_\beta, \qquad (5a)$$

$$[\nabla_\mu, \nabla_\nu](\psi) = K_{\mu\nu}^{\ \ \alpha\beta} \Gamma_\alpha \psi \Gamma_\beta. \qquad (5b)$$

This poses several difficulties in formulation which have yet to be fully addressed. For example, eq. (1d) will not give the correct field strength tensor of eq. (5b) because of the necessity of operating on both sides of the wavefunction. However, it is apparent that far more number of gauge fields can be represented in this way for a given dimension Clifford algebra than for the spin-gauge type theory. For example, in the particular case of the multivector Dirac equation in Majorana algebra, it was earlier argued that spin gauge theory would yield 10 fields (from the bar-negative vectors and bivectors of the Clifford group). The general bilateral connection of eq. (5a) implies 16x16=256 possible fields. It has been shown [7] that 136 of these can be interpreted as describing all the known spin 0 and 1 light unflavored mesons. Geometric constraints between spin and isospin suppresses the remaining 120 couplings, which exactly corresponds to those mesons forbidden by the standard model (which must appeal to the requirement of antisymmetry of a composite quark wavefunction).

4. Summary

The last category is the most general, containing all the previous categories as subsets. It is hence a prototype for the full exploitation of Clifford algebra in a physical formulation. Further it allows for a broadening of the postulates of physics (n.b. quantum mechanics) beyond mere reformulation.

Acknowledgements

Thanks to Professor Craig Harrison and Victoria Berdon of the Philosophy Department, and graduate student M. Enjalran of the Physics Department at San Francisco State University for helpful discussions. Also to Professor J.S.R. Chisholm (Univ. of

Kent, Canterbury) and J. Crawford (Pennsylvania State University) for educating the author on the subject of spin-gauge theories while providing hospitality at their respective institutions for the author during the summer of 1991. Finally we thank the University of Ghent, Belgium for the financial support which made it possible to attend the conference in Deinze.

References

1. Chisholm, J.R.S. and Farwell, R.:1989 *J. Phys.* **A22**, 1059.
2. Crawford, J.P.:1993, 'Local Automorphism Invariance: A Generalization of General Relativity', to appear in these proceedings.
3. Enjalran, M.:1993,'Spin Gauge theory of Pauli Particles', Master's thesis, San Francisco State University.
4. Greider, K.:1984, *Found. Phys.*14,467; :1980,*Phys. Rev. Lett.*44, 1718.
5. Pezzaglia, W.:1992, *Found. Phys. Lett.*5, 57.
6. Pezzaglia, W.:1992, 'Generalized Fierz Identities and the Superselection Rule for Geometric Multispinors', Proceedings of the Second Max Born Symposium, Univ. of Wroclaw, Poland, September 24-27.
7. Pezzaglia, W.:1993, 'Dextral and Bilateral Multivector Gauge Field Description of Light-Unflavored Mesonic Interactions',Preprint SFSU-TH-92-03.

THE "IDEAL" APPROACH TO SPINORS RECONSIDERED

F. PIAZZESE
Department of Physics, Politecnico.
Corso Duca degli Abruzzi 24, 10129 Torino, Italy

Abstract. The "ideal" approach to spinors is discussed and a comparison is performed with the classical one, as recently clarified by the author. The simplest case of non-relativistic rank-one spinors is enough to prove that, owing to inconsistent transformation laws, spinors can not be interpreted as the minimal ideals of Clifford's algebra.

Key words: spinors, Clifford's algebra, minimal ideals

1. Introduction

As well known, spinor fields were introduced in theoretical physics long ago to describe the wave functions of quantum systems with spin. However, the very existence of spinors appeared as a problem (cf., e.g., [1],[2]), and spinor calculus, as formalized by van der Waerden and Infeld between 1929 and 1933 (cf. [3]), was "not easy to learn" even in much more recent years [4]. Thus, several Authors even felt the need to *visualize* spinors one way or another (cf. [5],[6]). This state of things easily accounts for the following affirmation by Penrose [7]: "We have still not yet seen the full significance of spinors -particularly the 2-components ones- in the basic structure of physical laws". But, recently, the classical theory of 2- and 4-component spinors has been made entirely clear (cf. [8],[9],[10],[11],[12]).

A different approach was proposed about half a century ago by M. Riesz [13], who reinterpreted 2- and 4-component rank-one spinors as particular elements of Clifford's algebra, represented with square matrices, in which the only non-zero elements are equal to the spinor components. As a result, spinors were identified with the minimal left ideals of Clifford's algebra, and this is, in fact, "the modern definition of spinors", as accepted by many Authors (cf., e.g., [14, page xi]), although it is far from resolving all difficulties (cf. [15]).

It appears now that a careful revision of the "ideal" approch is necessary. The purpose of this paper is double: to do such a revision (cf. Sections 2 and 3), and to make a comparison with the classical approach as recently clarified (cf. Sections 4 and 5).

2. Interpreting the spinor

In the classical approach, spinors are elements of vector spaces, characterized by transformation laws induced by Lorentz transformations and modelled on the pattern of tensor transformations. Thus, one can search for an interpretation of spinors as elements of an isomorphic space, *with* a suitable transformation law. To this

F. Brackx et al. (eds.), Clifford Algebras and their Applications in Mathematical Physics, 325–332.

end, consider any vector space S. Employing the so-called "active" point of view (the "passive" one being unsuitable), a transformation law can be regarded as an endomorphism in S (cf. [16, pages 135–137])

$$\forall s \in S: \; s \to s' \tag{1}$$

induced by Lorentz transformations. Let Ω be any space isomorphic to S. Owing to the isomorphism, the following endomorphism is induced, in Ω

$$\forall s, s' \in S: \; \omega(s) \to \omega(s') \tag{2}$$

where ω denotes any one-to-one application of S on to Ω. On the other hand, Lorentz transformations may induce *directly* another endomorphism, in Ω too

$$\forall \omega(s) \in \Omega: \; \omega(s) \to \omega'(s) \tag{3}$$

If this is the case, a *self-consistent* interpretation of the elements of S through the elements of Ω requires that endomorphisms (2),(3) are consistent, i.e., the following condition is fulfilled

$$\forall s, s' \in S: \; \omega(s') = \omega'(s) \tag{4}$$

In this case, we call $\omega(S)$ a "representation" of S in Ω. If the elements of S are spinors, any representation offers a self-consistent interpretation for them.

In particular, any rank-one 2-component spinor ψ is defined as an element of 2-dimension complex vector space C^2 fulfilling the following transformation law

$$\psi'^a = M^a{}_b \psi^b \tag{5}$$

where $\psi^1, \psi^2 \in C$ are the (contravariant) components and matrix $M = \{M^a{}_b\}$ belongs to the unimodular group $SL(2, C)$, in particular to the unitary group $SU(2, C)$ $\subset SL(2, C)$, if the Lorentz transformation reduces to a space rotation. (The Latin indices a, b have the values $1, 2$; the other ones the values $1, 2, 3$. All the Greek indices have the values $0, 1, 2, 3$. The Einstein sum convention is understood.) Transformation (5) can be written in equivalent forms, in terms of covariant components $\psi_1 = \psi^2, \psi_2 = -\psi^1$, and "dotted" contravariant and covariant components (defined as the complex conjugates of the "undotted" ones), cf. [12].

The following representations of rank-one 2-component spinors are considered in this paper.

a) The sets of matrices

$$\psi = \begin{bmatrix} \psi^1 \\ \psi^2 \end{bmatrix} \; ; \; \tilde{\psi} = \begin{bmatrix} -\psi^{2*} \\ \psi^{1*} \end{bmatrix} \; ; \; \psi^* = [\, \psi^{1*} \;\; \psi^{2*} \,] \; ; \; \tilde{\psi}^* = [\, -\psi^2 \;\; \psi^1 \,] \tag{6}$$

(where the star denotes the matrix Hermitian conjugate, in particular the conjugate of a complex number) here regarded as elements of C^2 and the dual space C^{2*}, respectively, with transformation laws

$$\psi' = M\psi \; ; \; \tilde{\psi}' = (M^{-1})^* \tilde{\psi} \; ; \; \psi^{*\prime} = \psi^* M^* \; ; \; \tilde{\psi}^{*\prime} = \tilde{\psi}^* M^{-1} \tag{7}$$

We call (6), fulfilling (7), the "fundamental representations" of spinor ψ.

b) The minimal left and right ideals of the second order matrix algebra $M_2(C)$ (here regarded as vector spaces), with elements

$$J_{(1)} = \begin{bmatrix} \psi^1 & 0 \\ \psi^2 & 0 \end{bmatrix}; \ J_{(2)} = \begin{bmatrix} 0 & -\psi^{2*} \\ 0 & \psi^{1*} \end{bmatrix}; \ J^*_{(1)} = \begin{bmatrix} \psi^{1*} & \psi^{2*} \\ 0 & 0 \end{bmatrix}; \ J^*_{(2)} = \begin{bmatrix} 0 & 0 \\ -\psi^2 & \psi^1 \end{bmatrix} \quad (8)$$

and transformation laws

$$J'_{(1)} = MJ_{(1)} \ ; \ J'_{(2)} = (M^{-1})^* J_{(2)} \ ; \ J^{*\prime}_{(1)} = J^*_{(1)} M^* \ ; \ J^{*\prime}_{(2)} = J^*_{(2)} M^{-1} \quad (9)$$

c) The sets of matrices

$$N = \begin{bmatrix} \psi^1 & -\psi^{2*} \\ \psi^2 & \psi^{1*} \end{bmatrix} \ ; \ N^* = \begin{bmatrix} \psi^{1*} & \psi^{2*} \\ -\psi^2 & \psi^1 \end{bmatrix} \quad (10)$$

with transformation laws

$$N' = MN \ ; \ N^{*\prime} = N^* M^* \quad (11)$$

The representations defined by (10),(11) only hold in the *non-relativistic* case ($M^{-1} = M^*$). In this case, the columns of N and the rows of N^* coincide with the fundamental representations (cf. (6),(7)).

3. The "ideal" approach

Since matrix algebra $M_2(C)$ is isomorphic to Clifford's algebra $Cl_{3,0}$ (cf., e.g., [17]), is tempting, to take advantage of the geometrical features of the latter, regarding (8) as the "matrix representations" of some particular elements of $Cl_{3,0}$ as well. To discuss this point, we must go into the transformation laws.

As is known, from any orthonormal basis $\{e_i\}$ of R^3 a basis $\{e_A\}$ of $Cl_{3,0}$ can be constructed, with the elements

$$e_0 = 1, \ e_1, \ e_2, \ e_3, \ e_{12} = e_1 e_2, \ e_{13} = e_1 e_3, \ e_{23} = e_2 e_3, \ e_{123} = e_1 e_2 e_3 \quad (12)$$

Thus, any element (or, "Clifford number") $z \in Cl_{3,0}$ can be written as follows

$$z = s + x^i e_i + b^{hk} e_{hk} + t e_{123} \quad (h < k) \quad (13)$$

being s, x^i, b^{hk}, t its scalar, vector, bivector and three-vector components, respectively. Any proper rotation of basis $\{e_i\}$ in R^3 induces a corresponding transformation on $\{e_A\}$ in $Cl_{3,0}$. The elements of the new basis $\{e'_A\}$ are given by

$$e'_0 = 1, \ e'_i = A_i{}^j \ e_j, \ e'_{hk} = A_h{}^m A_k{}^n \ e_{mn}, \ e'_{123} = e_{123} \quad (h < k, m < n) \quad (14)$$

being $A = \{A_i{}^j\} \in SO(3, R)$. If the *passive* point of view is employed, any Clifford number must be *invariant* with respect to the rotations. As a consequence, the components of z must be transformed as follows

$$s' = s \ ; \ x'^i = A^i{}_j \ x^j \ ; \ b'^{hk} = A^h{}_m A^k{}_n \ b^{mn} \ ; \ t' = t \quad (h < k, m < n) \quad (15)$$

where $\{A^i{}_j\} = A^{-1}$. On the contrary, in the *active* point of view, z can be regarded as fixed in "rotating" basis $\{e'_A\}$, i.e. its components with respect to such a basis are constant. A non-rotating observer sees the components of z with respect to "fixed" basis $\{e_A\}$ *changing* because of the rotations in R^3 in accordance with (15) with $B^i{}_j = A_j{}^i$ instead of $A^i{}_j$. With the aid of Clifford's algebra, the transformation can be rewritten as the following endomorphism, in $Cl_{3,0}$

$$z \to z' = uzu^{-1} \tag{16}$$

where u belongs to the group $Spin(3,0)$, the even subgroup of group $Pin(3,0)$ (cf., e.g., [17]).

Owing to the isomorphism between $Cl_{3,0}$ and $M_2(C)$, any element z of $Cl_{3,0}$ can be one-to-one associated with the following matrix belonging to $M_2(C)$

$$Z = s\sigma_0 + x^i\sigma_i + b^{hk}\sigma_h\sigma_k + t\sigma_1\sigma_2\sigma_3 \quad (h < k) \tag{17}$$

where σ_0 denotes the identity matrix, and

$$\sigma_1 = \begin{bmatrix} 0 & 1 \\ 1 & 0 \end{bmatrix} \; ; \; \sigma_2 = \begin{bmatrix} 0 & -i \\ i & 0 \end{bmatrix} \; ; \; \sigma_3 = \begin{bmatrix} 1 & 0 \\ 0 & -1 \end{bmatrix} \tag{18}$$

are the well known Pauli matrices. In particular, matrices (18) represent the unit vectors e_i. Thus, the behaviour of matrices (17) is analogous to the one of the Clifford numbers they represent: i) in the *passive* point of view, they are *invariant*; ii) in the *active* one, owing to (16), are *transformed* according to the following endomorphism, in $M_2(C)$

$$Z \to Z' = UZU^{-1} \tag{19}$$

where $U \in SU(2,C)$ is the representation (17) of $u \in Cl_{3,0}$. As a result, matrices (17) with transformation law (19) are "representations" of (13) (and conversely), in the sense of Section 2, in the non-relativistic case.

We easily recognize that matrices (8) with transformation law (19) (with $Z = J_{(1)}$, or $J_{(2)}$) are *not* spinor representations, because condition (4) is not fulfilled (the minimal ideals of $M_2(C)$ are not even close with respect to endomorphism (19)!). Moreover, no element of $Cl_{3,0}$ can be one-to-one associated with any of matrices (8) with transformation law (9). *We conclude that rank-one 2-component spinors can* not *be interpreted as the minimal ideals of Clifford's algebra.*

Remark

When z reduces to a vector of R^3, we get from the above equations the following endomorphism in the 3-dimensional real vector space of the second-order traceless Hermitian matrices H_0^2

$$x^i\sigma_i \to B^i{}_j\, x^j\sigma_i = x^iU\sigma_iU^{-1} \tag{20}$$

The above equation is an expression of the well known homomorphism of $SO(3,R)$ on to $SU(2,C)$ (cf. e.g., [16, Section 4.5]).

4. The classical approach clarified

In this Section the main results given in [12] are summarized. The notations introduced in Section 2 are employed.

1) The real non-negative quantity

$$||\psi|| = +(\psi^1\psi^{1*} + \psi^2\psi^{2*})^{1/2} = +(\psi^*\psi)^{1/2} \tag{21}$$

only vanishing for $\psi^1 = \psi^2 = 0$, is invariant *only* in the non-relativistic case ($M^* = M^{-1}$). In this case, as customary *out* of spinor theory, $||\psi||$ can be called the "modulus" of ψ, and any spinor with modulus equal to 1 can be called "normalized". For any non-zero spinor ψ, a normalized spinor $\hat{\psi}$ with the components

$$\hat{\psi}^a = \psi^a/||\psi|| \tag{22}$$

can be defined. Thus, the following factorization holds

$$\psi^a = ||\psi||\ \hat{\psi}^a \tag{23}$$

In the relativistic case, we consider any ordered pair $\{\psi, \phi\}$ of non-zero 2-component spinors, whose transformation laws involve the same matrices. As follows from (7), scalar product

$$\psi^b\phi_b = \psi^1\phi^2 - \psi^2\phi^1 = \tilde{\psi}^*\phi \tag{24}$$

is a (complex) invariant quantity. When $\psi^b\phi_b = 1$, we call the pair a "normalized pair". For any "non-singular" pair $\{\psi, \phi\}$ (i.e., $\psi^b\phi_b \neq 0$), we define a normalized pair $\{\breve{\psi}, \breve{\phi}\}$ with the components

$$\breve{\psi}^a = \psi^a/(\psi^b\phi_b)^{1/2}\ \ ;\ \ \breve{\phi}^a = \phi^a/(\psi^b\phi_b)^{1/2} \tag{25}$$

As a result, the following factorization, involving spinors in pairs, holds

$$\psi^a = (\psi^b\phi_b)^{1/2}\breve{\psi}^a\ \ ;\ \ \phi^a = (\psi^b\phi_b)^{1/2}\breve{\phi}^a \tag{26}$$

2) Consider the group $L_+^\uparrow$ of the restricted (i.e., homogeneous orthochronous) Lorentz transformations

$$x^\mu \rightarrow x'^\mu = L^\mu{}_\nu x^\nu \tag{27}$$

($\{L^\mu{}_\nu\} \in L_+^\uparrow$). Owing to the homomorphism of $L_+^\uparrow$ on to $SL(2,C)$, transformation (27) induces the following endomorphism, in the 4-dimensional real vector space of the second-order Hermitian matrices H^2

$$X = x^\mu\sigma_\mu \rightarrow X' = x'^\mu\sigma_\mu = L^\mu{}_\nu x^\nu\sigma_\mu = x^\mu P\sigma_\mu P^* \tag{28}$$

being $P \in SL(2,C)$ (cf., e.g., [19, page 24]). If $detX \neq 0$, i.e. the considered events do not lie on the light cone, (28) can be written also as follows

$$X^{-1}detX = x^0\sigma_0 - x^h\sigma_h \rightarrow X'^{-1}detX = x'^0\sigma_0 - x'^h\sigma_h = (P^{-1})^*(x^0\sigma_0 - x^h\sigma_h)P^{-1} \tag{29}$$

The following proposition is easily proved.

The columns of matrices $P, (P^{-1})^*$ *and the rows of matrices* P^*, P^{-1} *are given by the fundamental representations of a normalized pair of spinors* $\{\chi, \eta\}$, *as follows*

$$P = [\chi \mid \eta\,] \,;\; (P^{-1})^* = [-\tilde{\eta} \mid \tilde{\chi}\,] \,;\; P^* = \left[\frac{\chi^*}{\eta^*}\right] \;;\; P^{-1} = \left[\frac{-\tilde{\eta}^*}{\tilde{\chi}^*}\right] \tag{30}$$

The proof follows from the group properties of $SL(2,C)$, when the product of any two restricted Lorentz transformations is considered. The unimodular property of $SL(2,C)$ assures the normalization of pair $\{\chi, \eta\}$. Notice that any normalized pair depend on *six* independent real parameters, which is exactly the number of parameters needed to describe any restricted Lorentz transformation.

With the aid of (28) the matrix elements $\{L^\mu{}_\nu\}$ can be written as the following bilinear spinor functions

$$\begin{cases} L^\mu{}_0 = \frac{1}{2}\delta^{\nu\rho}(\chi^*\sigma_\rho\chi \;+\eta^*\sigma_\rho\eta\;) \\ L^\mu{}_1 = \frac{1}{2}\delta^{\nu\rho}(\chi^*\sigma_\rho\eta \;+\eta^*\sigma_\rho\chi\;) \\ L^\mu{}_2 = \frac{i}{2}\delta^{\nu\rho}(\chi^*\sigma_\rho\eta \;-\eta^*\sigma_\rho\chi\;) \\ L^\mu{}_3 = \frac{1}{2}\delta^{\nu\rho}(\chi^*\sigma_\rho\chi \;-\eta^*\sigma_\rho\eta\;) \end{cases} \tag{31}$$

being $\delta^{\nu\rho}$ the Kronecker symbol.

3) Owing to **2)**, *A one-to-one correspondence exists between the normalized pairs of 2-component spinors defined up to sign and* $L_+^\uparrow$.

4) Owing to **3)** and factorization (26), *a one-to-one correspondence exists between the non-singular ordered pairs of 2-component spinors and Cartesian product* $(C - \{0\}) \times L_+^\uparrow$, *being* $C - \{0\}$ *invariant with respect to* $L_+^\uparrow$.

5) When a restricted Lorentz transformation reduces to a proper space rotation, endomorphism (28) reduces to (20) ($P = U \in SU(2,C)$). As a result, $\eta = \tilde{\chi}$, and, matrix P depends on only *one* 2-component non-relativistic spinor. The unimodular condition reduces to the normality condition for χ. Owing to such a condition, χ depends on *three* real independent parameters, which is the number of parameters needed to describe any proper space rotation. Propositions 3), 4) become:

3') *A one-to-one correspondence exists between the non-relativistic 2-component normalized spinors defined up to sign and the proper rotation group* $SO(3,R)$.

4') *A one-to-one correspondence exists between the non-relativistic non-zero 2-component spinors and the Cartesian product* $R_+ \times SO(3,R)$, *being the positive real axis* R_+ *invariant with respect to* $SO(3,R)$.

6) Since any ordered pair of 2-component spinors makes a "4-component spinor" (or a "bispinor"), all the above results can be easily restated with the 4-component spinor formalism.

7) In accordance with the active point of view, any real non-light-like 4-vector can be described by performing on any 4-vector of the same type (assumed as a "reference 4-vector") the following commuting operations: i) a suitable restricted Lorentz transformation, and ii) the product by a suitable positive factor (cf. [20, Chapter 4]). Any non-singular pair of 2-component spinors (or the equivalent 4-

component spinor) provides both a restricted Lorentz transformation and a non-zero complex invariant quantity, which can be employed to get such a factor. The resulting connections between such spinor fields and 4-vector fields (in particular, between non-relativistic 2-component spinor fields and 3-vector fields) are investigated in [12], where a number of Dirac covariant expressions are limpidly interpreted. The interpretation of spinors as describing a rotation and a dilatation is similar to the one suggested once by Hestenes [18]. However, in that Reference, spinors are defined as the even elements of the Clifford algebra, which is not standard. This use of spinors supports their interpretation as some "square root" of the vector (cf. [21, page 90]). Notice, however, that the invariant quantity $\psi^a\phi_a$ (or $||\psi||^2$ in the non-relativistic case) can find also other interpretations: for instance, in the Pauli theory $||\psi||^2$ describes a probability density (cf. [22]).

8) The above results, obtained in the context of Special Relativity, can be, locally, extended to the Affine Theories of Gravitation (including, in particular, General and Special Relativity) (cf. [23]). In this case, space-time is modelled as a 4-dimensional manifold $\mathcal{F}$, endowed with a Lorentz metric (cf., e.g., [24]). Since transformation (27) can be defined at any event $q \in \mathcal{F}$, the mathematical meaning of any 4-component spinor field, or the corresponding pair of 2-component spinor fields, is made entirely clear, in any 4-dimensional manifold having a global field of orthonormal tetrads (cf. [25]).

5. Concluding remarks

As shown in Section 3, rank-one spinors can *not* be interpreted as minimal ideals of Clifford's algebra. On the other hand, *no* Clifford algebra is required to get a clear and fruitful interpretation of rank-one 2- and 4-component spinors, in both the non-relativistic and relativistic case (cf. Section 4).

Finally, notice that an interpretation of spinors in the Clifford algebra framework *can* be given. In fact, assuming the active point of view, consider the product of any two proper rotations in R^3. Let $U, V \in SU(2, C)$ be the matrices describing the induced endomorphisms (19) in $M_2(C)$. In the first rotation, matrix (17), representing any element $z \in Cl_{3,0}$, is transformed in accordance with (19). In the second rotation, we have (again from (19), with V instead of U)

$$U \to VUV^{-1} \;\; ; \;\; Z \to VZV^{-1} \;\; ; \;\; U^{-1} \to VU^{-1}V^{-1} \tag{32}$$

The resulting transformation is

$$Z \to VUV^{-1}VZV^{-1}VU^{-1}V^{-1} = VUZU^{-1}V^{-1} \tag{33}$$

Thus, matrix U can be regarded as spinor representation **c)** (cf. (10),(11) with $U = N, V = M$). This suggests that the correct Clifford algebra counterpart of rank-one spinors is the *even Clifford group* (in particular, the spin group for the normalized spinors or the normalized pairs of spinors), as proposed in [18]. But there is little to be gained by calling "spinors" the elements of such a group.

Acknowledgements

This work has been produced under the auspices of the Italian Council for Research, C. N. R. (G. N. F. M.), with the support of the Italian Ministry of Research.

References

1. C.G. Darwin, *The Wave Equations of the Electron*, Proc. Royal Soc. London A **118** (1928) 654–680.
2. B.L. Van der Waerden, *Spinor Analysis*, in M.Fierz and V.F. Weisskopf (eds.), *Theoretical Physics in the XXth. Century* New York: Interscience, 1958, page 236.
3. W. L. Bade and H. Jehle, *An introduction to spinors*, Rev. Mod. Phys. **25** (1953) 714–728.
4. P. Rastall, *A new spinor calculus*, Gen. Rel. Grav. **3** (1972) 281–304.
5. R. Kjellander, *A geometrical definition of spinors from 'orientations' in three-dimensional space leading to a linear spinor visualization*, J. Phys. A: Math. Gen. 14 (1981) 1863–1885.
6. R. Ablamowicz, P. Lounesto and J. Maks, *Conference report. Second workshop on "Clifford algebras and their applications in mathematical physics," Université des Sciences et Techniques du Languedoc, Montepellier, France, September 17-30, 1989*, Found. Phys. **21** (1991) 735–748.
7. R. Penrose, *Spinors and torsion in general relativity*, Found. Phys. **13** (1983) 325–339.
8. F. Piazzese, *Spinors, rotations, vectors and Clifford's numbers in three-dimensional space*, Rend. Seminario Matematico Brescia **6** (1981) 60–74.
9. F. Piazzese, *On spinors in Minkowski space-time*, Rend. Seminario Matematico Brescia **7** (1984) 1–16.
10. F. Piazzese, *On the classical theory of elementary spinors*, in Proc. 6th Haifa Matrix Theory Conference (M. Goldberg, D. Hershkowitz, H. Schneider eds.), Lin. Alg. Appl. **167** (1992) 242–247.
11. F. Piazzese, *On 2- and 4-component spinor fields in curved space-times*, Proc. Physical Interpretation of Relativity Theory III London (1992) 236–239.
12. F. Piazzese, *2- and 4-component spinors: what, in fact, are they?*, J. Nat. Geom. **3** (1993) 59–79.
13. M. Riesz, in *C. R. Dixième Congrès Math. Scandinaves 1946*, Copenhagen: Jul. Gjellerup Forlag (1947) 123–148.
14. A. Crumeyrolle, *Orthogonal and Symplectic Clifford Algebras*, Dordrecht: Kluwer Academic Publ., 1990.
15. A. Crumeyrolle, *The primitive idempotents of the Clifford algebras and the amorphic spinor fibre bundles*, Reports on Math. Phys. **25** (1987) 305–328.
16. H. Goldstein, *Classical mechanics*, Reading: Addison-Wesley, 1980 *(2nd Edition)*.
17. P. Lounesto, *Scalar products of spinors and an extension of Brauer-Wall groups*, Found. Phys. 11 (1981) 721–740.
18. D. Hestenes, *Vectors, spinors, and complex numbers in classical and quantum physics*, Am. J. Phys. **39** (1971) 1013–1027.
19. A. O. Barut, *Electrodynamics and classical theory of fields* (Macmillan, New York, 1964).
20. J. L. Synge, *Relativity (The Special Theory)* (North-Holland, Amsterdam, 1964).
21. J. A. Wheeler, *Geometrodynamics* (Academic Press, New York, 1962).
22. E. Persico, *Fundamentals of quantum mechanics* (Prentice-Hall, Englewood Cliffs, 1950).
23. E. Schrödinger, *Space-time Structure* (Cambridge Univ., Cambridge, 1950).
24. S. W. Hawking, and G. F. R. Ellis, *The large scale structure of space-time* (Cambridge Univ., Cambridge, 1973).
25. R. Geroch, *Spinor structure of space-times in general relativity. I*, J. Math. Phys. **9** (1968) 1739–1744.

GEOMETRIC ASPECTS OF SPINORS

A Short Review

ANDRZEJ TRAUTMAN
Instytut Fizyki Teoretycznej UW
Hoża 69, 00-681 Warszawa, Poland

Abstract. Geometric properties of spinors are reviewed in connection with their role in complex and optical geometry. According to Cartan and Chevalley, a Weyl spinor $\varphi \neq 0$, associated by the Dirac representation γ with a complex, $2m$-dimensional vector space W, is called *pure* if the vector subspace $N(\varphi)$ consisting of all elements w of W such that $\gamma(w)\varphi = 0$ is maximal, i.e. m-dimensional. If W is the complexification of a real space V with a scalar product of signature $(2p+\epsilon, 2q+\epsilon)$, where $\epsilon = 0$ or 1 and $p+q+\epsilon = m$, then the *real index* of φ, $r = \dim(N(\varphi) \cap \overline{N(\varphi)})$, in the generic case equals ϵ. Therefore, the direction of a generic pure spinor defines in V a complex ($\epsilon = 0$) or an optical ($\epsilon = 1$) structure. These observations are applied to a smooth, orientable $2m$-dimensional spin manifold $\mathcal{M}$ with a bundle of directions of generic pure spinors. A section of this bundle – if it exists – defines an almost complex or an almost optical geometry, depending on whether $r = 0$ or 1. With such a section one associates a bundle $\mathcal{N}$ of maximal, totally null subspaces of the complexified tangent spaces to $\mathcal{M}$. Denoting by $\mathcal{Z}$ the module of sections of the bundle $\mathcal{N}$, one considers the integrability conditions $[\mathcal{Z}, \mathcal{Z}] \subset \mathcal{Z}$. In the pseudo-Euclidean case ($\epsilon = 0$), the condition is equivalent to the vanishing of the Nijenhuis tensor of the almost complex structure; in the Lorentzian, 4-dimensional case, it is related to the geodetic, shear-free properties of the trajectories of the real line bundle $\mathrm{Re}\,(\mathcal{N} \cap \overline{\mathcal{N}}) \to \mathcal{M}$.

Key words: Pure spinors, optical and CR geometries, algebraically special gravitational fields

1. Introduction: a Touch of History

Spinors may be considered to appear, for the first time, in slightly disguised form, in the work of L. Euler (1770) and O. Rodrigues (1840), who discovered a rational representation of rotations. The latter wrote equations equivalent to

$$Z' = UZU^\dagger / (1 + \tfrac{1}{4}(m^2 + n^2 + p^2)), \tag{1}$$

where $U^\dagger$ is the Hermitian conjugate of the 2 by 2 matrix

$$U = I + \tfrac{1}{2} i\,(m\sigma_x + n\sigma_y + p\sigma_z),$$

I and the sigmas are the unit and the Pauli matrices, respectively,

$$Z = x\sigma_x + y\sigma_y + z\sigma_z, \tag{2}$$

and similarly for Z'.

Spinors are even more explicit in the work of W. R. Hamilton (1844), A. Cayley (1845), W. K. Clifford (1878) and R. O. Lipschitz (1880). Their ideas led to many

F. Brackx et al. (eds.), *Clifford Algebras and their Applications in Mathematical Physics*, 333–344.

developments presented also at the two previous conferences on *Clifford algebras and their applications in mathematical physics* (Canterbury 1985 and Montpellier 1989). Élie Cartan (1913) discovered what are now called spinor representations of the complex Lie algebras $so(n)$, $n > 2$. According to B. L. van der Waerden (1960), the name *spinor* is due to P. Ehrenfest who suggested, on a visit to Göttingen, to develop a spinor analysis analogous to tensor calculus (Van der Waerden, 1929).

In atomic and particle physics, spinor-valued functions are used to describe the quantum-mechanical behaviour of fermions. Most of the time, there appear complex, four-component Dirac spinors over Minkowski space-time. Two-component Weyl spinors are used in the context of parity violating weak interactions.

Weyl (reduced, semi– or half–)spinors are used by Roger Penrose and his school (Penrose and Rindler, 1984 and 1986) in the theory of general relativity and twistors. As emphasized by P. Budinich, Weyl spinors occur there because they are simple (pure) in the sense of Cartan (Chevalley) and not because of their relevance in the theory of the neutrino. Incidentally, the germ of the notion of a pure spinor can be found in a paper by J. W. Givens (1937).

2. The Square Root Ideas

2.1. The Clifford Idea

If Z is as in (2), then $Z^2 = (x^2+y^2+z^2)\,I$ so that Z is a *linear form of the square root of the quadratic form* $x^2+y^2+z^2$. Clifford generalized this and similar observations; to fix the notation and terminology, I state the relevant definitions and theorems.

Let V be an n-dimensional vector space over the field $F = \mathbf{R}$ or $\mathbf{C}$; and let $g : V \times V \to F$ be a bilinear and symmetric map; if it is, moreover, non-degenerate, then it is a scalar product in V and defines an isomorphism $g : V \to V^*$. Recall that, denoting by $\mathrm{T}(V) = \bigoplus \bigotimes^k V$, where $\bigotimes^0 V = F$, the tensor algebra of V and by $\mathrm{I}(V, g)$ its bilateral ideal generated by all elements of the form $v \otimes v - g(v, v)$, one defines the *Clifford algebra* as $\mathrm{Cl}(V, g) = \mathrm{T}(V)/\mathrm{I}(V, g)$ and finds it to be associative, with unit element, and to have the following universal property: if $\mathcal{A}$ is an algebra over F, with unit element $1_{\mathcal{A}}$, and $f : V \to \mathcal{A}$ is a linear map such that

$$f(v)^2 = g(v, v)\, 1_{\mathcal{A}} \quad \text{for every } v \in V,$$

then there is a homomorphism of algebras with units $\tilde{f} : \mathrm{Cl}(V, g) \to \mathcal{A}$ extending f. In particular, $\mathrm{Cl}(V, 0) = \bigwedge V$ is the exterior algebra of V. For $v \in V$, let $e(v)$ and $c(v) : \bigwedge V \to \bigwedge V$ be the exterior multiplication by v and the contraction with $g(v) \in V^*$, respectively. The map $f : V \to \mathrm{End} \bigwedge V$ given by $f(v) = e(v) + c(v)$ has the properties described above, where now $\mathcal{A} = \mathrm{End} \bigwedge V$. The homomorphism $\tilde{f} : \mathrm{Cl}\,(V, g) \to \mathrm{End} \bigwedge V$ is injective and the map $\mathrm{Cl}\,(V, g) \to \bigwedge V$ given by $a \mapsto \langle \tilde{f}(a), 1_{\wedge V} \rangle$ is an isomorphism of vector spaces. Since it is natural, one can identify these two vector spaces; this will be done in the sequel without further comments; the Clifford product of $v \in V$ by $a \in \bigwedge V$ is then given by the Chevalley-Kähler formula (Chevalley, 1954; Kähler, 1962),

$$va = e(v)a + c(v)a.$$

2.2. The Cartan Idea

The Pythagorean equation

$$x^2 + z^2 = y^2$$

is equivalent to each of the statements:
(i) (x, y, z) is a null (isotropic, optical, light-like) vector in $\mathbf{R}^3$ with a scalar product of signature $(2, 1)$;
(ii) one has $\det X = 0$, where $X = x\sigma_x + iy\sigma_y + z\sigma_z$ is a real matrix;
(iii) there exists a *spinor* $\varphi \in \mathbf{R}^2$ such that

$$BX = \varphi \otimes \varphi, \tag{3}$$

where $B = \sigma_y/i$ is a matrix such that

$$X^t = -BXB^{-1}.$$

Informally, equation (3) can be interpreted to mean that the *spinor φ is a square root of the null vector* (x, y, z), represented by the symmetric matrix BX. Cartan (1938) found a generalization of this observation to vector spaces of any dimension, with a scalar product, admitting totally null subspaces of maximal dimension.

2.3. The 'Double-valuedness' of the Spinor Representations

2.3.1. The Algebraic Aspect

Let V and g be as in §2.1. If $u \in V$ is non-null, $u^2 = g(u, u) \neq 0$, then u is invertible, as an element of $\mathbf{Cl}(V, g)$, and the map $v \mapsto -uvu^{-1}$, where $v \in V$, is a reflection in the hyperplane orthogonal to u. The vector u can be normalized, but there is always an *ambiguity of sign*: the vectors u and $-u$ represent the same reflection. The group $\mathbf{Pin}(V, g)$ is defined as the subset of $\mathbf{Cl}(V, g)$ consisting of products of all finite sequences of unit vectors; if $a \in \mathbf{Pin}(V, g)$ and α is the *main automorphism* of the Clifford algebra, i.e. such that $\alpha(v) = -v$ for $v \in V$, then the linear map $v \mapsto \rho(a)v = \alpha(a)va^{-1}$ is an orthogonal transformation of V and there is the exact sequence of group homomorphisms

$$1 \to \mathbf{Z}_2 \to \mathbf{Pin}(V, g) \to \mathbf{O}(V, g) \to 1. \tag{4}$$

The main automorphism α defines the $\mathbf{Z}_2$-grading of the Clifford algebra,

$$\mathbf{Cl}\,(V, g) = \mathbf{Cl}_0(V, g) \oplus \mathbf{Cl}_1(V, g).$$

In general, the group $\mathbf{Pin}\,(V, g)$ has two distinguished subgroups:

$$\mathbf{Spin}\,(V, g) = \mathbf{Pin}(V, g) \cap \mathbf{Cl}_0(V, g)$$

and its connected component $\mathbf{Spin}_0(V, g)$. There is an exact sequence similar to (4),

$$1 \to \mathbf{Z}_2 \to \mathbf{Spin}(V, g) \to \mathbf{SO}(V, g) \to 1. \tag{5}$$

If V is a complex vector space, $V = \mathbf{C}^n$, and g is non-degenerate, then one writes $\mathbf{Cl}\,(n, \mathbf{C})$ and $\mathbf{Pin}\,(n, \mathbf{C})$ instead of $\mathbf{Cl}\,(V, g)$ and $\mathbf{Pin}\,(V, g)$, respectively. A similar notation is used for the groups $\mathbf{Spin}(n, \mathbf{C})$ which are connected for $n > 1$. If g is a scalar product of signature (s, t) in $\mathbf{R}^{s+t}$, then one writes $\mathbf{Cl}\,(s, t)$ instead of $\mathbf{Cl}\,(V, g)$, etc. The groups $\mathbf{Spin}(n, 0)$ and $\mathbf{Spin}(0, n)$ are isomorphic and abbreviated as $\mathbf{Spin}(n)$; similarly for $\mathbf{Spin}_0$. One has $\mathbf{Spin}(n) = \mathbf{Spin}_0(n)$ for $n > 1$.

2.3.2. The Topological Aspect

The kernel of ρ, $\mathbf{Z}_2 = \{1, -1\}$, appearing in (4) and (5), has an algebraic origin, described in the preceding paragraph. The case when $V = \mathbf{R}^2$ and g is of signature (1,1), is exceptional: the special Lorentz group $\mathbf{SO}_0(1,1)$ and the group $\mathbf{Spin}_0(1,1)$ are both isomorphic to the additive group $\mathbf{R}$. The generic case is illustrated by $s = 3$ and $t = 0$. On the basis of (1) one sees that the map $Z \mapsto Z'$ represents a rotation by the angle $\omega = 2\arctan\frac{1}{2}\sqrt{m^2+n^2+p^2}$ around an axis defined by the vector (m, n, p). Introducing the 'Cayley–Klein parameters', $a = (1 + \frac{1}{2}ip)\cos\frac{1}{2}\omega$, $b = \frac{1}{2}(im - n)\cos\frac{1}{2}\omega$, and allowing them to assume all values compatible with $|a|^2 + |b|^2 = 1$, one obtains $\mathbf{Spin}(3) = \mathbf{SU}(2)$ and

$$\rho\begin{pmatrix} e^{\frac{1}{2}i\omega} & 0 \\ 0 & e^{-\frac{1}{2}i\omega} \end{pmatrix} = \begin{pmatrix} \cos\omega & \sin\omega & 0 \\ -\sin\omega & \cos\omega & 0 \\ 0 & 0 & 1 \end{pmatrix}.$$

One sees that the restriction of ρ to the circle $\mathbf{U}(1) \subset \mathbf{SU}(2)$, defined by $0 \leq \omega \leq 4\pi$, is the 'squaring map', $z \mapsto z^2$, $z \in \mathrm{U}(1)$. This property of ρ is general in the following sense: if s or $t > 1$, then there are the exact sequences

$$1 \to \mathbf{Z}_2 \to \mathbf{Spin}_0(s,t) \to \mathbf{SO}_0(s,t) \to 1 \tag{6}$$

and

$$1 \to \pi_1(\mathbf{Spin}_0(s,t)) \to \pi_1(\mathbf{SO}_0(s,t)) \to \mathbf{Z}_2 \to 1. \tag{7}$$

The latter sequence defines a non-trivial $\mathbf{Z}_2$–grading of the fundamental group $\pi_1(\mathbf{SO}_0(s,t))$. Incidentally, there is a little 'experiment' closely related to the Hopf map $\mathbf{SU}(2) \to \mathbf{SO}(3) \to \mathbf{S}_2$. Take two coins of one ecu each and put them on a table. Holding one coin, roll the other around it so that their rims touch and there is no slipping. Children, who perform the experiment for the first time, are astonished when they realize that the moving coin rotates by 720^0 after completing one turn around the fixed coin.

3. Pure Spinors

3.1. Representations of Clifford Algebras Associated with Even-dimensional Vector Spaces

In this section, one considers only even-dimensional vector spaces with a scalar product. These assumptions imply that the corresponding Clifford algebras are simple and, as such, have only one, up to equivalence, faithful and irreducible representation.

3.1.1. Complex Spaces

Let (e_μ), $\mu = 1, \ldots, n$, be an orthonormal basis in the complex vector space $W = \mathbf{C}^n$, where n is even, $n = 2m$. One chooses the basis so that $g_{\mu\nu} = g(e_\mu, e_\nu)$ is given by $g_{\mu\nu} = 0$ for $\mu \neq \nu$ and $g_{\nu\nu} = (-1)^{\nu+1}$. The corresponding volume element $\eta = e_1 \ldots e_n \in \mathrm{Cl}(n)$ satisfies $\eta^2 = 1$ and can be used to define the Hodge dual $*a$ of $a \in \bigwedge W$ by means of the Kähler formula $*a = \eta a$ so that $**a = a$.

Recall (see e.g. (Benn and Tucker, 1987; Budinich and Trautman, 1988a) and the references given there) that, for every $m = 1, 2, \ldots$, there is a faithful and irreducible representation γ of the Clifford algebra $\mathrm{Cl}\,(2m)$ in a complex, 2^m- dimensional vector space S of 'Dirac spinors',

$$\gamma : \mathrm{Cl}\,(2m) \to \mathrm{End}\, S. \tag{8}$$

The automorphisms of S defined by $\gamma_\mu = \gamma(e_\mu)$ are called 'Dirac matrices' and

$$\Gamma = \gamma(\eta) = \gamma_1 \ldots \gamma_{2m}$$

is the 'helicity' automorphism, satisfying $\Gamma\gamma_\mu + \gamma_\mu\Gamma = 0$ and $\Gamma^2 = I$. If φ is an eigenvector of Γ,

$$\Gamma\varphi = (-1)^{\lambda(\varphi)}\varphi, \quad \text{where} \quad \lambda(\varphi) = 0 \text{ or } 1, \tag{9}$$

then it is called a 'Weyl spinor'. The transposed matrices γ_μ^t correspond to the contragredient representation of $\mathrm{Cl}\,(2m)$ in the dual space S^* . From the simplicity of $\mathrm{Cl}\,(2m)$ there follows the existence of an isomorphism $B : S \to S^*$ such that

$$\gamma_\mu^t = (-1)^m B\gamma_\mu B^{-1}, \quad \Gamma^t = (-1)^m B\Gamma B^{-1} \quad \text{and} \quad B^t = (-1)^{\frac{1}{2}m(m+1)}B. \tag{10}$$

Since Γ anticommutes with the Dirac matrices, the sign in the first equation above can be changed by replacing B with $B\Gamma$. Following Cartan, it is chosen in such a way that the equation holds also for the matrix $\gamma_{2m+1} = \Gamma$, appearing in the representation of the algebra $\mathrm{Cl}\,(2m+1)$.

3.1.2. Real Spaces

Let g be a scalar product of signature $(2p+\varepsilon, 2q+\varepsilon)$ in the real vector space $V = \mathbf{R}^{2m}$, where $m = p+q+\varepsilon$ and $\varepsilon = 0$ or 1. Since $\mathbf{C}\otimes\mathrm{Cl}(2p+\varepsilon, 2q+\varepsilon) = \mathrm{Cl}(2m)$, the Dirac representation (8) gives, by restriction, a representation

$$\gamma : \mathrm{Cl}\,(2p+\varepsilon, 2q+\varepsilon) \to \mathrm{End}\, S, \tag{11}$$

where the same letter γ is used for the restricted representation. Since now the Clifford algebra is defined over the reals, the complex conjugate $\bar{\gamma}$ of the representation (11) is also a representation; it follows from the simplicity of the algebra that these representations are equivalent: there exists an inertwining linear isomorphism $C : S \to \bar{S}$ such that

$$\bar{\gamma}_\mu = C\gamma_\mu C^{-1}, \tag{12}$$

where the Dirac matrices $\gamma_\mu = \gamma(e_\mu)$ correspond to an orthonormal basis in V such that $g_{\mu\nu} = 0$ for $\mu \neq \nu$, $g_{\mu\mu} = 1$ for $\mu = 1, \ldots, 2p+\varepsilon$ and -1 for $\mu = 2p+1+\varepsilon, \ldots, 2m$; the volume element $\eta = e_1 \ldots e_{2m}$ satisfies $\eta^2 = (-1)^{m+\varepsilon}$. The helicity automorphism is now defined by $\Gamma = i^{m+\varepsilon}\gamma(\eta)$ so that $\Gamma^2 = I$ still holds and

$$\bar{\Gamma} = (-1)^{m+\varepsilon} C\Gamma C^{-1}.$$

The defining properties of the maps B and C determine them only up to non-zero complex factors which can be chosen so that

$$\bar{C}C = (-1)^{\frac{1}{2}(q-p)(q-p+1)} I \qquad \text{and} \qquad C^t\bar{B}C = B. \tag{13}$$

The 'charge conjugate' of $\varphi \in S$ is the spinor

$$\varphi_c = C^{-1}\bar{\varphi}. \tag{14}$$

If $q-p \equiv 0$ or 3 mod 4, then the representation is real: the space S decomposes into the direct sum of two real spaces of 'Majorana spinors' characterized by $\varphi_c = \varphi$ or $-\varphi$. If φ is a Weyl spinor, then so is φ_c and

$$\lambda(\varphi) + \lambda(\varphi_c) \equiv m + \varepsilon \bmod 2. \tag{15}$$

When $p \equiv q$ mod 4 — and only in this case — there are non-zero spinors which are simultaneously Weyl and Majorana.

3.2. Multivectors Associated with Pairs of Spinors

Consider a pair (φ, ψ) of spinors. One associates with that pair the endomorphism $\varphi \otimes B\,\psi$ of S such that, for every $\chi \in S$, one has

$$(\varphi \otimes B\,\psi)(\chi) = \langle B\,\psi, \chi\rangle\, \varphi \quad \text{so that} \quad \mathrm{Tr}\,(\varphi \otimes B\psi) = \langle B\psi, \varphi\rangle.$$

Let γ be the isomorphism (8); there is a multivector (see e.g. (Pais, 1962) or (Budinich and Trautman, 1988b))

$$B(\psi,\varphi) = \sum_{k=0}^{2m} B_k(\psi,\varphi), \quad \text{where} \quad B_k(\psi,\varphi) \in \bigwedge^k \mathbf{C}^{2m}, \tag{16}$$

such that

$$\varphi \otimes B\,\psi = \gamma\,(B(\psi,\varphi)) \tag{17}$$

and

$$B_k(\psi,\varphi) = 2^{-m} \sum\nolimits_{\mu_1 < \ldots < \mu_k} \langle B\,\psi, \gamma^{\mu_k} \ldots \gamma^{\mu_1}\varphi\rangle e_{\mu_1} \ldots e_{\mu_k}. \tag{18}$$

The following theorem summarizes well-known results, which are obtained from the preceding definitions and the properties of the Dirac representation.

Theorem 1 *Let γ be the representation (8), let $B(\psi,\varphi)$ be the multivector defined by (16–18) and $v \in \mathbf{C}^{2m}$. Then*

(i) $$B(\psi, \gamma(v)\varphi) = (c(v) + e(v))B(\psi,\varphi);$$

(ii) $$B_k(\psi,\varphi) = (-1)^{\frac{1}{2}(k-m)(k-m-1)} B_k(\varphi,\psi);$$

(iii) $$*B(\psi,\varphi) = B(\psi, \Gamma\varphi);$$

(iv) $$B_k(\psi, \Gamma\varphi) = (-1)^{k-m} B_k(\Gamma\psi, \varphi);$$

(v) if φ and ψ are Weyl spinors, then

$$\lambda(\varphi) + \lambda(\psi) + k + m \equiv 1 \bmod 2 \quad \textit{implies} \quad B_k(\psi,\varphi) = 0;$$

(vi) in particular, if φ is a Weyl spinor, then

$$k - m \not\equiv 0 \bmod 4 \quad \textit{implies} \quad B_k(\varphi,\varphi) = 0.$$

The representation (11) of the real algebra is faithful (injective), but γ is not surjective: the multivector defined by (17) is not real, in general. There holds, by virtue of (13), the equality

$$\overline{B(\psi,\varphi)} = B(\psi_c,\varphi_c). \tag{19}$$

Therefore, if $q-p\equiv 0$ or 3 mod 4, and $\varphi_c=\varphi$, $\psi_c=\psi$, then the multivector $B(\psi,\varphi)$ is real. In general, the multivector $B_k(\varphi_c,\varphi)$ is real or pure imaginary, depending on whether $\frac{1}{2}(q-p)(q-p+1)+\frac{1}{2}(k-m)(k-m-1)$ is even or odd, as may be seen by referring to part (ii) of Theorem 1 and (13).

3.3. Pure spinors associated with $\mathbf{C}^{2m}$

Let again γ be the Dirac representation of $\mathrm{Cl}(2m)$ in S and $W=\mathbf{C}^{2m}$; if $\varphi\in S$ and $\varphi\neq 0$, then the vector space

$$N(\varphi)=\{v\in W:\gamma(v)\varphi=0\}$$

depends only on the *direction* of the spinor φ and is *totally null*, i.e. $v^2=0$ for every $v\in N(\varphi)$. If $N(\varphi)$ is maximal (in this case: m-dimensional), then φ is said to be *pure*. Let $W=N\oplus P$ be a decomposition of W into a direct sum of maximal, totally null (mtn) subspaces and let $(n_1,\ldots,n_m,p_1,\ldots,p_m)$ be the corresponding *Witt basis*,

$$n_\mu n_\nu+n_\nu n_\mu=p_\mu p_\nu+p_\nu p_\mu=0,\quad n_\mu p_\nu+p_\nu n_\mu=\delta_{\mu\nu},$$

for $\mu,\nu=1,\ldots,m$. Since the representation γ is faithful, there exists $\omega\in S$ such that $\varphi=\gamma(n_1\ldots n_m)\omega\neq 0$ and then $N(\varphi)=N$ so that φ is pure. Conversely, given a pure spinor φ, one can find an mtn subspace P complementary to $N=N(\varphi)$. In terms of the Witt basis, $\eta=[n_1,p_1]\ldots[n_m,p_m]$ and one sees that a pure spinor is Weyl. The collection of 2^m pure spinors $\gamma(p_{\mu_1}\ldots p_{\mu_k})\varphi$, where $1\leq\mu_1<\ldots<\mu_k\leq 2m$, is a 'Fock basis' in S. There is a bijective correspondence between the set of all mtn subspaces of W and the set of all directions of pure spinors. If v is a non-null vector, $v^2\neq 0$, and φ is a pure spinor, then $\gamma(v)\varphi$ is a pure spinor of opposite helicity and $N(\gamma(v)\varphi)=vN(\varphi)v^{-1}$. The group $\mathbf{O}(2m)$ acts transitively on the space of directions of pure spinors and $\mathbf{SO}(2m)$ acts transitively on the manifold of directions of pure spinors of one helicity: this is the complex, $\frac{1}{2}m(m-1)$–dimensional symmetric space $\mathcal{Q}_m=\mathbf{SO}(2m)/\mathbf{U}(m)$ (Ehresmann, Porteous).

Using Theorem 1 one proves

Theorem 2 *I. Let φ be a non-zero Weyl spinor associated with $W=C^{2m}$. The following conditions are equivalent:*

(i) φ *is pure and* $N=N(\varphi)=span\{n_1,\ldots,n_m\}$*;*

(ii) $B_m(\varphi,\varphi)=n_1\wedge n_2\wedge\ldots\wedge n_m$ *and* $B_k(\varphi,\varphi)=0$ *for* $k\neq m$*;*

(iii) *the vector space* $N=\{B_1(\psi,\varphi)\in W:\psi\in S\}$ *is mtn.*

II. If φ and ψ are pure spinors, then the dimension of $N(\varphi)\cap N(\psi)$ is the least integer r such that $B_r(\psi,\varphi)\neq 0$; moreover, $B_r(\psi,\varphi)=n_1\wedge\ldots\wedge n_r$ and $N(\varphi)\cap N(\psi)=span\{n_1,\ldots,n_r\}$, so that *(iv)* $\lambda(\varphi) + \lambda(\psi) + r + m \equiv 0 \bmod 2.$

If the pure spinors φ and ψ are linearly independent, then

(v) $\varphi + \psi$ *is pure iff* $r = m - 2$ *and then* $N(\varphi+\psi) \cap N(\psi) = N(\varphi) \cap N(\psi)$.

III. Consider now a real vector space V with a scalar product of signature $(2p+\varepsilon, 2q+\varepsilon)$ and let $W = \mathbb{C}\otimes V$. For every spinor φ one has

(vi) $N(\varphi_c) = \overline{N(\varphi)}$.

If φ is pure, as a spinor associated with W, then so is φ_c; its real index r is the dimension of the intersection $N(\varphi) \cap N(\varphi_c)$. There is a decomposition of the space $\mathcal{Q}_m$ of directions of pure spinors of one helicity into submanifolds $\mathcal{Q}_{p,q,\varepsilon,r}$ of directions of pure spinors having the same helicity and real index,

(vii) $\mathcal{Q}_m = \bigcup_r \mathcal{Q}_{p,q,\varepsilon,r}$, *where* $r = \varepsilon, \varepsilon+2, \ldots, \varepsilon + 2\min\{p,q\}$.

Pure spinors with the least value ε of the real index are generic in the sense that $\mathcal{Q}_{p,q,\varepsilon,\varepsilon}$ is open and dense in $\mathcal{Q}_m$. The action of the group $SO(2p+\varepsilon, 2q+\varepsilon)$ is transitive on every $\mathcal{Q}_{p,q,\varepsilon,r}$, a manifold of real dimension $m(m-1) - \frac{1}{2}r(r-1)$.

According to part (ii) of Theorem 2 *a pure spinor associated with $\mathbb{C}^{2m}$ is the square root of a self- (or anti-self-)dual and decomposable m-vector.* A proof of the equivalence of parts (i) and (iii) is based on the equation (Budinich and Trautman, 1988b)

$$c(v)B_1(\psi, \varphi) = B_0(\psi, \gamma(v)\varphi).$$

Another proof was given by Hughston and Mason (1988). Most of the statements in Parts I and II are due to Cartan and Chevalley, see also Benn and Tucker. The real case (Part III) is taken from Kopczyński and Trautman.

4. Analogies between Complex and Optical Geometries

4.1. The Algebraic Aspect

Consider a real, n-dimensional vector space V. A complex vector subspace N of $W = \mathbb{C}\otimes V$ defines a *complex flag* (K, L, J) in V, i.e. a pair (K, L) of subspaces of V such that $K \subset L$ and J is a complex structure in L/K, i.e. a linear map such that $J^2 = -\mathrm{id}$. Namely, given N, one puts $K = \mathrm{Re}(N \cap \bar{N})$, $L = \mathrm{Re}(N + \bar{N})$ and $J(n + \bar{n} \bmod K) = i(n - \bar{n}) \bmod K$, where $n \in N$. One easily sees that there is a bijective and natural correspondence between the set of all complex flags in V and the set of all complex subspaces of W. Denote by r the dimension of K. If, in particular, V and N are $2m$- and m-dimensional, respectively, and $r = 0$, then $L = V$ and J is a complex structure in V. When V is given a scalar product g, then a natural question, in relation to a complex flag, is to ask whether g *descends* to L/K and makes J orthogonal. This is answered by the following (Nurowski and Trautman, 1993)

Theorem 3 *Let (K, L, J) be a complex flag in an even-dimensional real vector space V with a scalar product g and let N be the corresponding subspace of the complexification W of V. The following two conditions are equivalent:*

(i) N is maximal among totally null subspaces of W;
(ii) g descends to a scalar product in L/K, the complex structure J is orthogonal and $K = L^{\perp}$.

The case of $m = 2$, g of signature $(3, 1)$ and $r = 1$ leads to an 'optical structure': the line K is interpreted as corresponding to a ray of light and L/K is a 2-dimensional 'screen space'; see (Trautman, 1984 and 1985; Robinson and Trautman, 1988) and the references given there.

4.2. Complex, Cauchy–Riemann and Optical Geometries

4.2.1. Geometries Defined by Subbundles of the Complexified Tangent Bundle

Consider an n-dimensional, paracompact, connected, smooth manifold $\mathcal{M}$ and let $\mathcal{W}$ denote the complexification of its tangent bundle $\mathcal{V} = T\mathcal{M}$. The preceding algebraic constructions can be applied, 'pointwise', to the fibres of the bundles. In particular, a *complex flag geometry* $(\mathcal{K}, \mathcal{L}, \mathcal{J})$ on $\mathcal{M}$ consists of two smooth subbundles $\mathcal{K}$ and $\mathcal{L}$ of $\mathcal{V}$, such that $\mathcal{K} \subset \mathcal{L} \subset \mathcal{V}$, and an automorphism $\mathcal{J}$ of the vector bundle $\mathcal{L}/\mathcal{K}$ such that $\mathcal{J}^2 = -\mathrm{id}$. By Theorem 3, such a geometry is equivalent to giving a smooth, complex vector subbundle $\mathcal{N}$ of the complex vector bundle $\mathcal{W}$. Denoting by $\mathcal{K}_x$ the fibre of $\mathcal{K}$ over $x \in \mathcal{M}$, and similarly for other bundles, one has $\mathcal{K}_x = \mathrm{Re}\,(\mathcal{N}_x \cap \overline{\mathcal{N}_x})$, etc.

Let $\mathcal{S}(\mathcal{N})$ be the module of sections of the vector bundle $\mathcal{N} \to \mathcal{M}$; a similar notation is used for other vector bundles over $\mathcal{M}$ and, in particular, $\mathcal{S}(M)$ is the algebra of smooth functions on $\mathcal{M}$. One says that the complex flag geometry defined by $\mathcal{N}$ is *integrable* if

$$[\mathcal{S}(\mathcal{N}), \mathcal{S}(\mathcal{N})] \subset \mathcal{S}(\mathcal{N}). \tag{20}$$

In the general case, when $0 < r < m < \frac{1}{2}(n + r)$, condition (20) implies

$$[\mathcal{S}(\mathcal{K}), \mathcal{S}(\mathcal{K})] \subset \mathcal{S}(\mathcal{K}), \tag{21}$$

and

$$[\mathcal{S}(\mathcal{K}), \mathcal{S}(\mathcal{L})] \subset \mathcal{S}(\mathcal{L}). \tag{22}$$

Therefore, the bundle $\mathcal{K}$ defines a *foliation* on $\mathcal{M}$, which is assumed to be *regular* in the sense that the set $\mathcal{M}' = \mathcal{M}/\mathcal{K}$ of all its leaves has the structure of a smooth, $(n-r)$-dimensional manifold such that the canonical map $\mathcal{M} \to \mathcal{M}'$ is a submersion. The following theorem rephrases well-known facts concerning complex and Cauchy–Riemann (CR) geometries (Wells, 1983; Yano and Kon, 1984).

Theorem 4 *Let $\mathcal{N} \to \mathcal{M}$ be a complex vector subbundle of the complexified tangent bundle of an n-dimensional manifold $\mathcal{M}$; denote by $(\mathcal{K}, L, J)$ the associated complex flag geometry, and let m and r be the dimensions, complex in the first and real in the second case, respectively, of the fibres of the bundles $\mathcal{N}$ and $\mathcal{K}$. Then*

(i) If $r = 0$ and $n = 2m$, then $\mathcal{J}$ defines an almost complex structure on $\mathcal{M}$ and (20) is its classical integrability condition, equivalent to the vanishing of the Nijenhuis tensor of $\mathcal{J}$;

(ii) If $r = 0$, but $n > 2m > 0$, then (20) is the integrability condition of a (non-trivial) CR geometry defined by $\mathcal{J}$ on the fibres of $\mathcal{L} \subset \mathcal{V}$;

(iii) In the general case, when $0 < r < m \leq \frac{1}{2}(n+r)$, condition (20) implies the Frobenius integrability condition (21) of $\mathcal{K}$; by virtue of (22) the quotient manifold $\mathcal{M}' = \mathcal{M}/\mathcal{K}$ inherits, by projection, a subbundle $\mathcal{L}'$ of its tangent bundle and $\mathcal{J}$ descends to $\mathcal{M}'$, defining there a CR geometry. The latter geometry is simply a complex geometry when $n = 2m - r$.

4.2.2. Hermitian and Optical Geometries

Let $\mathcal{M}$ be now a $2m$-dimensional manifold satisfying, in addition to what is stated at the beginning of §4.2.1, the following assumptions: $\mathcal{M}$ has a metric tensor g of signature $(2p+\varepsilon, 2q+\varepsilon)$, is orientable and has a *spin structure*, i.e. a principal bundle $\mathcal{P}$ over $\mathcal{M}$, with structure group $G = \mathbf{Spin}\,(2p+\varepsilon, 2q+\varepsilon)$, doubly-covering the bundle of orthonormal frames on $\mathcal{M}$ of coherent orientation; see Lawson and Michelsohn for details. The last assumption is, in fact, not essential because all that is needed here are bundles of 'projectivized' spinors which are associated with the bundles of orthonormal frames. It is often, however, convenient to work with spinor fields and to represent them by the corresponding maps from $\mathcal{P}$ to the typical fibre S equivariant with respect to the action of the group G: a spinor field is a map $\varphi : \mathcal{P} \to S$ such that $\varphi(\xi a) = \gamma(a^{-1})\varphi(\xi)$ for every $\xi \in \mathcal{P}$ and $a \in G$; similar representations are applied to vector fields and differential forms. For the sake of clarity, the same symbols will be used to represent such fields, irrespectively of whether they are considered as equivariant maps from $\mathcal{P}$ to a typical fibre or as sections of the corresponding associated bundle. The metric tensor determines a Levi-Civita connection and ∇ denotes the corresponding covariant derivative. With the pair (φ, ψ) of spinor fields one associates the multivector-valued field $\mathcal{B}(\psi, \varphi)$, given by

$$\mathcal{B}(\psi, \varphi)(\xi) = B(\psi(\xi), \varphi(\xi)).$$

The manifold $\mathcal{M}$ being orientable, it is meaningful to consider spinors of the same heliciy over $\mathcal{M}$. Generalizing the constructions due to Penrose, Atiyah *et al.* and O'Brian–Rawnsley, one defines the total space of the *twistor bundle* $\mathcal{T}_r$ of *real index* r as consisting of directions of all pure spinors on $\mathcal{M}$ of one, say positive, helicity and of real index r. According to Theorem 2, if a section of the twistor bundle exists, then it can be represented (in 'homogeneous coordinates') by a nowhere vanishing field $\varphi : \mathcal{P} \to S$ of pure spinors; such a section defines a complex vector bundle $\mathcal{N} \to \mathcal{M}$; its fibres are *mtns*, $\mathcal{N}_x = N(\varphi(\xi))$, where $\xi \in \mathcal{P}_x$. With the field φ one associates the self-dual and decomposable complex m-form

$$\Phi = g(\mathcal{B}_m(\varphi, \varphi)), \tag{23}$$

so that

$$u \in S(\mathcal{N}) \quad iff \quad c(u)\Phi = 0 \quad iff \quad g(u) \wedge \Phi = 0. \tag{24}$$

Theorem 5 *Let φ be a nowhere vanishing field of pure spinors on $\mathcal{M}$, let $\mathcal{N}$ be the corresponding bundle of* mtn *subspaces of the complexified tangent bundle of $\mathcal{M}$ and let Φ be the m-form defined by (23). The integrability condition (20) is equivalent to each of the following:*

(i) the Penrose–Sommers equation $\varphi \wedge \nabla_u \varphi = 0$ holds for every $u \in S(\mathcal{N})$;

(ii) there exists a 1-form μ such that $d\Phi = \mu \wedge \Phi$.

A spinorial proof of this theorem is given by Hughston and Mason who provide references to Penrose and Sommers; they also point out that, locally, by rescaling Φ, one can reduce the 1-form μ to 0 ; this is a 'generalized Robinson theorem'.

If (20) holds and the dimension r of the fibres of the bundle $\mathcal{K}$ is > 0, then the leaves of the corresponding foliation are r-dimensional *null geodetic manifolds*; a similar result was obtained by Plebański and Hacyan in the context of complex, four-dimensional Riemannian geometry.

From now on it is assumed that the field φ of pure spinors is generic, i.e. $r = \varepsilon$. There are two cases to consider: either (i) g is *pseudo-Euclidean*, $\varepsilon = 0$, or (ii) g is *pseudo-Lorentzian*, $\varepsilon = 1$. The prefix *pseudo* is dropped whenever p or $q = 0$.

(i) Hermitian Geometry. In the pseudo-Euclidean case, a section of the twistor bundle $\mathcal{T}_0$ defines an orthogonal almost complex structure $\mathcal{J}$ on $\mathcal{M}$; the orthogonality property is equivalent to the statement that the bilinear map $j : \mathcal{S}(V) \times \mathcal{S}(\mathcal{V}) \to \mathcal{S}(\mathcal{M})$ given by

$$j(u, v) = \langle g(u), \mathcal{J}(v) \rangle$$

defines a *two-form* on $\mathcal{M}$; using $\mathcal{J}$ to give a complex structure in the fibres of $\mathcal{V}$, one introduces in this bundle, with fibres of complex dimension m, the pseudo-Hermitian form $g + ij$ of signature (p, q). There is a convenient, spinorial expression of j in terms of the field φ. Since $r = 0$ and φ is nowhere zero, so is the function $\mathcal{B}_0(\varphi_c, \varphi)$ and $j = \mathrm{i}g(\mathcal{B}_2(\varphi_c, \varphi))/\mathcal{B}_0(\varphi_c, \varphi)$. Moreover, the form Φ can be scaled so that $\Phi \wedge \overline{\Phi} = \wedge^m j$. A manifold $\mathcal{M}$ with g of signature $(2p, 2q)$ and an orthogonal almost complex structure $\mathcal{J}$ is called an *almost pseudo-Hermitian* space; if $\mathcal{J}$ is integrable, then $\mathcal{M}$ is *pseudo-Hermitian*; it is a *pseudo-Kähler* space if, in addition, one of the following equivalent conditions is satisfied: (a) $\mathrm{d}j = 0$, (b) $\nabla \mathcal{J} = 0$, (c) there exists a field of 1-forms μ such that $\nabla \varphi = \mu \otimes \varphi$.

(ii) Optical Geometry. In the pseudo-Lorentzian case, a section of the twistor bundle $\mathcal{T}_1$ defines an *almost optical geometry*, i.e. a complex flag geometry $(\mathcal{K}, L, J)$ on $\mathcal{M}$ with a metric tensor g of signature $(2p+\varepsilon, 2q+\varepsilon)$ such that $\mathcal{K} = \mathcal{L}^\perp$ is a real line bundle and $\mathcal{J}$ is orthogonal with respect to the metric induced by g in the fibres of $\mathcal{L}/\mathcal{K}$. An almost optical geometry satisfying the integrability condition (20) is called *optical*: the trajectories of $\mathcal{K}$ are then null geodesics and the quotient $\mathcal{M}' = \mathcal{M}/\mathcal{K}$ is a $(2m-1)$-dimensional CR manifold. In the proper Lorentzian case $(p = 1, q = 0)$ the integrability condition is equivalent to the statement that the trajectories of $\mathcal{K}$ form a congruence of *null geodesics without shear*, cf. p. 193 in vol. 2 of Penrose and Rindler. Such congruences play a fundamental role in the study of *algebraically special* gravitational fields; see e.g. Kramer *et al.*, Robinson and Trautman (1986 and 1988) and the numerous references given there.

Acknowledgements

This work has been influenced by my collaboration with P. Budinich, W. Kopczyński, P. Nurowski and I. Robinson; I thank them for discussions. It was supported in part by the Polish Committee for Scientific Research under Grant no. 2-0430-9101.

References

Atiyah, M. F., Hitchin, N. J. and Singer, I. M.: 1978, Self-duality in four-dimensional Riemannian geometry, *Proc. Roy. Soc. Lond.* **A 362**, 425–461.

Benn, I. M. and Tucker, R. W.: 1987, *An Introduction to Spinors and Geometry with Applications to Physics* (Hilger, Bristol).

Budinich, P. and Trautman, A.: 1988a, *The spinorial chessboard*, Trieste Notes in Physics (Springer–Verlag, Berlin).

Budinich, P. and Trautman, A.: 1988b, Fock space description of simple spinors, *J. Math. Phys.* **30**, 2125–2131.

Cartan, É.: 1913, Les groupes projectifs qui ne laissent invariante aucune multiplicité plane, *Bull. Soc. Math. France* **41**, 53–96.

Cartan, É.: 1938, *Leçons sur la théorie des spineurs*, vols 1 and 2, Exposés de Géométrie (Hermann, Paris).

Chevalley, C.: 1954, *The algebraic theory of spinors* (Columbia U. P., New York).

Ehresmann, C.: 1950, Sur les variétés presque complexes, pp. 412–419 in *Proc. of the Intern. Congress of Math.*, Cambridge, MA, 1950 (Amer. Math. Soc. Publ., Providence, RI, 1952).

Givens, J. W.: 1937, Tensor coördinates of linear spaces, *Ann. of Math.* **38**, 355–385.

Hughston, L. P. and Mason, L. J.: 1988, A generalised Kerr–Robinson theorem, *Class. Quantum Grav.* **5**, 275–285.

Kähler, E.: 1962, Der innere Differentialkalkül, *Rend. Matematica* **21**, 425–523.

Kopczyński, W. and Trautman, A.: 1992, Simple spinors and real structures, *J. Math. Phys.* **33**, 550–559.

Kramer, D., Stephani, H., MacCallum, M. and Herlt, E.: 1980, *Exact Solutions of Einstein's Field Equations* (VEB Deutscher Verlag der Wissen., Berlin).

Lawson, H. B., Jr. and Michelsohn, M. L.: 1989, *Spin Geometry* (Princeton U. P., Princeton).

Nurowski, P. and Trautman, A.: 1993, Analogies between complex and optical geometries (in preparation).

O'Brian, N. R. and Rawnsley, J. H.: 1985, Twistor spaces, *Ann. Global Anal. Geom.* **3**, 29–58.

Pais, A.: 1962, On spinors in *n* dimensions, *J. Math. Phys.* **3**, 1135–1139.

Penrose, R.: 1978, The complex geometry of the natural world, pp. 189–194 in *Proc. of the Intern. Congress of Math.*, Helsinki, 1978.

Penrose, R.: 1983, Physical space-time and nonrealizable CR-structures, *Bull. Amer. Math. Soc. (N. S.)* **8**, 427–448.

Penrose, R. and Rindler, W.: 1984 and 1986, *Spinors and Space-time*, vols 1 and 2 (Cambridge U. P., Cambridge).

Plebański, J. F. and Hacyan, S.: 1975, Null geodesic surfaces and the Goldberg–Sachs theorem in complex Riemannian spaces, *J. Math. Phys.* **16**, 2403–2407.

Porteous, I. R.: 1969, *Topological Geometry* (Van Nostrand Reinhold Co., London).

Robinson, I. and Trautman, A.: 1986, Cauchy–Riemann structures in optical geometry, pp. 317–324 in: *Proc. of the Fourth Marcel Grossmann Meeting on General Relativity*, R. Ruffini, ed. (Elsevier Science Publ.).

Robinson, I. and Trautman, A.: 1988, Optical geometry, pp. 454–497 in *New Theories in Physics*, Proc. XI Warsaw Symp. Elem. Particle Phys., Kazimierz, 23–27 May 1988, Z. Ajduk *et al.*, eds (World Scientific, Singapore, 1989).

Rodrigues, O.: 1840, Des lois géométriques qui régissent les déplacements d'un système solide, *Journal de Math. (Liouville)* **5**, 380.

Trautman, A.: 1984, Deformations of the Hodge map and optical geometry, *J. Geom. Phys.* **1**, 85–95.

Trautman, A.: 1985, Optical structures in relativistic theories, in "Élie Cartan et les mathémathiques d'aujourd'hui", Proc. Conf., Lyon, 25–29 juin 1984, *Astérisque*, numéro hors série, 401–420.

Waerden, B. L. van der: 1929, Spinoranalyse, *Nach. Ges. Wiss. Göttingen, math.–phys.*, 100–115.

Waerden, B. L. van der: 1960, Exclusion principle and spin, pp. 199–244 in *Theoretical physics in the twentieth century*, M. Fierz and W. F. Weisskopf, eds (Interscience Publ., New York).

Wells, R. O., Jr.: 1983, The Cauchy–Riemann equations and differential geometry, *Proc. Symp. Pure Math.* **39**, Part I, 423–435.

Yano, K. and Kon, M.: 1984, *Structures on Manifolds* (World Scientific, Singapore).

A BASIS FOR DOUBLE SOLUTION THEORY

JAYME VAZ, JR. and WALDYR A. RODRIGUES, JR.
Department of Applied Mathematics
State University at Campinas
13081-970, Campinas, S.P., Brazil

Abstract. Some recent studies concerning the possible equivalence of Dirac and Maxwell equations and its geometrical interpretation by means of a Riemann-Cartan-Weyl geometry are exploit as a possible basis for de Broglie's Double Solution theory.

Key words: Double Solution – Dirac – Maxwell – Riemann-Cartan-Weyl

1. Introduction

Louis de Broglie, like Einstein and Schrödinger (among others), was never satisfied with the orthodox interpretation of quantum mechanics. This led him to revisite in the 50's some of his original ideas developed in the 20's which are the basis of his Double Solution theory (DST) (de Broglie, 1960).

DST supposes the existence of two "waves" associated to quantum phenomena: ψ and ϕ – the former with a statistical meaning and satisfying a linear equation, while the latter is an objective description of a particle by means of its singularities (indeed a region with very high amplitude) and obeying a non-linear equation. It remains, however, a still open question: what is the non-linear equation satisfied by ϕ, and what is the relation between ψ and ϕ ?

We shall exploit in this paper our recent studies (Rodrigues et al. 1993; Vaz and Rodrigues 1993, 1993a) concerning the possible equivalence of Dirac and Maxwell equations as a possible basis for DST. Of course, those "waves" ψ and ϕ are spinor fields in our approach, and our objective is to establish a non-linear equation for ϕ, show the relation between ψ and ϕ, and interpret the results in such a way to suggest the "reality" of the ϕ spinor field.

In order to implement our ideas, we exploit the fact (Rapoport et al. 1993) that a Dirac-Hestenes spinor field (DHSF) can be viewed as generating an effective Riemann-Cartan-Weyl (RCW) geometry in spacetime (ST). One can use the amplitude of DHSF to define a conformal transformation (CT) of the metric, and such that the trace of the torsion plays the role of Weyl's vector field and the metric is compatible with the connection, in distinction to the usual Weyl geometries (Obukhov 1982, Rapoport 1991, Rapoport et al. 1993). Our basic assumption will be that under that CT one can define two different DHSF. We shall take these fields as ψ and ϕ, and we show from the equation for the original DHSF – which is DH equation in Minkowski ST (MST) – we get for ψ DH equation in MST while for ϕ we get a non-linear equation where non-linearity is given by Weyl's vector field. Finally we show, by means of a theorem of Rainich-Misner-Wheeler (Rainich 1925, Misner

F. Brackx et al. (eds.), Clifford Algebras and their Applications in Mathematical Physics, 345–351.

and Wheeler 1957), how that non-linear equation becomes just the free Maxwell equation (for non-null electromagnetic field).

The fact that we have two spinor fields in close relationship, one satisfying Dirac equation and the other satisfying a non-linear equation, perfectly fits de Broglie's ideas in DST. Moreover, the fact that the non-linear equation is just a *spinorial representation* of Maxwell equations indicates an objective meaning to that spinor field which is usually hidden and suggests the existence of something "real" behind it. Indeed, we use de Broglie's ideas in order to try to interpret our mathematical results; we have some intriguing and unexpected results looking for an interpretation, and DST seems to be an adequated schema for it.

In this paper we shall use the following notation: $R_{1,3}$ is the spacetime algebra (STA) (Hestenes 1966) and $R^{+}_{1,3}$ its even subalgebra; the Clifford product is denoted by justaposition, while the interior and exterior products are denoted by $\cdot$ and $\wedge$, respectively; $\{\gamma_\mu\}$ is an orthonormal basis for TM (the tangent bunble over MST M) and $\{\gamma^\mu\}$ is the reciprocal basis: $g(\gamma^\mu,\gamma^\nu) = \eta^{\mu\nu}$ with $\eta^{\mu\nu} = diag(1,-1,-1,-1)$, $g \in T^{0,2}(M)$ and $\gamma^\mu \cdot \gamma_\nu = \delta^\mu_\nu$.

2. The RCW Geometry

Consider a DHSF Ψ on MST M, where $\Psi(x) \in R^{+}_{1,3}, \forall x \in M$. It has the canonical decomposition (Hestenes 1990, 1991):

$$\Psi = \sqrt{\rho} e^{\gamma^5 \beta/2} R, \tag{1}$$

where R represents a Lorentz rotation, $\sqrt{\rho}$ a dilatation and $e^{\gamma^5\beta/2}$ a duality rotation ($\gamma^5 = \gamma^0\gamma^1\gamma^2\gamma^3$ is the volume element and β is the Takabayasi angle, which we assume $\beta = 0$ in this paper). Now, Ψ acts on vectors as $\Psi\gamma^\mu\tilde{\Psi} = \rho\lambda^\mu$, where $\lambda^\mu = R\gamma^\mu\tilde{R}$ (the tilde denotes reversion). The amplitude of the DHSF Ψ can be seen as the *generator* of the CT $C_\rho : \gamma^\mu \mapsto \bar{\gamma}^\mu = \sqrt{\rho}\gamma^\mu\sqrt{\rho} = \rho\gamma^\mu$. Since the reciprocal basis is defined by $\bar{\gamma}^\mu \cdot \bar{\gamma}_\nu = \delta^\mu_\nu$ we must have $C_\rho(\gamma_\mu) = \bar{\gamma}_\mu = \rho^{-1}\gamma_\mu$. This CT introduces a Weyl vector field which is the trace of the torsion tensor, and it defines an effective RCW geometry which is metric compatible. Let us see how: for that CT define a "new" metric $\bar{g}$ by $\bar{g}(\gamma^\mu,\gamma^\nu) = g(\rho\gamma^\mu,\rho\gamma^\nu)$ and a "new" connection $\bar{\nabla}$ by $\rho\bar{\nabla}_\mu\gamma^\nu = \nabla_\mu(\rho\gamma^\nu)$. This gives:

$$\bar{\Gamma}^\nu_{\mu\sigma} = \Gamma^\nu_{\mu\sigma} - (\partial_\mu \ln\rho)\delta^\nu_\sigma \tag{2}$$

for the connection coeficients defined by $\nabla_\mu\gamma^\nu = -\Gamma^\nu_{\mu\sigma}\gamma^\sigma$, etc. It is easy to see that $\bar{\nabla}\bar{g} = 0$, which proves the metric compatibility of the connection $\bar{\nabla}$. Now, take orthonormal vectors $\gamma^a = h^a_\mu\gamma^\mu$, that is: $\bar{g}(\gamma^a,\gamma^b) = h^a_\mu h^b_\nu \bar{g}^{\mu\nu} = \eta^{ab} = diag(1,-1,-1,-1)$; of course that in our case $h^a_\mu = \rho^{-1}\delta^a_\mu$. It is easy to see that in this case we have:

$$\bar{\Gamma}^a_{\mu b} = \Gamma^a_{\mu b}. \tag{3}$$

Moreover, for the Dirac operator $\partial = \gamma^\mu\nabla_\mu$ we have $\partial \mapsto \bar{\partial} = \bar{\gamma}^\mu\bar{\nabla}_\mu = \rho\gamma^\mu\nabla_{\rho^{-1}\gamma_\mu} = \gamma^\mu\nabla_\mu = \partial$, since we always work in terms of orthonormal vectors.

In spite of the fact that $\bar{\Gamma}^a_{\mu b} = \Gamma^a_{\mu b}$, curvature and torsion do not vanish anymore. This is because the structure coeficients of the orthonormal basis are non-vanishing now. In order to calculate curvature and torsion, it is easy to use eq.(2), which gives for the components $\bar{T}^{\nu}_{\mu\sigma}$ of the torsion tensor and $\bar{R}^{\tau}_{\mu\nu\sigma}$ of the curvature tensor:

$$\bar{T}^{\nu}_{\mu\sigma} = (\partial_\sigma \ln\rho)\delta^{\nu}_{\mu} - (\partial_\mu \ln\rho)\delta^{\nu}_{\sigma}, \tag{4}$$

$$\bar{R}^{\tau}_{\mu\nu\sigma} = ([\partial_\sigma, \partial_\nu]\ln\rho)\delta^{\tau}_{\mu}. \tag{5}$$

Note that curvature is null away of the singularities of ρ. Moreover, the trace of the torsion is:

$$\bar{T}_\mu = \bar{T}^{\nu}_{\mu\nu} = -3(\partial_\mu \ln\rho) \tag{6}$$

and the torsion vector $\bar{T} = \bar{T}_\mu\gamma^\mu = -3(\partial\ln\rho)$ plays the role of the Weyl vector field in this geometry, which contrary to other Weyl's ones, is metric compatible and does not meet the kind of problems we have when connection is not metric compatible.

Now, for spaces with curvature and torsion, DH equation is (Vaz and Rodrigues 1993a):

$$\left[\partial\Psi + \frac{1}{2}T\Psi\right]\gamma^1\gamma^2 + \frac{mc}{\hbar}\Psi\gamma^0 = 0 \tag{7}$$

where $\{\gamma^a\}$ $(a = 0,1,2,3)$ are orthonormal vectors and $\partial\Psi = \gamma^\mu\nabla_\mu\Psi = \gamma^\mu[\partial_\mu\Psi + \frac{1}{2}\Gamma_\mu\Psi]$, where $\Gamma_\mu = \frac{1}{2}\Gamma^{ab}_{\mu}\gamma_a\gamma_b$. A thoughtful discussion about DHSF and DH equation in arbitrary manifolds will be given elsewhere, but one can be easily convinced that eq.(7) is just the generalization of Dirac equation given by Hehl et al. (1976) and Ivanenko and Obukhov (1985), now written in terms of STA (except for the factor $\frac{1}{2}$ multiplying T due to our definition of torsion as $T^{\sigma}_{\mu\nu} = \Gamma^{\sigma}_{\mu\nu} - \Gamma^{\sigma}_{\nu\mu} - C^{\sigma}_{\mu\nu}$).

Finnaly, for the case we considered above, DH equation assumes the form:

$$\left[\partial\Psi - \frac{3\partial\ln\rho}{2}\Psi\right]\gamma^1\gamma^2 + \frac{mc}{\hbar}\Psi\gamma^0 = 0. \tag{8}$$

3. Towards DST

We shall start by supposing the geometry of ST to be the RCW one discussed in the preceeding section. The natural, but speculative, way of interpretating that CT is to think of a change in the density of a "fluid" due to the presence of a "particle". DH equation is given by eq.(8), but it is appropriate to define two other DHSF, namely:

$$\phi = \rho^{-1}\Psi \tag{9}$$

and

$$\psi = \rho^{-3/2}\Psi \tag{10}$$

such that they are related by:

$$\phi = \rho^{1/2}\psi. \tag{11}$$

Finally, if we use eq.(9) and eq.(10) into eq.(8) we arrive at the following equations:

$$\partial\psi\gamma^1\gamma^2 + \frac{mc}{\hbar}\psi\gamma^0 = 0, \tag{12}$$

$$\partial\phi\gamma^1\gamma^2 + \frac{mc}{\hbar}\phi\gamma^0 = \frac{1}{2}(\partial\ln\rho)\phi\gamma^1\gamma^2, \tag{13}$$

the former being just DH equation in MST for ψ, while the latter is interpreted as a non-linear DH equation in MST for ϕ.

It is time to note that the above scenary perfectly fits the one of DST. We have two different spinor fields ψ and ϕ related by eq.(11), the former obeying a linear equation and the latter a non-linear one. It remains to show that these two spinor fields have a *different interpretation*, for on the contrary one can simply argue about the artificiality of our schema.

4. Interpretation of the Non-Linear Equation

In order to interpret the non-linear equation (13) and the DHSF ϕ, we need first two other results.

The first result concerns the origin of mass. Hestenes has discussed in (Hestenes 1990, 1991) how one can interpret mass as kinetic energy of rotation, such that for $p = mcv$ where $v = R\gamma^0\tilde{R}$ we have

$$p_\mu = S \cdot \Omega_\mu \tag{14}$$

where $\Omega_\mu = 2(\partial_\mu R)\tilde{R}$ and $S = \frac{\hbar}{2}R\gamma^2\gamma^1\tilde{R}$ is the spin bivector. We shall not discuss here "what is rotating" - see (Barut 1983, Barut and Bracken 1981, Barut and Zanghi 1984, Hestenes 1990, 1991) - but we remember that already Lorentz (1915) wrote: "Something invisible is rotating when we have a magnetic field" - afterwards we shall see the appropriateness of this quotation!

The second result is the theorem of Rainich-Misner-Wheeler (Rainich 1925, Misner and Wheeler 1957) - a simple proof of it using STA is given in (Vaz and Rodrigues 1993). It asserts: "Let an "extremal field" be any electromagnetic field for which the magnetic [electric] field is zero and the electric [magnetic] field is parallel to one spatial axis. Then at any point of MST any non-null electromagnetic field can be reduced to an extremal field by a Lorentz rotation and a duality rotation". Then $F_{ext} = e^{\gamma^5\alpha}LF\tilde{L}$ is an extremal field, which we take for convenience to be a *magnetic* one along the z-direction, i.e., $F_{ext} = h\gamma^2\gamma^1$ where $h > 0$ is its magnitude; and if we take $\beta = -\alpha$, $R = \tilde{L}$, we can write

$$F = \Phi\gamma^2\gamma^1\tilde{\Phi}, \tag{15}$$

where

$$\Phi = \sqrt{h}e^{\gamma^5\beta/2}R. \tag{16}$$

The role of the duality rotation is to change the invariants of the electromagnetic field; we assume $\beta = 0$, which means that we suppose the electric and magnetic fields to be orthogonal.

We finally observe that for the free particle case, Dirac theory gives (Hestenes 1990, 1991) that R describes rotations in the spin plane, so that the product $S\Omega_\mu$ has only scalar part. We can write therefore $S\Omega_\mu = S \cdot \Omega_\mu$, and from eq.(9) and eq.(14):

$$\frac{mc}{\hbar}\phi\gamma^0 = \frac{1}{\hbar}p\phi = \frac{1}{\hbar}\gamma^\mu S\Omega_\mu\phi = \gamma^\mu R\gamma^1\gamma^2\partial_\mu\tilde{R}\phi, \tag{17}$$

If we put eq.(17) into eq.(13) and define

$$h = \frac{1}{\rho}, \tag{18}$$

we have

$$\partial\phi\gamma^1\gamma^2 + \gamma^\mu R\gamma^1\gamma^2\partial_\mu\tilde{R}\phi = -(\partial \ln\sqrt{h})\phi\gamma^1\gamma^2, \tag{19}$$

or after multiplying on the right by $\tilde{\phi}$:

$$\partial\phi\gamma^1\gamma^2\tilde{\phi} + \gamma^\mu\phi\gamma^1\gamma^2\sqrt{h}\partial_\mu\tilde{R} + \gamma^\mu\phi\gamma^1\gamma^2(\partial_\mu \ln\sqrt{h})\tilde{\phi} = 0, \tag{20}$$

where we used $\phi\tilde{\phi} = h$, since using eq.(18) $\phi = \rho^{-1}\Psi = \rho^{-1/2}R = \sqrt{h}R$. Note that with the definition (18) ϕ has exactly the form given by eq.(16), and we can therefore identify ϕ and Φ and interpret $\phi\gamma^2\gamma^1\tilde{\phi}$ as an electromagnetic field F. Since $\partial_\mu\phi = (\partial_\mu \ln\sqrt{h})\phi + \sqrt{h}\partial_\mu\tilde{R}$ we have from eq.(20):

$$\partial F = 0 \tag{21}$$

for $F = \phi\gamma^2\gamma^1\tilde{\phi}$. This is just the *free Maxwell equation*! Note that from eq.(11) we have

$$F = \rho\psi\gamma^2\gamma^1\tilde{\psi} = \rho\mathcal{M}, \tag{22}$$

which suggestus to interpret $\mathcal{M} = \psi\gamma^2\gamma^1\tilde{\psi}$ as a "density" of F; and in fact M is interpreted in Dirac theory as density of magnetization (remember that ψ satisfies DH equation).

5. Concluding Remarks

The results of the preceeding section suggest a different meaning to those spinor fields ψ and ϕ, and consequently subsidy our conjectures made on sec.3. DST forms a schema naturally adapted for the interpretation of our results, which can therefore be viewed as a basis for it.

We observe that a spinorial representation of Maxwell equations like ours has already been found by Campolattaro (1980) – the equivalence of our representation with his one (which uses the traditional tensor and spinor calculus) has been proved in (Vaz and Rodrigues 1993).

Finally, it is important to consider the Klein-Gordon-type equation associated to our non-linear DH equation (13). If we apply the Dirac operator ∂ to eq.(13) we arrive at ($\Box = \partial^2$)

$$\Box\phi + \left(\frac{mc}{\hbar}\right)^2\phi = \frac{\Box\sqrt{\rho}}{\sqrt{\rho}}\phi + 2\left[\frac{\partial^\mu\sqrt{\rho}}{\sqrt{\rho}}\partial_\mu\phi - \left(\frac{\partial\sqrt{\rho}}{\sqrt{\rho}}\right)^2\phi\right] \tag{23}$$

which is a non-linear Klein-Gordon equation. If we neglect the term in brackets we get

$$\Box\phi + \left(\frac{mc}{\hbar}\right)^2\phi = \frac{\Box\sqrt{\rho}}{\sqrt{\rho}}\phi, \tag{24}$$

which is an equation similar to the one of Guerra and Pusterla (1982) (see also Vigier (1991)), except for the different meaning we gave to ρ. In terms of the amplitude $\sqrt{h}$ of ϕ we have

$$\Box\phi + \left(\frac{mc}{\hbar}\right)^2 \phi = -\frac{\Box\sqrt{h}}{\sqrt{h}}\phi - 2\partial^\mu \ln\sqrt{h}\partial_\mu\phi, \tag{25}$$

and neglecting the last term,

$$\Box\phi + \left(\frac{mc}{\hbar}\right)^2 \phi = -\frac{\Box\sqrt{h}}{\sqrt{h}}\phi. \tag{26}$$

The terms in this equation and in that of Guerra and Pusterla (1982) have the same meaning, but our non-linear equation differs from their one by the sign of the non-linear term. It is interesting to note the possibility of interpreting the non-linear term in brackets in eq.(23) as an irreversibility term in the stochastic interpretation of quantum mechanics (Vigier 1991).

There remains several other questions to be answered, of course; some have already been considered in (Vaz and Rodrigues 1993a) – like the question of the two additional degrees of freedom that the DHSF ϕ has (since only 6 are needed to represent an electromagnetic field according to eq.(15)) – and others like the study of cases with interactions are to be considered elsewhere.

Acknowledgements

The authors are grateful to Professor E. Recami, Dr. Q.A.G. de Souza, Dr. M.A.F. Rosa and Dr. J.R.R. Zeni for several important discussions, and to Professor J.P. Vigier for pointing out important references related to the issues addressed in this work, which has been partially supported by CNPq, CAPES and FAEP-UNICAMP.

References

Barut, A.O. (1983), "What is an Electron? Relativistic Electron Theory and Radiative Processes", in *Quantum Optics, Relativity and Theory of Measurement*, P. Meystre (ed.), Plenum Press, New York, pg. 155.

Barut, A.O. and Bracken, A.J. (1981), Phys. Rev. D **23**, 2454.

Barut, A.O. and Zanghi, N. (1984), Phys. Rev. Lett. **52**, 2009.

de Broglie, L. (1960), *Non-Linear Wave Mechanics – A Causal Interpretation*, Elvesier Publ. Co., Amsterdam.

Campolattaro, A.A. (1980), Int. J. Theor. Phys. **19**, 99.

Guerra, F. and Pusterla, M. (1982), Lett. Nuovo Cimento **34**, 351.

Hehl, F.W., von der Heyde, P., Kerlick, G.D. and Nester, J.M. (1976), Rev. Mod. Phys. **48**, 393.

Hestenes, D. (1966) *Space-Time Algebra*, Gordon & Breach, New York.

Hestenes, D. (1990), Found. Phys. **20**, 1213.

Hestenes, D. (1991) "Zitterbewegung in Radiative Processes", in *The Electron*, D. Hestenes and A. Weingartshofer (eds.), Kluwer Acad. Publ. , Dordrecht, pg. 21.

Ivanenko, D. and Obukhov, Yu. N. (1985), Ann. Physik **7**, 59.

Lorentz, H.A. (1915), *Theory of Electrons*, 2nd. ed., Verlag von B.G. Teubner, Leipzig und Berlin.

Misner, C.W. and Wheeler, J.A. (1957), Ann. Phys. **2**, 525.

Obukhov, Yu. N. (1982), Phys. Lett. A **90**, 13.

Rainich, G.Y. (1925), Trans. Am. Math. Soc. **27**, 106.

Rapoport-Campodonico, D.L. (1991), Int. J. Theor. Phys., **30**, 1497.

Rapoport-Campodonico, D.L., Rodrigues, Jr., W.A., Souza, Q.A.G. and Vaz, Jr., J., (1993) "The Riemann-Cartan-Weyl Geometry Generated by a Dirac-Hestenes Spinor Field", RP 56/92 IMECC-UNICAMP, submitted for publication.

Rodrigues, Jr., W.A., Vaz, Jr., J. and Recami, E. (1993) *Courants, Amers, Écueils en Microphysique*, G. & P. Lochak (ed.), Fondation Louis de Broglie (to appear); RP 29/92 IMECC-UNICAMP, INFN/AE-92/25.

Vaz, Jr., J. and Rodrigues, Jr., W.A. (1993), Int. J. Theor. Phys. (to appear); RP 57/92 IMECC-UNICAMP.

Vaz, Jr., J. and Rodrigues, Jr., W.A. (1993a) "A Geometrical Interpretation of the Equivalence of Dirac and Maxwell Equations", RP 06/93 IMECC-UNICAMP, submitted for publication.

Vigier, J.P. (1991), Found. Phys. **21**, 125.

SPATIAL INVERSION AND SPINORS

J. RICARDO ZENI
Depto. Ciências Naturais - FUNREI
36300-000 São João del Rei - MG - Brazil

Abstract. In this article we shed some light on spatial inversion and its action on spinors. We based our discussion on a geometrical background. We remark two points in our discussion: a "natural" way to define spatial inversion on Weyl spinors and so on Dirac spinors; and we are able to show that the two common eigenstates (spin up and spin down) of a spin 1/2 operator are related between themselves through spatial inversion.

Key words: Spatial Inversion - CPT Operations - Weyl Spinors - Dirac Equations - Spin Eigenstates

1. Introduction

Spatial inversion, as well as CPT operations, play a important role in quantum electrodynamics. However, we find some difficulties and ambiguities to define it on spinors [Costa de Beauregard; Feinberg and Weinberg]. We think this situation is based on the fact we have no 2×2 matrix related to spatial inversion [Barut], i.e., if $X = X^j \sigma_j$ represents an euclidean vector, where σ_j $(j = 1, 2, 3)$ are the Pauli matrices, we can prove that there is no $A \in C(2)$ satisfying:

$$AXA^\dagger = -X \tag{1}$$

where $\dagger$is the hermitian conjugate.

This fact has disconnected the definition of spatial inversion from its geometrical interpretation on vectors. So that, the usual definition of spatial inversion on spinors follows from its consequences, not from its causes. This situation results in an uncomfortable state of art, in particular we remark the discussion done in [Lifshitz et al] to justify the fact that spatial inversion relates a dotted to an undotted Weyl spinor.

In this article we expect to give a well reasoned definition to spatial inversion on spinors basing our discussion on the formalism of Clifford Algebras. Others researchers have already studied the subject of spatial inversion on this way [Hestenes 1975; Greider], but we think they have not disclosed all points on the subject, in particular two facts were not discussed in their works: that spatial inversion must relate a dotted to an undotted Weyl spinor; and that a double inversion (that can be interpreted as a 360° rotation) of a spinor (spin 1/2) must change the sign of the spinor.

At last, we are able to show from our definition of spatial inversion that the two spin eigenstates (Pauli or Weyl spinors) of the same operator of spin are related between themselves through spatial inversion, a result that furnishes a mathematical

F. Brackx et al. (eds.), Clifford Algebras and their Applications in Mathematical Physics, 353–358.

background to the usual interpretation of spin 1/2 eigenstates, i.e. those eigenstates are the projection in the forward (spin up) and backward (spin down) direction of the spatial axis defined by the spin operator [Zeni 1991, 1992].

2. Pauli and Weyl Spinors

In this section we work with the formalism of the Clifford Algebra of Euclidean 3-Space, also called Pauli Algebra, where the basic elements (which can be identified with the Pauli matrices) are interpreted as Euclidean vectors [Hestenes, 1966; Zeni and Rodrigues 1990, 1992]. In this formalism we have a well defined operation, called principal automorphism [Porteous], which can be identified with spatial inversion, because of course, vectors change sign under that operation. To carry on this operation on spinors, we take care in order for the spatially inverted spinor to belong to the same ideal than the original spinor, so we introduce one more step in our definition to pull back the inverted spinor to the original ideal, as described bellow.

We choose the following basis for the Pauli algebra, indicated hereafter by R_3,

$$\{1, \sigma_j, i\sigma_j, i\}, \quad (j = 1, 2, 3) \tag{2}$$

where $i = \sigma_1\sigma_2\sigma_3$ is interpreted as the volume element of the euclidean space. It is worth to remark that i commutes with every Pauli number and also $i^2 = -1$, so we can formally identify i with the imaginary unit of complex numbers.

The main automorphism of the Pauli algebra, indicated by $^\square$, also called spatial conjugation, is defined by:

$$\sigma_j^\square = -\sigma_j, \quad (\sigma_i\sigma_j)^\square = \sigma_i^\square\sigma_j^\square, \tag{3}$$

We remark that spatial conjugation changes sign of the euclidean volume element, $i^\square = -i$, as it follows from eq.(3). This fact is significant in our approach, because it shows that spatial conjugation is related to complex conjugation, once we identify i with the imaginary unit, as we said before.

Spinors are included in the discussion through the algebraic theory of spinors, which is based on the subject of ideals [Figueiredo et al, Maiorino et al; Zeni 1992]. Ideals are subspaces of a given algebra which are invariant by multiplication by every element of algebra. The elements of a given ideal are the algebraic spinors, which can be identified with the usual spinors (or covariant spinors), which are defined through group theory [Barut; Lifshitz et al].

We choose the ideal in R_3 related to the idempotent

$$e = \frac{1}{2}(1 + \sigma_3) \tag{4}$$

The basis (complex) for our left ideal is given by

$$I = R_3 e = \{e, \sigma_1 e\} \tag{5}$$

Our spinors are elements of the above ideal that can be written

$$\Psi = (\psi_1 + \psi_2\sigma_1)e \tag{6}$$

where ψ_i can be identified with complex numbers, the imaginary unit being $i = \sigma_1\sigma_2\sigma_3$. In matrix representation the ψ_i are identified with the matrix components of spinor Ψ.

We define spatial inversion on spinors, indicated by ^{P}, as shown bellow:

$$\Psi^P = \Psi^\Box \sigma_1 \tag{7}$$

We have two reasons for the above definition. First, Ψ^P lies in the same ideal that Ψ, whereas $\Psi^\Box$ does not. We remark that $\Psi^\Box \in R_3(1-e)$, i.e., the ideal generated by the idempotent $e^\Box = 1 - e$ so we have introduced σ_1 to the right side of $\Psi^\Box$ to pull back the spatially inverted spinor to the ideal containing the original Ψ, because

$$e^\Box \sigma_1 = \sigma_1 e$$

The fact that $\Psi^P \in I = R_3 e = \{e, \sigma_1 e\}$ avoids any worries in the calculations and interpretation of our results.

Second, we have that

$$(\Psi^P)^P = -\Psi \tag{8}$$

what is natural to happen if we wish, as it is usual to interpret the spatial inversion as a 180° rotation so that a double inversion can be identified with a 360° rotation which changes sign of spinors (spin 1/2).

The components of spatially inverted spinor defined by eq.(7) are given by:

$$\psi_1^P = -\psi_2^\dagger, \quad \psi_2^P = \psi_1^\dagger \tag{9}$$

where $\psi_i^\dagger$ is formally the complex conjugate of ψ_i.

We see from eq.(9) that if Ψ is an eigenstate of an operator of spin, then Ψ^P is the other eigenstate of the same operator [Sakurai]. Therefore we get an alternative, and we think much more natural, geometrical interpretation to the transformation relating eigenstates of spin down and up: it is spatial inversion. It furnishes a mathematical background to the usual interpretation of spin 1/2 eigenstates, i.e., the spin up is the eigenstate that is in the forward direction while the spin down is the eigenstate that is in the backward direction. In this common description of spin eigenstates it is natural to think that the spin eigenstates are related through spatial inversion. However, we learn from textbooks on Quantum Mechanics that the mathematical operation relating eigenstates of spin 1/2 is the time inversion [Sakurai].

Now, we wish to point a difficulty with the usual way to define spatial inversion on Weyl spinors: there are two spinors, say η and χ, which are related themselves through spatial inversion (one of them is a dotted Weyl spinor). After a critical examination of the reasoning exposed in the textbooks it becomes clear that η and χ are arbitrarily related themselves. Behind all that we have the fact that these spinors must satisfy the Dirac (2-spinor) equation. But, do we need Dirac equation to define spatial inversion?

We remark that the definition of spatial inversion given by eq.(7) involves only one spinor. So that we have a well defined rule, eq.(9), to transform the components of a spinor in another ones. We will show bellow that spatial inversion relates two different Weyl spinors when we work with them as components of Dirac spinors. However, as we criticize before, spatial inversion can be defined only on one Weyl spinor.

At this point we observe that if Ψ is a Weyl undotted spinor, then Ψ^P (or $\Psi^\square$) is a Weyl dotted spinor.

3. Dirac Spinors

Our results can be extended to include Dirac spinors, based on the fact that Dirac spinors can be written as a combination of two Weyl spinors, one undotted and other dotted. In the Clifford Algebras formalism, Dirac spinors are found in the Clifford Algebra of Space-Time, generated by the unit scalar 1 and the vectors γ_μ, $\mu = 0,1,2,3$ (that can be represented by a convenient choice of Dirac's matrices). We choose the Minkowski metric with signature $(+,-,-,-)$.

A basis for the space-time algebra, indicated by $R_{1,3}$, is given by:

$$\{1, \gamma_\mu, \sigma_j, \gamma_5\sigma_j, \gamma_5\gamma_\mu, \gamma_5\} \tag{10}$$

where $\gamma_5 = \gamma_0\gamma_1\gamma_2\gamma_3$ and $\sigma_j = \gamma_j\gamma_0$ $(j = 1,2,3)$.

The operation of spatial conjugation introduced in the Pauli algebra, eq.(3), can be extended to the space-time algebra through the following definition, if $A \in R_{1,3}$ we have that [Hestenes 1966]:

$$A^\square = \gamma_0 A \gamma_0 \tag{11}$$

It is convenient to decompose the space-time algebra $R_{1,3}$ in subspaces of even grade, indicated by $R^+_{1,3} = \{1, \sigma_j, \gamma_5\sigma_j, \gamma_5\}$, and odd grade, indicated by $R^-_{1,3} = \{\gamma_\mu, \gamma_5\gamma_\mu\}$. It is easy to see that every element of the odd subalgebra can be written as an element of the even subalgebra multiplied by γ_0. Symbolically we have:

$$R_{1,3} = R^+_{1,3} \oplus R^-_{1,3} = R^+_{1,3} \oplus \left(\gamma_0 R^+_{1,3}\right) \tag{12}$$

where $\oplus$ indicates the direct sum of linear spaces. We remark that the even subalgebra, $R^+_{1,3}$, is identified with the Pauli algebra, R_3.

Considering our previous discussion about ideals and spinors in the Pauli algebra, we get that every spinor on space-time algebra can be obtained multiplying a space-time number by an idempotent, like e given by eq.(4), so we have that a Dirac spinor can be formally written as [Hestenes 1966]:

$$\phi = \eta + \gamma_0\chi \tag{13}$$

where η and χ are Weyl spinors, i.e. they are elements of ideal $I = R_3 e$ (see eq.(5)). The above result is also guaranteed by the isomorphism between $R_{1,3}$ and the algebra of quaternionic matrices 2×2 [Figueiredo et al].

Eq.(13) describes algebraically a set of four complex variables (cf. eq.(6)) that can be identified with a Dirac spinor. However it does not have the adequate transformation law under the Lorentz group, $SL(2,C)$ or its counterpart in Clifford algebras $Spin_+(1,3)$. The transformation law shows that the space of Dirac spinors is the carrier space of the $D^{1/2,0} \oplus D^{0,1/2}$ representations of the Lorentz group, i.e. a Dirac spinor must be written as the sum of a dotted and an undotted Weyl spinor.

For the apropriate description of Dirac spinors in terms of transformation law, we remark that the spatial conjugation relates the two fundamentals representations of the Lorentz group (see the comments bellow eq.(3)). Therefore, we must write a Dirac spinor as follows [Figueiredo et al]:

$$\phi = \eta + \gamma_0 \chi^P \tag{14}$$

We remark that χ^P lies in the same ideal that η (see comments bellow eq.(7)), so eq.(14) is algebraically equivalent to eq.(13). However, eq.(14) shows explicitly the transformation law obeyed by a Dirac spinor, because if χ transforms like $u\chi$ where $u \in D^{1/2,0}$ then $\chi^\Box$ and also χ^P transform like $u^\Box \chi^P$, where $u^\Box \in D^{0,1/2}$ [Figueiredo et al].

We define spatial inversion on Dirac spinors as follows (cf. [Hestenes 1966; Greider]):

$$\phi^P = \phi \gamma_0 \sigma_1 \tag{15}$$

It is easy to show that spatial inversion of Dirac spinors must relate their components, Weyl dotted and undotted spinors:

$$\phi^P = -\chi + \gamma_0 \eta^P \tag{16}$$

If we look at a Dirac spinor as a pair of Weyl spinors, $\phi = (\eta, \chi^P)$, we get that $\phi^P = (-\chi, \eta^P)$. We interpret this result as a consequence of spatial inversion on Dirac spinors, and not on Weyl spinors. Therefore, the usual belief that spatial inversion of Weyl spinors must relate two arbitrary spinors is not true, it is necessary only when we are working with Dirac spinors.

Moreover, we get that eq.(8) is also satisfied to our definition of spatial inversion on Dirac spinors.

4. Conclusions

The theory of algebraic spinors is somewhat in the counterwise of the usual presentation of spinors, because in that conception the spinors can be constructed from antisymmetric tensors, the basic elements of a Clifford Algebra [Figueiredo et al; Hestenes and Sobczyck]. This fact and its consequence to physics theories, mainly Dirac theory, has been carefully examined in recent years [Hestenes (1990); Daviau; Rodrigues and Vaz].

The fact that spinors can be written as linear combination of tensors is essential to understand how spatial inversion acts under spinors. Based on these results we construct our definition of spatial inversion that has its advantages: it is a natural one because it is founded on spatial inversion of vectors; it satisfies the usual requirements of spatial inversion (see eq.(8) and eq.(16)); it relates the spin up and spin down eigenstates, giving a mathematical background to the usual interpretation of the spin eigenstates as the projection in the forward and backward directions. Moreover, our definition of spatial inversion shed some light on the fact that spatial inversion must relate the two fundamentals representations of the Lorentz group, or better, it must relate a Weyl undotted spinor to a Weyl dotted spinor. We hope to exploit this point in our future research on Dirac equations, in special working on CPT transformations.

Acknowledgements

The author is grateful to Professors Waldyr Rodrigues, Juan Mignaco, Guillermo Cabrera and Jayme Vaz for some useful discussions on the subject. This work was partially supported by FAPEMIG.

References

A.O. BARUT, "Classical Electrodynamics of Fields and Particles", Macmillan (1964).
O. COSTA DE BEAUREGARD, Found. Phys., 12, 861 (1982).
C.DAVIAU, Ann. Fond. Louis de Broglie, 14, 373 (1989).
G. FEINBERG and S. WEINBERG, Il Nuovo Cim., XIV, 571 (1959).
V.L. FIGUEIREDO, E.C. OLIVEIRA and W.A. RODRIGUES, Alg. Groups and Geom., 7, 153 (1990).
J.R. GREIDER, Found. Phys., 14, 467 (1984).
D. HESTENES, "Space-Time Algebra", Gordon & Breach (1966).
D. HESTENES, J. Math. Phys., 16, 556 (1975).
D. HESTENES, Found. Phys., 20, 1213 (1990).
D. HESTENES and G. SOBCZYCK, "Clifford Algebras to Geometric Calculus", D. Reidel Publ. Co. (1984).
E. LIFSHITZ, L. PITAIEVSKI and V. BERESTETSKI, "Relativistic Quantum Mechanics", Addison Wesley Publ. Co. (1971).
J.E. MAIORINO, W.A. RODRIGUES and J.R. ZENI, to appear in Alg., Groups and Geom. (1993).
I. PORTEOUS, "Topological Geometry", Cambridge Univ. Press, 1981.
W.A RODRIGUES and J. VAZ, "Non-Dispersive de Broglie Wave Packets From the Free Maxwell Equations", to appear: H. Journal (1993).
J.J. SAKURAI, "Modern Quantum Mechanics", World Scientific (1985).
J.R. ZENI and W.A. RODRIGUES, Int. J. Mod. Phys. A7, 1793 (1992).
J.R. ZENI and W.A. RODRIGUES, Rev. Brasileira Fisica, 20, 373 (1990).
J.R. ZENI, "Geometry of Spin 1/2 Eigenstates", to appear in Annals from XII Encontro Nac. Física Part. e Campos, Brazil (1991).
J.R. ZENI, "Spin 1/2 Eigenstates and Spatial Inversion", to appear in Annals from XIII E.N.F.P.C., Brazil (1992).

PHYSICAL MODELS

NON ABELIAN GAUGE FIELDS IN THE REAL CLIFFORD ALGEBRA OF SPACE TIME

ROGER BOUDET
Université de Provence
Pl. V. Hugo, 13331 Marseille, France

Abstract. One proposes a model of a $SU(2) \times U(1)$ gauge field, written in $Cl(1,3)$ and one studies its eventual conformity with the model of Weinberg-Salam.

Key words: Moving frames, infinitesimal rotations.

1. INTRODUCTION

We adopt here the same point of view as the one of D. Hestenes [1]. All the entities one can find in Physics are elements of the Clifford algebra $C(M) = Cl(1,3)$ of the Minkowski spacetime $M = R(1,3)$, or Real Space Time Algebra [1], because they are geometrical objects (as oriented parts of straight lines, planes, etc..) of M, and also they may express euclidean transformations of these objects.

We propose here a $SU(2) \times U(1)$ gauge model in which $SU(2)$ corresponds to a group of rotations of an effective geometrical three-dimensional space of M instead of some abstract set of transformations. This model differs from the one described in [2] to the extent that, here, $U(1)$ is explicitely interpreted as the goup of rotations of a plane of spacetime, instead as the group of the "duality rotations" (see Nota).

Let us recall how the electromagnetic $U(1)$ gauge group may be interpreted in the Dirac theory of the electron.

One considers the unit timelike vector v which defines the direction of the current at each point x of M.

Two orthonormal spacelike vectors n_1, n_2, orthogonal to v, allow one to define the spacetime intrinsic spin, represented by the antisymmetric tensor of rank two $(\hbar c/2)\sigma^{\mu\nu}$ such that

$$\sigma^{\mu\nu} = n_1^\mu n_2^\nu - n_2^\mu n_1^\nu. \tag{1}$$

The couple (n_1, n_2) determines a plane $P(x)$, the "spin plane" [1], whose role is fundamental as well for the definition of the spin, as for the one of the momentum-energy and the construction of the electromagnetic gauge.

Indeed, one has

$$p_\mu = \frac{\hbar c}{2}\omega_\mu - eA_\mu, \tag{2}$$

F. Brackx et al. (eds.), Clifford Algebras and their Applications in Mathematical Physics, 361–366.

where p_μ is the momentum-energy spacetime vector, A_μ is the electromagnetic potential acting on the electron and

$$\omega_\mu = (\partial_\mu n_1).n_2 = -n_1.(\partial_\mu n_2) \tag{3}$$

is nothing else but a spacetime vector, representing the infinitesimal rotation of the "spin plane" $P(x)$ on itself [3], that the relations $n_1.n_2 = n_1^\nu n_{2\nu} = 0$, $n_k^2 = n_k^\nu n_{k\nu} = -1$ allows one to define.

The replacement of the couple (n_1, n_2) by another couple (n_1', n_2') defining the same plane $P(x)$ and making, inside $P(x)$, with the first couple, an angle φ (the parameter $\varphi/2$ is nothing else but the "phase"), involves the change $\omega_\mu \to \omega_\mu' = \omega_\mu - \partial_\mu\varphi$.

Accompanied by the change $A_\mu \to A_\mu' = A_\mu + (\hbar c/2e)\partial_\mu\varphi$, it constitutes the electromagnetic gauge invariance. The gradient $(\hbar c/2)\partial_\mu\varphi$ is a "gauge photon". (This geometrical interpretation of the electromagnetic gauge has been clearly noticed, to my knolewdge, for the first time in [4]).

The generalization of this point of view to the $SU(2)$ non abelian gauge may be obtained by replacing the rotation of a spacelike plane $P(x)$ on itself by the rotation on itself of a three-dimensional space $E(v)$ orthogonal at each point x of spacetime, to a timelike unit vector v depending on x.

Moreover, if there exists a geometrical link between the energy and the *infinitesimal* rotation on itself of the plane $P(x)$, in the theory of the electron, surely, a similar link between the energy and the *infinitesimal* rotation on itself of the space $E(v)$ must exist in the physical theories which imply the $SU(2)$ gauge. It is not possible to make this link in a conspicuous position in the theories in which $SU(2)$ appears as acting on some abstract space.

2. THE YANG-MILLS SU(2) GAUGE IN Cl(1,3)

Let $B_0 = \{\gamma_\mu \in M, \mu = 0, 1, 2, 3\}$ be a fixed frame of M (laboratory frame). We denote $\gamma^\mu = g^{\mu\nu}\gamma_\nu \in M$ and $i = \gamma^0\gamma^1\gamma^2\gamma^3 \in \wedge^4 M$ (see [1]).

We emphasize that here the γ_μ are spacetime vectors (not matrices) and, though one has $i^2 = -1$ in $C(M)$, i is the unit pseudo-scalar of M (not some imaginary number $\sqrt{-1}$).

We consider a Dirac spinor ψ, depending on the point x of M. The spinor ψ is considered in real algebra as beeing an element of the even subalgebra of $C(M)$ (or ring of the biquaternions) [1].Then, ψ may be written in the form $\psi = (\rho e^{i\beta})^{1/2}R$ where $\rho > 0, \beta$ are real scalars and $R \in \mathrm{Spin}(M)$ corresponds to a Lorentz rotation [1]. Such a rotation defines an orthonormal frame $B(x) = \{n_\mu\}$, such that $n_\mu = R\gamma_\mu R^{-1}$, that we will consider as invariant in all change of the frame B_0 (thus, R varies in such a way that $R' = RL^{-1}$ if $\gamma_\mu' = L\gamma_\mu L^{-1}$).We will call $B(x)$ the intrinsic frame, and denote $E(v)$ the three-dimensional space, subspace of $B(x)$, orthogonal at x to the unit timelike vector $v = n_0$, .

We are here only interested by properties of gauge. So, we will not take into account the terms which could depend on the scalars ρ (the density) or β (the Yvon-Takabayasi angle) though they play a fundamental role in Physics. The part of ψ which is relative to gauge properties is only R. So we will use in what follows the spinor R, which represents a Lorentz rotation, instead of the spinor ψ.

We consider the three timelike orthonormal bivectors of $B(x)$, $\tau_k = n_k \wedge v = n_k v$ (we recall that the spacelike vectors n_k span the space $E(v)$), and we associate with each τ_k the real scalars $W^k_\mu(x)$, in such a way that each spacetime vector $W^k = W^k_\mu \gamma^\mu$ is invariant. As well one could define the timelike bivectors of $B(x)$, $W_\mu = W^k_\mu \tau_k$.

Let us consider the invariant (which is an element of $\wedge^1 M \bigoplus \wedge^3 M$)

$$I = \gamma^\mu(\partial_\mu R - \frac{g}{2} W^k_\mu i\tau_k R)R^{-1} \tag{4}$$

(with a summation over the index k) where g is a scalar constant, or, as well

$$I = \gamma^\mu(\partial_\mu R - \frac{g}{2} iW_\mu R)R^{-1}. \tag{5}$$

Denoting

$$\Omega_\mu = 2(\partial_\mu R)R^{-1}, \quad \Omega = \gamma^\mu \Omega_\mu, \tag{6}$$

(the Ω_μ represent the bivectors which define the infinitesimal rotation of the intrinsic frame $B(x)$), one has

$$I = \frac{1}{2}(\Omega - gW^k i(n_k \wedge v)) = \frac{1}{2}(\Omega - g\gamma^\mu iW_\mu). \tag{7}$$

We construct a $SU(2)$ change of gauge in the following way. We consider a rotation on itself on the three-dimensional space $E(v)$ of the intrinsic frame $B(x)$, which changes the spacelike vectors n_k in the vectors $n'_k = Sn_kS^{-1}$, where $S(x) \in$ Spin(M). Such a transformation leaves the timelike vector v invariant (and so we have $SvS^{-1} = v$), and changes the bivectors τ_k in the bivectors $\tau'_k = n'_k \wedge v$. The invariant I becomes I' such that

$$I' = \gamma^\mu S(\partial_\mu(S^{-1}R) - \frac{g}{2} iW_\mu(S^{-1}R))R^{-1}, \tag{8}$$

or

$$I' = \gamma^\mu(\partial_\mu R - \frac{g}{2} iW'_\mu R)R^{-1}, \tag{9}$$

where

$$W'_\mu = SW_\mu S^{-1} - \frac{i}{g}\hat{\Omega}_\mu, \quad \hat{\Omega}_\mu = 2(\partial_\mu S)S^{-1} = -2S\partial_\mu S^{-1}. \tag{10}$$

The bivectors $\hat{\Omega}_\mu$ are the bivectors of the infinetisimal rotation on itself of the frame $\{n_1, n_2, n_3\}$, in the change of this frame into the frame $\{n'_1, n'_2, n'_3\}$. The absolute infinitesimal rotation of this last frame is defined by the bivectors $\hat{\Omega}_\mu + S\Omega_\mu S^{-1}$. If one would want to make an analogy with the theory of the electron, the bivectors $\hat{\Omega}_\mu$ would correspond to the scalars $\partial_\mu \varphi$ which are the components of the gauge photon spacetime vector. Furthermore, the Ω_μ would allow one to define the momentum-energy of the particle which is in interaction with the field represented by the spacetime vectors W^k, in a same way that in equation (2), the

ω_μ give the components of the momentum-energy of the electron in interaction with the electromagnetic potential A.

If one imposes on the field $W_{\mu\nu}$, associated with the bivectors W_μ to verify the relation$W'_{\mu\nu} = SW_{\mu\nu}S^{-1}$ in the change of gauge, then one has

$$W_{\mu\nu} = \partial_\nu W_\mu - \partial_\mu W_\nu + ig(W_\mu W_\nu - W_\nu W_\mu), \tag{11}$$

which shows that the components $W_{\mu\nu}$ of the field are spacetime bivectors. The proof of eq. (11) call in the relation

$$\partial_\nu \hat{\Omega}_\mu - \partial_\mu \hat{\Omega}_\nu + \frac{1}{2}(\hat{\Omega}_\mu \hat{\Omega}_\nu - \hat{\Omega}_\nu \hat{\Omega}_\mu) = 0, \tag{12}$$

particular to the set of bivectors defining the infinitesimal rotation associated with a rotation, depending on x. It may be deduced from the equality $\partial^2_{\mu\nu} S = \partial^2_{\nu\mu} S$.

Certainly, what is above is a faithful translation, in the real Clifford algebra of spacetime, of the Yang-Mills $SU(2)$ gauge model, and does not bring anything new, except the possibility of interpreting what we call energy in terms of infinitesimal rotations of moving frames of spacetime (see [5]).

However an important point is to be emphasized for what follows. The symbol i which appears in the usual presentation of the theory [6] corresponds to the unit pseudo-scalar of spacetime (this i is the same as the one which is used in the writing $E + iH$ of the electromagnetic field). It has a meaning quite different from the one of the i which is used in the Dirac theory (see [1] and par. 4) and which corresponds to a bivector of M.

3. A SU(2)×U(1) GAUGE MODEL

In this model, the invariant I of eq. (4) is replaced by the invariant J

$$J = \gamma^\mu(\partial_\mu R - \frac{g}{2} iW_\mu R - \frac{g'}{2} B_\mu iIR)R^{-1} \tag{13}$$

where g' is a second scalar constant and the real scalars $B_\mu(x)$ are such that the spacetime vector $B = B_\mu \gamma^\mu$ is invariant. I is the bivector $I = RI_0R^{-1}$, $I_0 = \nu \wedge \gamma_0$, where ν is a unit fixed spacelike vector orthogonal to γ_0. Denoting $n = R\nu R^{-1}$, one can write

$$J = \frac{1}{2}(\Omega - gW^k i(n_k \wedge v) - g'Bi(n \wedge v)). \tag{14}$$

The bivector $iI = i(n \wedge v)$ is a spacelike bivector whose direction is the one of a plane $\Pi(x)$, orthogonal at x to the timelike vector v. The part $U(1)$ of a $SU(2)\times U(1)$ change of gauge will be constructed in such a way that it induces rotations of the plane $\Pi(x)$ on itself.

Such a change is achieved by writing $R' = S^{-1}RU$, where $U(x) \in \text{Spin}(M)$ is such that $UI_0U^{-1} = I_0$, and so I and the direction of the plane $\Pi(x)$ remain unchanged. As a consequence, U is in the form $U = exp(-iI_0\chi/2)$ where $\chi(x)$ is a real scalar, and one has $2(\partial_\mu U)U^{-1} = -iI_0\partial_\mu\chi$. J becomes

$$J' = \gamma^{\mu} S(\partial_{\mu} R' - \frac{g}{2} i W_{\mu} R' - \frac{g'}{2} B_{\mu} i R' I_0) U^{-1} R^{-1} \tag{15}$$

or

$$J' = \gamma^{\mu} (\partial_{\mu} R - \frac{g}{2} i W'_{\mu} R - \frac{g'}{2} B'_{\mu} i I R) R^{-1} \tag{16}$$

in which W'_{μ} is given by eq. (10), and where

$$B'_{\mu} = B_{\mu} + \frac{1}{g'} \partial_{\mu} \chi. \tag{17}$$

If one uses the relations of Weinberg

$$W^3 = \sin\theta\ A - \cos\theta\quad Z, B = \cos\theta\ A + \sin\theta\ Z,\ \ g\sin\theta = g'\cos\theta, \tag{18}$$

(from which one deduces $W^3 \wedge B = A \wedge Z$), one can write

$$\frac{1}{2}(gW^3 n_3 + g' B n) = g\sin\theta (A \frac{1}{2}(n + n_3) + Z \frac{1}{2}(\mathrm{tg}\theta\ n - \mathrm{cotg}\theta\ n_3)) \tag{19}$$

which introduces in eq. (13) the bivector

$$\frac{1}{2}(n + n_3) \wedge v = \frac{1}{2}(I + \tau_3). \tag{20}$$

If $W^1 = W^2 = Z = 0$, if A is the electromagnetic spacetime potential vector and $g\sin\theta = e/\hbar c$, if one has $n = n_3 = s$, where s is the intrinsic spin unit vector of the electron (orthogonal to the current and "the spin plane"), the invariant J becomes an exact term of the part of the intrinsic Dirac equation (whose values are in $\wedge^1 M \bigoplus \wedge^3 M$) [3] which does not involve the density ρ [7]. Then, the plane $\Pi(x)$ becomes the "spin plane" $P(x)$ and the gauge is reduced to the $U(1)$ electromagnetic gauge.

4. COMPARISON WITH THE WEINBERG-SALAM GAUGE MODEL

The interpretation of the above model as a traduction in Clifford algebra of the W.S. gauge requires two remarks.

a) Here, the symbols τ_k, I represent bivectors of M. In the W.S. model they correspond to mixing matrices acting on the components of a doublet in such a way that, in particular, the field that multiplies the matrix $(I + \tau_3)/2$ concerns only the charged part of the doublet. As it has been done in [8] for the doublet proton-neutron, a convenient choice of the vector n (more precisely, of the plane $\Pi(x)$ whose direction and orientation is defined by the spacelike bivector $iI = i(n \wedge v)$) will make or will not make this field taken into account: one will write $n = n_3$ or $n = -n_3$ following that the spinor R is or is not relative to the charged term of the doublet.

b) The conformity of our model with the one of W.S. implies that, in the usual presentation of this model ([9],[10]), the symbol $i = \sqrt{-1}$ which is used in factor of g on one side, of g', on the other, has two quite different meanings. The first i must have the meaning of a pseudo-scalar of M, as in the Yang-Mills theory, in such a way that the spacelike bivectors $i\tau_k$ are the generators of rotations in different planes. The second i must have the same meaning as in the Dirac theory of the electron (see [1]), i.e. also, the one of a generator of a rotation in a plane, but in this case the plane is unspecified. As in the presentation of the Dirac theory of the electron made in [1], the ambiguity is absent of our model, in which the i which multiplies g' in the usual presentation is replaced by the bivector iI.

Nota. It is not algebraicaly incoherent to concede to the i of the usual presentation of the W.S. model, the same meaning as in the Dirac equation, in both terms containing g and g'.

As it is shown in [1], the symbol $i\Psi$, where Ψ is a usual Dirac column spinor must be represented in real algebra by $\psi i\sigma_3$ where ψ is a biquaternion and σ_3 is a fixed timelike bivector (called I_0 here) of the frame B_0, such that $\sigma_3^2 = 1$.

Furthermore, because $\gamma_5\Psi$ becomes $i\psi i\sigma_3 = -\psi\sigma_3$, a left particle, represented by the column spinor $f_L = ((1-\gamma_5)/2)\Psi$ becomes the biquaternion ψu, where $u = (1+\sigma_3)/2$. So, if_L is to be represented by $\psi i\sigma_3 u = \psi i u$.

Denoting $\tau_k^0 = R^{-1}\tau_k R$, one sees that $g\tau_k if_L$ and $g' if_L$ becomes in real algebra $g\psi\tau_k^0 i\sigma_3 u = g\psi i\tau_k^0 u$, $g'\psi i\sigma_3 u = g'\psi i u$. Then, one may consider a group of transformations of the space B_0 on itself whose generators are $\{i\tau_1^0, i\tau_2^0, i\tau_3^0; i\}$ as in [2], instead of the set $\{i\tau_1^0, i\tau_2^0, i\tau_3^0; i\sigma_3\}$ (in which $i\sigma_3$ is to be considered as independant of the other generators, though τ_k^0, σ_3 are bivectors of the same frame B_0). The second set only allows one to consider explicitely the $U(1)$ of the $SU(2)\times U(1)$ group as representing the rotations in a plane of M.

5. CONCLUSION

The traduction in real Clifford algebra of the gauge theories goes beyond the question of the choice of a unambiguous and clear algebraic language, directly related to the euclidean stucture of the Minkowski spacetime. In particular it gives the possibility to interprete the energy as being associated with the infinitesimal rotations of moving frames in spacetime, in the way of the geometrical methods of G. Darboux and E. Cartan, instead with the Lie algebras of some abstract groups.

References

1. D. Hestenes, *J. Math. Phys.* **8**, 798 (1967).
2. D. Hestenes, *Found. of Phys.* **12**, 153 (1982).
3. R. Boudet, *C.R. Ac. Sc. (Paris)* **272** A, 767 (1971).
4. F. Halbwachs, J.M. Souriau and J.R. Vigier, *J. Phys. et le Radium* **22**, 293 (1961).
5. R. Boudet, in *Clifford algebras and their applications in mathematical physics*, A. Micali, R. Boudet and J. Helmstetter, eds. (Kluwer, Dordrecht, 1992), p. 343.
6. M. Carmeli, Kh. Huleihil and E. Leibowitz, *Gauge Fields*, (World Sc. Pub., Singapore, 1989).
7. R. Boudet, *J. Math. Phys.* **26**, 718 (1985).
8. D. Hestenes, *J. Math. Phys.* **8**, 809 (1967).
9. E. Leader and E. Predazzi, *Gauge theories*, (Cambridges University Press, London, 1982).
10. E. Elbaz, *De l'électromagnétisme à l'électrofaible*, (Ed. Marketing, Paris, 1989).

SPIN GAUGE THEORIES: PRINCIPLES AND PREDICTIONS

J.S.R. CHISHOLM*
Institute of Mathematics and Statistics
University of Kent, Canterbury, Kent, U.K.

and

R.S. FARWELL
Faculty of Information Technology
University of Brighton, Brighton, East Sussex, U.K.

1. Introduction

The paper provides an overview of spin gauge theories by first describing the principles on which they are based. These principles are grouped together in four basic categories and are illustrated by referring to a particular spin gauge theory model. The principles have evolved over a number of years and are considerably more sophisticated now compared with those used in our first models (see for example, [1-5]).

In the third section there is a description of our most recent model which unifies the four fundamental forces of elementary particles. This model is the object of current attention since it produces a mass formula which leads to a prediction for the mass of the top quark. The top quark has hitherto been unobserved although current estimates for its mass from Fermilab show that our prediction is well within the allowable range, which is gradually narrowing.

Finally we consider a more speculative suggestion for a way in which the different families could be incorporated into our unified model.

2. Principles of Spin Gauge Theories

2.1. The Geometrisation of Particle Interactions

By geometrisation we mean the unification of space-time and the space describing the particle interactions within a single geometric entity. Clifford, or Geometric[1], Algebras provide a natural setting to achieve such an objective, as we shall illustrate.

Spin gauge theories are Lagrangian field theories describing fundamental fermions and their interactions. The Lagrangian density is given in terms of a Clifford Algebra

* UKC/IMS/A93/5b

[1] The term Geometric Algebra was used by W.K. Clifford in his paper *On the Classification of Geometric Algebras* [6] the abstract of which was presented to the London Mathematical Society on 10 March 1876. In it he generalises Grassmann's extensive quantities a to the cases where $aa \neq 0$.

F. Brackx et al. (eds.), Clifford Algebras and their Applications in Mathematical Physics, 367–374.

$R_{p,q}$ associated with an n-dimensional manifold M with $n = p + q$. In our models M consists of two parts:

- 4 dimensional curved space-time;
- an $(n-4)$ dimensional flat space corresponding to the "higher dimensions".

We emphasise that the two parts are not combined by a direct product, but rather are different subspaces of the larger manifold M. For any point x on the manifold, the tangent space $T(x)$ is spanned by the vectors $\{e_i\}$ of $R_{p,q}$

$$\{e_i, e_j\} = 2I\eta_{ij} \tag{1}$$

where I is the unit of the algebra and the metric is

$$\eta_{ij} = diag(+,\ldots,+,-,\ldots,-)\,, \tag{2}$$

with p pluses and q minuses.

The vectors $\Gamma_\mu(x)$ on M are defined by

$$\Gamma_\mu = \begin{cases} h^i_\mu(x)e_i & \mu = 1,2,3,4 \\ \delta^i_\mu e_i & \mu = 5,\ldots,n \end{cases} \tag{3}$$

where $h^i_\mu(x)$ is the vierbein field and δ^i_μ is the Kronecker delta.

Then

$$\{\Gamma_\mu(x), \Gamma_\nu(x)\} = 2Ig_{\mu\nu}(x)\,, \tag{4}$$

where

$$g_{\mu\nu}(x) = h^i_\mu h^j_\nu \eta_{ij}\,. \tag{5}$$

In spin gauge theories the fermions are described by spinors $\psi(x)$ which are taken to be elements of minimal left ideals of $R_{p,q}$. The bar conjugate, or dual, spinors are elements of minimal right ideals

$$\bar\psi(x) = \psi^+(x)\Gamma\,, \tag{6}$$

where Γ is the conjugation operator and $+$ represents an automorphism within $R_{p,q}$.

For example in a model based on $R_{1,6}$ the elements of the algebra may be minimally represented by 8×8 matrices. In this model spinors will thus also be represented by 8×8 matrices which are equivalent to 8 component spinors that only have non-zero entries in a single column of the matrix. These can be used to model two 4 component Dirac spinors which may represent the electron and its neutrino, that is, the lepton pair l in the first family of fermions; similar spinors can be used to represent the three colours of quark pairs q_{red}, q_{blue} and q_{green} in the first family:

$$l = \begin{pmatrix} \epsilon \\ v \end{pmatrix} \quad q_{red} = \begin{pmatrix} d_{red} \\ u_{red} \end{pmatrix} \quad q_{blue} = \begin{pmatrix} d_{blue} \\ u_{blue} \end{pmatrix} \quad q_{green} = \begin{pmatrix} d_{green} \\ u_{green} \end{pmatrix} \tag{7}$$

In this and our other models the neutrino is taken to have 4 components, but the form of the gauge transformations ensures that there are no right handed neutrino interactions. The model based on $R_{1,6}$ can be used to describe the electroweak interactions of either the lepton pair or the quark pair.

2.2. Gauge Transformations and Symmetries

In gauge theories the particle interactions are associated with local symmetries of the spinors in the Lagrangian density. We choose a transformation of the spinors which preserves the ideal structure of the spinors and conjugate spinors, namely:

$$\begin{aligned}\psi &\to Q(x)\psi \\ \bar{\psi} &\to \bar{\psi}Q^{-1}(x)\end{aligned} \tag{8}$$

where the transformation $Q(x)$ belongs to the algebra $R_{p,q}$. The transformation (8) implies that

$$\psi\bar{\psi} \to Q(x)\psi\bar{\psi}Q^{-1}(x)\,. \tag{9}$$

Now the product $\psi(x)\bar{\psi}(x)$ is not necessarily an element of an ideal of the algebra; it is an arbitrary element of $R_{p,q}$. Thus, for consistency, when the spinors are transformed by (8) then all the elements A in the algebra $R_{p,q}$ should also be transformed according to

$$A \to Q(x)AQ^{-1}(x)\,. \tag{10}$$

One of the implications of the full transformations (8,10) is that the Dirac matrices in the kinetic energy term of the electron, say, would transform at the same time as the electron spinor. It is the transformation (10) which distinguishes spin gauge theories from standard gauge theories since, in the latter, the Dirac matrices would be treated as constant. One of the advantages of viewing all the terms in the Lagrangian density as elements of a single Clifford algebra and thus by using the combined transformations (8,10) is that a greater variety of terms can be built into the Lagrangian. This comes about since all terms of the form $\bar{\psi}A\psi$ are spin gauge invariant; fermion mass terms can thus be built into the Lagrangian in an invariant way.

In summary, in spin gauge theories we gauge the freedom to choose a different representation of the spinors *and* the spin basis $\{\Gamma_\mu(x)\}$ at each point x on the manifold M.

Invariance of the Lagrangian under the local transformation (8,10) is achieved by introducing a covariant derivative D_μ. The (spin) connection terms in D_μ define the interactions of the fermions via intermediate vector bosons. The free Lagrangian for these bosons is generated in a standard way by using the self-commutator squared of the covariant derivative

$$KTr\{g^{\mu\sigma}g^{\nu\tau}[D_\mu, D_\nu][D_\sigma, D_\tau]\}\,. \tag{11}$$

K is taken to be an arbitrary normalisation factor.

2.3. The Frame Field and the Extended Covariant Derivative

The principles behind the frame field and the extended covariant derivative are best illustrated within a particular model based on $R_{1,3}$, with M corresponding to space-time only [3]. In such a model $\psi(x)$ may be taken to represent a 4 component spinor describing an electron with mass m. The Lagrangian density representing the interaction of the massive electron with the electromagnetic field is given by

$$\mathcal{L} = \tfrac{1}{2}[\bar{\psi}i\gamma^\mu(\partial_\mu - ieA_\mu)\psi + h.c.] + m\bar{\psi}\psi\,, \tag{12}$$

where $\{\gamma_\mu(x)\}$ is a vector basis for $R_{1,3}$. In any representation of $R_{1,3}$ we have the identity

$$\gamma^\mu(x)\gamma_\mu(x) = 4U\ , \tag{13}$$

where U is the unit of $R_{1,3}$. Thus we can rewrite the Lagrangian as follows

$$\Delta_\mu = D_\mu - \tfrac{1}{4}im\gamma_\mu(x)\ . \tag{14}$$

In factorising the mass term in this way we can recognise analogies between it and the electromagnetic interaction term. There are parallels between the electromagnetic field $A_\mu(x)$ and the electromagnetic coupling constant e and the terms $\gamma_\mu(x)$ and $m/4$ respectively. We can thus consider $\gamma_\mu(x)$ to be a field, the "frame field". Fermion mass is then interpreted as an interaction with the frame field: it is the strength of the interaction and is no longer an intrinsic property of fermions.

Pursuing the analogy between $A_\mu(x)$ and $\gamma_\mu(x)$ further, it is possible to "combine" the fields together by treating $\gamma_\mu(x)$ as an additional term in a covariant derivative. We thus define an extended covariant derivative Δ_μ which in the $R_{1,3}$ model is given by

$$\Delta_\mu = D_\mu - \tfrac{1}{4}im\gamma_\mu(x)\ . \tag{15}$$

It is perfectly sensible within the context of the Clifford Algebra to define such a quantity since we are forming a linear combination of two algebraic expressions D_μ and $-im\gamma_\mu(x)/4$.

In our models we impose a further condition that the frame field be covariantly constant:

$$D_\mu\gamma_\nu = 0\ . \tag{16}$$

By considering the Clifford anti-commutator (4) we can interpret the frame field as the "square root" of the metric tensor, and thus the condition (16) ensures that the manifold is Riemannian since (16) implies that the metric is covariantly constant. Effectively (16) also places a condition on the form of connection in the covariant derivative D_μ; in $R_{1,3}$ it implies that bivector and scalar terms only are possible. The bivector terms represent local Lorentz transformations (in space-time).

The mechanism of factorising the mass term to define the frame field and the definition of the extended covariant derivative can be generalised to any *spin* gauge theory model. The emphasis here is on spin since the critical ingredient is the Clifford Algebra perspective: without it the extended covariant derivative is not well-defined.

A general question then arises by extension of an earlier principle, namely what is produced from the term

$$KTr\{g^{\mu\sigma}g^{\nu\tau}[\Delta_\mu, \Delta_\nu][\Delta_\sigma, \Delta_\tau]\}\ . \tag{17}$$

We shall see that remarkable things happen when we consider the self-commutator squared of the extended covariant derivative. For example, by including the frame field principle in the model based on $R_{1,6}$ we can unify the space-time local Lorentz transformations with the electroweak symmetry transformation $SU(2) \times U(1)$. The Lagrangian density model in the models describes the lepton pair l and its interactions, including a lepton mass matrix term M. By "factorising" the mass term as above we define an extended covariant derivative

$$\Delta_\mu = D_\mu - \tfrac{1}{4}iM\Gamma_\mu(x)\ . \tag{18}$$

The terms arising by using (18) in (17) are quite astonishing; we obtain:

- the free Lagrangian density for the bosons plus a "spin gravity term" $R_{\mu\nu\sigma\tau}R^{\mu\nu\sigma\tau}$ where $R_{\mu\nu\sigma\tau}$ is the Riemannian curvature tensor;
- the boson mass terms with the correct W/Z mass ratio and zero mass photons;
- the Einstein-Hilbert gravitational Lagrangian $R/16\pi G$, where R is the curvature scalar;
- a constant term.

All but the first terms here arise from the inclusion of the frame field.

One of the objectives of spin gauge theories is to provide a framework for the unification of the four fundamental forces, with different interactions associated with different aspects of the same symmetry principle. This is where the importance of the form of the manifold M emerges. There is no distinction between space-time and the internal interaction space. Thus by using a unified space and symmetry transformations within it, we can realise these objectives. Within such a picture it is possible to build in gravity on the same footing as the other interactions.

2.4. The Family Tetrahedron

The last principle has a different focus from the others above: we turn to the algebra $R_{2,2}$ in which the idempotents may be represented by monomial matrices:

$$P_1 = \begin{pmatrix} 1&0&0&0\\0&0&0&0\\0&0&0&0\\0&0&0&0 \end{pmatrix} \quad P_2 = \begin{pmatrix} 0&0&0&0\\0&1&0&0\\0&0&0&0\\0&0&0&0 \end{pmatrix} \quad P_3 = \begin{pmatrix} 0&0&0&0\\0&0&0&0\\0&0&1&0\\0&0&0&0 \end{pmatrix} \quad P_4 = \begin{pmatrix} 0&0&0&0\\0&0&0&0\\0&0&0&0\\0&0&0&1 \end{pmatrix}. \tag{19}$$

Within the algebra $R_{2,2}$ it is possible to define an operator B_4 such that

$$B_4 P_r B_4^{-1} = P_{r+1} \quad B_4 P_4 B_4^{-1} = P_4 \quad B_4^3 = I\,, \tag{20}$$

where $r = 1,2,3$ and $(r+1)$ is evaluated modulo 3. Analogous operators B_1, B_2 and B_3 can also be defined which leave P_1, P_2 and P_3 invariant respectively [7]. Thus a consistent geometrical representation of the idempotents $\{P_i; i = 1,2,3,4\}$ is as vertices of a tetrahedron[2] with the operators $\{B_i\}$ representing the generators of the rotational tetrahedral symmetries. Tetrahedral symmetry of quarks and leptons has been suggested before by Glashow [9] for example. We propose that the vertices of the tetrahedron, and hence the idempotents P_4 and $\{P_i; i = 1,2,3\}$ be identified with the lepton and quark pairs l and $\{q_c; c = red,\ blue,\ green\}$ respectively referred to above. Then B_4 is reminiscent of the colour symmetry operator. Moreover $SU(3)$ can be generated from the generators of the tetrahedral group expressed as linear combinations of the elements of $R_{2,2}$. This last principle provides the ingredient to build the strong interactions into a spin gauge theory model.

[2] This extends an idea of Greider and Weideman [8] that the idempotents P_1, P_2, P_3 may be interpreted geometrically as the vertices of an equilateral triangle which in turn may be identified with quarks of three colours.

3. The Unified Model and a Prediction of the Mass of the Top Quark

A spin gauge theory model which incorporates all of the principles described in section 2 can be built using the algebra $R_{3,8}$. It provides a model of the four fundamental interactions of the first family of fermions: the electron and its neutrino and the up and down quarks of three colours.

The algebra $R_{3,8}$ has a number of interesting and useful features.

- It contains a pseudoscalar which has square $-I$ and which commute with all other elements in the algebra and can thus be used to represent the imaginary unit i.
- By partitioning the basis vectors in such a way that the vectors which span space-time are separated from the rest, we obtain the basis vectors which span a space of signature (2,5). The associated algebra $R_{2,5}$ has a pseudoscalar with square I and thus within it a helicity projection operator can be defined which can be used to split the algebra into left and right handed parts.[3]
- The spinors are minimally represented by 32×32 matrices and we take a block form for the spinor matrix consisting of the sum of two ideals:

$$\psi = \left(\begin{array}{ccc|c} q_{red} & 0 & 0 & 0 \\ q_{blue} & 0 & 0 & 0 \\ q_{green} & 0 & 0 & 0 \\ \hline 0 & 0 & 0 & l \end{array}\right) , \tag{21}$$

Each term in this matrix representation of ψ is an 8×8 matrix representing an ideal in $R_{1,6}$. The form of the gauge transformations is deliberately selected so that this structure is preserved. In particular a representation of $SU(3)$ can be found from the elements of $R_{2,2}$ which does not mix the quarks with the leptons and does not give the lepton any strong interactions, as required.

The four fundamental interactions can be incorporated into this model by using gauge transformations generating local Lorentz transformations in space-time and $SU(2) \times U(1)$ and $SU(3)$ transformations in the higher dimensions.

The factorisation of the mass term to produce the frame field may be used in this model and the extended covariant derivative thus defined used to generate the boson Lagrangian according to (17). The remarkable terms generated by this process in the model $R_{1,6}$ described above are duplicated here, and there is just one additional feature of this unified model: the masses of the gluons are automatically zero as required.

Equal normalisation of the boson kinetic and mass terms gives a mass formula

$$m_e^2 + m_\mu^2 + 3(m_u^2 + m_d^2) = \frac{32}{3} M_W^2 \tag{22}$$

where m_f denotes the mass of the fermion f and M_W is the mass of the W boson. The formula (22) clearly does not hold for the the first family of fermions. However if it is extended to three families then we obtain instead

$$\sum_{all\ f} m_f^2 = \frac{32}{3} M_W^2 \tag{23}$$

[3] We are grateful to Li Deming for pointing this out to us [10].

where we neglect family mixing and there is a multiplicity of three for the quarks. The mass term for the top quark will dominate the left-hand-side of (23) and thus

$$m_{top}^2 \approx \frac{4}{3}\sqrt{2}M_W \tag{24}$$

or

$$m_{top} \approx 151.7\, GeV\ . \tag{25}$$

Family mixing might lead to a correction of about 0.1 GeV.

4. The Introduction of Families

In our model the spinor representing a single family is given by the matrix (21) which has a particular block form. To introduce more than one family would require multiple copies of this representation. Experimental results suggests that maximum number of families is three. It is difficult to see how a factor of three might arise in our model since all basis elements of Clifford Algebras have order two or four.

Yet copies of the matrix representation (21) arise quite naturally in the tetrahedral group and we shall see that in turn this can be related to the algebra $R_{2,2}$ which is central to one of our spin gauge theory principles.

In the tetrahedral group there are 12 elements and they have a regular representation in terms of 12×12 matrices. In group representation theory the regular representation breaks into *three* equivalent 3×3 irreducible representations and *three* inequivalent 1×1 irreducible representations. In other words we have three copies of the required family spinor (21) provided naturally:

$$\begin{pmatrix} \boxed{3\times 3} & & & & & \\ & \boxed{3\times 3} & & & & \\ & & \boxed{3\times 3} & & & \\ & & & \boxed{1} & & \\ & & & & \boxed{1} & \\ & & & & & \boxed{1} \end{pmatrix} . \tag{26}$$

In particular it is interesting to note that the three "quark" representations are equivalent which potentially allows for quark mixing among the families.

A second interesting pointer arises if we try to realise the elements of the tetrahedral group from the elements of $R_{2,2}$. The algebra $R_{2,2}$ has 16 basis elements. If we distinguish the $\pm$ elements in the algebra then we have 32 different basis elements which are generate a group G_{32} of order 32. In this group there is an automorphism of order *three* defined by

$$e_3 \rightarrow e_4 \rightarrow e_3e_4 \rightarrow e_3 \tag{27}$$

$$e_1e_3e_4 \rightarrow e_1e_2 \rightarrow e_2e_3e_4 \rightarrow e_1e_3e_4\ , \tag{28}$$

where $\{e_i; i = 1, 2, 3, 4\}$ represent the basis vectors in $R_{2,2}$. The two sets of elements in (27) and (28) each generate two quaternionic subgroups of G_{32} which are fixed by the automorphism.

The automorphism (27) can be represented[4] as a similarity transformation by introducing an element ω with

$$\omega^3 = I, \quad \omega e_3 \omega^{-1} = e_4, \quad etc. \tag{29}$$

Then we may define *three* sets

$$IG_{32}, \quad \omega G_{32}, \quad \omega^2 G_{32}$$

or, in terms of the algebra basis,

$$\pm R_{2,2}, \quad \pm\omega R_{2,2}, \quad \pm\omega^2 R_{2,2}\,. \tag{30}$$

These three sets together form a group G_{96} of order 96 in which there are 12 equivalent *tetrahedral* subgroups. This provides a mathematical relation between the algebra $R_{2,2}$ and the tetrahedral group. Also $G_{96} = H \times T$, where H is one of the quaternionic subgroups of G_{32} and T is the tetrahedral group.

Given the importance we attach in section 2.4 to the tetrahedral group, since it enables us to build the strong interactions into our model, it appears to be more than a mere coincidence that the tetrahedral group might provide us also with the opportunity to introduce exactly three families of fermions.

References

1. Farwell, R.S. and Chisholm, J.S.R.: 1986, 'Unified Spin Gauge Theory Models', in *Clifford Algebras and their Applications in Mathematical Physics*, eds. J.S.R. Chisholm and A.K. Common, Reidel, Dordrecht, pp. 363-370.
2. Chisholm, J.S.R. and Farwell, R.S.: 1987, 'Electroweak spin gauge theories and the frame field' *J. Phys. A: Math. Gen.* **20**, 6561-6580.
3. Chisholm, J.S.R. and Farwell, R.S.: 1988, 'Gravity and the frame field', *Gen. Rel. and Grav.* **20**, 371.
4. Chisholm, J.S.R. and Farwell, R.S.: 1989, 'Unified spin gauge theory of electroweak and gravitational interactions' *J. Phys. A: Math. Gen.* **22**, 1059-1071.
5. Chisholm, J.S.R. and Farwell, R.S.: 1990, 'Unified spin gauge theories of the four fundamental forces' in *Proceedings of the IMA Conferecnce: The Interface of Mathematics and Physics*, eds. D.G. Quillen, G.B. Segal and Tsou, S.T. Oxford University Press, Oxford, pp. 193-202.
6. W.K. Clifford: 1876, 'On the Classification of Geometric Algebras', in *Mathematical Papers*, Reprint, Chelsea, New York.
7. Chisholm, J.S.R. and Farwell, R.S.: 1992, 'Tetrahedral Structure of Idempotents of the Clifford Algebra $C_{3,1}$' in *Clifford Algebras and their Applications in Mathematics Physics*, eds. A. Micali, R. Boudet and J. Helmstetter, Kluwer, Dordrecht, pp. 27-32.
8. Greider, K. and Weiderman, T.: 1988, 'Generalised Clifford Algebras as Special Cases of Standard Clifford Algebras I' *UCD Preprint 16.*
9. Glashow, S.: 1980, in *1979 Cargese Summer Institute Lectures.*
10. Li, Deming: 1991, private communication.
11. Belch, R.: 1993, private communication.

[4] We are grateful to R. Belch for noting this representation [11].

GRAVITY AS A GAUGE THEORY IN THE SPACETIME ALGEBRA

CHRIS DORAN*
DAMTP, Silver Street, Cambridge, CB3 9EW, UK

and

ANTHONY LASENBY and STEPHEN GULL
Cavendish Laboratory, Madingley Road, Cambridge CB3 0HE, UK

Abstract. We outline a theory of gravitational interactions utilising the spacetime algebra – the geometric algebra of spacetime. The theory arises by demanding invariance under active Poincaré transformations. Making this symmetry local results in a first-order theory with 40 degrees of freedom. The matter-free field equations are presented, and are solved for radially-symmetric static fields. We discuss the behaviour of point particles under the fields described by these solutions, and compare and contrast the results with those of general relativity.

Key words: gravity, gauge theories, spacetime algebra

1. Introduction

For some time we have been convinced that geometric algebra is the best available mathematical tool for physics [1]. This is particularly true for relativistic physics, in which the spacetime algebra, or STA [2], offers great advantages over conventional 4-vector techniques. A major stumbling block to wider acceptance of the STA approach has been the inability to formulate general relativity satisfactorily. The spacetime algebra is, after all, the geometric (Clifford) algebra of *flat* spacetime. Points in this spacetime are represented as *vectors*, and the Clifford algebra is built on this vector space. It therefore seems at first sight to be impossible to formulate a theory of gravity within the STA, since the structure of general relativity is intrinsically related to the notion of spacetime curvature.

Hestenes & Sobczyk [3] have developed a beautiful theory of curved manifolds within geometric algebra by viewing the manifold as a surface embedded in a larger 'ambient' flat space. The properties of the manifold are determined solely by the pseudoscalar (the volume element of tangent space) as it moves through the embedding space, and all the standard (intrinsic) results of Riemannian geometry are recovered by the projection of multivector quantities onto the manifold. This approach is useful both for concrete calculations and for working abstractly, when results can be generated without specifying the dimension of the embedding space. Nevertheless, it is not clear that this approach is the one required for general relativity. Einstein's equations are entirely local and do not predict any global features of a manifold, so that they do not contain information about how the pseudoscalar

* Supported by a SERC studentship.

F. Brackx et al. (eds.), Clifford Algebras and their Applications in Mathematical Physics, 375–385.

moves through an embedding space. In formulating general relativity within the framework of geometric calculus we are therefore presented with two alternatives. We must either modify the Einstein equations so that they specify some extrinsic properties, or we must assume that the pseudoscalar is constant, in which case we are explicitly working in a flat spacetime.

In this paper we adopt the second of these two approaches, and formulate a theory of gravity in terms of multilinear functions defined in the algebra of flat spacetime. That this is possible may seem surprising at first, since we are traditionally taught that incorporating gravity into special relativity leads inexorably to the concept of a curved spacetime [4, 5]. Yet our theory does reproduce the *experimentally-verified* predictions of general relativity. At the very least, this means that physicists have a choice between formulating gravity in terms of spacetime curvature or in terms of forces in a flat spacetime. A detailed discussion of the issues involved in this choice is given in a forthcoming paper [6], in which it is argued that the existence of torsion generated by quantum spin strongly favours the flat-space approach.

In the present paper we provide an introduction to our theory, concentrating on the matter-free field equations. Our theory is a gauge theory based on the gauge group of active Poincaré transformations of spacetime fields. Geometric algebra is a coordinate-free language, so that the passive coordinate transformations of general relativity have no place. We discover a class of radially-symmetric static solutions and discuss the relationship between these and the Schwarzschild metric of general relativity. By considering the motion of a test particle, we recover the standard modification of the Newtonian radial acceleration of the particle under the influence of these radial fields. A horizon still exists, but particles can now cross the horizon in a finite external coordinate time. This is illustrated with some simple diagrams. We conclude by discussing some implications for black-hole physics.

We follow throughout the conventions of [1, 7, 8, 9]; thus (Clifford) vectors are written in lower case Roman (a) or Greek (γ_μ) and general multivectors in upper case Roman (A) or Greek (ψ). The symbol $\langle A\rangle_r$ denotes the projection onto the grade-r components of A, and the scalar (grade-0) part is written as $\langle A\rangle$. We define the interior, exterior, scalar and commutator products as follows:

$$\begin{aligned} A_r \cdot B_s &= \langle A_r B_s\rangle_{|r-s|} & A_r \wedge B_s &= \langle A_r B_s\rangle_{r+s} \\ A*B &= \langle AB\rangle & A\times B &= \tfrac{1}{2}(AB-BA), \end{aligned} \tag{1}$$

and reversion as:

$$\begin{aligned} (AB)^{\sim} &= \tilde{B}\tilde{A} \\ \tilde{a} &= a \qquad \text{for any vector } a. \end{aligned} \tag{2}$$

Upon introducing an orthonormal frame $\{\gamma_\mu\}$ ($\mu = 0\ldots3$), satisfying

$$\gamma_\mu \cdot \gamma_\nu = \eta_{\mu\nu} = \mathrm{diag}(+\,-\,-\,-), \tag{3}$$

the full STA is spanned by the quantities

$$1, \qquad \{\gamma_\mu\}, \qquad \{\sigma_k,\, i\sigma_k\}, \qquad \{i\gamma_\mu\}, \qquad i, \tag{4}$$

where

$$i \equiv \gamma_0\gamma_1\gamma_2\gamma_3, \qquad \sigma_k \equiv \gamma_k\gamma_0. \tag{5}$$

Linear functions are written as $\underline{h}(a)$ and are extended via outermorphism [3, 10] to act on the entire algebra. The adjoint to a linear function is defined by

$$\overline{h}(a) = \partial_b \langle a\underline{h}(b)\rangle \tag{6}$$

where

$$\partial_b = \gamma^\mu \frac{\partial}{\partial b^\mu}, \tag{7}$$

and $b^\mu = \gamma^\mu \cdot b$ is the (scalar) component of b along the γ_μ axis. An important class of linear functions is obtained from differentiating a (possibly non-linear) transformation $f(x)$. For these we write

$$\underline{f}(a) = \underline{f}_x(a) = a\cdot\nabla_x f(x) \tag{8}$$

where ∇_x is the vector derivative with respect to x. The dependence on spacetime position of $\underline{f}$ is indicated by a subscript, to distinguish it from the argument of the linear function, but we will drop this subscript whenever no confusion can arise. The adjoint to $\underline{f}$ (defined by (6)) satisfies the integrability condition that

$$\nabla\wedge\overline{f}(a) = 0, \quad \text{for all constant } a. \tag{9}$$

2. The Matter-free Field Equations

We are concerned in this paper with the gravitational field equations in the absence of matter. In the full treatment [6] the field equations are derived by demanding invariance of the Dirac equation under local, active Poincaré transformations. For our present purpose it suffices to consider the simpler equation

$$\nabla\psi = \nabla_x\psi(x) = 0, \tag{10}$$

where ψ is some arbitrary multivector function. This equation encompasses the neutrino equation (ψ = even multivector) and the free-field Maxwell equations (ψ = bivector). Given a solution $\psi(x)$ to (10), we can obtain a new solution by a translation; that is, if we define

$$\psi(x) \mapsto \psi(x') \equiv \psi'(x), \quad \text{where } x' = x + a \tag{11}$$

for some constant vector a, then $\psi'(x)$ also satisfies (10). To make this symmetry local we let a become an arbitrary function of position, which is done by replacing the translation with

$$x \mapsto f^{-1}(x) \equiv x' \tag{12}$$

so that the field transforms to

$$\psi'(x) \equiv \psi(x'). \tag{13}$$

Here f is an arbitrary non-linear mapping of spacetime vectors. Now $\psi'(x)$ no longer satisfies (10), but satisfies the new equation

$$\overline{f}_{x'}(\nabla)\psi' = \nabla_{x'}\psi(x') = 0. \tag{14}$$

This immediately tells us how to generalise (10). We introduce an arbitrary position-dependent linear function $\overline{h}$, and replace (10) by

$$\overline{h}(\nabla)\psi = 0. \tag{15}$$

If $\overline{h}$ is now transformed according to

$$\overline{h}_x \mapsto \overline{h}_{x'}\overline{f}_{x'} \equiv \overline{h}'_x, \tag{16}$$

then the transformed functions $\overline{h}'$ and ψ' together satisfy (15), provided that $\overline{h}$ and ψ do so.

Equation (15) is also invariant under rotations, although not in the conventional manner in which the spacetime dependence of ψ is also rotated (with the opposite orientation to the rotation of the fields). The gauging of translations has already allowed for the most general type of transformation of position dependence, so rotational invariance is instead achieved by transforming both ψ and $\overline{h}$:

$$\begin{aligned} \psi &\mapsto R\psi\tilde{R} \\ \overline{h}(a) &\mapsto R\overline{h}(a)\tilde{R}, \end{aligned} \tag{17}$$

where R is a constant rotor ($R\tilde{R} = 1$) and we have assumed a double-sided transformation law for ψ. (If ψ is a spinor function, which has a single-sided transformation law, the analysis is similar and the resultant field equations are the same [6].) To make this symmetry local, we write $\overline{h}(\nabla)$ as $\overline{h}(\partial_a)a\cdot\nabla$ and replace the directional derivative $a\cdot\nabla$ by a directional coderivative defined by

$$\mathcal{D}_a \equiv a\cdot\nabla + \Omega(a)\times. \tag{18}$$

Here $\Omega(a)$ is a position-dependent bivector-valued linear function of the vector a, which behaves under local rotations as

$$\Omega(a) \mapsto R\Omega(a)\tilde{R} - 2a\cdot\nabla R\tilde{R} \tag{19}$$

and under translations as

$$\Omega_x(a) \mapsto \Omega_{x'}\underline{f}_{x'}^{-1}(a). \tag{20}$$

The full generalisation of equation (10) now reads

$$\overline{h}(\partial_a)\mathcal{D}_a\psi \equiv \mathcal{D}\psi = 0, \tag{21}$$

and is invariant under both local translations and rotations. Local Poincaré invariance has been achieved at the expense of introducing two gauge fields, $\overline{h}$ and Ω, with a total of $(4\times4)+(4\times6) = 40$ degrees of freedom. (This is precisely the number expected from gauging the 10-dimensional Poincaré group.) The advantage of this approach over previous formulations of gravity as a gauge theory [11, 12] is that the freedom from coordinates has shown us exactly how the gauge fields enter into the field equations.

We must now construct an action integral for the gauge fields $\overline{h}$ and Ω. We first define the field-strength tensor $R(a\wedge b)$ by

$$\begin{aligned} &[\mathcal{D}_a, \mathcal{D}_b]\psi = R(a\wedge b)\times\psi \\ \Rightarrow\ &R(a\wedge b) = a\cdot\nabla\Omega(b) - b\cdot\nabla\Omega(a) + \Omega(a)\times\Omega(b), \end{aligned} \tag{22}$$

so that $R(B)$ is a bivector-valued function of the bivector B. This transforms under local translations as

$$R_x(a\wedge b) \mapsto R_{x'}\underline{f}_{x'}^{-1}(a\wedge b), \tag{23}$$

and under rotations as

$$R(a\wedge b) \mapsto RR(a\wedge b)\tilde{R}. \tag{24}$$

¿From $R(a\wedge b)$ we define the contractions

$$R(b) = \overline{h}(\partial_a)\cdot R(a\wedge b) \tag{25}$$
$$\mathcal{R} = \overline{h}(\partial_b\wedge\partial_a)\cdot R(a\wedge b), \tag{26}$$

and of these the ('Ricci') scalar $\mathcal{R}$ has the simple transformation property $\mathcal{R}(x) \mapsto \mathcal{R}(x')$ under local translations. It follows that the action integral

$$S = \int |d^4x|(\det h)^{-1}\mathcal{R} \tag{27}$$

is invariant under local Poincaré transformations, as will be the field equations derived from S. The derivation of the field equations from S is carried out in full in [6]; this derivation is similar to the Palatini formulation of general relativity, but there is no need for the concept of a metric associated with a curved space. Here we simply quote the required equations [6]:

$$\mathcal{D}\wedge\overline{h}(a) \equiv \overline{h}(\partial_b)\wedge(\mathcal{D}_b\overline{h}(a)) = \overline{h}(\nabla\wedge a) \quad \text{for all } a, \tag{28}$$

and

$$R(a) = 0 \quad \text{for all } a. \tag{29}$$

These constitute our flat-space matter-free gravitational field equations.

3. Radially-symmetric Static Solutions

In order to find radially-symmetric static solutions to the field equations (28) and (29) we introduce a set of polar coordinates (t, r, θ, ϕ), and define the coordinate frame

$$\begin{aligned} e_t &= \partial_t x = \gamma_0 \\ e_r &= \partial_r x = \sin\theta\cos\phi\gamma_1 + \sin\theta\sin\phi\gamma_2 + \cos\theta\gamma_3 \\ e_\theta &= \partial_\theta x = r(\cos\theta\cos\phi\gamma_1 + \cos\theta\sin\phi\gamma_2 - \sin\theta\gamma_3) \\ e_\phi &= \partial_\phi x = r\sin\theta(-\sin\phi\gamma_1 + \cos\phi\gamma_2). \end{aligned} \tag{30}$$

For our initial ansätz we choose $\overline{h}$ to be of the form

$$\begin{aligned} \overline{h}(e_t) &= f_1 e_t + f_2 e_r \qquad & \overline{h}(e_\theta) &= e_\theta \\ \overline{h}(e_r) &= g_1 e_r + g_2 e_t \qquad & \overline{h}(e_\phi) &= e_\phi, \end{aligned} \tag{31}$$

where f_i and g_i are functions of r only. We can write $\overline{h}$ in a more compact form as

$$\overline{h}(n) = n + n\cdot e_t\left((f_1-1)e_t + f_2 e_r\right) - n\cdot e_r\left((g_1-1)e_r + g_2 e_t\right). \tag{32}$$

We also take a trial form for the bivector field Ω; abbreviating $\Omega(e_\mu)$ to Ω_μ, this is

$$\begin{aligned} \Omega_t &= a e_r e_t & \Omega_\theta &= (b_1 e_r + b_2 e_t) e_\theta / r \\ \Omega_r &= 0 & \Omega_\phi &= (b_1 e_r + b_2 e_t) e_\phi / r, \end{aligned} \tag{33}$$

where a and b_i are also scalar functions of r only. This is not the most general form of radially-symmetric Ω, but the general form can be generated from (33) by local gauge transformations.

The first of the field equations (28) can be written as

$$\overline{h}(e^\mu)\wedge(\mathcal{D}_\mu \overline{h}(e^\nu)) = 0, \tag{34}$$

where $\{e^\mu\}$ is the frame reciprocal to the $\{e_\mu\}$. Inserting (31) and (33) into (34) generates the four equations

$$g_2 f_2' - g_1 f_1' - a f_1^2 + a f_2^2 = 0 \tag{35}$$
$$g_1 g_2' - g_1' g_2 + a f_1 g_2 - a f_2 g_1 = 0 \tag{36}$$
$$g_1 = b_1 + 1 \tag{37}$$
$$g_2 = b_2, \tag{38}$$

where the primes denote differentiation with respect to r. We use (37) and (38) to eliminate b_1 and b_2. Next, calculating the 6 quantities $R_{\mu\nu} \equiv R(e_\mu \wedge e_\nu)$, yields

$$R_{tr} = -a' e_r e_t \tag{39}$$
$$R_{t\theta} = a(g_1 e_t + g_2 e_r) e_\theta / r \tag{40}$$
$$R_{r\theta} = (g_1' e_r + g_2' e_t) e_\theta / r \tag{41}$$
$$R_{\theta\phi} = (g_1^2 - g_2^2 - 1) e_\theta e_\phi / r^2, \tag{42}$$

with $R_{t\phi}$, $R_{r\phi}$ having the same form as $R_{t\theta}$, $R_{r\theta}$ respectively. By forming the contraction $\overline{h}(e^\mu)\cdot R_{\mu\nu}$ and setting the result equal to zero, we find that

$$2a + a'r = 0 \tag{43}$$
$$2g_1' + f_1 a' r = 0 \tag{44}$$
$$2g_2' + f_2 a' r = 0 \tag{45}$$
$$ar(f_1 g_1 - f_2 g_2) + r(g_1 g_1' - g_2 g_2') + g_1^2 - g_2^2 - 1 = 0. \tag{46}$$

Equation (43) yields a immediately, and equations (44) and (45) define f_1 and f_2 in terms of g_1 and g_2. Upon combining these with (35) we find that

$$\det h \equiv \overline{h}(i) i^{-1} = f_1 g_1 - f_2 g_2 = \text{constant}, \tag{47}$$

and we set this constant equal to 1, since $\overline{h}$ is required to reduce to the identity at large distances. All that remains is a simple equation for $g_1^2 - g_2^2$, and the full solution to our field equations is

$$\begin{aligned} a &= GM/r^2 \\ g_1^2 - g_2^2 &= 1 - 2GM/r \\ GM f_1 &= r^2 g_1' \\ GM f_2 &= r^2 g_2', \end{aligned} \tag{48}$$

subject to the boundary conditions that

$$\left.\begin{array}{l} f_1, g_1 \to 1 \\ f_2, g_2 \to 0 \end{array}\right\} \quad \text{as} \quad r \to \infty. \tag{49}$$

These boundary conditions guarantee that at large distances the effects of the fields fall away to zero. The solution (48) contains a single arbitrary function, g_2 say, subject to the condition that

$$g_2{}^2(r) \geq 2GM/r - 1, \tag{50}$$

together with the boundary conditions. From our initial restricted choice of $\overline{h}$ and Ω we have found a one-parameter family of solutions. This is extended to a four-parameter family by considering radially-symmetric gauge transformations. The four classes of transformation which preserve radial symmetry are

$$\begin{array}{rl} \text{Radial boost}: & R = \exp(\alpha(r) e_r e_t/2) \\ \text{Rotation}: & R = \exp(\alpha(r) i e_r e_t/2) \\ \text{Time translation}: & f(x) = x + \alpha(r) e_t \\ \text{Radial translation}: & f(x) = x + \alpha(r) e_r. \end{array} \tag{51}$$

These transformations induce more general forms of $\overline{h}$ and Ω, and combinations of the first and third can be used to move within the one-parameter family described by (48). All transformations in (51) leave $g_1{}^2 - g_2{}^2$ and Ω_t unchanged.

We are now in a position to compare our solutions with the Schwarzschild metric of general relativity. The 'line element' in our flat-space theory is given by:

$$\begin{aligned} ds^2 &= \underline{h}^{-1}(e_\mu)\underline{h}^{-1}(e_\nu) dx^\mu dx^\nu \\ &= (1 - 2GM/r)dt^2 - (f_1 g_2 - f_2 g_1) 2dr\, dt - (f_1{}^2 - f_2{}^2) dr^2 \\ &\quad - r^2(d\theta^2 + sin^2\theta\, d\phi^2). \end{aligned} \tag{52}$$

We see that the exterior Schwarzschild metric is recovered by setting $g_2 = 0$, which can only be done outside the horizon ($r > 2GM$). If we wish to extend the same line element inside the horizon, we must set $g_1 = 0$ for $r < 2GM$. However this solution is then strongly discontinuous at the horizon: something has gone wrong! To see exactly what, we focus on the cross-term in the line element, since standard treatments of the Schwarzschild solution always assume that this term can be transformed away. The coefficient of this term is $f_1 g_2 - f_2 g_1$, and since our solutions have $g_1 = \pm g_2$ at the horizon and $f_1 g_1 - f_2 g_2 = 1$ everywhere, we find that

$$f_1 g_2 - f_2 g_1 = \pm 1 \quad \text{at } r = 2GM. \tag{53}$$

The assumption that $f_1 g_2 - f_2 g_1$ can be transformed away fails at the horizon, so that the standard form of the Schwarzschild solution is not admissible in our theory. We shall shortly see that this removes much of the pathological behaviour of test particles at the horizon of a Schwarzschild black hole. The reason for our more restrictive class of solutions is that we retain a notion of position in a flat spacetime, and demand that the $\overline{h}$ and Ω functions be well-defined throughout this spacetime (except possibly where a point source is present). General relativity, by contrast, does not place such restrictions on the components of the metric. These components are scalar functions which can be transformed by a coordinate transformation. General relativity admits coordinate transformations which result in patches of spacetime not being covered; such transformations have no counterpart in our theory.

The shift from the Schwarzschild solution, with $f_1g_2 - f_2g_1 = 0$, to two distinct families of solutions, with $f_1g_2 - f_2g_1 = \pm 1$ at the horizon, is characteristic of the transition from second-order to first-order theories. A similar phenomenon is seen in the theory of propagation of electromagnetic waves, for example, where the first-order formulation correctly fixes the obliquity factors which have to be put in by hand in the second-order theory [7]. Furthermore, the appearance of two disconnected families of solutions is precisely as expected from a gauge theory of the Poincaré group, since the disconnected families are related by the discrete symmetry of time-reversal, which switches the sign of $f_1g_2 - f_2g_1$.

4. Test-particle Motion under Radial Gravitational Forces

To find the equations of motion for a test particle we need the version of the geodesic equation appropriate to our flat-space theory. The required equation is

$$\mathcal{D}_v v = 0, \tag{54}$$

and, on defining

$$\dot{x} = \underline{h}(v) \tag{55}$$

$$\dot{v} = \partial_\tau v = \dot{x}\cdot\nabla v, \tag{56}$$

our geodesic equation (54) becomes

$$\dot{v} = -\Omega(\dot{x})\cdot v. \tag{57}$$

In relativistic physics, accelerations should be thought of as bivectors [13] and in (57) we can identify the bivector $\Omega(\dot{x})$ as the acceleration of the test particle.

If we assume that all motion takes place in the azimuthal plane ($\theta = \pi/2$), we can write

$$\dot{x} = \dot{t}e_t + \dot{r}e_r + \dot{\phi}e_\phi \tag{58}$$

and the equations of motion (57) give

$$r^2\dot{\phi} = L \quad \text{(constant)} \tag{59}$$

$$\dot{r}^2 = A^2 - 1 - (1-2GM/r)L^2/r^2 + 2GM/r \tag{60}$$

$$\dot{t}(1-2GM/r) = \dot{r}(f_1g_2 - g_2f_1) + A. \tag{61}$$

Equations (59) and (60) agree with those found in general relativity using the Schwarzschild metric, and (60) can be differentiated to give

$$\ddot{r} - r\dot{\phi}^2 = -(1 + 3L^2/r^2)GM/r^2. \tag{62}$$

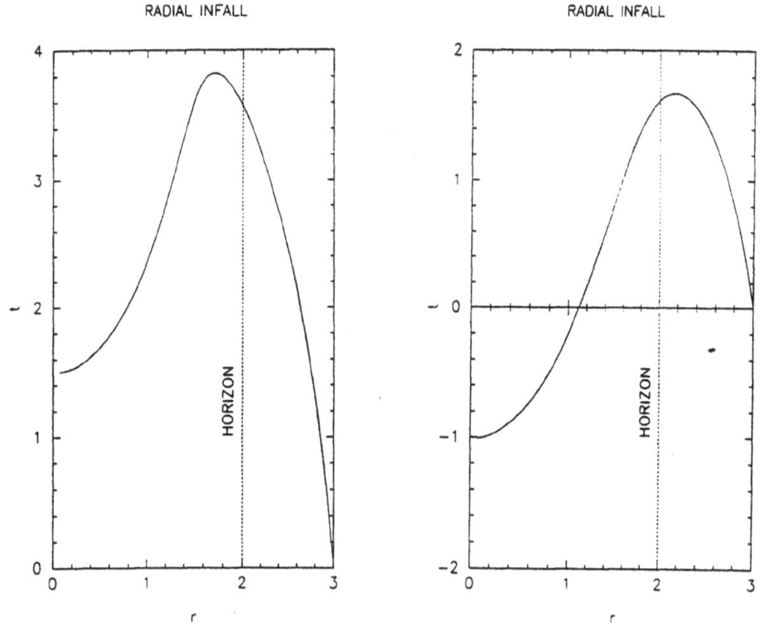

Fig. 1. 1 *Radial infall for* $g_2 = 1.4/r^2$ *(left) and* $g_2 = 2.0/r^2$ *(right).*

This will reproduce the standard results for the shape of the orbit and the precession rate per orbit, which have been checked to high accuracy by measurements on binary pulsar systems [14]. Equation (61), which describes the rate of change of coordinate time with respect to proper time, differs from that of the Schwarzschild metric through the inclusion of the $\dot{r}(f_1g_2 - g_2f_1)$ term. We saw in Section 3 that $f_1g_2 - g_2f_1$ must equal ± 1 at the horizon, so, taking the positive sign and radial infall ($\dot{r} < 0$, $A > 0$), we find that

$$\dot{r}(f_1g_2 - g_2f_1) + A = 0 \quad \text{at } r = 2GM. \tag{63}$$

This result removes the pole present in the corresponding equation for the Schwarzschild metric, and allows particles to cross the horizon in *finite* external coordinate time. This is demonstrated in Figure 1, which plots radial infall of a particle for two distinct choices of the function g_2. Different field configurations have different $\dot{t}$ equations, and lead to different trajectories relative to the flat background. The trajectories are mapped onto each other by gauge transformations.

Solutions with $f_1g_2 - g_2f_1 = 1$ at the horizon act as 'one-way valves', in that particles can cross from the outside in finite coordinate time, but once inside can never get back out again. No part of an outgoing trajectory from the past is inside the horizon. The solutions for which $f_1g_2 - g_2f_1 = -1$ at the horizon have the reverse properties — matter can escape in finite coordinate time, but can never pass through the horizon from the outside (without going to infinite coordinate time). A geodesic for an escaping particle is plotted in Figure 2.

The pictures presented by these two (separate) types of solution have a counterpart in general relativity as the extension of the Schwarzschild metric written

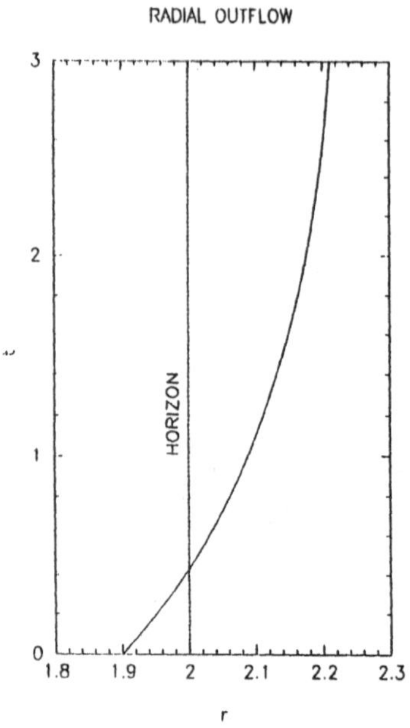

Fig. 2. 2 *Radial outflow for* $g_2 = -1.4/r^2$.

using advanced- and retarded-time Eddington-Finkelstein coordinates. General relativity understands these as representing the same solution, both obtained from the Schwarzschild solution by (passive) coordinate transformations. In our theory, however, they are ***different*** solutions, with different physical properties. They are related to each other via the active transformation of time-reversal. These differences from general relativity are seen most starkly when we consider the maximal analytic extension of the Schwarzschild metric found by Kruskal (see Hawking and Ellis [15], for example). The Kruskal metric mixes advanced and retarded coordinates and therefore has no counterpart in our theory, since it would combine elements of distinct solutions. The formulation of gravity as a first-order theory in flat spacetime rules out Kruskal's extension. Further, related differences from general relativity are dealt with in greater detail in a forthcoming paper [6].

5. Conclusions

Gravity can be formulated in geometric algebra as a first-order gauge theory of multilinear functions in a flat spacetime. The dynamical variables are the functions $\bar{h}$ and Ω, and these are arrived at by gauging the Poincaré group of transformations of spacetime fields. All the dynamical quantities have coordinate-free definitions, and become much easier to manipulate than in tensor analysis. The insistence on a globally-defined underlying flat space has implications for cosmology which are discussed in the following paper [16].

The search for radially-symmetric static solutions has led us naturally to two distinct families of solutions, both different from the standard Schwarzschild solution of general relativity. This in turn has produced a new picture of geodesic motion around massive bodies, in which particles cross the horizon in finite external coordinate time. Implications for the formation of horizons will be discussed in future work, in which models of collapsing matter will be presented.

References

1. S.F. Gull, A.N. Lasenby, and C.J.L. Doran. Imaginary numbers are not real — the geometric algebra of spacetime. To appear in: *Foundations of Physics.*, 1993.
2. D. Hestenes. *Space-Time Algebra.* Gordon and Breach, 1966.
3. D. Hestenes and G. Sobczyk. *Clifford Algebra to Geometric Calculus.* D. Reidel Publishing, 1984.
4. H. Stephani. *General Relativity.* Cambridge University Press, 1982.
5. W. Misner, K.S. Thorne, and J.A. Wheeler. *Gravitation.* W.H. Freeman and Company, 1973.
6. A.N. Lasenby, C.J.L. Doran, and S.F. Gull. Gravity, gauge theory and geometric algebra. In Preparation, 1993.
7. S.F. Gull, A.N. Lasenby, and C.J.L. Doran. Electron paths, tunnelling and diffraction in the spacetime algebra. To appear in: *Foundations of Physics.*, 1993.
8. C.J.L. Doran, A.N. Lasenby, and S.F. Gull. States and operators in the spacetime algebra. To appear in: *Foundations of Physics.*, 1993.
9. A.N. Lasenby, C.J.L. Doran, and S.F. Gull. A multivector derivative approach to Lagrangian field theory. To appear in: *Foundations of Physics.*, 1993.
10. D. Hestenes. The design of linear algebra and geometry. *Acta. Appli. Math.*, 23:65, 1991.
11. R. Utiyama. Invariant theoretical interpetation of interaction. *Phys. Rev.*, 101(5):1597, 1956.
12. T.W.B. Kibble. Lorentz invariance and the gravitational field. *J. Math. Phys.*, 2(3):212, 1961.
13. D. Hestenes. Proper particle mechanics. *J. Math. Phys.*, 15(10):1768, 1974.
14. D. Kleppner. The gem of general relativity. *Physics Today*, 46(4):9, 1993.
15. S.W. Hawking and G.F.R. Ellis. *The Large Scale Structure of Space-Time.* Cambridge University Press, 1973.
16. A.N. Lasenby, C.J.L. Doran, and S.F. Gull. Cosmological consequences of a flat-space theory of gravity. *These proceedings.*

COSMOLOGICAL CONSEQUENCES OF A FLAT-SPACE THEORY OF GRAVITY

ANTHONY LASENBY
Cavendish Laboratory, Madingley Road, Cambridge CB3 0HE, UK

CHRIS DORAN*
DAMTP, Silver Street, Cambridge, CB3 9EW, UK

and

STEPHEN GULL
Cavendish Laboratory, Madingley Road, Cambridge CB3 0HE, UK

Abstract.
In the preceding paper [1] we described some aspects of a new theory of gravity, and found radially-symmetric static solutions to the free-field equations. Here we apply the theory to the universe on the largest scales by investigating the consequences of spatial homogeneity. A guiding principle for our theory is that spacetime itself does not play an active role in physics. This means that spacetime cannot be thought of as expanding, and we show also that material test particles do not expand away from each other. Nevertheless, we do predict the observed galactic redshifts. Furthermore, we find that the only cosmological models compatible with complete spatial homogeneity are those at critical density.

Key words: cosmology, spacetime algebra, gravitation, Friedmann-Robertson-Walker metric

1. Introduction

In the preceding paper [1] (henceforth Paper I) we outlined a theory of gravity based on the requirement of invariance under local, active Poincaré transformations. Gravitational interactions were described by gauge fields in a flat spacetime, and the theory contained only first-order derivatives of these fields. Here, we apply our theory to the universe on the largest scales. The guiding principle is spatial homogeneity, and we find that this requirement is more restrictive in our theory than in general relativity (GR). In particular, a testable consequence of our theory is that the density parameter Ω should be 1; that is, the universe is at critical density. This is quite different from general relativity, which in principle allows for any value of Ω. That we can make this prediction might seem surprising, since it is generally thought that first-order theories of gravity (such as ours) have identical outcomes to their second-order counterparts if the effects of quantum spin are neglected. Here we do indeed neglect the effects of spin, but nevertheless obtain several results not accessible to standard GR.

Some of the philosophy behind our approach was discussed in Paper I, and a fuller presentation, together with the detailed mathematics, will be given in a forthcoming paper [2]. An essential point to stress, however, is that in our approach spacetime

* Supported by a SERC studentship.

F. Brackx et al. (eds.), *Clifford Algebras and their Applications in Mathematical Physics*, 387–396.

is an entirely *passive* participant in physics. That is, we reject the notion of a *dynamical* spacetime, and deal with gravitational effects solely in terms of forces in flat spacetime. We regard this rejection of curved spacetime as representing a very conservative approach to gravity, although some of its consequences may seem unconventional. In this paper we consider the cosmological implications of working within this fixed Minkowski space which has no evolution in time and no variation in space. In particular, since spacetime itself cannot be expanding, the observed galactic redshifts must have a different origin. Perhaps even more surprisingly, we show that the cosmological fluid which models the material content of the universe is not expanding either – its energy density is simply decaying with time!

We start by exploring the consequences of imposing spatial homogeneity on the $\overline{h}$ and Ω functions introduced in Paper I. The only preferred direction imposed is one in time. The resulting theory is then compared with standard Friedmann-Robertson-Walker (FRW) models. We confirm that our model reproduces some basic experimental facts, in particular the observed galactic redshifts. This entails discussing the motion of test particles, both with and without mass. A major prediction of this theory is that the universe is at critical density. We discuss whether this prediction is likely to be testable in the near future. We end by discussing whether any of our conclusions may be altered when multiparticle and spin effects are incorporated.

The conventions, notations and results of Paper I are assumed throughout, and references to equations in Paper I are made in the form (I.n). The standard notations of cosmology are also adopted. These can be found in most textbooks on the subject (see [3, 4, 5] for example).

2. A Simple Homogeneous Model

In Paper I we found a class of radially-symmetric static solutions to the free-field equations. These provide a model of a system in which the only preferred direction is the spatial vector e_r, pointing towards some chosen origin. The analogous problem for cosmological models is to find solutions of the field equations in which the only preferred direction is one in *time* rather than space. Following the example of Paper I, we take a trial form for the $\overline{h}$ function as

$$\overline{h}(a) = f(t)a\cdot e_t e_t + \alpha(t)a\wedge e_t e_t, \tag{1}$$

in which f and α are functions of time only. The form of the Ω function must also be consistent with spatial homogeneity, so that

$$\Omega(a) = A(t)a\wedge e_t. \tag{2}$$

The first of the field equations (I.28) can be written, in the absence of spin, as

$$\mathcal{D}\wedge\overline{h}(a) = 0, \quad \text{for all constant } a. \tag{3}$$

This implies the single equation

$$f\dot{\alpha} + \alpha^2 A = 0. \tag{4}$$

For the second of the field equations (I.29) we must couple to a source – the cosmological fluid. In this case (I.29) is modified, to become

$$G(a) \equiv R\underline{h}(a) - \tfrac{1}{2}a\mathcal{R} = 8\pi GT(a), \tag{5}$$

where $T(a)$ is the stress-energy tensor generated by the matter fields. For a 'dust' model (that is, one with no pressure), $T(a)$ takes the form

$$T(a) = \rho a \cdot e_t e_t, \tag{6}$$

where $\rho = \rho(t)$ is the matter density. Assuming such a model, we find that the remaining field equations give the pair of equations

$$3\alpha^2 A^2 = 8\pi G\rho, \tag{7}$$
$$\alpha^2 A^2 + 2\dot{A}\alpha f = 0. \tag{8}$$

Equations (4) and (8) immediately imply that A^2/α is a constant. If we make the further asumption that α is positive (so that the spatial part of $\overline{h}$ is simply-connected to the identity function) then we can write the solution in the form

$$A = -C\alpha^{1/2}, \tag{9}$$
$$f = C\alpha^{5/2}/\dot{\alpha}, \tag{10}$$

and

$$\alpha = \left(\frac{8\pi G\rho}{3C^2}\right)^{1/3}, \tag{11}$$

where C is an arbitrary constant of integration. From equation (10) we see that we must have $\dot{\alpha} \neq 0$, so that this model universe is necessarily dynamic.

The time history of α is determined completely by the time history of the density ρ – but what determines this density? It seems that we have an under-determined theory. At this stage a parallel analysis in general relativity has usually found that the scale factor R (related to α in our theory by $R = R_0/\alpha$) is proportional to $t^{2/3}$. In fact, the conventional approach has tacitly assumed $f = 1$ from the outset, which would imply that

$$\dot{\alpha} = C\alpha^{5/2}$$
$$\Rightarrow \alpha \propto t^{-2/3}. \tag{12}$$

This assumption is made when the FRW metric is written as

$$ds^2 = \lambda dt^2 - \frac{R^2(t)}{c^2} \times \text{spatial part}, \tag{13}$$

and λ is taken equal to 1. This implies that $f^2 = 1$ in equation (1), which corresponds to the assumption that 'cosmic time' (dt) is the proper time (ds) measured by observers comoving with the Hubble flow of galaxies. (An observer can determine whether they are comoving by measuring a zero dipole component in the cosmic microwave background anisotropy — this provides a preferred time axis e_t which is the one we are using in our theory.)

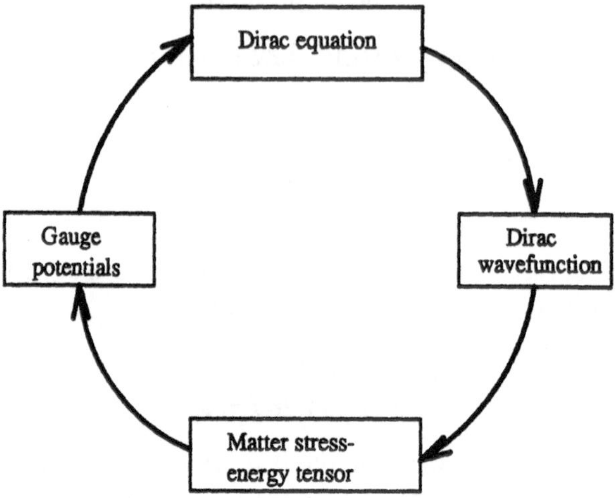

Fig. 1. 1 *The self-consistency loop for self-gravitating matter.*

In our theory, however, we cannot simply assume that $\overline{h}(e_t) = e_t$. Suppose, for example, that our universe is in a phase of 'expansion', so that $\dot{\alpha} < 0$. If we choose $f > 0$ it follows from (10) that the constant C must be negative. If the universe were subsequently to recontract, then f would have to ***flip sign*** since C is fixed. This argument is only schematic, since we have yet to establish the proper time history of α, but it suggests that time would indeed run backwards during a contracting phase. This 'solves' a long-standing problem (see [6] for a popular account), a conclusion which is only made possible by keeping track of the sign of f. In a metric theory such as GR this sign would have been lost, since the theory can only access the variable f^2.

A significant problem remains; that is, how to determine the time dependence of the energy density ρ (or equivalently α)? To find this we need to take account of a matter field satisfying its own field equations. This coupling is carried out explicitly in [2], where the matter is taken initially to satisfy the free-particle Dirac equation (expressed in real spacetime algebra form [7]). Gravitational fields are introduced by gauging local Poincaré transformations of the spinor field. In the resulting theory, matter generates gravitational fields (at least, at the level of classical fields) which themselves feed back into the matter equations. In principle, therefore, one would like to find self-consistent solutions of the 'loop' of equations illustrated in Figure 1. This has not yet proved possible, essentially because of the complication of quantum spin. Spin generates 'torsion', which alters the form of the first of the field equations (3).

On the other hand, it is possible to 'close the loop' of Figure 1 if the effects of spin can be ignored. This would be the case if the spin contributions to T average to zero, or if the matter sector consisted of spin-zero sources. Our approach requires that spin-zero sources be built from pairs of spin-1/2 particles, thus the second possibility would involve a multiparticle theory of gravity – a goal towards which we are currently working. Assuming that the spin can be ignored, we find that the Dirac wavefunction is of the form

$$\psi = e^{-mti\sigma_3 - \xi(t)}, \tag{14}$$

and that the Dirac equation fixes f to be $+1$. The form of the stress-energy tensor implies that

$$\rho = me^{-2\xi(t)}, \tag{15}$$

(ignoring the spin contribution) and the remaining content of the Dirac equation reduces to the conservation of the current $mJ = \rho e_t$. In other words, we must have

$$\begin{aligned} \mathcal{D}\cdot J &= 0 \\ \Rightarrow \nabla\cdot(\overline{h}(J) \det h^{-1}) &= 0, \end{aligned} \tag{16}$$

and this holds provided that ρ/α^3 is a constant – which is indeed consistent with equation (11). The full solution now takes the form

$$\alpha = \left(\frac{t}{t_0}\right)^{2/3} \tag{17}$$

and

$$\rho = \rho_0 \alpha^3 = \rho_0 \left(\frac{t_0}{t}\right)^2, \tag{18}$$

where $Ct_0 = -2/3$, and $\rho_0 = 3C^2/(8\pi G)$ is the density at the present epoch ($t = t_0$).

The Hubble parameter $H(t)$ is defined by

$$H(t) \equiv \frac{\dot{R}}{R}, \tag{19}$$

where R is the scale factor ($= R_0/\alpha$), and we find that $H(t)$ is equal to $2/(3t)$. The Hubble constant, $H(t_0)$, is therefore given by $2/(3t_0)$, and we now recognise that

$$C = -\text{Hubble constant} = -H_0. \tag{20}$$

The present energy density is therefore

$$\rho_0 = \frac{3H_0^2}{8\pi G}, \tag{21}$$

which is equal to the *critical density*. In conventional terms this is the maximum density that allows continued expansion, and this model universe will continue expanding forever. It is usual to talk in terms of the dimensionless parameter Ω, which is defined as (actual density)/$(3{H_0}^2/8\pi G)$ [5]. This model therefore predicts an Ω of 1.

In our current approach, however, does it make any sense to talk about expansion? Conventionally, it is spacetime itself that is held to be expanding, but for our theory this is impossible – the flat background Minkowski spacetime is unchanging.

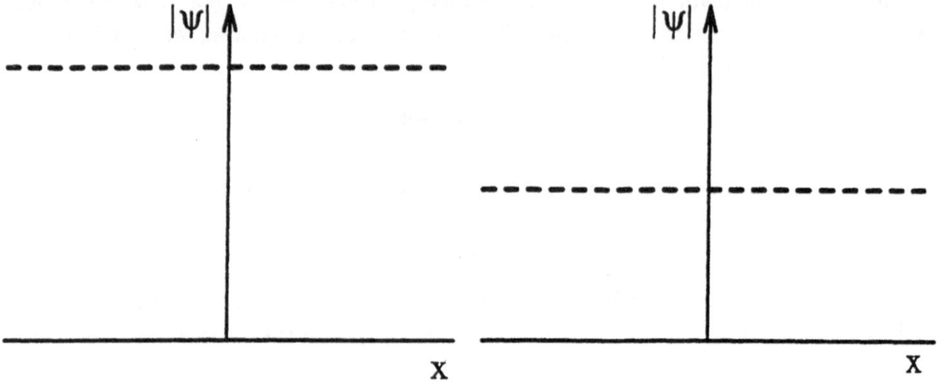

Fig. 2. 2 *The amplitude of the wavefunction ψ, plotted against one spatial coordinate. At left, an early time; right, a later time.*

One might then suppose that it is the particles within the space that are expanding away from each other. We investigate this by examining the motion of test particles in this universe. Suppose that we launch a particle with

$$v = \cosh u\, e_t + \sinh u\, \hat{n}, \tag{22}$$

where $\hat{n}$ is a unit spatial vector. The geodesic equation $\mathcal{D}_v v = 0$ simplifies to

$$\dot{v} = \sinh u\, \alpha^{3/2} C(\sinh u\, e_t + \cosh u\, \hat{n}), \tag{23}$$

and this implies a *deceleration* in a straight line, since $C < 0$. It can easily be shown that the de Broglie wavelength, λ, satisfies $\lambda \propto t^{2/3}$, and the particle therefore gets *redshifted*. A similar analysis for massless particles reveals the same relationship between wavelength and time, so photons are redshifted in the same way. We note, however, that a test particle released at rest *remains* at rest ($u = 0$, where u is the rapidity), rather than expanding away radially from some fixed point. We are therefore led to some startling conclusions:

- the universe is *not* expanding;
- the redshift is not Doppler in origin, but is due to a loss of energy to the gravitational field.

However, the universe we are considering is certainly dynamic, so what is it that is changing with time? In fact it is the *energy density* that changes. The amplitude of the wavefunction $|\psi|$ is decreasing with time, as shown in Figure 2, and therefore the stress-energy tensor derived from ψ also decreases in value, reflecting the fact that the matter field is losing energy to the gravitational field.

3. Models with $\Omega \neq 1$

This completes our analysis of the simplest model for the universe, which has turned out to be one at critical density. We now investigate the existence of $\Omega \neq 1$ (non-critical density) models. In order to simulate standard FRW $\Omega \neq 1$ models we have

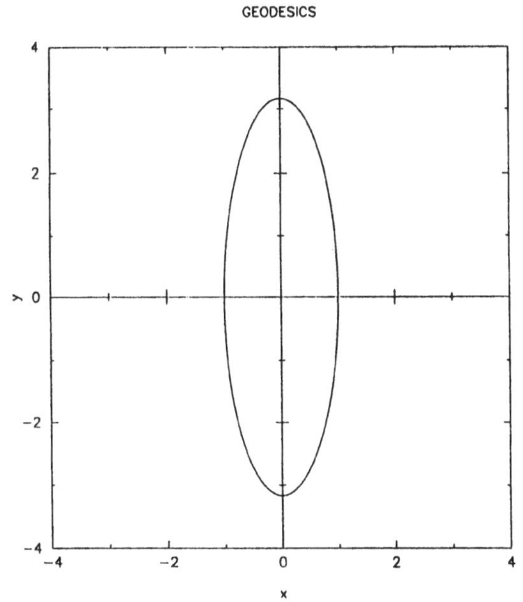

Fig. 3. 3 *A particle projected parallel to the y-axis in a universe with* $K = 0.1$ *(corresponding to* $\Omega > 1$*) and* $C = -0.5$.

to recast the acceleration bivectors as

$$\begin{aligned} \Omega_t &= 0 & \Omega_\theta &= (be_r + Be_t)e_\theta/r \\ \Omega_r &= Ae_r e_t & \Omega_\phi &= (be_r + Be_t)e_\phi/r, \end{aligned} \tag{24}$$

where $b = \sqrt{1 - Kr^2} - 1$ and K measures the spatial curvature. The form of $\overline{h}$ also changes to include terms with an explicit r-dependence. This is conventionally thought to be unproblematic, since the Einstein tensor remains spatially homogeneous. But a series of worrying points now emerges when one repeats the analysis of Section 2:

1. The 'b' terms imply a preferred spatial direction, in the same manner as the corresponding terms for radially-symmetric solutions [1].
2. We find that the implied Dirac wavefunction ψ must have a spatial variation given by
$$\psi \propto \frac{1}{1 + \sqrt{1 - Kr^2}}, \tag{25}$$
so that it is no longer homogeneous.
3. The stress-energy tensor derived from the Dirac field is also no longer spatially homogeneous, so we cannot close the loop of Figure 1 and find a self-consistent solution.

We must therefore conclude that $\Omega \neq 1$ FRW models do not work, since it does not seem possible to find *any* self-consistent solution to the coupled field equations.

Further support for this conclusion arises when considering the motion of a test particle in an $\Omega \neq 1$ universe. Test particles no longer move in a straight line, but orbit the origin, as shown in Figure 3. In conventional approaches to cosmology the non-radial motions of a test particle are seldom plotted, but if they were, the interpretation of Figure 3 would be that one is looking down at the North Pole of a sphere, on which the particle is executing great-circle (*i.e.* geodesic) motion. This motion is projected into an ellipse when one dimension is supressed. This missing dimension is neither the z- nor t-axes, which are indeed missing in our figure, but is the extra dimension of an embedding picture in which the 3 dimensions of ordinary space become the surface of a 3-sphere (*c.f.* box 27.2 of reference [4]). In a flat spacetime this trick of reinterpretation is simply not available to us; instead, Figure 3 is that of a test particle revealing by its motion that a preferred direction e_r has been introduced.

The above points all lead us to the conclusion that $\Omega \neq 1$ FRW universes are not admisssible, since they break spatial homogeneity. The universe is therefore required to be at $\Omega = 1$ (critical density). This fits in well with some current observations, which give $\Omega \approx 0.85 \pm 0.2$ from results on large-scale streaming motions, and $\Omega = 1$ from measurements of the angular diameter of sub-kiloparsec-scale jets in a sample of radio sources [8] (see [9] for a recent discussion). These results notwithstanding, the value of Ω remains a matter of great debate and controversy, and it will probably require several further years of observations to settle the matter.

We have considered several more tests and predictions of our model and find that in general, once one has accepted that the density is at the critical value, there is a fairly straightforward mapping between (a) a conventional viewpoint in which the material particles expand away from each other and the energy density decreases because the comoving volume is increasing, and (b) our current viewpoint, in which the particles are fixed, and the energy density of the cosmological fluid decreases due to interaction with the gravitational field. In particular, the prediction for angular size versus redshift remains the same as in the conventional (GR) $\Omega = 1$ model. Discussions of the physics of the 'big bang' are also largely unaltered, since the relevant factor is the energy density as a function of time. Our theory will therefore agree with the current predictions of big-bang cosmology, for example calculations of nucleosynthesis rates in the early universe. More details of these comparisons and predictions will be presented in a forthcoming paper.

4. Discussion and Conclusions

In connection with self-consistent solutions to the coupled Einstein-Dirac equations we should mention the work of Isham & Nelson [10], who studied the quantization of a Dirac field in FRW universes. They solved for a classical Dirac field, and found that they could not obtain a self-consistent solution to the GR field equations, with the Dirac field providing the stress-energy tensor, except for the case where $\Omega = 1$. This reinforces the conclusions of Section 3, but we should note that Isham & Nelson claimed that the system was underdetermined even after the inclusion of the Dirac equation, and that a choice of time coordinate (equivalent to a free choice of f) still remained. Our analysis disagrees with this. In addition, they did not have available to them the further arguments presented in Section 3, which we believe provide additional, compelling reasons for excluding non-critical density models.

For the reasons discussed in the Introduction, it may seem surprising that our current theory is able to make such a strong prediction, at variance with GR. A point that increases the plausibility of our conclusions is that the theory of gravity discussed here is first-order in derivatives, and with complete spatial homogeneity contains only one adjustable free parameter when integrated in time. In our case this parameter was taken as the Hubble constant, $-C$. Alternatively, we could have characterized the free parameter by the present density, ρ_0, in which case H_0 would have been fixed. This is to be contrasted with GR, which is explicitly a second-order theory, so both the 'velocity' (H_0) and 'acceleration' ($q_0 = \frac{1}{2}\Omega$) must be specified in terms of their values at the current epoch.

We can summarise the conclusions arising from the approach in this paper as follows:

- Spacetime is not curved.
- The universe is not expanding.
- The redshift is a loss of energy to the gravitational field, not a Doppler effect.
- The universe is at critical density, and so its energy density will continue decreasing forever.
- $\Omega \neq 1$ FRW models are *not* spatially homogeneous.

A further conclusion, not fully addressed in the text, is that proper time would run backwards for observers in any 'recontraction' phase of the universe. Of course, the arguments presented in this paper 'censor out' such a phase by requiring that $\Omega = 1$.

These conclusions have been presented in unqualified form so far, so we end with some necessary provisos. Firstly, we have not considered the whole class of spatially-homogeneous models. It is possible that a Bianchi model will work in the 'self-consistent' sense of Figure 1, although it seems unlikely that it could meet the other objections raised to non-spatially-flat FRW models. Secondly, the spin-torsion sector of our theory has not yet been properly considered. To address this fully, we must provide a spinless source for the stress-energy tensor which is compatible with the rest of the theory. This requires a *multiparticle* Dirac theory, incorporating gravitational interactions. Although this is an ambitious undertaking, we are confident that the flat-space, gauge-force approach will work here as well, and could provide a fruitful approach to a full theory of quantum gravity.

Acknowledgements

The authors thank the organisers for arranging such an enjoyable conference. We also thank Anton Garrett for his careful reading of these manuscripts, and his numerous suggestions for improvements.

References

1. C.J.L. Doran, A.N. Lasenby, and S.F. Gull. Gravity as a gauge theory in the spacetime algebra. *These proceedings.*
2. A.N. Lasenby, C.J.L. Doran, and S.F. Gull. Gravity, gauge theory and geometric algebra. In Preparation, 1993.
3. S. Weinberg. *Gravitation and Cosmology.* John Wiley and Sons, 1972.
4. C.W. Misner, K.S. Thorne, and J.A. Wheeler. *Gravitation.* W.H. Freeman and Company, 1973.
5. M. Berry. *Principles of Cosmology and Gravitation.* Cambridge University Press, 1976.

6. S.W. Hawking. *A Brief History of Time.* Bantam Press, 1988.
7. C.J.L. Doran, A.N. Lasenby, and S.F. Gull. States and operators in the spacetime algebra. To appear in: *Foundations of Physics.*, 1993.
8. K.I. Kellermann. The cosmological deceleration parameter estimated from the angular size/redshift relation for compact radio sources. *Nature*, 361:134, 1993.
9. P. Scheuer. Weighing the universe (News and Views). *Nature*, 361:112, 1993.
10. C.J. Isham and J.E. Nelson. Quantization of a coupled Fermi field and Robertson-Walker metric. *Phys. Rev. D*, 10(10):3226, 1974.

ZITTERBEWEGUNG AND ELECTRON STRUCTURE

WALDYR A. RODRIGUES, JR. and JAYME VAZ, JR.
Department of Applied Mathematics
State University at Campinas
13081-970, Campinas, S.P., Brazil

Abstract. In this work we analyse the conjecture that Zitterbewegung could appear as the basis of electron spin. Our studies are based on the Barut-Zanghi model, translated in terms of spacetime algebra.

Key words: Zitterbewegung – Barut-Zanghi model – Electron spin

1. Introduction

In recent times various different, independent approaches led to the conclusion that it exists a classical limit of spin (at least in the case of the electron) and that it is related to a helical motion. It suggests that the Zitterbewegung (ZBW) motion, associated since long with the electron structure, seems to be responsible for the electron spin. Indeed, Schrödinger (1930) proposed that the electron spin is a consequence of a local circulatory motion, constituting the ZBW and resulting from the interference between positive and negative energy solutions of Dirac equation. Such an issue turned out to be of renewed interest, following works of Barut et al. (1981, 1983, 1984, 1987) and Hestenes (1990, 1991). We shall consider this issue by studing the model proposed by Barut and Zanghi (BZ) (1984).

BZ model is a classical model for the Dirac electron. BZ considered a classical electron as characterized, besides by the usual pair of conjugate variables (x^μ, p_μ), by a second pair of conjugate classical spinor variables $(z, i\bar{z})$ representing internal degrees of freedom. BZ model is based on the lagrangian

$$\mathcal{L} = \frac{1}{2}i(\dot{\bar{z}}z - \bar{z}\dot{z}) + p_\mu(\dot{x}^\mu - \bar{z}\gamma^\mu z) + eA_\mu\bar{z}\gamma^\mu z, \qquad (1)$$

where $z = z(\tau) \in C^4$ is a Dirac spinor, $\bar{z}$ its Dirac adjoint, γ^μ are the Dirac matrices and τ is an invariant time parameter to be identified with the proper time of the "center of mass" motion of the electron and A_μ is the electromagnetic potential. This model provides a classical action from which one can define a path integral giving the Dirac propagator (Barut and Duru 1984) and gives the helical motion as the classical analogue of ZBW.

BZ model has many interesting properties which we now study using the space-time algebra (STA) (Hestenes 1966). We shall give (sect.2) a STA approach to BZ model as a hamiltonian system (Vaz and Rodrigues 1993), and then we study some general properties of it (sect.3), and in particular we present solutions corresponding to a helical motion as the analogue of ZBW. We obtain BZ solution corresponding

F. Brackx et al. (eds.), Clifford Algebras and their Applications in Mathematical Physics, 397–404.

to a time-like helical path and also a solution corresponding to a light-like path. This kind of solution has been conjectured by Hestenes (1990, 1991) in the context of Dirac theory in his ZBW interpretation of quantum mechanics. Here, however, we look to that conjecture at a "classical level", taking BZ model as a "classical analogue" of Dirac theory. Finally, if ZBW is responsible for some properties of the electron, then we expected its electromagnetic field to oscillate with ZBW frequency, and in sect.4 we present some conjectures about the electron's oscillating field.

In the following we adopt the notation: $R_{1,3}$ is STA, $R^{+}_{1,3}$ its even subalgebra, the Clifford (or geometrical) product is denoted by justaposition, while the interior and exterior products are denoted by $\cdot$ and $\wedge$, respectively; tilde denotes reversion, i.e., $(AB)\tilde{} = \tilde{B}\tilde{A}$ with $\tilde{A} = A$ for A scalar or vector. We remark that in STA $\{\gamma^{\mu}\}$ are interpreted as vectors of TM (the tangent bundle over Minkowski spacetime), in contrast to those $\{\gamma^{\mu}\}$ in eq.(1) which are matrices.

2. STA Approach to BZ Model

The lagrangian (1) in terms of STA reads (Gull 1991, Pavšič et al. 1992):

$$\mathcal{L} = \langle \dot{\tilde{\psi}}\dot{\psi}\gamma_2\gamma_1 + p(\dot{x} - \psi\gamma_0\tilde{\psi}) + eA\psi\gamma_0\tilde{\psi}\rangle_0, \tag{2}$$

where $\psi = \psi(\tau)$ is a Dirac-Hestenes (DH) spinor field ($\psi \in R^{+}_{1,3}$) and $\langle\ \rangle_0$ means "the scalar part". A non-singular DH spinor ψ has a beautiful geometrical interpretation in terms of the canonical decomposition:

$$\psi = \sqrt{\rho}e^{\gamma_5\beta/2}R, \tag{3}$$

that is: R represents a Lorentz rotation ($R^{-1} = \tilde{R} \in R^{+}_{1,3}$), $\sqrt{\rho}$ a dilatation and $e^{\gamma_5\beta/2}$ a duality rotation by the Takabayasi angle β (we shall assume $\beta = 0$). Since for a Dirac spinor z we have $z = \psi f$, where f is an approapriate idempotent (that is: a Dirac spinor can be defined as an element of the ideal $C \otimes R_{1,3}f$; see Lounesto 1993), one can now understand the presence of z in BZ lagrangian, for a spinning particle must be individuated, besides by (x,p), by the Frenet tetrad $\{e_{\mu}\}$ given by $e_{\mu} = R\gamma_{\mu}\tilde{R}$ (Hestenes and Sobczyk 1984), or $\rho e_{\mu} = \psi\gamma_{\mu}\tilde{\psi}$, for $\psi = \psi(\tau)$.

We remark that in the lagrangian formulation of BZ model, those p_{μ} are introduced as lagrangian multipliers. However, in a hamiltonian formulation those p_{μ} appears as canonical momentum. The fact that BZ model is a hamiltonian system was proved by Rawnsley (1992) (using an exact-symplectic manifold more general than the cotangent bundle), and the STA approach to it as a hamiltonian system was given by Vaz and Rodrigues (1993).

We take as BZ hamiltonian $H = H(x, p, \psi, \bar{\psi})$:

$$H = \langle (p - eA)\psi\gamma_0\tilde{\psi}\rangle_0 = \langle (p - eA)\psi\gamma_0\gamma_1\gamma_2\bar{\psi}\rangle_0, \tag{4}$$

where p is the canonical conjugate momentum to x and $\bar{\psi}$ is the canonical conjugate momentum to the DH spinor ψ, defined by

$$\bar{\psi} = \gamma_2\gamma_1\tilde{\psi}. \tag{5}$$

Hamilton equations are:

$$\dot{x} = \partial_p H, \quad \dot{p} = -\partial_x H, \tag{6}$$

$$\dot{\psi} = \partial_{\bar{\psi}} H, \quad \dot{\bar{\psi}} = -\partial_{\psi} H, \tag{7}$$

where ∂_X denotes the multivector derivative in relation to the multivector X (Hestenes and Sobczyk 1984). We note that a symplectic structure to the phase space of BZ model can be given (Vaz and Rodrigues 1993) according to methods given by Hestenes (1992). Take a basis $\{E_0, E_1, E_2, E_3\}$ of R^4 such that $E_a \cdot E_b = \delta_{ab}$ $(a = 0, \ldots, 3)$ and define $X = \sum_a x_a E_a$; take another copy of R^4 and a basis $\{E'_0, E'_1, E'_2, E'_3\}$ with $E'_a \cdot E'_b = \delta_{ab}$ and define $P' = \sum_a p_a E'_a$. Since $R^+_{1,3} \ni \psi = <\psi>_0 + <\psi>_2 + <\psi>_4$ (where $<>_a$ means "the a-vector part") we can take a basis $\{F_0, F_1, \ldots, F_7\}$ of R^8 such that $F_m \cdot F_n = \delta_{mn}$ $(m, n = 0, 1, \ldots, 7)$ and define $\phi = <\psi>_0 F_0 + (<\psi>_2)_{01} F_1 + \cdots + (<\psi>_2)_{23} F_6 + <\psi>_4 F_7$; take also another copy of R^8 with a basis $\{F'_0, F'_1, \ldots, F'_7\}$ such that $F'_m \cdot F'_n = \delta_{mn}$ and define $\bar{\phi}' = <\bar{\psi}>_0 F'_0 + (<\bar{\psi}>_2)_{01} F'_1 + \cdots + (<\bar{\psi}>_2)_{23} F'_6 + <\bar{\psi}>_4 F'_7$. Finally, take the space $(R^4 \oplus R^8) \oplus (R^4 \oplus R^8)$ with a basis $\{E_0, \ldots, E_3; F_0, \ldots, F_7; E'_0, \ldots, E'_3; F'_0, \ldots, F'_7\}$ with $E_a \cdot F_m = E_a \cdot F'_m = E'_a \cdot F_m = E'_a \cdot F'_m = 0$ $(\forall a, m)$ and $E_a \cdot E'_b = 0$ $(\forall a, b)$, $Fm \cdot F'_n = 0$ $(\forall m, n)$ as the phase space of the BZ model. The symplectic structure is given by the symplectic bivector $\mathcal{J}$:

$$\mathcal{J} = J + K = \sum_a E_a \wedge E'_a + \sum_m F_m \wedge F'_m. \tag{8}$$

After defining

$$\Pi = X' + P + \phi' + \bar{\phi}, \tag{9}$$

$$\partial'_\Pi = \partial'_P - \partial_X + \partial'_{\bar{\phi}} - \partial_\phi \tag{10}$$

where $X' = -\mathcal{J} \cdot X$, $\partial'_P = -\mathcal{J} \cdot \partial_P$, etc ..., we can write Hamilton equations as

$$\dot{\Pi} = \partial'_\Pi H. \tag{11}$$

From the hamiltonian (4) we derive the equations

$$\dot{\psi}\gamma_1\gamma_2 + \pi\psi\gamma_0 = 0, \tag{12}$$

$$\dot{x} = \psi\gamma_0\tilde{\psi}, \tag{13}$$

$$\dot{\pi} = eF \cdot \dot{x}, \tag{14}$$

where $F = \partial \wedge A$ and

$$\pi = p - eA \tag{15}$$

is the kinetic momentum. From our approach to BZ model as a hamiltonian system it is clear that p is the canonical momentum and π is the kinetic one. This is the interpretation given by Pavšič et al. (1992), but Lasenby et al. (1993) interpreted π as canonical momentum. Pavšič et al. (1992) made also the following remark: let $x \in \mathcal{M}$ ($\mathcal{M}$ is Minkowski spacetime) and $\Psi = \Psi(x)$ be such that its restriction to the world-line $\sigma : \tau \mapsto x(\tau)$ is $\psi(\tau)$, i.e., $\Psi(x)|_\sigma = \psi(\tau)$. Since $\dot{\psi} = v \cdot \partial\Psi$ the general eq.(12) can be written as a non-linear equation for $\Psi(x)$; we have

$$(\Psi\gamma_0\tilde{\Psi}) \cdot \partial\Psi + \pi\Psi\gamma_0 = 0, \tag{16}$$

and in the case where Ψ is a eigenspinor of the momentum operator $p_\mu \Psi = \partial_\mu \Psi \gamma_2 \gamma_1$ $(A = 0)$ we have

$$\partial \Psi \gamma_1 \gamma_2 + m \Psi \gamma_0 = 0, \tag{17}$$

which is just the DH equation (Hestenes 1990, 1991) (Pavšič et al. 1992) – this seems to be indeed a non trivial result.

3. Some Properties of BZ Model

One can extract several informations from BZ model by analysing the system (12-14). Let $\Omega = 2\dot{R}\tilde{R}$ and $S = \frac{\hbar}{2} R \gamma_2 \gamma_1 \tilde{R}$ (the spin bivector). Then, using eq.(3) (with $\beta = 0$) into eq.(12) and after splitting the resulting equation into its scalar, bivector and pseudo-scalar parts, we get $(\hbar = 1)$:

$$\dot{\rho} = 0, \tag{18}$$

$$\Omega = -4\pi \cdot (e_0 \wedge S), \tag{19}$$

$$\pi \wedge e_0 \wedge e_1 \wedge e_2 = 0. \tag{20}$$

From eq.(18) we conclude that $\rho =$ constant, which we take to be $\rho = 1$ for τ being the proper time along the center of mass. In this case $v = \dot{x} = e_0$, and since eq.(20) says that π is a linear combination of $\{e_0, e_1, e_2\}$, we conclude that we can have solutions of eq.(12) for which v and π are not parallel. The fact that velocity and momentum are non-parallel is an essential characteristic of Dirac theory, as discussed by Hestenes (1990, 1991).

In BZ model, the mass comes from kinetic energy of rotation (see also Hestenes 1990, 1991). In fact, from eq.(19) we can show that

$$\Omega \cdot S = \pi \cdot e_0 = m, \tag{21}$$

where the constant m is identified with mass.

Another interesting conclusion can be obtained from eq.(12) after multiplying it on the right by $\tilde{\psi}$ and subtracting from this result the reverse of eq.(12) multiplied on the left by ψ, which gives:

$$\dot{S} - \pi \wedge v = 0. \tag{22}$$

Now, from the STA formulation of Dirac theory one can show, from the definition of the energy-momentum and angular momentum tensors, that

$$\rho[\dot{S} - \pi \wedge v] = N^\mu \wedge \gamma_\mu - \partial_\mu M^\mu, \tag{23}$$

where N_μ describes the flow of energy-momentum normal to the velocity streamline and M_μ describes the flow of angular momentum normal to that streamline, that is: $T_\mu = \rho v_\mu p + N_\mu$, $M_\mu = \rho(S \cdot \gamma_\mu) v$, where $T_\mu = T_{\mu\nu}\gamma^\nu$ with $T_{\mu\nu}$ being the energy-momentum tensor and S is the spin bivector (see Hestenes 1973 for details). But now, for a Weyssenhoff fluid (Weyssenhoff and Raabe 1947), the net flux of energy-momentum and angular momentum through the walls of a comoving volume element vanishes, i.e. (Hestenes 1973)

$$\partial_\mu N^\mu = 0, \ \ N^\mu \wedge \gamma_\mu = \partial_\mu M^\mu, \tag{24}$$

and in this case eq.(23) reduces to eq.(22). We conclude therefore that a BZ fluid is an example of a Weyssenhoff fluid.

One can easily show from eq.(22) that

$$\frac{d}{d\tau}(S + x \wedge \pi) = e\, x \wedge (F \cdot v), \tag{25}$$

so that in the free particle case we have conservation of the angular momentum J given by

$$J = x \wedge p + S. \tag{26}$$

In the free particle case it is easy to solve system (12-14). One solution with particular interest is the one given by BZ (1984) which in terms of STA reads

$$\psi = \psi(\tau) = \cos m\tau \psi(0) + \sin m\tau \gamma_0 \psi(0) \gamma_0 \gamma_1 \gamma_2, \tag{27}$$

where m is given by eq.(21). We can also write eq.(27) as

$$\psi(\tau) = \psi_+(0) e^{\gamma_1 \gamma_2 m\tau} + \psi_-(0) e^{-\gamma_1 \gamma_2 m\tau}, \tag{28}$$

where $\psi_\pm(0) = \Lambda_\pm[\psi(0)]$, with $\Lambda_\pm(\psi) = \frac{1}{2}[\psi \pm \gamma_0 \psi \gamma_0]$ being the positive and negative energy projection operators (Lounesto 1993). It is interesting to remark that although eq.(28) is a superposition of positive and negative energy states of the free DH equation (restricted to the world-line σ), it is also a solution of the non-linear eq.(16) (which seems to be more important). This solution gives for the time-like velocity

$$v = v(\tau) = \frac{p}{m} + [v(0) - \frac{p}{m}] \cos 2m\tau + \frac{\dot{v}(0)}{2m} \sin 2m\tau, \tag{29}$$

and integration gives a world-line $x = x(\tau)$ which is a cilindrical helix.

As a concrete example, if we take $\psi(0) = \sqrt{\rho_+} + \sqrt{\rho_-}\gamma_1\gamma_0$ with $\rho_+ - \rho_- = 1$ for simplicity, we get $v(\tau) = (\rho_+ + \rho_-)\gamma_0 + 2\sqrt{\rho_+\rho_-}[\gamma_1 \cos 2m\tau + \gamma_2 \sin 2m\tau]$, which after a convenient rotation can be written as $v = (\rho_+ + \rho_-)e_0 - 2\sqrt{\rho_+\rho_-}e_2$. We note that $\dot{L} \neq 0$ for $L = x \wedge p$, but $\dot{J} = 0$ for $J = L + S = \frac{1}{2}(\rho_+ + \rho_-)\gamma_2\gamma_1$.

If, on the other hand, we take $\psi(0) = \frac{1}{\sqrt{2}}(1 + \gamma_1\gamma_0)$ (which is actually a Majorana spinor (according to Lounesto 1993)[1] since charge conjugation operator C is $C\psi = \psi\gamma_1\gamma_0$), then for the velocity we get the light-like vector $v(\tau) = \gamma_0 + \gamma_1 \cos 2m\tau + \gamma_2 \sin 2m\tau$, or, after a convenient rotation,

$$v = e_0 - e_2. \tag{30}$$

We also have $\dot{J} = 0$ with $J = \frac{1}{2}\gamma_2\gamma_1$ and the helix diameter is just the Comptom wavelength. Note that the velocity in this case is light-like, and expression (30) is just Hestenes' (1990, 1991) conjecture for the electron's velocity in his ZBW interpretation of quantum mechanics.

For cases with interactions the solution of system (12-14) is not so easy to find as in the free particle case, of course. The difficult arises because althought we can

[1] From Lounesto's definition we have that Majorana spinor is singular and the spin vector is zero, while the current and the spin-bivector are non-null; it is an eigenspinor of the charge conjugate operator, with eigenvalue 1 in our example.

write $\pi = me_0$ $(e_0 \cdot e_0 = 1)$, we cannot in general identify $e_0 = \gamma_0$. In the case of interaction with a magnetic field $F = B\gamma_1\gamma_2$ we can always find a stationary solution (Gull 1991) such that $v = \gamma_0$, $\pi = m\gamma_0$, $\psi = e^{-\gamma_1\gamma_2 m\tau}$ like in the free case, but such a solution does not have any interesting feature. This is a particularity that arises from choosing the magnetic field in the same direction of the spin vector.

4. Some Remarks about the Electric Field of the Electron

We remarked in sect.3 that expression (30) is just Hestenes' conjecture for the electron's velocity. According to his ZBW interpretation of quantum mechanics, an electron is supposed to move along a cilindrical helix whose diameter equals the Compton wavelength. Mass is explained as the kinetic energy of that circulation and the electron spin as its angular momentum. But, if this is the case, a kind of self-interaction must exists, and since the ZBW frequency is about 10^{21} Hz, we expect the electron to be the seat of an oscillating electromagnetic field of the same order, which when averaged over a ZBW period gives the usual Coulomb field. Let us see (Vaz and Rodrigues 1993a) how one can obtain this oscillating coulomb-like field by modifying Maxwell equations in order to account for that self-interaction. An alternative approach (see Vaz and Rodrigues 1993a for details) consists in exploiting the fact that gauge invariance is related to rotations on the spin plane in Dirac theory (Hestenes 1990, 1991), and in order to obtain this field we break gauge invariance for local rotations by introducing "defects" on the spin plane and representing the electromagnetic potential in terms of space-like vectors belonging to the spin plane. This representation is given by $A_\mu = -e(\pi_1 \cdot \partial_\mu \pi_2)$ where π_1, π_2 are orthogonal space-like vectors belonging to the spin plane.

First, let us remember that London (see Reitz et al. 1960) in his theory of supercondutivity, in order to explain the Meissner effect, introduced a supercurrent $\vec{j}_s$ which adds to the convection current $\vec{j}$ in Maxwell-Ampere law. This supercurrent is taken to be proportional to the vector potential $\vec{A}$. London take it as $\vec{j}_s = -\frac{1}{\lambda^2}\vec{A}$, with the minus sign needed to obtain an exponential decay behaviour for the solutions. Since we want an oscillatory behaviour, we take

$$\vec{j}_s = \frac{1}{\lambda^2}\vec{A}, \tag{31}$$

and Maxwell-Ampere law reads $\nabla \times \vec{B} = \partial\vec{E}/\partial t + \vec{j} + \vec{j}_s$. The so-called Maxwell-London equations can now be solved, and with $\vec{j} = 0$ we have for the electric field

$$\vec{E} = -2\sin^2\omega_{zbw}t\frac{e\vec{r}}{r^3}, \tag{32}$$

once we take[2] $A_0 = -\frac{e}{r}$, $\vec{A} = \frac{e}{\omega_{zbw}}\frac{\sin 2\omega_{zbw}t}{2}\frac{\vec{r}}{r^3}$ and identify $\lambda = \frac{c}{2\omega_{zbw}} = \frac{r_0}{2}$, where r_0 is half the Compton wavelenght and ω_{zbw} is the ZBW frequency, that is: $r_0 = \frac{\hbar}{2mc}$, $\omega_{zbw} = \frac{2mc^2}{\hbar}$. Note that we have $\vec{B} = 0$ in this case, but one can add the quantity

[2] The justification for taking these values for A_0 and $\vec{A}$ can be found in (Vaz and Rodrigues 1993a).

$\frac{1}{2}(y\sigma_1 - x\sigma_2)$ (where $\sigma_i = \gamma_i\gamma_0$) to $\vec{A}$, giving the same field $\vec{E}$ but $\vec{B} = \frac{1}{2}\sigma_3$. If we average the field $\vec{E}$ over a ZBW period we get

$$\langle\vec{E}\rangle = -\frac{e\vec{r}}{r^3}, \tag{33}$$

which is the usual Coulomb field.

We remark that taking the current proportional to the electromagnetic potential in order to account for that self-interaction is an assumption that already produced interesting results. We quote specifically Dirac (1951) non-linear electrodynamics and its generalization proposed by us (Rodrigues, Vaz and Recami 1993) in terms of which we discuss a purely electromagnetic model for a spinning charged particle.

Acknowledgements

The authors are grateful to Professor E. Recami, Dr. Q.A.G. de Souza and Dr. M.A.F. Rosa for several important discussions. This work has been partially supported by CNPq, CAPES and FAEP-UNICAMP.

References

Barut, A.O. (1983), "What is an Electron? Relativistic Electron Theory and Radiative Processes", in *Quantum Optics, Relativity and Theory of Measurement*, P. Meystre (ed.), Plenum Press, New York, pg. 155.

Barut, A.O. and Bracken, A.J. (1981), Phys. Rev. D **23**, 2454.

Barut, A.O. and Duru, I.H. (1984), Phys. Rev. Lett. **53**, 2355.

Barut, A.O. and Pavšič, M. (1987), Class. Quant. Grav. **4**, L131.

Barut, A.O. and Zanghi, N. (1984), Phys. Rev. Lett. **52**, 2009.

Dirac, P.A.M. (1951), Proc. Roy. Soc. London **A209**, 292.

Gull, S.F. (1991) "Charged Particles at Potential Steps", in *The Electron*, D. Hestenes and A. Weingartshofer (eds.), Kluwer Acad. Publ. , Dordrecht.

Hestenes, D. (1966) *Space-Time Algebra*, Gordon & Breach, New York.

Hestenes, D. (1973), J. Math. Phys. **14**, 893.

Hestenes, D. (1990), Found. Phys. **20**, 1213.

Hestenes, D. (1991) "Zitterbewegung in Radiative Processes", in *The Electron*, D. Hestenes and A. Weingartshofer (eds.), Kluwer Acad. Publ. , Dordrecht.

Hestenes, D. (1992) "Hamiltonian Mechanics with Geometric Calculus", preprint Arizona State University.

Hestenes, D. and Sobczyk, G. (1984) *Clifford Algebra to Geometric Calculus*, D. Reidel Publ. Co., Dordrecht.

Lasenby, A., Doran, C. and Gull, S.F. (1993) "A Multivector Derivative Approach to Lagrangian Field Theory", to appear in Found. Phys. (1993).

Lounesto, P. (1993) "Clifford Algebra and Hestenes Spinors", preprint Helsinki University of Technology, to appear in Found. Phys. (may 1993).

Pavšič, M., Recami, E., Rodrigues, Jr., W.A., Maccarrone, G.D., Raciti, F. and Salesi, G. (1992) "Spin and Electron Structure", preprint IMECC-UNICAMP 45/92, submitted for publication.

Rawnsley, J. (1992), Lett. Math. Phys. **24**, 331.

Reitz, J.R., Milford, F.J. and Christy, R.W. (1960) *Foundations of Electromagnetic Theory*, Addison-Wesley Publ. Co. Inc., Reading, Mass.

Rodrigues, Jr., W.A., Souza, Q.A.G., Vaz, Jr., J. and Maiorino, J.E. (1993) "A Thoughtful Study of Dirac-Hestenes Spinor Fields, its Covariant Derivative and Dirac-Hestenes Equation in Riemann-Cartan Spacetimes", preprint IMECC-UNICAMP, submitted for publication.

Rodrigues, Jr., W.A., Vaz, Jr., J. and Recami, E. (1993) Found. Phys. **23**, 469.

Vaz, Jr., J. and Rodrigues, Jr., W.A. (1993) "Clifford Algebra Approach to Barut-Zanghi Model as a Hamiltonian System", preprint IMECC-UNICAMP RP 11/93, submitted for publication.
Vaz, Jr., J. and Rodrigues, Jr., W.A. (1993a) "Zitterbewegung and the Electromagnetic Field of the Electron", preprint IMECC-UNICAMP RP 05/93, to appear in Phys. Lett. B.
Weyssenhoff, J. and Raabe, A. (1947), Acta Phys. Pol. 9, 7.

SEPARATION OF THE DIRAC EQUATION AND POSITIVE DEFINITENESS OF QUANTUM NUMBERS

MANFRED STEIN
Fachbereich Physik der Universitt
Postfach 3049
D-6750 Kaiserslautern
West Germany

Abstract. The Clifford valued form of the Dirac equation, in contrast to its matrix form, makes it possible to carry out a thorough discussion of the procedure of how to construct the most general solution for a given potential. In applying several separation steps to the Dirac equation in the case of a spherically symmetric time independent potential, not only the effectivity of the calculational tools provided by Clifford algebra becomes obvious, but it is also possible to replace the sign of the appearing quantum numbers by a geometrically more meaningful set of elements of the Pauli algebra. In the case of a sufficiently weak Coulomb potential, we can construct a theory which has only positive definite quantum numbers.

1. Introduction

The construction of solutions of the Dirac equation

$$\hbar\partial_{-}\Psi i_3 - \frac{q}{c}A_{-}\Psi = mc\Psi^*, \tag{1}$$

due to its high dimensionality, is a difficult task. In matrix formulation, operator calculus is the predominant tool. Basically however, people try to relate themselves to the results of nonrelativistic (one dimensional complex) Schrödinger theory, which in a somewhat modified form, are then applied to the four dimensional complex Dirac theory [1]. In contrast to this operator approach, the above given Clifford valued formulation of the Dirac equation [2] makes it possible to systematize the construction in a way, which makes use of only two basic principles, namely Clifford algebra itself and separation, so that now one can solve the Dirac equation using only results already inherent in the theory. This makes the solving procedure very transparent.

The importance of the separation techniques has also been realized by some people working with the matrix formulation of the Dirac equation [3, 4]. In addition to this, because of the presence of objects with a straigthforward geometric interpretation in the Pauli algebra $C_{3,0}$, the Clifford valued solutions have a strong coupling to geometry. This article will try to emphasize this aspect. I will try to develop a complete systematic approach to the separable solutions of the Dirac equation in the case of spherically symmetric potentials, which only relates to Clifford algebra. It will be seen, that the sign of the quantum numbers, appearing in the course of the separation procedure, can be made positive definite by admitting elements of a four dimensional subalgebra of constant end elements in the spinor solution of the Dirac

F. Brackx et al. (eds.), Clifford Algebras and their Applications in Mathematical Physics, 405–411.

equation. Thus, whats usually hidden in the sign of the quantum number, can be attributed with a geometrical interpretation.

The above given Dirac equation is the projection of the real Hestenes form into the Pauli algebra $C_{3,0}$, i. e. that real Clifford algebra, which is connected to the euclidean three form

$$\begin{pmatrix} 1 & 0 & 0 \\ 0 & 1 & 0 \\ 0 & 0 & 1 \end{pmatrix}$$

Its eight basis elements shall be denoted as follows:

Skalar 1 , Vectors $\sigma_1, \sigma_2, \sigma_3$, Bivectors i_1, i_2, i_3, Pseudoscalar i

$\partial_- \equiv \frac{1}{c}\frac{\partial}{\partial t} + \sigma_k \frac{\partial}{\partial r^k} \equiv \frac{\partial}{\partial r^0} + \partial$ projects the four gradient into the Pauli algebra,
$A_- \equiv A^0 - A^k\sigma_k \equiv A^0 - A$ projects the four potential into the Pauli Algebra [5].

The star operation $\left(A^* \equiv \gamma_0^{-1} A \gamma_0 \text{ in the Dirac algebra } C_{1,3}\right)$ can be defined in $C_{3,0}$ as follows:

$$1^* \equiv 1; \; \sigma_k^* \equiv -\sigma_k \;\; i_k^* \equiv i_k, \; i^* \equiv -i; \; (A\,B)^* \equiv A^*\,B^*.$$

The separation steps following in the next chapters shall be treated as general as possible, i. e. the restrictions imposed onto the four potential, will be formulated as weak as possible and all possibilities to solve the resulting equations shall be considered.

2. Separation of time (time independent potentials)

Be A^0, A^k functions of r^k, $k \in \{1, \; 2 \;\; 3\}$ only.

Ansatz: $\Psi(r) \equiv \psi(r) \;\; \tau\left(r^0\right)$

$$\Rightarrow \psi(r)\; \hbar\partial_0 \tau i_3 + \hbar\partial\psi(r)\, i_3\, i_3^{-1} \tau i_3 - \frac{q}{c} A_-(r)\, \psi(r)\, \tau = mc\, \psi(r)^* \,\tau^*$$

This leads to the following set of separation conditions

$$\tau 1 : \hbar\partial_0 \tau\, i_3 = \frac{E}{c}\tau$$

$$\tau 2 : i_3^{-1} \tau\, i_3 = C_1 \tau$$

$$\tau 3 : \tau^* = C_2 \tau$$

These three conditions must be evaluated to obtain the (bound state) solution of the time dependent spinor part. Starting with the most general form of τ , namely $\tau \in C_{3,0}$, conditions $\tau 2$ and $\tau 3$ divide the possible solutions into four subsets of the form $\tau\left(r^0\right)\; \tau_0$ with $\tau\left(r^0\right) = \tau_1\left(r^0\right) + i_3\, \tau_2\left(r^0\right)$ and $\tau_0 \in \{1, \sigma_1, \sigma_3, i_2\}$. $C_{1,2}$ can only take the values ± 1 (scalar), according to the choice of the constant end element τ_0. This dependence is given in the following table:

C_1	C_2	τ_0
+1	+1	$1\ exp(i_3\alpha)$
+1	−1	$\sigma_3\ exp(i_3\alpha)$
−1	+1	$i_2\ exp(i_3\alpha)$
−1	+1	$\sigma_1\ exp(i_3\alpha)$

Multiplying $\tau 1$ to the left with $i_3^{-1} = -i_3$, we get the equivalent equation

$$\partial_0\, i_3^{-1} \tau\, i_3 = -i_3 \frac{C_1 E}{\hbar c}\tau$$

$$\Leftrightarrow \partial_0\ \ln\tau = -i_3 \frac{C_1 E}{\hbar c}$$

$$\Leftrightarrow \ln(\tau/\tau_0) = -i_3 \frac{C_1 E}{\hbar c}\left(r^0 - r^0_0\right)$$

$$\Leftrightarrow \tau = \exp\left\{-i_3 \frac{C_1 E}{\hbar c}\left(r^0 - r^0_0\right)\right\}\tau_0$$

$$r^0_0 \equiv 0 \Rightarrow \tau\left(r^0\right) = \exp\left\{-i_3\, \frac{E C_1}{\hbar c} r^0\right\}\ \tau_0 \tag{2}$$

The phase factor $\exp(i_3\alpha)$, which leaves equation (1) invariant, is usually interpreted as a free gauge factor [6] and will be suppressed in the following. We can further observe, that the energy constant E appears together with the constant C_1, the sign of E is connected to the choice of τ_0 . It is interesting to observe, that in the case of a sufficiently weak Coulomb potential, the resulting differential equation for the radial spinor part admits normalizable solutions only for one sign of the energy. Using the final form (2) of the solution τ, we obtain the time separated Dirac equation in the case of time independent potentials:

$$C_1\, \hbar\, \partial\, \psi\, i_3 + \left(\frac{E}{c} - \frac{q}{c} A_-\right)\, \psi = C_2\, mc\, \psi^* \tag{3}$$

3. Separation of azimuth angle in the case of polar coordinates

I will now show, by applying the same separation techniques, how the successive steps to find the most general separable solution of the Dirac equation for spherically symmetric potentials can be taken. To achieve this, we first start with a spinor representation of polar coordinates. It is possible to interpret polar coordinates as an application of two rotations, specified by two rotation axes and two rotation angles, and a (commutative) stretching factor ρ. The spatial vector $\boldsymbol{r}$ is then given by means of the following spinorial representation [7]:

$$\boldsymbol{r} = r^k \sigma_k \equiv \rho\, \tilde{\Omega}\, \sigma_3\, \Omega; \quad \Omega = \exp\left(i_2\, \frac{\theta}{2}\right) \exp\left(i_3 \frac{\varphi}{2}\right), \qquad \tilde{}\ \text{Reversion}$$

This spinor representation defines a bijective (almost everywhere) mapping from ρ, θ, φ onto r^1, r^2, r^3. Using the chain rule and the theory of outermorphisms, as developped in [9], this form of r can be used very effectively to give a spinor representation of the gradient $\partial = \sigma_k \frac{\partial}{\partial r^k}$. For the following calculations it is useful to factor out a unit of length, which will be chosen as the Compton wavelength $\lambda\!\!\!^- = \frac{\hbar}{mc}$ here. We get

$$\partial = \sigma_k \frac{\partial}{\partial r^k} \equiv \sigma_k \, \partial_k = \lambda\!\!\!^{-\,-1} S^{-1} \, \partial' \, S$$

Denotations: $S = \left(\rho^2 \sin\theta\right)^{1/2} \Omega$; $\partial' = \frac{\sigma_1}{x} \partial_\theta + \frac{\sigma_2}{x \sin\theta} \partial_\varphi + \sigma_3 \, \partial_x$; $x \equiv \frac{\rho}{\lambda\!\!\!^-}$
Definition: $\eta\,(x,\,\theta,\,\varphi) \equiv S\,\psi\,(r)$
Using this definition and the spinor form of ∂, we get the following equation, equivalent to (3)

$$C_1 \, \partial' \eta \, i_3 + \left(\varepsilon - A'_-\right) \eta = C_2 \, \eta^* \tag{4}$$

where the following denotations were used: $A'_- = \frac{\alpha \lambda\!\!\!^-}{q_{ee}} \Omega A_- \tilde{\Omega}$; $\alpha = \frac{q \, q_{ee}}{\hbar c}$; $\varepsilon = \frac{E}{mc^2}$.

We now restrict ourselves to the case of potentials A'_- independent of φ. Using the Ansatz $\eta\,(x,\,\theta,\,\varphi) = \chi\,(x,\,\theta)\, f\,(\varphi)$, we arrive at

$$C_1 \left(\sigma_3 \partial_x + \frac{\sigma_1}{x} \partial_\theta\right) \chi\,(x,\,\theta)\, i_3 \, i_3^{-1} \, f \, i_3 + \frac{C_1 \, \sigma_2}{x \sin\theta} \chi \, \partial_\varphi f \, i_3 + \left(\varepsilon - A'_-\right) \chi f = C_2 \, \chi^* \, f^*$$

The separation conditions have the same form as those in the case of time separation:

$$\varphi 1 : \partial_\varphi f \, i_3 = \mu f$$

$$\varphi 2 : i_3^{-1} f \, i_3 = C_3 \, f$$

$$\varphi 3 : f^* = g_0^{-1} f \, \gamma_0 = C_4 \, f$$

and we can immediately give the most general solution to be

$$f\,(\varphi) = \exp\left\{-i_3 \, \mu \, C_3 \, \varphi\right\} \; f_0 \tag{5}$$

C_3	C_4	f_0
+1	+1	$1 \exp(i_3\alpha)$
+1	−1	$\sigma_3 \; exp(i_3\alpha)$
−1	+1	$i_2 \; exp(i_3\alpha)$
−1	+1	$\sigma_1 \; exp(i_3\alpha)$

The Dirac equation now takes the form (Denotation: $C_{ij\ldots k} \equiv C_i C_j \ldots C_k$)

$$C_{13} \sigma_3 \left[\partial_x \chi + \frac{i_2}{x}\left(\partial_\theta \chi + \frac{\mu C_3}{\sin\theta} i_3^{-1} \chi \, i_3\right)\right] i_3 + \left(\varepsilon - A'_-\right)\chi = C_{24} \, \chi^* \tag{6}$$

Together with the four possibilities from time separation, we altogether have sixteen solutions. From these sixteen however, only four are found to be linear independent. They can be chosen as:

$$1. \quad \chi(++++)\exp\left[-i_3\left(\mu\varphi+\frac{E}{\hbar c}r^0\right)\right] 1$$

$$2. \quad \chi(+++-)\exp\left[-i_3\left(\mu\varphi+\frac{E}{\hbar c}r^0\right)\right] \sigma_3$$

$$3. \quad \chi(-+++)\exp\left[i_3\left(-\mu\varphi+\frac{E}{\hbar c}r^0\right)\right] i_2$$

$$4. \quad \chi(-++-)\exp\left[i_3\left(-\mu\varphi+\frac{E}{\hbar c}r^0\right)\right] \sigma_3$$

where e. g. $\chi(-++-)$ means, that χ is a solution of equation (6) with $C_1 = -1 = C_4$ and $C_2 = +1 = C_3$. As we have in all of the four solutions $C_2 = +1 = C_3$, these two constants can be suppressed. Examining the form of equation (6), another linear dependence can be found:

$$\chi(++++)\exp\left[-i_3\left(\mu\varphi+\frac{E}{\hbar c}r^0\right)\right] 1 = \chi(-+++)\exp\left[i_3\left(\mu\varphi+\frac{E}{\hbar c}r^0\right)\right] i_2$$

$$\chi(+++-)\exp\left[-i_3\left(\mu\varphi+\frac{E}{\hbar c}r^0\right)\right] \sigma_3 = \chi(-++-)\exp\left[i_3\left(\mu\varphi+\frac{E}{\hbar c}r^0\right)\right] \sigma_1$$

We now have two alternatives to avoid this dependence:
a) If we admit only $\{1,\ \sigma_3\}$ as constant end elements, then μ can take positive or negative values.
b) Admitting $\{1,\ \sigma_3,\ i_2,\ \sigma_1\}$, μ must be chosen to be positive.

4. Separation of polar angle and radial spinor parts

The last restriction imposed on A'_- is it to be independent of θ. We then can transform the eight-dimensional equation (6) to a set of two equivalent coupled quaternionic equations by means of the projectors $P_\pm \equiv \frac{1}{2}(1 \pm \sigma_3)$. They have the properties: $P_\pm = \pm\sigma_3 P_\pm$; $P_+ + P_- = 1$ $P_\pm^* = P_\mp$; $P_\pm^n = P_\pm$
Introducing $\chi(x,\theta) = \chi\,(P_+ + P_-) = \chi P_+ + \chi P_- \equiv \chi_+ P_+ + \chi_- P_-$ with $\chi_\pm^* = \chi_\pm$ being elements of the quaternion subalgebra, we get

$$+C_1 i_3 \left[\partial_x \chi_+ + \frac{i_2}{x}\left(\partial_\theta \chi_+ + \frac{\mu}{\sin\theta} i_3^{-1}\chi_+ i_3\right)\right] + (\varepsilon - A'_-)\,\chi_+ = C_4 \chi_- \qquad (7)$$

$$-C_1 i_3 \left[\partial_x \chi_- + \frac{i_2}{x}\left(\partial_\theta \chi_- + \frac{\mu}{\sin\theta} i_3^{-1}\chi_- i_3\right)\right] + (\varepsilon - A'_-)\,\chi_- = C_4 \chi_+ \qquad (8)$$

If we choose $\chi_- = i_2^{-1}\chi_+ i_2$, the equations become identical and we get

$$C_1 i_3 \left[\partial_x \chi_+ + \frac{i_2}{x}\left(\partial_\theta \chi_+ + \frac{\mu}{\sin\theta} i_3^{-1}\chi_+ i_3\right)\right] + (\varepsilon - A'_-)\,\chi_+ = C_4 i_2^{-1}\chi i_2 \quad (9)$$

The separation ansatz: $\chi_+ = a(x)\, g(\theta)$ with $i_3^{-1}\, a\, i_3 = C_5\, a$ $(C_5 = \pm 1)$ then leads to

$$C_1\, i_3 \left[\partial_x\, a\, g + \frac{i_2\, a}{x}\left(\partial_\theta g + \frac{\mu C_5}{\sin\theta} i_3^{-1} g\, i_3\right)\right] + \left(\varepsilon - A'_-\right) a\, g = C_4\, i_2^{-1}\, a\, g\, i_2$$

which gives the following separation conditions:

$$\theta 1\ :\ i_2^{-1}\, g\, i_2 = C_6\, g \ \ (C_6 = \pm 1)$$

$$\theta 2\ :\ \partial_\theta\, g + \frac{\mu C_5}{\sin\theta}\, i_3^{-1} g\, i_3 = \kappa\, i_2\, g$$

Finally we have arrived at the form of the radial Dirac equation in the case of spherically symmetric potentials:

$$C_1\, i_3 \left[\partial_x\, a + \frac{\kappa}{x} i_2^{-1}\, a\, i_2\right] + \left(\varepsilon - A'_-\right) a = C_{46}\, i_2^{-1}\, a\, i_2 \tag{10}$$

As the solutions of equation $\theta 2$ and (10) are well known [7, 8], we will not deal with them here. To complete the examination of linear dependence, we have to consider only the symmetries of the resulting two systems of linear differential equations.

The symmetries of equation $\theta 2$ are:
Be $g(+\kappa, +\mu) = g_1 + i_2 g_2$ a solution of $\theta 2$ with values $(+\kappa, +\mu) = g_1 + i_2 g_2$ fixed. Then: $\tilde{g}(+\kappa, +\mu) = g(-\kappa, +\mu)$ and $i_2\, g(+\kappa, +\mu) = g(\kappa, -\mu)$.

Symmetries of equation (10):
Be $a = a_1 + i_3\, a_2$ a solution of (10). Then: $a\, i_2$ is also a solution of (10) and $a\, i_3$ is a solution of (10) and $a\, i_3$ is a solution of (10) where κ is replaced by $-\kappa$ and C_{46} by $-C_{46}$. Using these symmetries and the form of the solution

$$\left[a\, g\, P_+ + i_2^{-1}\, a\, g\, P_+\, i_2\right] \exp\left[-i_3\left(\mu\varphi + \frac{EC_1}{\hbar c} r^0\right)\right] \tau_0 \tag{11}$$

we find, that putting $C_5 = 1 = C_6$ is no restriction and that there is a last linear dependence: If we admit $C_4 = \pm 1$, then κ has to be chosen positive to avoid linear dependence, choosing $C_4 = +1$, we must admit positive as well as negative values for κ.

5. Summary

To finish, the results of the separation procedure in the case of spherically symmetric potentials shall be summarized. The complete set of linear independent separable solutions of the Dirac equation in the case of spherically symmetric potentials can be given in two alternative ways.

a) If we admit as constant end elements the elements of the set $\{1,\ \sigma_3,\ i_2,\ \sigma_1\}$, μ and κ have to be positive to avoid linear dependence of the solutions.

b) If we admit only one constant end element, e. g. *1*, the sign of μ and κ cannot be restricted, if we want to construct all possible solutions of the Dirac equation.

Analyzing the square integrable solutions of the Coulomb potential $A'_- = \frac{1}{x}$, we find [1], that only for positive values (depending on the sign of q_{ex}) solutions with suitable boundary conditions can be constructed. In this case, we can construct a complete set solutions using only positive definite quantum numbers, if we admit the elements of $\{1, \sigma_3, i_2, \sigma_1\}$ as end elements of the spinor solution. All states of the theoretical spectrum can be identified with experimentally observed ones, there is no need for additional antiparticle solutions. As the Coulomb problem is more reasonable than the free particle problem, it seems to be only natural to use its properties to interpret the physical contents of Dirac theory.

As a short application of the results just derived let us discuss charge conjugation, which is usually introduced in the following manner: Instead of the particle spinor ψ_p consider the antiparticle spinor $\psi_{ap} \equiv \psi_p \sigma_2$. Now, if ψ_p is a solution of (1), then ψ_{ap} is a solution of (1) where the charge q of the Dirac particle is replaced by $-q$. If we now consider the antiparticle hydrogen problem, we find, that there is another minus sign in A_- due to the negative charge of the antiproton, so that we have an overall of two minus signs and the resulting anti particle problem becomes identical to the particle hydrogen problem. However, all of the solutions we have derived in the previous chapters, can be uniquely mapped onto states from the experimentally observed spectrum, so there is no room for additional linear independent anti particle solutions. If we introduce another change in the sign of $\frac{q}{c}A_-$, we observe, that we get linearly independent solutions with a negative sign of the energy. This change in the sign can be either achieved by considering the sign of the anti particle a second time, i. e. by putting $q_{ap} \equiv -q_p$, or by not applying the charge conjugation at all and by putting $q_{ap} \equiv -q_p$.

The fact, that not the sign of q_p has to be changed to $-q_p$, but the total sign of $q_p A_p$ has to be reversed to get negative energy solutions has already been observed by Dirac [10]. Especially in a self consistent theory, as it has for instance been formulated by Barut and his co-workers [11], and which works with bound state solutions instead of solutions of the free particle Dirac equation, this observation should be of importance.

References

1. H. A. Salpeter, E. E. Salpeter: Quantum Mechanics Of One- And Two-Electron Atoms, 1. paperback ed. (Plenum Publishing Corporation, New York) (1977)
2. D. Hestenes, R. Gurtler, J. Math. Phys. 16 (no. 3), pp. 572-584 (1975)
3. G.V. Shishkin, V.M. Villalba, J. Math. Phys. 30 (no. 9), pp. 2132-2142 (1989 a)
4. G.V. Shishkin, V.M. Villalba, J. Math. Phys. 30 (no. 10), pp. 2373-2381 (1989 b)
5. D. Hestenes, Spacetime Algebra (Gordon and Breach, New York) (1966)
6. D. Hestenes, Found. Phys. 12 (2), 153 (1982)
7. H. Krüger in The Electron (pp. 49-81) ed D. Hestenes, A, Weingartshofer (Kluwer Acad. Publ. Dordrecht) (1991)
8. M. Stein, J. Phys. B to be published
9. D. Hestenes, G. Sobczyk: Clifford Algebra to Geometric Calculus (D. Reidel Publ. Comp., Dordrecht) (1984)
10. P. A. M. Dirac, Proc. Roy. Soc. Lond., Ser. A 133, p. 80 (1931)
11. A.O. Barut, J.F. van Huele, Phys. Rev. A 32 (no. 6), pp.3187-3195 (1985)